Quantum Groups, Hopf Algebras, and Applications

In Memory of Earl Jay Taft

CONTEMPORARY MATHEMATICS

814

Quantum Groups, Hopf Algebras, and Applications

In Memory of Earl Jay Taft

AMS Special Session in Honor of Earl J. Taft
Quantum Groups, Hopf Algebras, and Applications
October 22–23, 2022
University of Utah, Salt Lake City, Utah

Susan Montgomery
Siu-Hung Ng
Editors

AMS AMERICAN MATHEMATICAL SOCIETY

2020 *Mathematics Subject Classification.* Primary 14L15, 16T05, 16T20, 16T99, 17A70, 17B37, 17B62, 18M15, 18M20, 20G42.

Library of Congress Cataloging-in-Publication Data

Names: AMS Special Session in Honor of Earl J. Taft : Quantum Groups, Hopf Algebras, and Applications (2022 : Salt Lake City, Utah) | Taft, Earl J., honoree. | Montgomery, Susan, editor. | Ng, Siu-Hung, editor.

Title: Quantum groups, Hopf algebras, and applications : in memory of Earl Jay Taft : AMS Special Session in Honor of Earl J. Taft : Quantum Groups, Hopf Algebras, and Applications, October 22–23, 2022, University of Utah, Salt Lake City, Utah / Susan Montgomery, Siu-Hung Ng, editors.

Description: Providence, Rhode Island : American Mathematical Society, 2025. | Series: Contemporary mathematics, 0271-4132 ; volume 814 | Includes bibliographical references.

Identifiers: LCCN 2024035051 | ISBN 9781470473624 (paperback) | ISBN 9781470478650 (ebook)

Subjects: LCSH: Hopf algebras–Congresses. | Quantum groups–Congresses. | AMS: Algebraic geometry – Algebraic groups – Group schemes. | Associative rings and algebras – Hopf algebras, quantum groups and related topics – Hopf algebras and their applications. | Associative rings and algebras – Hopf algebras, quantum groups and related topics – Ring-theoretic aspects of quantum groups. | Associative rings and algebras – Hopf algebras, quantum groups and related topics – None of the above, but in this section. | Nonassociative rings and algebras – General nonassociative rings – Superalgebras. | Nonassociative rings and algebras – Lie algebras and Lie superalgebras – Quantum groups (quantized enveloping algebras) and related deformations. | Nonassociative rings and algebras – Lie algebras and Lie superalgebras – Lie bialgebras; Lie coalgebras. | Group theory and generalizations – Linear algebraic groups and related topics – Quantum groups (quantized function algebras) and their representations.

Classification: LCC QA613.8 .A48 2022 | DDC 512/.55–dc23/eng/20241226

LC record available at https://lccn.loc.gov/2024035051

DOI: https://doi.org/10.1090/conm/814

---

10 9 8 7 6 5 4 3 2 1   30 29 28 27 26 25

Contents

Preface

The current definition of Hopf algebras was motivated by the development of algebraic topology as well as algebraic groups. The progress on the theory of Hopf algebras in the last few decades has provided the foundation for their applications in other areas of mathematics such as combinatorics, low-dimensional topology, and mathematical physics. We are in debt to many pioneers of the subject. Professor Earl Jay Taft was certainly one of them. In memory of Professor Taft, the AMS special session was organized to provide a forum for different elements of development initiated by the theory of Hopf algebras in the last few decades.

A Hopf algebra H over a field $\Bbbk$ is a unital $\Bbbk$-algebra endowed with two algebra maps $\Delta : H \to H \otimes H$, $\epsilon : H \to \Bbbk$ and an algebra antihomomorphism $S : H \to H$, respectively, called comultiplication, counit, and antipode of H, satisfying some compatibility conditions. The antipode S is uniquely determined by H, and it plays the role of the inverse of an element in a group. Hopf algebras have been considered as generalizations of groups and Lie algebras, and they are indeed group objects in the category of vector spaces. For this reason, they were believed to be commutative or cocommutative (which means $\tau \circ \Delta = \Delta$ where τ is the flip map of the tensor product $H \otimes H$).

In a ground-breaking article of Professor Earl Jay Taft in 1971, a family of finite-dimensional noncommutative, noncocommutative, and nonsemisimple Hopf algebras was discovered, which are called Taft algebras $T_q(n)$ of dimension n^2 for any primitive nth root of unity $q \in \mathbb{C}$. The Taft algebras show that the orders of antipodes of finite-dimensional Hopf algebras can be arbitrarily large. Noncommutative and noncocommutative Hopf algebras are often referred to as quantum groups; they can be constructed from deformations of enveloping algebras of Lie algebras, and of coordinate rings of algebraic groups. However, the Taft algebras were considered by some as pathological before the introduction of quantum groups.

The comultiplication of a Hopf algebra H provides a tensor product on the category $\operatorname{Rep}(H)$ of finite-dimensional representations of H. It was shown by Sweedler in the 70s that the antipode S of any finite-dimensional Hopf algebra H is bijective. The invertibility of S implies $\operatorname{Rep}(H)$ is rigid, which means every finite-dimensional representation V of H admits a left dual and a right dual in $\operatorname{Rep}(H)$. The category of finite-dimensional representations of a finite-dimensional Hopf algebra is a prototype of finite tensor categories.

The Drinfeld double $D(H)$ construction for any finite-dimensional Hopf algebra H systematically provides examples of noncommutative and noncocommutative Hopf algebras. Moreover, each finite-dimensional representation V of $D(H)$ yields a solution of the quantum Yang–Baxter equation, via the notion of braiding on the representations of $D(H)$. In short, $\operatorname{Rep}(D(H))$ is a braided tensor category.

The small quantum group $u_q(sl_2)$ is a Hopf algebra quotient of the Drinfeld double $D(T_q(n))$ of the Taft algebra $T_q(n)$, and its representation category $\text{Rep}(u_q(sl_2))$ is also a braided tensor category with its braiding inherited from $\text{Rep}(D(T_q(n)))$.

A semisimple finite tensor category with finitely many simple objects and a nondegenerate braiding is often called a *modular tensor category* (MTC), which was originally used in conformal field theory. MTCs form the mathematical foundations of topological quantum computations and topological phase of matter, and they play an integral role in topological quantum field theory. The representation category $\text{Rep}(D(H))$ of $D(H)$ for any finite-dimensional semisimple Hopf algebra H over $\mathbb{C}$ or the semisimplification of $\text{Rep}(u_q(\mathfrak{g}))$ for any simple Lie algebra $\mathfrak{g}$ are typical examples of MTCs. The notion of Frobenius–Schur indicators introduced for finite groups and semisimple Hopf algebras can be extended to the context of MTCs. This generalized notion from Hopf algebras leads to affirmative answers for the congruent character conjecture in conformal field theory and the rank finiteness conjecture motivated from quantum computations.

Contemporary Mathematics
Volume **814**, 2025
https://doi.org/10.1090/conm/814/16299

On the finite-dimensional representations of the double of the Jordan plane

Nicolás Andruskiewitsch and Héctor Martín Peña Pollastri

In memoriam Earl Taft

ABSTRACT. We continue the study of the Drinfeld double of the Jordan plane, denoted by $\mathcal{D}$ and introduced in (Andruskiewitsch and Peña Pollastri, J. Algebra Appl. **20** (2021), no. 1, Paper No. 2140012, 26). The simple finite-dimensional modules were computed in (Andruskiewitsch, et al., Ark. Mat. **60** (2022), no. 2, 213–229); it turns out that they factorize through $U(\mathfrak{sl}_2(\Bbbk))$. Here we introduce the Verma modules and the category $\mathfrak{O}$ for $\mathcal{D}$, which have a resemblance to the similar ones in Lie theory but induced from indecomposable modules of the 0-part of the triangular decomposition. Accordingly, there is the notion of highest weight rank (hw-rk). We classify the indecomposable modules of hw-rk one and find families of hw-rk two. The Gabriel quiver of $\mathcal{D}$ is computed implying that it has a wild representation type.

1. Introduction

Let $\Bbbk$ be an algebraically closed field of characteristic 0. There were substantial advances in the ongoing classification of Nichols algebras over abelian groups with finite Gelfand–Kirillov dimension (abbreviated as GKdim), an important step towards the classification of Hopf algebras with finite GKdim, see [**AAH, AAM, AGI**]. Recall that Nichols algebras of diagonal type correspond to infinitesimal braidings that are semisimple as Yetter–Drinfeld modules (over abelian groups). In the context of finite GKdim, Nichols algebras with nonsemisimple infinitesimal braidings appear naturally. The well-known Jordan plane, denoted by J, is a paradigmatic example.

Quantized enveloping algebras can be defined via Drinfeld doubles of (bosonizations of) suitable Nichols algebras of diagonal type. A natural question is to investigate the behavior of the Drinfeld doubles of other Nichols algebras, specifically those with nonsemisimple infinitesimal braidings. In this direction, we studied in

---

2020 *Mathematics Subject Classification.* Primary 16T05, 16T20, 17B37, 17B62.
Key words and phrases. Hopf algebras, Nichols algebras, Gelfand–Kirillov dimension.
Both authors were partially supported by CONICET (PIP 11220200102916CO), FONCyT-ANPCyT (PICT-2019-03660) and Secyt (UNC). The second author was also supported by the Simons Foundation Award 889000 as part of the Simons Collaboration on Global Categorical Symmetries.

[**AP**, **ADP**] a Hopf algebra $\mathcal{D}$ thought of as the Drinfeld double of the bosonization $J\#\Bbbk\mathbb{Z}$, which differs from quantized enveloping algebras for some distinctive features:

- [**AP**] The Hopf algebra paired with $J\#\Bbbk\mathbb{Z}$ in order to build the double has the form $J^d\#U(\mathfrak{h})$ where J^d is the graded dual of J and $\dim\mathfrak{h} = 1$.
- [**AP**] The Hopf algebra $\mathcal{D}$ is not simple and, in fact, it fits into an *abelian* exact sequence of Hopf algebras

$$\mathcal{O}(\mathbf{G}) \hookrightarrow \mathcal{D} \xrightarrow{\pi} U(\mathfrak{sl}_2(\Bbbk))$$

  where $\mathcal{O}(\mathbf{G})$ is the algebra of regular functions on $\mathbf{G} = (\mathbf{G}_a \times \mathbf{G}_a) \rtimes \mathbf{G}_m$.
- [**AP**] The analogue of the Cartan subalgebra of $\mathcal{D}$ splits as $\mathcal{D}^0 \simeq U(\mathfrak{h}) \otimes \Bbbk\Gamma$.
- [**ADP**] The finite-dimensional simple modules over $\mathcal{D}$ are in bijective correspondence with those of $U(\mathfrak{sl}_2(\Bbbk))$ via π.
- [**BS**] The prime and primitive spectra of $\mathcal{D}$ are completely determined; from this information, it follows that $\mathcal{D}$ satisfies the Dixmier–Moeglin equivalence.

The next step is to understand the indecomposable modules. We observe that the action of $\mathcal{D}^0$ is not semisimple, what leads us to introduce Verma modules inducing from indecomposable $\mathcal{D}^0$-modules; highest weight modules and the category $\mathfrak{O}$ they belong to. The notions of hw-rk and hw-series arise naturally. See Section 3. In Sections 4 and 5 we study the indecomposable modules with hw-rk one and two respectively. In the case of hw-rk $= 1$ we get the complete classification by the family of uniserial indecomposable modules $\mathtt{T}(n,m)$; see Theorem 4.7. In Section 6 we classify extensions of simple modules using the families described previously. Thus we compute the Gabriel quiver of $\mathcal{D}$ and deduce that $\mathcal{D}$ has wild representation type; see Proposition 6.5. Section 2 is devoted to preliminaries on $\mathcal{D}$.

There are convincing reasons to believe that the representation theory of the Drinfeld doubles of many of the Nichols algebras discovered in [**AAH**, **AAM**] will have the features described above.

Conventions. If $\ell < n \in \mathbb{N}_0$, we set $\mathbb{I}_{\ell,n} = \{\ell, \ell+1, \dots, n\}$, $\mathbb{I}_n = \mathbb{I}_{1,n}$. If Y is a subobject of an object X in a category $\mathcal{C}$, then we write $X \leq Y$.

Let A be an algebra. Given $a_1, \dots, a_n \in A$, $n \in \mathbb{N}$, $\Bbbk\langle a_1, \dots, a_n\rangle$ denotes the subalgebra generated by $a_1, \dots, a_n$. Let ${}_A\mathrm{Mod}$ (respectively, ${}_A\mathrm{mod}$, $\operatorname{irrep} A$) denote the category of left A-modules (respectively, the full subcategory of finite-dimensional ones, the set of isomorphism classes of simple objects in ${}_A\mathrm{mod}$). Often we do not distinguish a class in $\operatorname{Irrep} A$ and one of its representatives. If $M \in {}_A\mathrm{Mod}$ and $m_1, \dots, m_n \in M$, $n \in \mathbb{N}$, then $\langle m_1, \dots, m_n\rangle$ denotes the submodule generated by $m_1, \dots, m_n$. Given a subalgebra B, of A, $\operatorname{Ind}_B^A : {}_B\mathrm{Mod} \to {}_A\mathrm{Mod}$ and $\operatorname{Res}_A^B : {}_A\mathrm{Mod} \to {}_B\mathrm{Mod}$ denote the induction and restriction functors, e.g., $\operatorname{Ind}_B^A(M) = A \otimes_B M$.

Let L be a Hopf algebra. The kernel of the co-unit ε is denoted L^+, the antipode (always assumed bijective) by $\mathcal{S}$, the space of primitive elements by $\mathcal{P}(L)$ and the group of group-likes by $G(L)$. The space of (g,h)-primitives is $\mathcal{P}_{g,h}(L) = \{x \in L : \Delta(x) = x \otimes h + g \otimes x\}$ where $g, h \in G(L)$. The category of Yetter–Drinfeld modules over L is denoted by ${}^L_L\mathcal{YD}$. We refer to [**R**] for unexplained terminology on Hopf algebras.

2. Preliminaries

2.1. The double of the Jordan plane. The well-known Jordan plane, the quadratic algebra $J = \Bbbk\langle x, y | xy - yx - \frac{1}{2}x^2\rangle$, is a braided Hopf algebra where x and y are primitive. The Hopf algebra $\mathcal{D}$ was introduced in [**AP**, Proposition 2.3], where it is denoted by $\widetilde{D}$. See [**AP**, **ADP**] for properties of $\mathcal{D}$, some of which are listed below. The algebra $\mathcal{D}$ is presented by generators u, v, ξ, $g^{\pm 1}$, x, y and relations

$$g^{\pm 1} g^{\mp 1} = 1, \quad \xi g = g\xi, \tag{2.1}$$

$$\begin{aligned} gx &= xg, \quad gy = yg + xg, \quad \xi y = y\xi - 2y, \quad \xi x = x\xi - 2x, \\ ug &= gu, \quad vg = gv + gu, \quad v\xi = \xi v - 2v, \quad u\xi = \xi u - 2u, \end{aligned} \tag{2.2}$$

$$\begin{aligned} yx &= xy - \frac{1}{2}x^2, \quad vu = uv - \frac{1}{2}u^2, \\ ux &= xu, \quad vx = xv + (1-g) + xu, \\ uy &= yu + (1-g), \quad vy = yv + \frac{1}{2}g\xi + yu. \end{aligned} \tag{2.3}$$

Notice that we have replaced the generator ζ by $\xi = -2\zeta$ in the presentation of [**AP**] and adjusted the relations accordingly. The Hopf algebra structure is determined by $g \in G(\mathcal{D})$, $u, \xi \in \mathcal{P}(\mathcal{D})$, $x, y \in \mathcal{P}_{g,1}(\mathcal{D})$ and

$$\Delta(v) = v \otimes 1 + 1 \otimes v - \frac{1}{2}\xi \otimes u.$$

The following set is a PBW-basis of $\mathcal{D}$:

$$B = \{x^n\, y^r\, g^m\, \xi^k\, u^i\, v^j : i, j, k, n, r \in \mathbb{N}_0, \quad m \in \mathbb{Z}\}.$$

We shall consider the following subalgebras of $\mathcal{D}$:

$$\begin{aligned} &\mathcal{D}^{<0} := \Bbbk\langle x, y\rangle, && \mathcal{D}^{0} := \Bbbk\langle g^{\pm 1}, \xi\rangle, && \mathcal{D}^{\leq 0} := \Bbbk\langle g^{\pm 1}, \xi, x, y\rangle, \\ &\mathcal{D}^{>0} := \Bbbk\langle u, v\rangle && \text{and} && \mathcal{D}^{\geq 0} := \Bbbk\langle g^{\pm 1}, \xi, u, v\rangle. \end{aligned}$$

The algebra $\mathcal{D}$ has a $\mathbb{Z}$-grading $\mathcal{D} = \oplus_{n\in\mathbb{Z}} \mathcal{D}^{[n]}$ given by

$$\deg x = \deg y = -2, \qquad \deg u = \deg v = 2, \qquad \deg g = \deg \xi = 0. \tag{2.4}$$

The algebra $\mathcal{D}$ has a triangular decomposition $\mathcal{D} \simeq \mathcal{D}^{<0} \otimes \mathcal{D}^{0} \otimes \mathcal{D}^{>0}$.

2.2. An exact sequence of Hopf algebras. Let $\mathcal{O} := \Bbbk\langle x, u, g^{\pm 1}\rangle$; this is a commutative Hopf subalgebra of $\mathcal{D}$, hence $\mathcal{O} \simeq \mathcal{O}(\mathbf{G})$, where $\mathbf{G}$ is the algebraic group as in Section 1. Let e, f, h be the Chevalley generators of $\mathfrak{sl}_2(\Bbbk)$, i.e., $[e, f] = h$, $[h, e] = 2e$, $[h, f] = -2f$. The Hopf algebra map $\pi\colon \mathcal{D} \to U(\mathfrak{sl}_2(\Bbbk))$ determined by

$$\pi(v) = \tfrac{1}{2}e, \qquad \pi(y) = f, \qquad \pi(\xi) = h, \qquad \pi(u) = \pi(x) = \pi(g-1) = 0, \tag{2.5}$$

induces an isomorphism of Hopf algebras $\mathcal{D}/\mathcal{D}\mathcal{O}^+ \simeq U(\mathfrak{sl}_2(\Bbbk))$. Thus we have an exact sequence as mentioned in Section 1.

2.3. Simple modules. The simple objets in ${}_{\mathcal{D}}\mathrm{mod}$ are classified in [**ADP**]; the starting point is the following fact that we will use later.

PROPOSITION 2.1 ([**ADP**, Proposition 3.9]). *If $M \in {}_{\mathcal{D}}\mathrm{mod}$, then $g-1$, x and u act nilpotently on M.* □

Let $\mathtt{L}(n)$ be the simple $\mathfrak{sl}_2(\Bbbk)$-module with highest weight n. Then $\mathtt{L}(n)$ becomes a simple $\mathcal{D}$-module via the projection π in (2.5). Precisely, $\mathtt{L}(n)$ has a basis $\{z_{n,0}, \dots, z_{n,n}\}$ where the action is given by

$$\begin{aligned} &\xi \cdot z_{n,i} = (n-2i)z_{n,i}, && v \cdot z_{n,i} = \tfrac{i(n-i+1)}{2} z_{n,i-1}, && y \cdot z_{n,i} = z_{n,i+1}, \\ &x \cdot z_{n,i} = 0, && u \cdot z_{n,i} = 0, && g \cdot z_{n,i} = z_{n,i}, \end{aligned} \tag{2.6}$$

$i \in \mathbb{I}_{0,n}$ where $z_{n,-1} = z_{n,n+1} = 0$ by convention.

THEOREM 2.2 ([**ADP**, 3.11]). *The family* $(\mathtt{L}(n))_{n\in\mathbb{N}_0}$ *parametrizes* $\operatorname{irrep}\mathcal{D}$. □

REMARK 2.3. If $M \in {}_{\mathcal{D}}\mathrm{mod}$, then y and v act nilpotently on M and the eigenvalues of the action of ξ are integers, but the action of ξ is not necessarily semisimple (consider a Jordan–Hölder series of M and apply Theorem 2.2).

3. Highest weight modules

3.1. Weight decompositions. Proposition 2.1 and Remark 2.3 lead us to the following considerations. Given $P \in {}_{\mathcal{D}^0}\mathrm{Mod}$ and $n \in \mathbb{Z}$, we set

$$P^{(n)} = \{m \in P : (\xi - n)^a \cdot m = 0,\ (g-1)^b \cdot m = 0 \text{ for some } a, b \in \mathbb{N}_0\}.$$

By a standard argument (see, e.g., [**D**, 1.2.13]) the sum $\sum_{n\in\mathbb{Z}} P^{(n)}$ is direct. Let ${}_{\mathcal{D}^0}\mathcal{SG}$ be the full subcategory of ${}_{\mathcal{D}^0}\mathrm{Mod}$ consisting of those $P \in {}_{\mathcal{D}^0}\mathrm{Mod}$ such that $P = \oplus_{n\in\mathbb{Z}} P^{(n)}$. Notice that ${}_{\mathcal{D}^0}\mathrm{mod}$ is a subcategory of ${}_{\mathcal{D}^0}\mathcal{SG}$.

Given $M \in {}_{\mathcal{D}}\mathrm{Mod}$, we set by abuse of notation

$$M^{(n)} = \left(\operatorname{Res}_{\mathcal{D}}^{\mathcal{D}^0}(M)\right)^{(n)}.$$

The weights of $M \in {}_{\mathcal{D}}\mathrm{Mod}$ are the elements of $\Pi(M) := \{n \in \mathbb{Z} : M^{(n)} \neq 0\}$; the $M^{(n)}$'s are called weight subspaces even when they are 0.

DEFINITION 3.1. A module $M \in {}_{\mathcal{D}}\mathrm{Mod}$ is *suitably graded* if

$$M = \oplus_{n\in\mathbb{Z}} M^{(n)} \qquad \text{and} \qquad \dim M^{(n)} < \infty \qquad \text{for all } n \in \mathbb{Z}.$$

Let ${}_{\mathcal{D}}\mathcal{SG}$ be the full subcategory of ${}_{\mathcal{D}}\mathrm{Mod}$ of suitably graded modules.

REMARK 3.2.

(i) From the defining relations we get that

$$\begin{aligned} x \cdot M^{(n)} \subset M^{(n-2)} \supset y \cdot M^{(n)}, \quad g \cdot M^{(n)} = M^{(n)} \supset \xi \cdot M^{(n)}, \\ u \cdot M^{(n)} \subset M^{(n+2)} \supset v \cdot M^{(n)}, \quad n \in \mathbb{Z}. \end{aligned} \tag{3.1}$$

(ii) Morphisms of $\mathcal{D}$-modules preserve the weight subspaces. Thus submodules and quotients of suitably graded modules are suitably graded. Given an exact sequence of $\mathcal{D}$-modules $N \hookrightarrow M \twoheadrightarrow S$, the sequence

$$N^{(n)} \hookrightarrow M^{(n)} \twoheadrightarrow S^{(n)}$$

of $\mathcal{D}^0$-modules is exact for any $n \in \mathbb{Z}$.

(iii) Any $M \in {}_{\mathcal{D}}\text{mod}$ is suitably graded by Remark 2.3. If $n \in \Pi(M)$, then

$$\dim M^{(n)} = \dim M^{(-n)} \qquad \text{and} \qquad \{n, n-2, n-4, \dots, 2-n, -n\} \subseteq \Pi(M).$$

PROOF. Let $\lambda \in \Bbbk$. Then $(\xi - \lambda + 2)y = y(\xi - \lambda)$ by (2.2); hence by induction $(\xi - \lambda + 2)^a y = y(\xi - \lambda)^a$ for any $a \in \mathbb{N}$. Also, $(g-1)y = y(g-1) + gx$, hence by induction $(g-1)^a y = y(g-1)^a + axg(g-1)^{a-1}$ for any $a \in \mathbb{N}$. Therefore $y \cdot M^{(n)} \subset M^{(n-2)}$; the rest is similar. The proof of (iii) follows from the representation theory of $\mathfrak{sl}_2$. □

Let $M \in {}_{\mathcal{D}}\mathcal{SG}$. Given $n \in \mathbb{Z}$, one identifies the dual $\big(M^{(n)}\big)^*$ with the subspace $\big(\oplus_{n \neq a \in \mathbb{Z}} M^{(a)}\big)^{\perp} = \{f \in M^* : f_{|M^{(a)}} = 0, a \neq n\}$ of M^*. Clearly the sum of the various duals $\big(M^{(n)}\big)^*$ is direct. The graded dual of M is

$$M^{\vee} = \oplus_{n \in \mathbb{Z}} \big(M^{(n)}\big)^* \hookrightarrow M^*.$$

We need the formula for the antipode:

$$\begin{aligned} &\mathcal{S}(g) = g^{-1}, \quad \mathcal{S}(x) = -g^{-1}x, \quad \mathcal{S}(y) = -g^{-1}y, \\ &\mathcal{S}(\xi) = -\xi, \quad \mathcal{S}(u) = -u, \qquad \mathcal{S}(v) = -v - \frac{1}{2}\xi u. \end{aligned} \tag{3.2}$$

LEMMA 3.3. *If $M \in {}_{\mathcal{D}}\mathcal{SG}$, then $M^{\vee} \in {}_{\mathcal{D}}\mathcal{SG}$ and*

$$(M^{\vee})^{(n)} = \big(M^{(-n)}\big)^*, \quad n \in \mathbb{Z}. \tag{3.3}$$

PROOF. Fix $q \in \mathbb{Z}$. Notice that $\big(M^{(q)}\big)^*$ is stable under the action of g and ξ. We claim that

$$\begin{aligned} x \cdot \big(M^{(q)}\big)^* \subset \big(M^{(q+2)}\big)^* \supset y \cdot \big(M^{(q)}\big)^*, \\ u \cdot \big(M^{(q)}\big)^* \subset \big(M^{(q-2)}\big)^* \supset v \cdot \big(M^{(q)}\big)^*. \end{aligned}$$

Indeed, if $f \in \big(M^{(q)}\big)^*$, $p \in \mathbb{Z}$ and $m \in M^{(p)}$, then

$$\langle x \cdot f, m \rangle = -\langle f, x \cdot m \rangle = 0 \text{ unless } x \cdot m \in M^{(q)} \text{ that is } p - 2 = q.$$

The other inclusions are similar. Thus $M^{\vee}$ is a submodule of M^*. We next claim that $\big(M^{(q)}\big)^* \subset \big(M^*\big)^{(-q)}$. Since $\dim M^{(q)} < \infty$, there exist $a, b \in \mathbb{N}$ such that $(\xi - q)^a$ and $(g-1)^b$ act by 0 on $M^{(q)}$. Pick $f \in \big(M^{(q)}\big)^*$. By (3.2) we have, for any $p \in \mathbb{Z}$ and $m \in M^{(p)}$, that

$$\begin{aligned} &\langle (\xi + q) \cdot f, m \rangle = \langle f, (-\xi + q) \cdot m \rangle \\ &\qquad \Longrightarrow \langle (\xi + q)^a \cdot f, m \rangle = (-1)^a \langle f, (\xi - q)^a \cdot m \rangle = 0; \end{aligned}$$

$$\begin{aligned} &\langle (g-1) \cdot f, m \rangle = \langle f, (g^{-1} - 1) \cdot m \rangle \\ &\qquad \Longrightarrow \langle (g-1)^b \cdot f, m \rangle = (-1)^b \langle f, (g-1)^b \cdot (g^{-b} \cdot m) \rangle = 0 \end{aligned}$$

The claim follows. Together with the first claim, this implies (3.3). □

REMARK 3.4. If $M, N \in {}_{\mathcal{D}}\mathcal{SG}$, then $M \otimes N = \oplus_{n \in \mathbb{Z}} (M \otimes N)^{(n)}$ and

$$(M \otimes N)^{(n)} = \oplus_{p+q=n} M^{(p)} \otimes N^{(q)}.$$

Notice however that $\dim (M \otimes N)^{(p+q)}$ is not necessarily finite.

PROOF. It suffices to show that $M^{(p)} \otimes N^{(q)} \subset (M \otimes N)^{(p+q)}$ for any $p, q \in \mathbb{Z}$; this follows from the equalities

$$\Delta(\xi - \lambda)^a = \sum_{0 \le c \le a} \binom{a}{c} (\xi - \mu)^c \otimes (\xi - \nu)^{a-c}, \quad \mu + \nu = \lambda;$$

$$\Delta(g - 1)^b = \sum_{0 \le d \le b} \binom{b}{d} (g - 1)^d \otimes g^d (g - 1)^{b-d}, \quad a, b \in \mathbb{N}.$$

3.2. Highest weight modules. Let us now consider $P \in {}_{\mathcal{D}^0}\text{mod}$ such that $P = P^{(n)}$ for some $n \in \mathbb{Z}$. Then P becomes a $\mathcal{D}^{\geq 0}$-module by $u \cdot m = 0$, $v \cdot m = 0$, $m \in P$. We define the Verma module

$$\mathtt{M}(P) := \operatorname{Ind}_{\mathcal{D}^{\geq 0}}^{\mathcal{D}}(P) \simeq \mathcal{D} \otimes_{\mathcal{D}^{\geq 0}} P.$$

Thus $\mathtt{M}(P) \simeq \mathcal{D}^{<0} \otimes_{\Bbbk} P$ as vector spaces.[1]

REMARK 3.5. The Verma module $\mathtt{M}(P)$ is suitably graded. Indeed we have $P = P^{(n)} \subset \mathtt{M}(P)^{(n)}$ by definition, hence $y^i x^j P \subset \mathtt{M}(P)^{(n-2(i+j))}$ for any $i, j \in \mathbb{N}_0$ by (3.1). That is, $\mathtt{M}(P)^{(n-2k)} = \oplus_{i+j=k} y^i x^j P$ and

$$\mathtt{M}(P) = \bigoplus_{k \in \mathbb{N}_0} \mathtt{M}(P)^{(n-2k)}.$$

DEFINITION 3.6. Let $\mathfrak{O}$ be the full subcategory of ${}_{\mathcal{D}}\mathcal{SG}$ consisting of those M such that $\varPi(M)$ is bounded above. For $M \in \mathfrak{O}$, its highest weight is $\operatorname{hw} M := \sup \varPi(M)$; we introduce $\text{hw-rk}\, M = \dim M^{(\operatorname{hw} M)}$.

DEFINITION 3.7. If $M \in \mathfrak{O}$ is generated by $M^{(\operatorname{hw} M)}$, then we say that M is a *highest weight* module.

REMARK 3.8.

(i) If $M \in \mathfrak{O}$ has highest weight n and $P := M^{(n)}$, then there is a morphism of $\mathcal{D}$-modules $\varPhi : \mathtt{M}(P) \to M$ which is the identity on P; indeed, $u \cdot M^{(n)} = v \cdot M^{(n)} = 0$. Furthermore, $\operatorname{hw}(M/\operatorname{Im}\varPhi) < \operatorname{hw} M$.

(ii) The Verma modules $\mathtt{M}(P)$ with $P = P^{(n)} \in {}_{\mathcal{D}^0}\text{mod}$ are highest weight modules. Any highest weight module M is the quotient of a Verma module, namely of $\mathtt{M}(P)$ with $P = M^{(\operatorname{hw} M)}$.

(iii) The category ${}_{\mathcal{D}}\text{mod}$ is a subcategory of $\mathfrak{O}$. Thus any $M \in {}_{\mathcal{D}}\text{mod}$ has a unique series of submodules $0 = M_0 \subsetneq M_1 \subsetneq M_2 \cdots \subsetneq M_r = M$ (that we call the *hw-series*) such that M_i/M_{i-1} is a highest weigth module, $i \in \mathbb{I}_r$, and $\text{hw-rk}\, M_i/M_{i-1} > \text{hw-rk}\, M_{i+1}/M_i$, $i \in \mathbb{I}_{r-1}$.

(iv) If $M, N \in \mathfrak{O}$, then $M \otimes N \in \mathfrak{O}$. This follows from Remark 3.4; an elementary argument shows that the set of pairs $(p, q) \in \varPi(M) \times \varPi(N)$ such that $p+q = n$ for a given n, is finite when both $\varPi(M)$ and $\varPi(N)$ are bounded above (or below).

[1] Notice that the Verma modules in [**ADP**] are induced from one-dimensional modules where g and ξ act by arbitrary eigenvalues, thus more and less general than the previous definition.

3.3. Lowest weight modules. We also have the full subcategory $\mathfrak{o}$ of ${}_{\mathcal{D}}\mathcal{SG}$ consisting of those M such that $\Pi(M)$ is bounded below; for $M \in \mathfrak{o}$, we set $\operatorname{lw} M := \inf \Pi(M)$ and $\operatorname{lw-rank} M := \dim M^{(\operatorname{lw} M)}$. If $M \in \mathfrak{O}$, then $M^{\vee} \in \mathfrak{o}$. This gives a contravariant equivalence of categories between $\mathfrak{O}$ and $\mathfrak{o}$.

Given $P \in {}_{\mathcal{D}^0}\text{mod}$ such that $P = P^{(n)}$ for some $n \in \mathbb{Z}$, P becomes a $\mathcal{D}^{\leq 0}$-module by $x \cdot m = 0$, $y \cdot m = 0$, $m \in P$. The opposite Verma module is

$$\overline{\mathtt{M}}(P) := \operatorname{Ind}_{\mathcal{D}^{\leq 0}}^{\mathcal{D}}(P) \simeq \mathcal{D} \otimes_{\mathcal{D}^{\leq 0}} P.$$

Clearly, $\overline{\mathtt{M}}(P) \simeq \mathcal{D}^{>0} \otimes_{\Bbbk} P$ as vector spaces. Then $M \in \mathfrak{o}$ is a *lowest weight* module if it is generated by $M^{(\operatorname{lw} M)}$. We have the following properties.

(i) If $M \in \mathfrak{o}$ and $P := M^{(\operatorname{lw} M)}$, then there is a morphism of $\mathcal{D}$-modules $\varphi : \overline{\mathtt{M}}(P) \to M$, $\varphi_{|P} = \mathrm{id}_P$. Moreover, $\operatorname{lw}(M/\operatorname{Im}\varphi) > \operatorname{lw} M$.

(ii) The opposite Verma modules are lowest weight modules and any lowest weight module is the quotient of one of them.

(iii) The category ${}_{\mathcal{D}}\text{mod}$ is a subcategory of $\mathfrak{o}$ and any object belonging to $\mathfrak{o}$ and $\mathfrak{O}$ is in ${}_{\mathcal{D}}\text{mod}$. Thus any $M \in {}_{\mathcal{D}}\text{mod}$ has a unique series of submodules $0 = M_0 \subsetneq M_1 \subsetneq M_2 \cdots \subsetneq M_t = M$ (that we call the *lw-series*) such that M_i/M_{i-1} is a lowest weigth module, $i \in \mathbb{I}_t$, and $\operatorname{lw-rank} M_i/M_{i-1} < \operatorname{lw-rank} M_{i+1}/M_i$, $i \in \mathbb{I}_{t-1}$.

(iv) If $M, N \in \mathfrak{o}$, then $M \otimes N \in \mathfrak{o}$.

(v) Let $P \in {}_{\mathcal{D}^0}\text{mod}$ such that $P = P^{(n)}$ for some $n \in \mathbb{Z}$. Then P^* does not generate $\mathtt{M}(P)^{\vee}$.

(vi) Let $P \in {}_{\mathcal{D}^0}\text{mod}$ such that $P = P^{(n)}$ for some $n \in \mathbb{Z}$. Then the natural map of $\mathcal{D}$-modules $\varphi : \overline{\mathtt{M}}(P) \to \mathtt{M}(P^*)^{\vee}$ is not an isomorphism.

PROOF OF (v). Let $p \in P$, $f \in P^*$ and $a, b \in \Bbbk$. Because of (2.3), we have

$$\begin{aligned}\langle (au+bv)\cdot f, x\cdot p\rangle &= -\langle \cdot f, (au+bv)x\cdot p\rangle \\ &\quad - \langle f, (x(au+b(u+v)) + b(1-g))\cdot p\rangle \\ &\quad - \langle f, b(1-g)\cdot p\rangle;\end{aligned}$$

thus, $\langle (au+bv)\cdot f, x\cdot p\rangle = 0$ if $p \in \ker(1-g)$. Since $\ker(1-g) \neq 0$, we conclude that $(x\cdot P)^*$ is not contained in the submodule generated by P^*. □

PROOF OF (vi). The natural isomorphism $P \simeq P^{**}$ is of $\mathcal{D}^0$-modules because $\mathcal{S}^2_{\mathcal{D}^0} = \mathrm{id}$; thus φ exists. Now $\operatorname{Im}\varphi$ is generated by P^{**}, thus (v) applies. □

4. Modules of highest weight rank one

We aim to describe indecomposable finite-dimensional highest weight modules; we start in this section with those having highest weight 1. We shall define an indecomposable module $\mathtt{T}(n,m) \in {}_{\mathcal{D}}\text{mod}$ for any $(n,m) \in \mathbb{N}_0^2$.

The only Verma modules of hw-rk $= 1$, with the conventions of this article, are $\mathtt{M}(n) := \mathtt{M}(\lambda_n)$, $n \in \mathbb{Z}$, where $\lambda_n \in {}_{\mathcal{D}^{\geq 0}}\text{mod}$ has dimension 1, with basis $\{z_n\}$ and action

$$g\cdot z_n = z_n, \qquad \xi\cdot z_n = nz_n, \qquad u\cdot z_n = 0, \qquad v\cdot z_n = 0.$$

The elements $z_{n,(i,j)} := y^i x^j \cdot z_n$, $i,j \in \mathbb{N}_0$, form a basis of $\mathtt{M}(n)$. Recall that $[t]^{[k]}$ denotes the raising factorial $[t]^{[k]} := \prod_{i=1}^{k}(t+i-1)$ for $t \in \Bbbk$ and $k \in \mathbb{N}_0$. By

[**AP**, Lemma 2.5] the action of $\mathcal{D}$ is explicitly given by

$$
\begin{aligned}
x\cdot z_{n,(i,j)} &= \sum_{k=0}^{i} \binom{i}{k}\tfrac{k!}{2^k} z_{n,(i-k,j+k+1)},\\
y\cdot z_{n,(i,j)} &= z_{n,(i+1,j)},\\
g\cdot z_{n,(i,j)} &= \sum_{k=0}^{i} \binom{i}{k}\tfrac{[2]^{[k]}}{2^k} z_{n,(i-k,j+k)},\\
\xi\cdot z_{n,(i,j)} &= \big(n-2(i+j)\big) z_{n,(i,j)},\\
u\cdot z_{n,(i,j)} &= (-1)\sum_{k=1}^{i-1} \binom{i}{k+1}\tfrac{(k+1)!}{2^k} z_{n,(i-1-k,j+k)},\\
v\cdot z_{n,(i,j)} &= \tfrac{i(n-2j-i+1)}{2} z_{n,(i-1,j)} + \sum_{k=1}^{i-1} \tfrac{i!(n-j-i+1)}{(n-k-1)!2^{k+1}} z_{n,(i-1-k,j+k)}.
\end{aligned}
\tag{4.1}
$$

It was shown in [**ADP**] that the Verma module $\mathtt{M}(n)$ has a unique simple quotient $\mathtt{M}(n)/\langle z_{n,(0,1)}, z_{n,(n+1,0)}\rangle$ which is isomorphic to $\mathtt{L}(n)$, cf. Theorem 2.2. Now $\mathtt{M}(n)$ is presented by the generator z_n with defining relations

$$
g\cdot z_n = z_n, \qquad \xi\cdot z_n = nz_n, \qquad u\cdot z_n = 0, \qquad v\cdot z_n = 0. \tag{4.2}
$$

Similarly, $\mathtt{L}(n)$ is presented by z_n with defining relations (4.2) and

$$
x\cdot z_n = 0, \qquad y^{n+1}\cdot z_n = 0. \tag{4.3}
$$

DEFINITION 4.1. Let $(n,m)\in\mathbb{N}_0^2$. We define $\mathtt{T}(n,m) := \mathtt{M}(n+2m)/\mathtt{N}(n,m)$ where $\mathtt{N}(n,m)$ is the submodule of $\mathtt{M}(n+2m)$ generated by

$$
\begin{aligned}
z_{n+2m,(0,m+1)} &= x^{m+1}\cdot z_{n+2m},\\
z_{n+2m,(n+2(m-j)+1,j)} &= y^{n+2(m-j)+1}x^j\cdot z_n, \quad j\in\mathbb{I}_{0,m}.
\end{aligned}
\tag{4.4}
$$

Thus $\mathtt{T}(n,m)$ is presented by $z = z_{n+2m}$ with defining relations

$$
g\cdot z_{n+2m} = z_{n+2m}, \qquad \xi\cdot z_{n+2m} = (n+2m)z_{n+2m}, \tag{4.5}
$$

$$
u\cdot z_{n+2m} = 0, \qquad v\cdot z_{n+2m} = 0, \tag{4.6}
$$

$$
x^{m+1}\cdot z_{n+2m} = 0, \qquad y^{n+2(m-j)+1}x^j\cdot z_{n+2m} = 0, \qquad j\in\mathbb{I}_{0,m}. \tag{4.7}
$$

Let $\mathtt{z}_{n,m,(i,j)}$ be the image of $z_{n+2m,(i,j)}$ in $\mathtt{T}(n,m)$. For simplicity, we denote $z_{n+2m,(i,j)}$ by $z_{(i,j)}$ and $\mathtt{z}_{n,m,(i,j)}$ by $\mathtt{z}_{(i,j)}$.

LEMMA 4.2. *The set* $B = \{\mathtt{z}_{(i,j)} : j\in\mathbb{I}_{0,m}, i\in\mathbb{I}_{0,n+2(m-j)}\}$ *is a basis of* $\mathtt{T}(n,m)$*, thus*

$$
\dim\mathtt{T}(n,m) = (m+1)(n+m+1).
$$

PROOF. Let $\mathtt{N}$ be the vector subspace of $\mathtt{M}(n+2m)$ generated by the elements

$$
z_{(i,j)} \qquad \text{with either} \qquad \begin{cases} j\geq m+1 & \text{and } i\in\mathbb{N}_0, \text{ or else}\\ 0\leq j\leq m & \text{and } i\geq n+2(m-j)+1. \end{cases} \tag{4.8}
$$

We claim that $\mathtt{N}(n,m) = \mathtt{N}$; clearly this equality implies the Lemma. For this claim, since the generators (4.4) belong to $\mathtt{N}$, we are reduced to prove

(i) $\mathtt{N}(n,m) \supset \mathtt{N}$. (ii) $\mathtt{N}$ is a submodule of $\mathtt{M}(n+2m)$,

(i) It is clear: if $j \geq m+1$, then $z_{(i,j)} = y^i x^{j-m-1} \cdot z_{(0,m+1)}$; if $j \in \mathbb{I}_{0,m}$ and $i \geq n+2(m-j)+1$, then $z_{(i,j)} = y^{i-n-2(m-j)-1} \cdot z_{(n+2(m-j)+1,j)}$.

(ii): We use the formulas (4.1) to show that the generators leave $\mathtt{N}$ invariant. Fix $z_{(i_0,j_0)}$ as in (4.8). If $j_0 \geq m+1$, then by (4.1) the actions of the generators involve linear combinations of $z_{(i,j)}$ with $j \geq j_0$ hence they are in $\mathtt{N}$. So we assume that $0 \leq j_0 \leq m$ and $i_0 \geq n+2(m-j_0)+1$.

Action of x on $z_{(i_0,j_0)}$. Here, $x \cdot z_{(i_0,j_0)}$ is a linear combination of elements of the form $z_{(i_0-k,j_0+k+1)}$ with $k \in \mathbb{I}_{0,i_0}$. If $j_0+k+1 \geq m+1$, then we are done. Otherwise,

$$i_0 - k \geq n + 2m - 2j_0 + 1 - k \geq n + 2m - 2(j_0 + k + 1) + 1.$$

Action of y or ξ on $z_{(i_0,j_0)}$. This is clear.

Action of g on $z_{(i_0,j_0)}$. Here, $g \cdot z_{(i_0,j_0)}$ is a linear combination of elements of the form $z_{(i_0-k,j_0+k)}$ with $k \in \mathbb{I}_{0,i_0}$. If $j_0+k \geq m+1$, then we are done. Otherwise,

$$i_0 - k \geq n + 2m - 2j_0 + 1 - k \geq n + 2m - 2(j_0 + k) + 1.$$

Action of u on $z_{(i_0,j_0)}$. Here, $u \cdot z_{(i_0,j_0)}$ is a linear combination of elements of the form $z_{(i_0-k-1,j_0+k)}$ with $k \in \mathbb{I}_{i_0-1}$. If $j_0+k \geq m+1$, then we are done. Otherwise,

$$i_0 - k - 1 \geq n + 2m - 2j_0 + 1 - k - 1 \geq n + 2m - 2(j_0 + k) + 1.$$

Action of v on $z_{(i_0,j_0)}$. Here, $v \cdot z_{(i_0,j_0)}$ is a linear combination of $z_{(i_0-1,j_0)})$ and elements of the form $z_{(i_0-k-1,j_0+k)}$ with $k \in \mathbb{I}_{i_0-1}$. We begin with the latter case. If $j_0+k \geq m+1$, then we are done. Otherwise,

$$i_0 - 1 - k \geq n + 2m - 2j_0 + 1 - 1 - k \geq n + 2m - 2(j_0 + k) + 1.$$

Now for $z(i_0-1,j_0)$, if $i_0 = n+2(m-j_0)+1$ then the coefficient that goes with $z(i_0-1,j_0)$ is zero. So we can assume $i_0 > n+2(m-j_0)+1$. Then $i_0 - 1 \geq n+2(m-j_0)+1$ and we are done. □

Clearly, $\mathtt{T}(n,0) \simeq \mathtt{L}(n)$.

LEMMA 4.3. *The linear map $\psi : \mathtt{L}(n) \hookrightarrow \mathtt{T}(n,m)$ given by $z_{n,i} \mapsto \mathtt{z}_{(i,m)}$ for $i \in \mathbb{I}_{0,n}$ is a monomorphism of $\mathcal{D}$-modules. Let $\mathtt{L} := \operatorname{Im}\psi$. If $m \geq 1$, then*

$$\mathtt{T}(n,m)/\mathtt{L} \simeq \mathtt{T}(n+2,m-1).$$

PROOF. Using the formulas in (4.1), we see first that $\mathtt{z}_{(0,m)}$ satisfies the defining relations (4.2) and (4.3) of $\mathtt{L}(n)$, implying the existence of ψ with the desired properties. Second, we see that the class of $\mathtt{z}_{(0,0)}$ in $\mathtt{T}(n,m)/\mathtt{L}$ satisfies the defining relations (4.5), (4.6), and (4.7) of $\mathtt{T}(n+2,m-1)$. Thus we have an epimorphism $\mathtt{T}(n+2,m-1) \twoheadrightarrow \mathtt{T}(n,m)/\mathtt{L}$ which is an isomorphism because $\dim \mathtt{T}(n+2,m-1) = m(n+m+2) = \dim \mathtt{T}(n,m)/\mathtt{L}$. □

LEMMA 4.4. *Let N be a nonzero submodule of $\mathtt{T}(n,m)$. Then $\mathtt{L} \subseteq N$.*

PROOF. We first show that for every $j \in \mathbb{I}_m$ and $i \in \mathbb{I}_{0,n+2(m-j)}$ we have

$$v^i \cdot \mathtt{z}_{(i,j)} = \frac{(i!)^2}{2^i}\binom{n+2(m-j)}{i}\mathtt{z}_{(0,j)}. \tag{4.9}$$

We argue recursively on i. For $i=0$ the equality is clear. Suppose that

$$v^\ell \cdot \mathtt{z}_{(\ell,j)} = \frac{(\ell!)^2}{2^\ell}\binom{n+2(m-j)}{\ell}\mathtt{z}_{(0,j)}, \qquad \text{for } \ell \in \mathbb{I}_{0,i}.$$

Using (4.1) we see that $v^{i+1} \cdot \mathtt{z}_{(i+1,j)}$ is equal to

$$\begin{aligned}
&\tfrac{(i+1)(n+2(m-j)-i)}{2} v^i \mathtt{z}_{(i,j)} + \sum_{k=1}^{i} \tbinom{i+1}{k+1} \tfrac{(n+2m-j-i)(k+1)!}{2^{k+1}} v^k v^{i-k} \cdot \mathtt{z}_{(i-k,j+k)} \\
&= \tfrac{(i+1)(n+2(m-j)-i)(i!)^2}{2^{i+1}} \tbinom{n+2(m-j)}{i} \mathtt{z}_{(0,j)} \\
&\quad + \sum_{k=1}^{i} \tbinom{i+1}{k+1} \tfrac{(n+2m-j-i)(k+1)!((i-k)!)^2}{2^{i+1}} \tbinom{n+2(m-j-k)}{i-k} v^k \mathtt{z}_{(0,j+k)} \\
&= \tfrac{((i+1)!)^2}{2^{i+1}} \tbinom{n+2(m-j)}{i+1} \mathtt{z}_{(0,j)}.
\end{aligned}$$

Clearly, (4.9) implies that $v^k \cdot \mathtt{z}_{(i,j)} = 0$ for $k > i$. Now let $z \in N - 0$. Then $z = \sum_{i,j} c_{i,j}\mathtt{z}_{(i,j)}$ for some $c_{i,j} \in \Bbbk$. Let $i_0 = \max\{i\colon c_{i,j} \neq 0\}$. Then

$$v^{i_0} \cdot z = \sum_j c_{i_0,j} \frac{(i_0!)^2}{2^{i_0}} \binom{n+2(m-j)}{i_0} \mathtt{z}_{(0,j)} \in N.$$

Taking $j_0 = \max\{j\colon c_{i_0,j} \neq 0\}$ we get that

$$x^{m-j_0} v^{i_0} \cdot z = c_{i_0,j_0} \frac{(i_0!)^2}{2^{i_0}} \binom{n+2(m-j_0)}{i_0} \mathtt{z}_{(0,m)} \in N.$$

Hence $\mathtt{z}_{(0,m)} \in N$ and since $\mathtt{L}$ is simple, this shows that $\mathtt{L} \subseteq N$. □

PROPOSITION 4.5. *The module* $\mathtt{T}(n,m)$ *is uniserial and indecomposable.*

PROOF. If N is a simple submodule of $\mathtt{T}(n,m)$, then $N = \mathtt{L}$ by Lemma 4.4, thus $\mathtt{L}$ is the socle of $\mathtt{T}(n,m)$. We conclude from Lemma 4.3 that the socle series is a composition series, hence $\mathtt{T}(n,m)$ is uniserial and indecomposable. □

REMARK 4.6. The dual module $\mathtt{T}(n,m)^*$ is also uniserial and indecomposable but it is not a highest weight module; the subfactors of its hw-series are the simple modules $\mathtt{L}(n)$ etc.

THEOREM 4.7. *Let* $\mathtt{T}$ *be an indecomposable finite-dimensional highest weight module with* $\operatorname{hw-rk} \mathtt{T} = 1$. *Then* $\mathtt{T} \simeq \mathtt{T}(n,m)$ *for some* $(n,m) \in \mathbb{N}_0$.

PROOF. Assume that $\operatorname{hw} \mathtt{T} = p \in \mathbb{Z}$. Since $\mathtt{T}$ is generated by $\mathtt{T}^{(p)}$, $\mathtt{T}$ is a quotient of $\mathtt{M}(p)$. Fix $\mathtt{z} \in \mathtt{T}^{(p)} - 0$. Since $\operatorname{hw-rk} \mathtt{T} = 1$, $\mathtt{z}$ generates $\mathtt{T}$, $\xi \cdot \mathtt{z} = p\mathtt{z}$ and $g \cdot \mathtt{z} = \mathtt{z}$. Given a simple quotient L of $\mathtt{T}$, L is then a (finite-dimensional) simple quotient of $\mathtt{M}(p)$. Then $p \in \mathbb{N}_0$ and $L \simeq L(p)$ as in [**ADP**, 3.11].

By Proposition 2.1, $x^m \cdot \mathtt{z} \neq 0$ and $x^{m+1} \cdot \mathtt{z} = 0$ for some $m \in \mathbb{N}_0$. Let $n := p - 2m$. Applying the relations (4.1), we see that the submodule generated by $x^m \cdot \mathtt{z}$ is a quotient of $M(n)$; hence $n \in \mathbb{N}_0$ arguing as above.

Let $j \in \mathbb{I}_{0,m}$. By Remark 2.3 there exists $a_j \in \mathbb{N}_0$ such that $y^{a_j} x^j \cdot \mathtt{z} \neq 0$, $y^{a_j+1} x^j \cdot \mathtt{z} = 0$. Then the last commutation relation in (4.1) says that

$$\begin{aligned} 0 = v y^{a_j+1} x^j \cdot \mathtt{z} &= \tfrac{(a_j+1)(p-2j-a_j)}{2} y^{a_j} x^j \cdot \mathtt{z} \\ &+ \sum_{k=1}^{a_j} \tfrac{(a_j+1)!(p-j-a_j)}{(p-k-1)!2^{k+1}} y^{a_j-k} x^{j+k} \cdot \mathtt{z}. \end{aligned} \tag{4.10}$$

Let $I = \{k \in \mathbb{I}_{0,a_j} : y^{a_j-k} x^{j+k} \cdot \mathtt{z} \neq 0\}$. Then $\{y^{a_j-k} x^{j+k} \cdot \mathtt{z} : k \in I\}$ is linearly independent. Indeed if $\sum_{k \in I} c_k y^{a_j-k} x^{j+k} \cdot \mathtt{z} = 0$, then applying y enough times, one can show that each c_k should be zero. Then equation (4.10) tells us that $p - 2j - a_j = 0$ since $0 \in I$. Then $a_j = p - 2j = n - 2(m-j)$ by definition of n. Hence $\mathtt{T}$ is a quotient of $\mathtt{T}(n,m)$. Let $\mathtt{z}_{i,j} := y^i x^j \cdot \mathtt{z}$. By an argument similar to the one given in Lemma 4.2, one shows that the elements $\{\mathtt{z}_{i,j}\}$ form a basis of $\mathtt{T}$, hence $\mathtt{T} \simeq \mathtt{T}(n,m)$. □

REMARK 4.8. The previous result shows that the requirement in Definition 3.7 that highest weight modules are generated by their highest weight subspaces is necessary. Indeed, otherwise there would be many more indecomposables than in Proposition 4.7. For instance let $\widetilde{\mathtt{T}} = \mathtt{T}(n,m) \oplus S_\gamma(n)$, where $m \geq 1$ and $S_\gamma(n)$ is the indecomposable module defined in Proposition 5.2, let N be the submodule generated by the elements $s_i - \mathtt{z}_{i,m}$, and let $\mathtt{T} = \widetilde{\mathtt{T}}/N$. Then $\mathtt{T}$ is easily seen to be indecomposable since it satisfies a property similar to the one proven in Lemma 4.4. It satisfy hw-rk $\mathtt{T} = 1$, but it is not isomorphic to any $\mathtt{T}(n',m')$ since it has two copies of $\mathtt{L}(n)$ as composition factors.

5. Modules of highest weight rank two

In this section we introduce a family $\mathtt{S}_\gamma(n)$ of highest weight modules of hw-rk 2 and use it to classify self extensions of simple modules.

5.1. Highest weight rank 2. Let $n \in \mathbb{N}_0$ and $(\lambda,\mu) \in \Bbbk^2$. We consider the $\mathcal{D}^0$-module $\mathtt{P}_{\lambda,\mu}(n)$ with basis $\mathtt{s},\mathtt{w}$ and action

$$g \cdot \mathtt{s} = \mathtt{s}, \qquad g \cdot \mathtt{w} = \mathtt{w} + \lambda \mathtt{s}, \qquad \xi \cdot \mathtt{s} = n\mathtt{s}, \qquad \xi \cdot \mathtt{w} = n\mathtt{w} + \mu \mathtt{s}. \tag{5.1}$$

The module $\mathtt{P}_{\lambda,\mu}(n)$ is isomorphic to $\mathtt{P}_{t\lambda,t\mu}(n)$ for any $t \in \Bbbk^\times$ and is indecomposable whenever $(\lambda,\mu) \neq 0$; any indecomposable in ${}_{\mathcal{D}^0}\mathrm{mod}$ of dimension 2 has this shape. Let $\mathtt{M}_{\lambda,\mu}(n) = \operatorname{Ind}_{\mathcal{D}^{\geq 0}}^{\mathcal{D}}(\mathtt{P}_{\lambda,\mu}(n))$; it is presented by generators $\mathtt{s}$ and $\mathtt{w}$ with defining relations (5.1) and

$$u \cdot \mathtt{s} = 0, \qquad v \cdot \mathtt{s} = 0, \qquad u \cdot \mathtt{w} = 0, \qquad v \cdot \mathtt{w} = 0. \tag{5.2}$$

Here is a basis of $\mathtt{M}_{\lambda,\mu}(n)$:

$$\mathtt{s}_{i,j} := y^i x^j \cdot \mathtt{s}, \qquad \mathtt{w}_{i,j} := y^i x^j \cdot \mathtt{w}, \qquad i,j \in \mathbb{N}_0.$$

Let $\mathtt{M}_1$ be the span of $(\mathtt{s}_{i,j})_{i,j \in \mathbb{N}_0}$ and $\mathtt{M}_2 = \mathtt{M}_{\lambda,\mu}(n)/\mathtt{M}_1$. Thus we have a short exact sequence $\mathtt{M}(n) \simeq \mathtt{M}_1 \hookrightarrow \mathtt{M}_{\lambda,\mu}(n) \twoheadrightarrow \mathtt{M}_2 \simeq \mathtt{M}(n)$ in ${}_{\mathcal{D}}\mathrm{Mod}$.

In the rest of Section 5.1, N is a submodule of $\mathtt{M}_{\lambda,\mu}(n)$,

$$S := \mathtt{M}_{\lambda,\mu}(n)/N, \qquad S_1 := \mathtt{M}_1/\mathtt{M}_1 \cap N \hookrightarrow S, \qquad S_2 := S/S_1 \simeq \mathtt{M}_2/\left(N/\mathtt{M}_1 \cap N\right).$$

Let s, w, s_{ij}, w_{ij} be the images of $\mathtt{s}, \mathtt{w}, \mathtt{s}_{ij}, \mathtt{w}_{ij}$ in S; let $\overline{r} \in S_2$ be the image of $r \in S$ under the canonical projection.

5.1.1. *S_1 and S_2 are simple.*

LEMMA 5.1. *Keep the notation above.*

(i) *Assume that the following relations hold in S and S_2:*

$$x \cdot s = 0, \tag{5.3}$$

$$x \cdot \overline{w} = 0. \tag{5.4}$$

Then S_1 is spanned by $s_i := s_{i,0} = y^i \cdot s$, $i \in \mathbb{N}_0$; and S_2 is spanned by $\overline{w_i}$ where $w_i := w_{i,0} = y^i \cdot w$, $i \in \mathbb{N}_0$. Hence

$$S^{(n-2i)} = \Bbbk s_i + \Bbbk w_i, \quad i \in \mathbb{N}_0, \qquad \text{and} \qquad S = \oplus_{i\in\mathbb{N}_0} S^{(n-2i)}. \tag{5.5}$$

In addition, there exists $\gamma \in \Bbbk$ such that $x \cdot w = \gamma s_1$.

(ii) *The following relations hold for all $i \in \mathbb{N}_0$:*

$$x \cdot s_i = 0, \quad u \cdot s_i = 0, \quad v \cdot s_i = \tfrac{i(n-i+1)}{2} s_{i-1}, \tag{5.6}$$

$$x \cdot w_i = \gamma s_{i+1}, \tag{5.7}$$

$$\xi \cdot w_i = (n-2i) w_i + \mu s_i, \tag{5.8}$$

$$g \cdot w_i = w_i + (\lambda + i\gamma) s_i, \tag{5.9}$$

$$u \cdot w_i = -\left(i\lambda + \tfrac{i(i-1)}{2}\gamma\right) s_{i-1}, \tag{5.10}$$

$$v \cdot w_i = \tfrac{i(n-i+1)}{2} w_{i-1} + \left(\tfrac{i(n-2i+2)}{2}\lambda + \tfrac{i(i-1)(n-2i+2)}{4}\gamma - i\mu\right) s_{i-1}. \tag{5.11}$$

PROOF. (i) follows directly from (5.3) and (5.4), looking at the basis of $\mathtt{M}_{\lambda,\mu}(n)$. Then (3.1) implies (5.5). Now $x \cdot w \in S_1^{(n-2)}$, hence there exists $\gamma \in \Bbbk$ such that $x \cdot w = \gamma s_1$ by (5.4). The relations in (ii) are proved arguing recursively. For(5.6) the defining relations of $\mathcal{D}$ are used. For (5.7) we have

$$x \cdot w_{i+1} = xy \cdot w_i = (y + \tfrac{1}{2}x)x \cdot w_i = \gamma(y + \tfrac{1}{2}x) \cdot s_{i+1} \overset{(5.6)}{=} \gamma s_{i+2}.$$

The proof of (5.8) is direct starting from (5.1):

$$\xi \cdot w_{i+1} = \xi y \cdot w_i = (y\xi - 2y) \cdot w_i = (n - 2(i+1)) w_{i+1} + \mu s_{i+1}.$$

The proof of (5.9) also starts from (5.1):

$$\begin{aligned} g \cdot w_{i+1} &= gy \cdot w_i = (y+x) g \cdot w_i = (y+x) \cdot (w_i + (\lambda + i\gamma) s_i) \\ &\overset{(5.6)}{=} w_{i+1} + (\lambda + i\gamma) s_{i+1} + \gamma s_{i+1} = w_{i+1} + (\lambda + (i+1)\gamma) s_{i+1}. \end{aligned}$$

To prove (5.10) we start from (5.2) and argue:

$$\begin{aligned} u \cdot w_{i+1} &= uy \cdot w_i = (yu + 1 - g) \cdot w_i \overset{(5.9)}{=} \left(-(i\lambda + \tfrac{i(i-1)}{2}\gamma) - (\lambda + i\gamma)\right) s_i \\ &= -\left((i+1)\lambda x + \tfrac{i(i+1)}{2}\gamma\right) s_i \end{aligned}$$

as needed. Finally, we prove (5.11). First we observe that

$$-g\zeta \cdot w_i = g \cdot (\tfrac{(n-2i)}{2} w_i - \mu s_i) = \tfrac{(n-2i)}{2} w_i + (\tfrac{(n-2i)}{2}\lambda + \tfrac{(n-2i)i}{2}\gamma - \mu) s_i.$$

Then we proceed recursively:

$$\begin{aligned} v \cdot w_{i+1} =& vy \cdot w_i = (yv - g\zeta + yu) \cdot w_i \\ =& \tfrac{i(n-i+1)}{2} w_i + \left(\tfrac{i(n-2i+2)}{2} \lambda + \tfrac{i(i-1)(n-2i+2)}{4} \gamma - i\mu \right) s_i \\ &+ \tfrac{(n-2i)}{2} w_i + (\tfrac{(n-2i)}{2} \lambda + \tfrac{(n-2i)i}{2} \gamma - \mu) s_i - \left(i\lambda + \tfrac{i(i-1)}{2} \gamma \right) s_i; \end{aligned}$$

it remains to perform the routine verification of the following equalities:

$$\begin{aligned} \tfrac{i(n-i+1)}{2} + \tfrac{n-2i}{2} &= \tfrac{(i+1)(n-(i+1)+1)}{2}, \\ \tfrac{i(n-2i+2)}{2} + \tfrac{n-2i}{2} - i &= \tfrac{(i+1)(n-2(i+1)+2)}{2}, \\ \tfrac{i(i-1)(n-2i+2)}{4} + \tfrac{(n-2i)i}{2} - \tfrac{i(i-1)}{2} &= \tfrac{i(i+1)(n-2(i+1)+2)}{4}, \end{aligned}$$

a task left to the readers. □

Fix $\gamma \in \Bbbk$ and let N be the submodule of $\mathtt{M}_{\lambda,\mu}(n)$ generated by $x \cdot \mathtt{s}$, $y^{n+1} \cdot \mathtt{s}$, $x \cdot \mathtt{w} - \gamma y \cdot \mathtt{s}$ $y^{n+1} \cdot \mathtt{w}$. That is, the module $S = \mathtt{M}_{\lambda,\mu}(n)/N$ is presented by generators $\mathtt{s}$ and $\mathtt{w}$ with defining relations (5.1), (5.2),

$$x \cdot \mathtt{s} = 0, \qquad y^{n+1} \cdot \mathtt{s} = 0, \tag{5.12}$$

$$x \cdot \mathtt{w} = \gamma y \cdot \mathtt{s}, \qquad y^{n+1} \cdot \mathtt{w} = 0. \tag{5.13}$$

PROPOSITION 5.2. *Let S be as above.*

(i) *The module S has dimension n^2 if an only if*

$$\lambda + \tfrac{n}{2}\gamma = 0 \qquad \textit{and} \qquad \mu = 0. \tag{5.14}$$

(ii) *Set $\mathtt{S}_\gamma(n) := S$ when $(\lambda, \mu) = (-\frac{n}{2}\gamma, 0)$; then $\mathtt{S}_\gamma(n) \simeq \mathtt{S}_{\gamma'}(n)$ if and only if $\gamma = t\gamma'$ for some $t \in \Bbbk^\times$.*

(iii) *For any γ, $\mathtt{S}_\gamma(n)$ is an extension of L_n by L_n. Any extension is like this and $\dim \operatorname{Ext}^1_{\mathcal{D}}(\mathtt{L}(n), \mathtt{L}(n)) = 1$, $n \in \mathbb{N}_0$.*

For our conventions on extensions, see Section 6.1.

PROOF. We keep the notation above and apply Lemma 5.1; thus $\dim S \le n^2$. By (5.10) and (5.11), we see that

$$u \cdot w_{n+1} = -(n+1)\left(\lambda + \tfrac{n}{2}\gamma\right) s_n, \tag{5.15}$$

$$v \cdot w_{n+1} = -(n+1)\left(\tfrac{n}{2}\lambda + \tfrac{n^2}{4}\gamma + \mu\right) s_n. \tag{5.16}$$

If $\lambda + \frac{n}{2}\gamma \neq 0$, then (5.15) says that $\mathtt{s}_n \in N$. If $\lambda + \frac{n}{2}\gamma = 0$ but $\mu \neq 0$, then (5.16) says that $\mathtt{s}_n \in N$. In both cases, $\mathtt{s} \in N$ by (5.6), hence $\dim S \le n$. Conversely, assume that (5.14) holds. Then (3.1) implies that

$$N \subset \oplus_{k \in \mathbb{N}} \mathtt{M}_{\lambda,\mu}(n)^{(n-2k)}.$$

Then $N \cap \mathtt{M}_{\lambda,\mu}(n)^{(n)}$, in particular, $\mathtt{s} \notin N$. Now there are morphisms of $\mathcal{D}$-modules $\iota : \mathtt{L}(n) \to S_1$ and $\widetilde{\iota} : \mathtt{L}(n) \to S_2$ given by $\iota(z_{n,i}) = s_i$ and $\widetilde{\iota}(z_{n,i}) = \overline{w}_i$, $i \in \mathbb{I}_{0,n}$. Now $s \neq 0$ in S thus ι is injective (because $\mathtt{L}(n)$ is simple) and $\{s_0, \dots, s_n\}$ are linearly independent. Similarly $\overline{w} \neq 0$ in S_2, $\widetilde{\iota}$ is injective and $\{w_0, \dots, w_n\}$ are linearly independent. Hence $\dim S = n^2$ and (i) is proved.

(ii): The $\mathcal{D}^0$-module $\mathtt{S}_0(n)^{(n)} \simeq \mathtt{P}_{0,0}(n)$ is decomposable while $\mathtt{S}_\gamma(n)^{(n)} \simeq \mathtt{P}_{\gamma,0}(n)$ is indecomposable when $\gamma \neq 0$.

$\pi : \mathtt{S}_\gamma(n) \to \mathtt{L}(n)$ given by $\pi(s_i) = 0$ and $\pi(w_i) = z_{n,i}$, $i \in \mathbb{I}_{0,n}$. Now $s \neq 0$ in $\mathtt{S}_\gamma(n)$, thus ι is injective (because $\mathtt{L}(n)$ is simple) and $\{s_0, \dots, s_n\}$ are linearly independent; while $\{w_0, \dots, w_n\}$ are linearly independent because they project to a basis of $\mathtt{L}(n)$.

Indeed, let $\iota \colon L_n \longrightarrow \mathtt{S}_\gamma(n)$ given by $s_i \mapsto s_i$ and $\pi \colon \mathtt{S}_\gamma(n) \longrightarrow L_n$ given by $s_i \mapsto 0$, $w_i \mapsto s_i$. Then the following sequence is is exact

$$L_n \overset{\iota}{\hookrightarrow} \mathtt{S}_\gamma(n) \overset{\pi}{\twoheadrightarrow} L_n .$$

□

REMARK 5.3. In the context of Lemma 5.1, if $\dim S < \infty$ and $x \cdot s = 0$, then S_1 actually belongs to ${}_{U(\mathfrak{sl}_2)}\mathrm{mod}$, thus $y^{n+1} \cdot s = 0$; similarly $\overline{w} = 0$ implies that $y^{n+1} \cdot \overline{w} = 0$, hence $y^{n+1} \cdot w = 0$ by looking at the weights of S_1.

REMARK 5.4. The set $\{s_0, \dots, s_n, w_0, \dots, w_n\}$ is a basis of $\mathtt{S}_\gamma(n)$. Set $w_{n+1} = s_{n+1} = w_{-1} = s_{-1} = 0$ by convention. Then the action is given by

$$\begin{aligned}
x \cdot s_i &= 0, & x \cdot w_i &= \gamma s_{i+1}, \\
y \cdot s_i &= s_{i+1}, & y \cdot w_i &= w_{i+1}, \\
g \cdot s_i &= s_i, & g \cdot w_i &= w_i - \tfrac{n-2i}{2}\gamma\, s_i, \\
\xi \cdot s_i &= (n-2i)s_i, & \xi \cdot w_i &= (n-2i)\, w_i, \\
u \cdot s_i &= 0, & u \cdot w_i &= \tfrac{i(n+1-i)}{2}\gamma\, s_{i-1}, \\
v \cdot s_i &= \tfrac{i(n-i+1)}{2} s_{i-1}, & v \cdot w_i &= \tfrac{i(n-i+1)}{2} w_{i-1} - \tfrac{i(n-2i+2)(n+1-i)}{4}\gamma\, s_{i-1}.
\end{aligned} \tag{5.17}$$

6. Extensions of simple modules

6.1. Generalities. Let $n, m \in \mathbb{N}_0$. The goal of this Section is to compute the vector spaces $\mathrm{Ext}^1_{\mathcal{D}}(\mathtt{L}(n), \mathtt{L}(m))$. To fix the notation, an extension of $\mathtt{L}(n)$ by $\mathtt{L}(m)$ is a short exact sequence

$$\mathtt{L}(n) \overset{\iota}{\hookrightarrow} T \overset{\pi}{\twoheadrightarrow} \mathtt{L}(m). \tag{6.1}$$

By abuse of notation we say also that T is an extension of $\mathtt{L}(n)$ by $\mathtt{L}(m)$.

REMARK 6.1. Let H be a Hopf algebra and $M, N, P \in {}_H\mathrm{mod}$. Given an extension $M \overset{\iota}{\hookrightarrow} P \overset{\pi}{\twoheadrightarrow} N$, the sequence $N^* \overset{\pi^*}{\hookrightarrow} P^* \overset{\iota^*}{\twoheadrightarrow} M^*$ is exact. Then $\mathrm{Ext}^1_H(M, N) \simeq \mathrm{Ext}^1_H(N^*, M^*)$. Since $\mathtt{L}(p)^* \simeq \mathtt{L}(p)$, $p \in \mathbb{N}_0$, we have

$$\mathrm{Ext}^1_{\mathcal{D}}(\mathtt{L}(n), \mathtt{L}(m)) \simeq \mathrm{Ext}^1_{\mathcal{D}}(\mathtt{L}(m), \mathtt{L}(n)), \qquad m, n \in \mathbb{N}_0.$$

Fix an extension (6.1) and identify $\mathtt{L}(n)$ as a submodule of T via ι. Pick $w_m(0) \in T$ such that $\pi(w_m(0)) = z_m(0) \in \mathtt{L}(m)$ and set

$$w_m(i) := y^i \cdot w_m(0), \qquad i \in \mathbb{I}_{0,m}.$$

Then $\{z_{n,0}, \dots, z_{n,n}, w_m(0), \dots, w_m(m)\}$ is a linear basis of T. Also set

$$r_d := d \cdot w_m(0), \qquad d \in \mathcal{O} = \Bbbk\langle x, u, g^{\pm 1}\rangle.$$

Now $\mathcal{O}$ acts by 0 on $\mathtt{L}(m)$, hence $r_d \in \mathtt{L}(n)$ for $d \in \mathcal{O}$. Also $v \cdot w_m(0) \in \mathtt{L}(n)$.

We start giving restrictions on the existence of nontrivial extensions.

LEMMA 6.2. *If* $\mathrm{Ext}^1_{\mathcal{D}}(\mathtt{L}(n), \mathtt{L}(m)) \neq 0$, *then* $m - n \in \{2, 0, -2\}$.

PROOF. Let T be an extension of $\mathtt{L}(n)$ by $\mathtt{L}(m)$ and keep the notation above.

CLAIM. If $r_x = r_{g-1} = r_u = 0$, then the extension T is trivial.

We claim that x, $g-1$ and u act by 0 on T. Since they act trivially on $\mathtt{L}(n)$, it is enough to consider the action on the $w_m(i)$'s. We argue recursively, the case $i=0$ being the hypothesis. If x, $g-1$ and u act by 0 in $w_m(i)$, then

$$x \cdot w_m(i+1) = xy \cdot w_m(i) = (yx + \frac{1}{2}x^2)w_m(i) = 0,$$

$$u \cdot w_m(i+1) = uy \cdot w_m(i) = (yu + (1-g)) \cdot w_m(i) = 0,$$

$$(1-g) \cdot w_m(i+1) = (1-g)y \cdot w_m(i) = (y(1-g) - gx) \cdot w_m(i) = 0.$$

Hence $T \in {}_{\mathcal{D}/\mathcal{D}\mathcal{O}^+}\text{mod}$; since $\mathcal{D}/\mathcal{D}\mathcal{O}^+ \simeq U(\mathfrak{sl}_2(\Bbbk))$, the extension is trivial.

Thus, if T is a nontrivial extension, then at least one of r_x, r_{g-1}, r_u is not zero. Let $s \in \mathtt{L}(n)$ be such that $\zeta \cdot w_m(0) = mw_m(0) + s$.

CASE 1. $r_u \neq 0$. Since

$$v \cdot r_u = vu \cdot w_m(0) = (uv - \tfrac{1}{2}u^2) \cdot w_m(0) = 0,$$

there exists $a \in \Bbbk^\times$ such that $r_u = az_{n,0}$. We compute in two ways:

$$\xi \cdot r_u = \xi u \cdot w_m(0) = (u\xi + 2u) \cdot w_m(0) = mr_u + u \cdot s + 2r_u = (m+2)r_u$$

and also $\xi \cdot r_u = a\xi \cdot z_{n,0} = nr_u$. We conclude that $n = m+2$.

CASE 2. $r_u = 0$ and $r_{g-1} \neq 0$. Since

$$v \cdot r_{g-1} = v(g-1) \cdot w_m(0) = ((g-1)v + gu) \cdot w_m(0) = g \cdot r_u = 0,$$

there exists $a \in \Bbbk^\times$ such that $r_{g-1} = az_{n,0}$. We compute in two ways:

$$\xi \cdot r_{g-1} = (g-1)\xi \cdot w_m(0) = mr_{g-1} + (g-1) \cdot s = mr_{g-1}$$

and also $\zeta \cdot r_{g-1} = a\xi \cdot z_{n,0} = nr_{g-1}$. We conclude that $n = m$.

CASE 3. $r_u = r_{g-1} = 0$ and $r_x \neq 0$. Since

$$v \cdot r_x = vx \cdot w_m(0) = (xv + (1-g) + xu) \cdot w_m(0) = 0,$$

there exists $a \in \Bbbk^\times$ such that $r_x = az_{n,0}$. We compute in two ways:

$$\zeta \cdot r_x = \xi x \cdot w_m(0) = (x\xi - x) \cdot w_m(0) = mr_x + x \cdot s - 2r_x = (m-2)r_x$$

and also $\xi \cdot r_x = a\xi \cdot z_{n,0} = nr_x$. We conclude that $n = m-2$. □

6.2. Extensions of $\mathtt{L}(n)$ by $\mathtt{L}(n \pm 2)$. Let $n \in \mathbb{N}_0$. We next introduce the $\mathcal{D}$-module $\mathtt{T}(n,1)$ generated by z_n with defining relations

$$(4.2) \qquad g \cdot z_n = z_n, \qquad \zeta \cdot z_n = -\tfrac{(n+2)}{2} z_n, \qquad u \cdot z_n = 0, \qquad v \cdot z_n = 0,$$

$$(6.2) \qquad x^2 \cdot z_n = 0, \qquad y^{n+3} \cdot z_n = 0, \qquad y^{n+1}x \cdot z_n = 0.$$

This belongs to the family of $\mathcal{D}$-modules $\mathtt{T}(n,m)$ studied in Section 4. The set $\{\mathtt{z}_n(i,j) := y^i x^j \cdot z_n \colon j \in \mathbb{I}_{0,1}, i \in \mathbb{I}_{0,n+2-2j}\}$ is a basis of $\mathtt{T}(n,1)$, see Section 4 or prove it directly. Given $b \in \Bbbk^\times$, let $\mathcal{E}(b)$ be the exact sequence

$$\mathtt{L}(n) \xhookrightarrow{\ \iota_b\ } \mathtt{T}(n,1) \xrightarrow{\ \pi\ } \mathtt{L}(n+2)\ ,$$

where ι_b and π are determined by $\iota_b(z_{n,i}) = b\mathtt{z}_n(i,1)$ and $\pi(\mathtt{z}_n(i,1)) = 0$ for $i \in \mathbb{I}_{0,n}$; and $\pi(\mathtt{z}_n(k,0)) = z_{n+2}(k)$ for $k \in \mathbb{I}_{0,n+2}$.

PROPOSITION 6.3. *Any extension* $\mathtt{L}(n) \hookrightarrow T \twoheadrightarrow \mathtt{L}(n+2)$ *is either trivial or else isomorphic to* $\mathcal{E}(b)$ *for a unique* $b \in \Bbbk^{\times}$. *Hence* T *is isomorphic either to* $\mathtt{L}(n) \oplus \mathtt{L}(n+2)$ *or to* $\mathtt{T}(n,1)$ *and*

$$\dim \operatorname{Ext}^1_{\mathcal{D}}(\mathtt{L}(n), \mathtt{L}(n+2)) = 1. \tag{6.3}$$

PROOF. Identify $\mathtt{L}(n)$ as a submodule of T via ι and pick $\mathbf{w} \in T$ satisfying

$$\pi(\mathbf{w}) = z_{n+2}(0) \in \mathtt{L}(n+2).$$

Hence there exist $c_0, \dots, c_n \in \Bbbk$ such that

$$\zeta \cdot \mathbf{w} = (n+2)\mathbf{w} + \sum_{i=0}^{n} c_i z_{n,i}.$$

Let $w_{n+2}(0) := \mathbf{w} + \sum_{i=0}^{n} \frac{c_i}{2(i+1)} z_{n,i}$. Clearly $\pi(w_0) = z_{n+2}(0)$, but also

$$\begin{aligned}
\xi \cdot w_{n+2}(0) &= (n+2)\mathbf{w} + \sum_{i=0}^{n} c_i z_{n,i} + \sum_{i=0}^{n} \frac{c_i(n-2i)}{2(i+1)} z_{n,i} \\
&= (n+2)\mathbf{w} + \sum_{i=0}^{n} \frac{(2i+2+n-2i)c_i}{2(i+1)} z_{n,i} \\
&= (n+2)\left(\mathbf{w} + \sum_{i=0}^{n} \frac{c_i}{2(i+1)} z_{n,i}\right) = (n+2) w_{n+2}(0).
\end{aligned}$$

Hence $w_{n+2}(0) \in T^{n+2}$,

$$w_{n+2}(j) := y^j \cdot w_{n+2}(0) \in T^{n+2-2j}, \qquad \pi(w_{n+2}(j)) = z_{n+2}(j), \qquad j \in \mathbb{I}_{n+2},$$

and $\{z_{n,0}, \dots, z_{n,n}, w_{n+2}(0), \dots, w_{n+2}(n+2)\}$ is a basis of T. Thus

$$T^k = \begin{cases} \Bbbk z_n(\frac{n-k}{2}) \oplus \Bbbk w_{n+2}(\frac{n+2-k}{2}), & k \in \mathbb{I}_{-n,n}; \\ \Bbbk w_{n+2}(0) & k = n+2, \\ \Bbbk w_{n+2}(n+2) & k = -n-2. \end{cases}$$

Now $y^{n+3} \cdot w_{n+2}(0) \in T^{-n-4}$ and $u \cdot w_{n+2}(0), v \cdot w_{n+2}(0) \in T^{n+4}$ by (3.1), i. e.

$$y^{n+3} \cdot w_{n+2}(0) = 0, \qquad u \cdot w_{n+2}(0) = 0, \qquad v \cdot w_{n+2}(0) = 0. \tag{6.4}$$

Also $g \cdot w_{n+2}(0) \in T^{n+2}$, i. e. $g \cdot w_{n+2}(0) = a w_{n+2}(0)$ for some $a \in \Bbbk$; applying π to both sides of the last equality we get that $a = 1$, that is

$$g \cdot w_{n+2}(0) = w_{n+2}(0). \tag{6.5}$$

Finally $x \cdot w_{n+2}(0) \in \mathtt{L}(n) \cap T^n = \Bbbk z_{n,0}$, i. e. $x \cdot w_{n+2}(0) = c z_{n,0}$ for some $c \in \Bbbk$. If $c = 0$, then the subspace with basis $w_{n+2}(0), \dots, w_{n+2}(n+2)$ is a submodule isomorphic to $\mathtt{L}(n+2)$ and the extension is trivial. If $c \neq 0$, then the extension is isomorphic to $\mathcal{E}(c^{-1})$. □

PROPOSITION 6.4. *If* S *is an extension of* $\mathtt{L}(n)$ *by* $\mathtt{L}(n-2)$, *then* S *is isomorphic either to* $\mathtt{L}(n) \oplus \mathtt{L}(n-2)$ *or to* $\mathtt{T}(n-2,1)^*$. *Also,*

$$\dim \operatorname{Ext}^1_{\mathcal{D}}(\mathtt{L}(n), \mathtt{L}(n-2)) = 1. \tag{6.6}$$

PROOF. By Remark 6.1 and Proposition 6.3. □

6.3. The quiver and representation type. Propositions 5.2, 6.3, and 6.4 give us the *Gabriel quiver* of $\mathcal{D}$, i.e., the quiver $\operatorname{Ext} Q(\mathcal{D})$ with vertices $\mathbb{N}_0$ and $\dim \operatorname{Ext}^1_{\mathcal{D}}(S_i, S_j)$ arrows from the vertex i to the vertex j. That is,

(6.7)

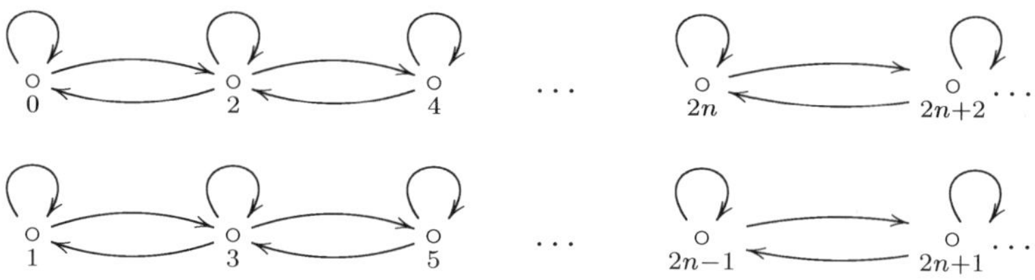

From the analysis of this quiver one concludes the following.

PROPOSITION 6.5. *The algebra $\mathcal{D}$ has wild representation type.*

PROOF. This is evident for experts in representation theory of artin algebras but we include a proof for completeness.

CLAIM 1. Let $A \twoheadrightarrow B$ be a surjective map of algebras and $M, N \in {}_B\mathrm{mod}$. Then the canonical map $\operatorname{Ext}^1_B(M,N) \to \operatorname{Ext}^1_A(M,N)$ is injective.

CLAIM 2. Let now A be a (possibly infinte-dimensional) algebra over a field $\Bbbk$ with Ext-quiver Q such thst $\dim_{\Bbbk} \operatorname{Ext}^1_A(L, L') < \infty$ for any $L, L' \in \operatorname{irrep} A$. Let F be a finite subset of $\operatorname{irrep} A$ and let Q_F be the (full) subquiver of Q spanned by F. Then there exists a finite-dimensional quotient algebra $A \twoheadrightarrow B$ such that the Ext-quiver of B is isomorphic to Q_F.

Given $L, L' \in F$, pick a basis (v_i) of $\operatorname{Ext}^1_A(L, L')$ and for each v_i an extension M_i of L by L' representing v_i. Let M be the direct sum of all L, L' in F and all the corresponding M_i. Clearly, $\dim M < \infty$ hence, so is the image B of the representation $A \to \operatorname{End} M$. By construction and Claim 1 the canonical map $\operatorname{Ext}^1_B(L, L') \to \operatorname{Ext}^1_A(L, L')$ is bijective, hence the Ext-quiver of B is isomorphic to Q_F.

By Claim 2 applied to $F = \{0, 2, 4\}$ there exists a surjective algebra map $\mathcal{D} \twoheadrightarrow B$ where $\dim B < \infty$ and the Ext-quiver of B is isomorphic to

(6.8)

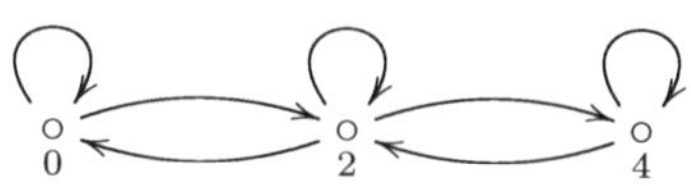

Let C be the basic algebra which is Morita equivalent to B and $\mathfrak{r} = \operatorname{rad} C$. Then $C/\mathfrak{r}^2$ has finite or tame representation type if and only if the separated quiver Γ_s of (6.8) is a disjoint union of Dynkin and affine Dynkin diagrams, see [**ARS**, Theorem X.2.6]. But Γ_s has the form

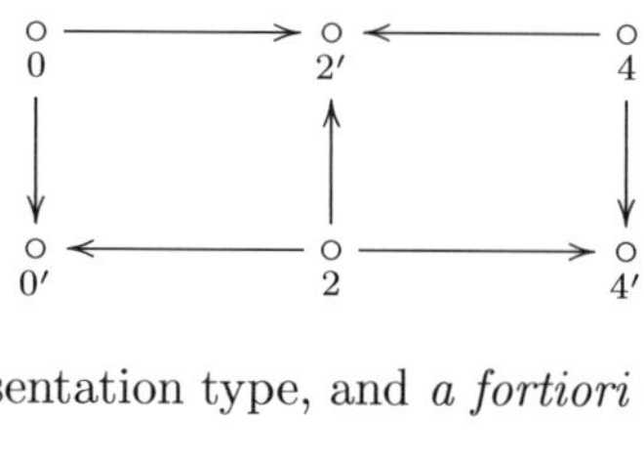

Thus $C/\mathfrak{r}^2$ has wild representation type, and *a fortiori* C, B and $\mathcal{D}$ also. $\square$

Acknowledgment

The first author thanks Henning Krause and Bernhard Keller for enlightening discussions.

References

[AAH] N. Andruskiewitsch, I. Angiono, and I. Heckenberger, *On finite GK-dimensional Nichols algebras over abelian groups*, Mem. Amer. Math. Soc. **271** (2021), no. 1329, ix+125, DOI 10.1090/memo/1329. MR4298502

[AAM] N. Andruskiewitsch, I. Angiono, and M. Moya Giusti, *Rank 4 Nichols algebras of pale braidings*, SIGMA Symmetry Integrability Geom. Methods Appl. **19** (2023), Paper No. 021, 41, DOI 10.3842/SIGMA.2023.021. MR4574403

[ADP] N. Andruskiewitsch, F. Dumas, and H. M. Peña Pollastri, *On the double of the Jordan plane*, Ark. Mat. **60** (2022), no. 2, 213–229, DOI 10.4310/arkiv.2022.v60.n2.a1. MR4500364

[AP] N. Andruskiewitsch and H. Peña Pollastri, *On the restricted Jordan plane in odd characteristic*, J. Algebra Appl. **20** (2021), no. 1, Paper No. 2140012, 26, DOI 10.1142/S0219498821400120. MR4209966

[AGI] I. Angiono and A. García Iglesias, *On finite GK-dimensional Nichols algebras of diagonal type: rank 3 and Cartan type*, Publ. Mat. **67** (2023), no. 2, 757–793, DOI 10.5565/publmat6722311. MR4609019

[ARS] M. Auslander, I. Reiten, and S. O. Smalø, *Representation theory of Artin algebras*, Cambridge Studies in Advanced Mathematics, vol. 36, Cambridge University Press, Cambridge, 1997. Corrected reprint of the 1995 original. MR1476671

[BS] K. A. Brown and J. T. Stafford, *The prime spectrum of the Drinfeld double of the Jordan plane*, Recent advances in noncommutative algebra and geometry, Contemp. Math., vol. 801, Amer. Math. Soc., [Providence], RI, [2024] ©2024, pp. 51–71, DOI 10.1090/conm/801/16036. MR4756377

[D] J. Dixmier, *Enveloping algebras*, Graduate Studies in Mathematics, vol. 11, American Mathematical Society, Providence, RI, 1996. Revised reprint of the 1977 translation, DOI 10.1090/gsm/011. MR1393197

[R] D. E. Radford, *Hopf algebras*, Series on Knots and Everything, vol. 49, World Scientific Publishing Co. Pte. Ltd., Hackensack, NJ, 2012. MR2894855

Facultad de Matemática, Astronomía, Física y Computación, Universidad Nacional de Córdoba. CIEM – CONICET. Medina Allende s/n (5000) Ciudad Universitaria, Córdoba, Argentina

Email address: nicolas.andruskiewitsch@unc.edu.ar

Department of Mathematics, Indiana University, Bloomington, Indiana 47405

Email address: hpenapol@iu.edu

Contemporary Mathematics
Volume **814**, 2025
https://doi.org/10.1090/conm/814/16300

2-unitary operads of GK-dimension 3

Yan-Hong Bao, Dong-Xing Fu, Yu Ye, and James J. Zhang

This paper is dedicated to the memory of Earl J. Taft.

ABSTRACT. We study and classify the 2-unitary operads of Gelfand–Kirillov (GK) dimension 3.

1. Introduction

Algebraic operads originated from homotopy theory in algebraic topology, and was first introduced by Boardman–Vogt [**BV**] and May [**Ma**] in the 1960s and 1970s. Throughout the past twenty years, operad theory has become an important tool in homological algebra, category theory, algebraic geometry, and mathematical physics. It is well known that every operad encodes an algebra system. For example, $\mathcal{A}ss$ encodes all unital associative algebras. Further, a $\Bbbk$-linear operad itself is an algebraic object similar to an associative algebra, and algebraic structures of operads have been widely investigated by many mathematicians; see [**BYZ, Dot, DK, DMR, DT, Fr1, Fr2, KP, LV, MSS, QXZZ**].

The Gelfand–Kirillov dimension of an associative algebra is a useful numerical invariant in ring theory and noncommutative algebraic geometry; see [**KL**]. In a similar way, the Gelfand–Kirillov dimension can be defined for other algebraic objects, including algebraic operads [**BYZ, Fi**]. Let $\Bbbk$ be a base field. An operad $\mathcal{P}$ is said to be *locally finite* if each $\mathcal{P}(n)$ is finite dimensional over $\Bbbk$. In this paper we only consider locally finite operads. The *Gelfand–Kirillov dimension* (or *GK-dimension* for short) of a locally finite operad $\mathcal{P}$ is defined to be

$$\operatorname{GKdim}\mathcal{P} := \limsup_{n\to\infty} \log_n \left(\sum_{i=0}^{n} \dim_{\Bbbk} \mathcal{P}(i)\right).$$

2020 *Mathematics Subject Classification.* Primary 18M05, 16D10.

Key words and phrases. 2-unitary operad, Gelfand–Kirillov dimension, truncation ideal, trident algebra.

The first author was partially supported by the National Natural Science Foundation of China (Grant No. 12371015) and the Science Fundation for Distinguished Young Scholars of Anhui Province (No. 2108085J01).

The third author was partially supported by the National Natural Science Foundation of China (Grant Nos. 12131015, 12161141001 and 12371042) and the Innovation Program for Quantum Science and Technology (Grant No. 2021ZD0302902).

The fourth author was partially supported by the US National Science Foundation (Grant Nos. DMS-2001015, and DMS-2302087).

We refer to [**BYZ**, **KP**, **QXZZ**] for more information related to the GK-dimension of an operad.

Recall that an operad $\mathcal{P}$ is *unitary* if $\mathcal{P}(0) = \Bbbk \mathbb{1}_0$ with a basis element $\mathbb{1}_0$ (which is called a 0-unit), see [**Fr2**, Section 2.2]. Denote by Op_+ the category of unitary operads, in which a morphism preserves the 0-unit. A *2-unitary operad* $\mathcal{P}$ is a unitary operad $\mathcal{P}$ equipped with a morphism $\mathcal{M}ag \to \mathcal{P}$ in Op_+, where $\mathcal{M}ag$ is the unital magmatic operad (see [**BYZ**, Section 8.4] or [**Lo**, Section 4.1.10]). The definition of a $2a$-unitary operad is given in Definition 2.5(4). In [**BYZ**], the authors proved that GK-dimension of a 2-unitary operad $\mathcal{P}$ is either an nonnegative integer or infinity and that the generating series of $\mathcal{P}$ is a rational function when $\operatorname{GKdim} \mathcal{P} < \infty$. The pattern of GK-dimension of a general non-2-unitary operad (or nonsymmetric operad) is very different; see Remark 7.5.

The only 2-unitary operad of GK-dimension 1 is $\mathcal{C}om$, which encodes all unital commutative algebras. All locally finite 2-unitary operads of GK-dimension 2 were classified in [**BYZ**, Theorem 0.6]. One way of viewing this classification is the following. We refer to [**BYZ**, Section 6] for the construction of 2-unitary operads of GK-dimension 2.

THEOREM 1.1 ([**BYZ**, Theorem 0.6]). *There are natural equivalences between*

(1) *the category of finite dimensional, not necessarily unital,* $\Bbbk$*-algebras,*
(2) *the category of* 2*-unitary operads of GK-dimension* ≤ 2,
(3) *the category of* $2a$*-unitary operads of GK-dimension* ≤ 2.

In the case of GK-dimension 3, our result is not as clean as Theorem 1.2. Nevertheless, we will provide a classification. Recall that an operad is called *$\mathcal{C}om$-augmented* if there is an operadic unit map $u_{\mathcal{P}} : \mathcal{C}om \to \mathcal{P}$. The morphisms in the category of $\mathcal{C}om$-augmented operads are supposed to be compatible with operadic unit maps. An algebra over a $\mathcal{C}om$-augmented operad is a commutative algebra together with additional operations on it. A typical example of $\mathcal{C}om$-augmented operad is the Poisson operad $\mathcal{P}ois$. Here is the main result of this paper.

THEOREM 1.2. *There is a natural equivalence between*

(1) *the category of finite dimensional trident algebras,*
(2) *the category of* $\mathcal{C}om$*-augmented operads of GK-dimension* ≤ 3.

If $\operatorname{char} \Bbbk \neq 2$, we also prove that every 2-unitary operad of GK-dimension 3 is equipped with a $\mathcal{C}om$-augmentation with possibly new 2-unit (Proposition 3.5). Combining Theorem 1.2 with Proposition 3.5, we obtain a classification of 2-unitary operads of GK-dimension 3 in terms of finite dimensional trident algebras. However, Theorem 1.2 fails if the condition "$\mathcal{C}om$-augmented" in part (2) is replaced by "2-unitary" (Example 7.4).

The definition of a trident algebra is given in Section 4. Roughly speaking, a trident algebra consists of a pair of $\Bbbk$-vector spaces (A, M) equipped with some algebraic structures and a pair of $\Bbbk$-linear maps (f, g) satisfying some identities. One special case is when $M = 0$, then both f and g must be zero maps. The corresponding operads have GK-dimension 2, which is in the situation of Theorem 1.1. Another special case is when $A = \Bbbk$ and M is a nonzero finite dimensional right $\mathbb{S}_2$-module (with both f and g being 0 automatically). Then $(\Bbbk, M)$ is a trident algebra. The trident algebras of the form $(\Bbbk, M)$ correspond exactly to connected 2-unitary (and $\mathcal{C}om$-augmented) operads of GK-dimension three. So we have the following corollary.

COROLLARY 1.3 (Corollary 6.1). *The category of connected 2-unitary operads of GK-dimension 3 is equivalent to the category of nonzero right $\mathbb{S}_2$-modules.*

Note that a seemingly technical result, Theorem 1.2, makes some questions easy to solve. For example, we prove Corollary 1.4.

COROLLARY 1.4 (Proposition 7.3). *There is no $\mathcal{C}om$-augmented Hopf operad of GK-dimension 2.*

Corollary 1.4 motivates the following question.

QUESTION 1.5. Is there a $\mathcal{C}om$-augmented Hopf operad of finite GK-dimension larger than 2?

This paper is organized as follows. In Section 2 we recall some basic definitions and properties of 2-unitary operads. We prove some properties of 2-unitary operads of GK-dimension 3 in Section 3. A key preliminary result is that a 2-unitary operad of GK-dimension 3 is $\mathcal{C}om$-augmented after changing the 2-unit (Proposition 3.5). We define a concept of a trident algebra in Section 4. In Section 5, we construct an operad from a trident algebra and complete the classification of 2-unitary operads of GK-dimension 3 by Theorem 1.2 and Proposition 3.5. In Section 6, we describe all connected 2-unitary operads of GK-dimension three. In Section 7, we give some comments, examples, and remarks.

2. Preliminaries

Throughout, let $\Bbbk$ be a base field, and every object is over $\Bbbk$. Let n be a nonnegative integer. Set $[n] = \{1, 2, \cdots, n\}$ for $n > 0$ and $[0] = \varnothing$. We use $\mathbb{S}_n$ to denote the symmetric group for $n \geq 0$. By convention, $\mathbb{S}_0$ is the trivial group. Following the notation introduced in [**BYZ**, Section 8.1], for each $\sigma \in \mathbb{S}_n$, we use the sequence $(i_1, i_2, \cdots, i_n)$ to denote a permutation $\sigma \in \mathbb{S}_n$ with $\sigma(i_k) = k$ for all $1 \leq k \leq n$. Denote by $\mathbb{S}$ the disjoint union of all symmetric group $\mathbb{S}_n$ for all $n \geq 0$. Recall that a $\Bbbk\mathbb{S}$-module (or $\mathbb{S}$-module) means a sequence $\{\mathcal{P}(n)\}_{n\geq 0}$ of right $\Bbbk\mathbb{S}_n$-modules, where the right $\mathbb{S}_n$-action on $\mathcal{P}(n)$ is denoted by $*$.

In this section, we retrospect some basic facts about operads.

2.1. Definitions. From different viewpoints, there are various equivalent definitions about operads. In this paper, we mainly use the *partial* definition and refer to [**LV**, Chapter 5] for other versions of the definition.

DEFINITION 2.1 ([**Fr2**, Section 2.1]). An *operad* $\mathcal{P}$ consists of the following data:

(i) a $\Bbbk\mathbb{S}$-module $\{\mathcal{P}(n)\}_{n\geq 0}$, where an element in $\mathcal{P}(n)$ is called an *n-ary operation.*

(ii) an element $\mathbb{1} \in \mathcal{P}(1)$, which is called the *identity,*

(iii) for all $m \geq 1, n \geq 0$ and $1 \leq i \leq m$, a *partial composition*:

$$- \underset{i}{\circ} -\colon \mathcal{P}(m) \otimes \mathcal{P}(n) \to \mathcal{P}(m+n-1)$$

satisfying the following coherence axioms:

(OP1) (Identity) for all $\theta \in \mathcal{P}(m)$ and all $1 \leq i \leq m$,

$$\theta \underset{i}{\circ} \mathbb{1} = \theta = \mathbb{1} \underset{1}{\circ} \theta;$$

(OP2) (Associativity) for all $\lambda \in \mathcal{P}(l), \mu \in \mathcal{P}(m)$ and $\nu \in \mathcal{P}(n)$,

(E2.1.1) $$(\lambda \underset{i}{\circ} \mu) \underset{i-1+j}{\circ} \nu = \lambda \underset{i}{\circ} (\mu \underset{j}{\circ} \nu), \quad 1 \leq i \leq l, 1 \leq j \leq m,$$

(E2.1.2) $$(\lambda \underset{i}{\circ} \mu) \underset{k-1+m}{\circ} \nu = (\lambda \underset{k}{\circ} \nu) \underset{i}{\circ} \mu, \quad 1 \leq i < k \leq l;$$

(OP3) (Equivariance) for all $\mu \in \mathcal{P}(m), \nu \in \mathcal{P}(n)$ and $\sigma \in \mathbb{S}_n$, $\phi \in \mathbb{S}_m$,

(E2.1.3) $$\mu \underset{i}{\circ} (\nu * \sigma) = (\mu \underset{i}{\circ} \nu) * \sigma',$$

(E2.1.4) $$(\mu * \phi) \underset{i}{\circ} \nu = (\mu \underset{\phi(i)}{\circ} \nu) * \phi'',$$

where $\sigma' = 1_m \underset{i}{\circ} \sigma$ and $\phi'' = \phi \underset{i}{\circ} 1_n$ are given by the partial composition in the associative algebra operad $\mathcal{A}ss$. We refer to [**BYZ**, Section 8] for more details concerning σ' and ϕ''.

Let $\mathcal{P}$ be an operad in the sense of Definition 2.1. Then one can define the composition map by

(E2.1.5) $$\lambda \circ (\mu_1, \cdots, \mu_n) = (\cdots((\lambda \underset{n}{\circ} \mu_n) \underset{n-1}{\circ} \mu_{n-1}) \underset{n-2}{\circ} \mu_{n-2} \cdots) \underset{1}{\circ} \mu_1$$

for all $\lambda \in \mathcal{P}(n)$ and $\mu_i \in \mathcal{P}$ and for $1 \leq i \leq n$ [**BYZ**, Remark 1.3].

Example 2.2 ([**LV**, Section 5.2.10]). Let $\mathcal{C}om$ denote the commutative algebra operad. The space of n-ary operations of $\mathcal{C}om$ is $\mathcal{C}om(n) = \Bbbk \mathbb{1}_n$ equipped with the trivial action of the symmetric group and the partial composition is given by $\mathbb{1}_m \underset{i}{\circ} \mathbb{1}_n = \mathbb{1}_{m+n-1}$ for all m, n, i. Note that $\mathbb{1}_1$ is the identity $\mathbb{1}$ of $\mathcal{C}om$.

Example 2.3 ([**LV**, Section 13.8]). Suppose that $\mathcal{M}ag$ is the operad generated by the $\mathbb{S}$-module

$$(\Bbbk\mu, \Bbbk\mathbb{1}, \Bbbk\mathbb{S}_2\nu, 0, 0, \cdots)$$

and subject to relations

$$\nu \underset{i}{\circ} \mu = \mathbb{1}, (i = 1, 2),$$

where $\Bbbk\mathbb{S}_2\nu$ is the regular $\Bbbk\mathbb{S}_2$-module with the basis ν. In this paper we use $\mathbb{1}_0$ for μ and $\mathbb{1}_2$ for ν.

Definition 2.4 ([**LV**, Chapter 5]). Let $\mathcal{P}$ and $\mathcal{P}'$ be operads. A *morphism* from $\mathcal{P}$ to $\mathcal{P}'$ is a sequence of $\Bbbk\mathbb{S}_n$-homomorphism $\gamma = (\gamma_n : \mathcal{P}(n) \to \mathcal{P}'(n))_{n \geq 0}$, satisfying

$$\gamma(\mathbb{1}) = \mathbb{1}' \quad \text{and} \quad \gamma(\mu \underset{i}{\circ} \nu) = \gamma(\mu) \underset{i}{\circ} \gamma(\nu),$$

where $\mathbb{1}$ and $\mathbb{1}'$ are identities of $\mathcal{P}$ and $\mathcal{P}'$, respectively, and $\mu, \nu \in \mathcal{P}$. Let Op denote the category of operads.

Next we collect some definitions given in [**BYZ, Fr2**].

Definition 2.5.

(1) An operad $\mathcal{P}$ is called *unitary* if $\mathcal{P}(0) = \Bbbk\mathbb{1}_0$, where $\mathbb{1}_0$ is a basis of $\mathcal{P}(0)$, and is called a 0-*unit*. The category of unitary operads is denoted by Op_+, in which morphisms are operadic morphisms preserving 0-units.

(2) A unitary operad is said to be *connected*, if $\mathcal{P}(1) = \Bbbk\mathbb{1}$ where $\mathbb{1}$ is the identity of $\mathcal{P}$. In this case we also use $\mathbb{1}_1$ for $\mathbb{1}$.

(3) A *2-unitary operad* is a unitary operad $\mathcal{P}$ equipped with a morphism $\mathcal{M}ag \to \mathcal{P}$ in Op_+, where $\mathcal{M}ag$ is the unital magmatic operad [Example 2.3], or equivalently, there is an element $\mathbb{1}_2 \in \mathcal{P}(2)$ (called a 2-unit) such that

(E2.5.1)
$$\mathbb{1}_2 \underset{1}{\circ} \mathbb{1}_0 = \mathbb{1}(= \mathbb{1}_1) = \mathbb{1}_2 \underset{2}{\circ} \mathbb{1}_0$$

where $\mathbb{1}_0$ is a 0-unit of $\mathcal{P}$.

(4) A *2a-unitary* operad is a unitary $\mathcal{P}$ equipped with a morphism $\mathcal{A}ss \to \mathcal{P}$ in Op_+, or equivalently, $\mathcal{P}$ is 2-unitary with a 2-unit $\mathbb{1}_2$ satisfying

(E2.5.2)
$$\mathbb{1}_2 \underset{1}{\circ} \mathbb{1}_2 = \mathbb{1}_2 \underset{2}{\circ} \mathbb{1}_2.$$

In this case $\mathbb{1}_2$ is called *2a-unit.*

(5) An operad $\mathcal{P}$ is called *$\mathcal{C}om$-augmented* if there is a morphism from $\mathcal{C}om \to \mathcal{P}$. It is clear that $\mathcal{C}om$-augmented operads are 2a-unitary. In this case $\mathbb{1}_2 * (2,1) = \mathbb{1}_2$ and $\mathbb{1}_2$ is called a *symmetric 2a-unit.*

Let dim denote $\dim_{\Bbbk}$.

DEFINITION 2.6. Let $\mathcal{P} = (\mathcal{P}(n))_{n\geq 0}$ be a locally finite operad, i.e. $\dim \mathcal{P}(n) < \infty$ for all $n \geq 0$.

(1) The *generating series* of $\mathcal{P}$ is defined to be

$$G_{\mathcal{P}}(t) = \sum_{n=0}^{\infty} \dim \mathcal{P}(n) t^n \in \mathbb{Z}[[t]].$$

(2) The *Gelfand-Kirillov dimension* (*GK-dimension* for short) of $\mathcal{P}$ is defined to be

$$\mathrm{GKdim}(\mathcal{P}) = \limsup_{n\to\infty} \log_n \Big(\sum_{i=0}^{n} \dim \mathcal{P}(i)\Big).$$

EXAMPLE 2.7.

(1) The Poisson operad $\mathcal{P}ois$ is $\mathcal{C}om$-augmented. In fact, the operad $\mathcal{P}ois$ fits into the sequences $\mathcal{C}om \hookrightarrow \mathcal{P}ois \twoheadrightarrow \mathcal{L}ie$ and $\mathcal{L}ie \hookrightarrow \mathcal{P}ois \twoheadrightarrow \mathcal{C}om$, where $\mathcal{L}ie$ is the Lie operad, see for instance [**LV**, 13.3.3]. Note that the Lie operad $\mathcal{L}ie$ is not $\mathcal{C}om$-augmented.

(2) A 2-unitary operad of GK-dimension 2 is always $\mathcal{C}om$-augmented, see [**BYZ**, Lemma 6.2]. Moreover, as we will show below, if $\mathrm{char}\,\Bbbk \neq 2$, then every 2-unitary operad of GK-dimension 3 is equipped with a $\mathcal{C}om$-augmentation with possibly new 2-unit [Proposition 3.5].

Let $\mathcal{P}$ and $\mathcal{P}'$ be 2-unitary operads. A morphism of 2-unitary operads is a morphism $\gamma : \mathcal{P} \to \mathcal{P}'$ in Op_+ satisfying the following commutative diagram

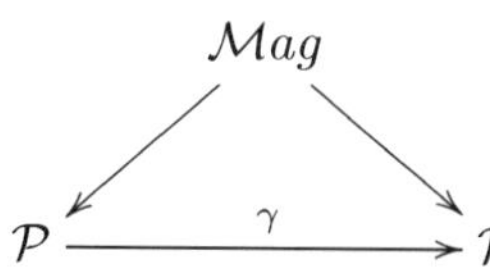

or equivalently, the operad morphism preserves the 2-unit. The categories of 2-unitary operads, 2a-unitary operads and $\mathcal{C}om$-augmented operads, are denoted by $\mathcal{M}ag \downarrow \mathrm{Op}_+$, $\mathcal{A}ss \downarrow \mathrm{Op}_+$ and $\mathcal{C}om \downarrow \mathrm{Op}_+$, respectively.

2.2. Truncation ideals. Let $\mathcal{P}$ be a unitary operad and I a subset of $[n]$. Recall that a *restriction operator* [**Fr2**, Section 2.2.1] means

$$\text{(E2.7.1)} \qquad \pi^I : \mathcal{P}(n) \to \mathcal{P}(s), \qquad \pi^I(\theta) = \theta \circ (\mathbb{1}_{\chi_I(1)}, \cdots, \mathbb{1}_{\chi_I(n)})$$

for all $\theta \in \mathcal{P}(n)$, where χ_I is the characteristic function of I, i.e. $\chi_I(x) = 1$ for $x \in I$ and $\chi_I(x) = 0$ otherwise. Note that $\circ$ is defined in (E2.1.5). If $I = \{i_1, \cdots, i_s\} \subset [n]$ with $1 \le i_1 < \cdots < i_s \le n$, we also denote π^I as $\pi^{i_1,\cdots,i_s}$. We refer to [**BYZ**, Section 2.3] for more details.

For integers $k \ge 1$, the kth truncation ideals ${}^k\Upsilon$ is defined by

$$\text{(E2.7.2)} \qquad {}^k\Upsilon_{\mathcal{P}}(n) = \begin{cases} \bigcap\limits_{I \subset [n],\, |I| = k-1} \ker \pi^I, & \text{if } n \ge k; \\ 0, & \text{otherwise.} \end{cases}$$

By convention, let ${}^0\Upsilon_{\mathcal{P}} = \mathcal{P}$. If no confusion, we write ${}^k\Upsilon = {}^k\Upsilon_{\mathcal{P}}$ for brevity.

For every subset $I = \{i_1, \cdots, i_s\} \subset [n]$ with $i_1 < \cdots < i_s$, we denote a permutation

$$c_I : = (i_1, \cdots, i_s, 1, \cdots, i_1 - 1, i_1 + 1, \cdots, i_s - 1, i_s + 1, \cdots, n) \in \mathbb{S}_n.$$

THEOREM 2.8 ([**BYZ**, Theorem 4.6]). *Let $\mathcal{P}$ be a 2-unitary operad. For each $k \ge 0$, let*

$$\Theta^k : = \{\theta_1^k, \cdots, \theta_{z_k}^k\}$$

be a $\Bbbk$-basis for ${}^k\Upsilon(k)$. Let $\mathbf{B}_k(n)$ be the set

$$\{\mathbb{1}_2 \circ (\theta_i^k, \mathbb{1}_{n-k}) * c_I \mid 1 \le i \le z_k, I \subset [n], |I| = k\}.$$

Then $\mathcal{P}(n)$ has a $\Bbbk$-basis

$$\bigcup_{0 \le k \le n} \mathbf{B}_k(n) = \{\mathbb{1}_n\} \cup \bigcup_{1 \le k \le n} \mathbf{B}_k(n),$$

and for every $k \ge 1$, ${}^k\Upsilon(n)$ has a $\Bbbk$-basis $\bigcup_{k \le i \le n} \mathbf{B}_i(n)$.

LEMMA 2.9 ([**BYZ**, Lemma 5.2]). *Let $\mathcal{P}$ be a 2-unitary operad and $f_{\mathcal{P}}(k) = \dim {}^k\Upsilon(k)$ for each $k \ge 0$. Then*

(1) $G_{\mathcal{P}}(t) = \sum\limits_{k=0}^{\infty} f_{\mathcal{P}}(k) \dfrac{t^k}{(1-t)^{k+1}}$.

(2) $\operatorname{GKdim} \mathcal{P} = \max\{k \mid f_{\mathcal{P}}(k) \ne 0\} + 1 = \min\{k \mid {}^k\Upsilon = 0\}$.

Combining Lemma 2.9(2) with [**BYZ**, Proposition 0.5], if $\mathcal{P}$ is 2-unitary, then there is a canonical morphism of 2-unitary operads

$$\text{(E2.9.1)} \qquad \epsilon : \quad \mathcal{P} \longrightarrow \mathcal{P}/{}^1\Upsilon = \mathcal{C}om.$$

3. Basic facts on 2-unitary operads of GK-dimension 3

Let $\mathcal{P}$ be a 2-unitary operad of GK-dimension 3. By Theorem 2.8 and Lemma 2.9, we have the following basic facts:

(1) ${}^2\Upsilon \ne 0$, and ${}^k\Upsilon = 0$ for all $k \ge 3$.

(2) $\dim \mathcal{P}(n) = 1 + dn + m\dfrac{n(n-1)}{2}$, where $d := f_{\mathcal{P}}(1) = \dim {}^1\Upsilon(1)$ and $m := f_{\mathcal{P}}(2) = \dim {}^2\Upsilon(2)$.

(3) $G_{\mathcal{P}}(t) = \dfrac{1}{1-t} + d\dfrac{t}{(1-t)^2} + m\dfrac{t^2}{(1-t)^3}$.

Based on the above facts, we have the following lemmas, which is useful to understand the structure of a 2-unitary operad of GK-dimension 3.

LEMMA 3.1. *Let $\mathcal{P}$ be a 2-unitary operad of GK-dimension 3 and $\mathbb{1}_2$ a 2-unit. Then for all $\tau, \mu \in {}^2\Upsilon(2)$ and $\delta \in {}^1\Upsilon(1)$,*

$$\tau \underset{1}{\circ} \mathbb{1}_2 = \mathbb{1}_2 \underset{2}{\circ} \tau + (\mathbb{1}_2 \underset{2}{\circ} \tau) * (2,1,3), \tag{E3.1.1}$$

$$\tau \underset{2}{\circ} \mathbb{1}_2 = \mathbb{1}_2 \underset{1}{\circ} \tau + (\mathbb{1}_2 \underset{1}{\circ} \tau) * (1,3,2). \tag{E3.1.2}$$

$$\tau \underset{i}{\circ} \mu = 0 \quad (i = 1, 2). \tag{E3.1.3}$$

$$\mathbb{1}_2 \circ (\mu, \delta) = 0, \tag{E3.1.4}$$

$$\mathbb{1}_2 \circ (\tau, \mu) = 0, \tag{E3.1.5}$$

PROOF. By a direct calculation, we have

$$(\tau \underset{1}{\circ} \mathbb{1}_2 - \mathbb{1}_2 \underset{2}{\circ} \tau - (\mathbb{1}_2 \underset{2}{\circ} \tau) * (2,1,3)) \underset{i}{\circ} \mathbb{1}_0 = 0$$

for all $1 \leq i \leq 3$. It follows that

$$\tau \underset{1}{\circ} \mathbb{1}_2 - \mathbb{1}_2 \underset{2}{\circ} \tau - (\mathbb{1}_2 \underset{2}{\circ} \tau) * (2,1,3) \in {}^3\Upsilon(3).$$

Since $\mathcal{P}$ is of GK-dimension 3 and ${}^3\Upsilon(3) = 0$, Equation (E3.1.1) holds. Similarly, (E3.1.2), (E3.1.3), (E3.1.4) and (E3.1.5) hold. □

Let $\mathcal{P}$ be a 2-unitary operad with a 2-unit $\mathbb{1}_2$. By convention, we define $\mathbb{1}'_n = \mathbb{1}_n$ for $n = 0, 1, 2$. Recall from [**BYZ**, Section 2] that, for every $n \geq 3$, one can define inductively that

$$\mathbb{1}_n = \mathbb{1}_2 \underset{1}{\circ} \mathbb{1}_{n-1}, \quad \text{and} \quad \mathbb{1}'_n = \mathbb{1}_2 \underset{2}{\circ} \mathbb{1}'_{n-1}.$$

By Definition 2.5(4) a 2-unitary operad is called 2a-unitary if $\mathbb{1}_2$ is associative, or equivalently $\mathbb{1}_3 = \mathbb{1}'_3$.

LEMMA 3.2. *Let $\mathcal{P}$ be a 2-unitary operad of GK-dimension 3 with 2-unit $\mathbb{1}_2$. Then $\mathbb{1}_2$ is a 2a-unit. Moreover, if $\mathbb{1}_2$ is a 2a-unit, then so is $\mathbb{1}_2 + \tau$ for any $\tau \in {}^2\Upsilon(2)$.*

PROOF. Suppose that $\mathbb{1}_2$ is a 2-unit of $\mathcal{P}$. By definition, $\mathbb{1}_3 = \mathbb{1}_2 \underset{1}{\circ} \mathbb{1}_2$ and $\mathbb{1}'_3 = \mathbb{1}_2 \underset{2}{\circ} \mathbb{1}_2$. One can easily check that $(\mathbb{1}_3 - \mathbb{1}'_3) \underset{i}{\circ} \mathbb{1}_0 = 0$ for all $i = 1, 2, 3$. This means that $\mathbb{1}_3 - \mathbb{1}'_3 \in {}^3\Upsilon(3)$. Since $\mathcal{P}$ is of GK-dimension 3, ${}^3\Upsilon(3) = 0$. Thus $\mathbb{1}_3 = \mathbb{1}'_3$ as required.

Clearly, for any $\tau \in {}^2\Upsilon(2)$, we have

$$(\mathbb{1}_2 + \tau) \underset{i}{\circ} \mathbb{1}_0 = \mathbb{1}_2 \underset{i}{\circ} \mathbb{1}_0 = \mathbb{1}_1$$

for $i = 1, 2$. So $\mathbb{1}_2 + \tau$ is a 2-unit. Moreover, by Lemma 3.1 (E3.1.3), we have

$$(\mathbb{1}_2 + \tau) \underset{i}{\circ} (\mathbb{1}_2 + \tau) = \mathbb{1}_2 \underset{i}{\circ} \mathbb{1}_2 + \tau \underset{i}{\circ} \mathbb{1}_2 + \mathbb{1}_2 \underset{i}{\circ} \tau$$

for $i = 1, 2$. Since $\mathbb{1}_2$ is a 2a-unit, $\mathbb{1}_2 \underset{1}{\circ} \mathbb{1}_2 = \mathbb{1}_2 \underset{2}{\circ} \mathbb{1}_2$. By direct computation, we have

$$(\mathbb{1}_2 \underset{1}{\circ} \tau + \tau \underset{1}{\circ} \mathbb{1}_2 - \mathbb{1}_2 \underset{2}{\circ} \tau - \tau \underset{2}{\circ} \mathbb{1}_2) \underset{i}{\circ} \mathbb{1}_0 = \tau - \tau = 0$$

for $i = 1, 2, 3$. Therefore, $(\mathbb{1}_2 \underset{1}{\circ} \tau + \tau \underset{1}{\circ} \mathbb{1}_2 - \mathbb{1}_2 \underset{2}{\circ} \tau - \tau \underset{2}{\circ} \mathbb{1}_2) \in {}^3\Upsilon(3)$. Since $\mathcal{P}$ is of GK-dimension 3 and ${}^3\Upsilon(3) = 0$, we know $\mathbb{1}_2 \underset{1}{\circ} \tau + \tau \underset{1}{\circ} \mathbb{1}_2 = \mathbb{1}_2 \underset{2}{\circ} \tau + \tau \underset{2}{\circ} \mathbb{1}_2$. It follows that $\mathbb{1}_2 + \tau$ is a 2a-unit. □

LEMMA 3.3 ([**BYZ**, Lemma 2.7]). *Let $\mathcal{P}$ be a 2a-unitary operad. Then the following hold.*

(1) *For every $n \geq 3$, $\mathbb{1}_n = \mathbb{1}'_n$.*
(2) *For every $n \geq 1$ and $k_1, \cdots, k_n \geq 0$, $\mathbb{1}_n \circ (\mathbb{1}_{k_1}, \cdots, \mathbb{1}_{k_n}) = \mathbb{1}_{k_1+\cdots+k_n}$.*

LEMMA 3.4. *Let $\mathcal{P}$ be a 2-unitary operad of GK-dimension ≤ 3. Suppose that ${}^1\Upsilon(1)$ has a $\Bbbk$-basis $\{\delta_j \mid j \in [d]\}$ and ${}^2\Upsilon(2)$ has a $\Bbbk$-basis $\{\tau_s \mid s \in [m]\}$. Then for every $n \geq 3$, $\mathcal{P}(n)$ has a $\Bbbk$-basis:*

(E3.4.1) $$\{\mathbb{1}_n\} \cup \{\delta^n_{(i),j} \mid i \in [n], j \in [d]\} \cup \{\tau^n_{(i_1 i_2),s} \mid 1 \leq i_1 < i_2 \leq n, s \in [m]\},$$

where $\delta^n_{(i),j} = (\mathbb{1}_n \underset{1}{\circ} \delta_j) * c_i$, $\tau^n_{(i_1 i_2),s} = (\mathbb{1}_{n-1} \underset{1}{\circ} \tau_s) * c_{i_1 i_2}$, *and* $c_i = (i, 1, \cdots, i-1, \hat{i}, i+1, \cdots, n)$ *and* $c_{i_1 i_2} = (i_1, i_2, 1, \cdots, i_1 - 1, \hat{i_1}, i_1 + 1, \cdots, i_2 - 1, \hat{i_2}, i_2 + 1, \cdots, n)$.

PROOF. Since $\mathcal{P}$ is a 2-unitary operad of GK-dimension 3, we have ${}^k\Upsilon(k) = 0$ for all $k \geq 3$. By Theorem 2.8, we can choose a basis of $\mathcal{P}(n)$ as follows

$$\{\mathbb{1}_n\} \cup \{\mathbb{1}_2 \circ (\delta_j, \mathbb{1}_{n-1}) * c_i \mid i \in [n], j \in [d]\}$$
$$\cup \{\mathbb{1}_2 \circ (\tau_s, \mathbb{1}_{n-2}) * c_{i_1 i_2} \mid 1 \leq i_1 < i_2 \leq n, s \in [m]\}.$$

By Lemmas 3.2 and 3.3

$$\mathbb{1}_n \underset{1}{\circ} \delta_j = \mathbb{1}'_n \underset{1}{\circ} \delta_j = (\mathbb{1}_2 \underset{2}{\circ} \mathbb{1}_{n-1}) \underset{1}{\circ} \delta_j = (\mathbb{1}_2 \underset{1}{\circ} \delta_j) \underset{2}{\circ} \mathbb{1}_{n-1} = \mathbb{1}_2 \circ (\delta_j, \mathbb{1}_{n-1}),$$

$$\mathbb{1}_{n-1} \underset{1}{\circ} \tau_s = \mathbb{1}'_{n-1} \underset{1}{\circ} \tau_s = (\mathbb{1}_2 \underset{2}{\circ} \mathbb{1}_{n-2}) \underset{1}{\circ} \tau_s = (\mathbb{1}_2 \underset{1}{\circ} \tau_s) \underset{3}{\circ} \mathbb{1}_{n-2} = \mathbb{1}_2 \circ (\tau_s, \mathbb{1}_{n-2}),$$

we immediately obtain basis (E3.4.1) of $\mathcal{P}(n)$. □

PROPOSITION 3.5. *Suppose* $\operatorname{char} \Bbbk \neq 2$. *Let $\mathcal{P}$ be a 2-unitary operad of GK-dimension 3 with a* 2-unit $\mathbb{1}_2$. *Let*

$$\bar{\mathbb{1}}_2 := \frac{1}{2}(\mathbb{1}_2 + \mathbb{1}_2 * (2, 1)).$$

Then $\bar{\mathbb{1}}_2$ is also a 2a-unit. *Consequently, $\mathcal{P}$ is $\mathcal{C}om$-augmented.*

PROOF. It is easy to check that $\bar{\mathbb{1}}_2$ is a 2-unit. By Lemma 3.2, $\bar{\mathbb{1}}_2$ is a 2*a*-unit, namely, $(\mathcal{P}, \mathbb{1}_0, \mathbb{1}, \bar{\mathbb{1}}_2)$ is a 2a-unitary operad.

After replacing $\mathbb{1}_2$ by $\bar{\mathbb{1}}_2$ we may assume that $\mathbb{1}_2 * (2, 1) = \mathbb{1}_2$. It follows from induction and Lemma 3.3(1) that $\mathbb{1}_n * \sigma = \mathbb{1}_n$ for all $\sigma \in \mathbb{S}_n$. Therefore there is a canonical morphism from $\mathcal{C}om$ [Example 2.2] to $\mathcal{P}$ sending $\mathbb{1}_n \mapsto \mathbb{1}_n$ for all $n \geq 0$. □

4. Trident algebras

Let R be a untial associative algebra over $\Bbbk$ with a right action of an abelian group G satisfying $(ab).g = (a.g)(b.g)$ and for all $a, b \in R$ and $g \in G$. Such an R is called a $\Bbbk G$-*module algebra* (or G-*module algebra*). Recall that the *skew group algebra* $R\#G$ is the vector space $R \otimes \Bbbk G$ with the multiplication

$$(a\#g)(b\#h) = ((a.h)b)\#(gh)).$$

Furthermore, a right module M over $R\#G$ means that M is a right R-module and a right $\Bbbk G$-module satisfying $(\mu a)g = (\mu g)(a.g)$ for all $\mu \in M, g \in G$ and $a \in R$. The following lemma is easy.

LEMMA 4.1. *Let G be an abelian group. Let A and B be G-module algebras. Suppose M, N are right modules over the skew group algebras $A\#G$ and $B\#G$, respectively. Then $M \otimes N$ is a right $(A \otimes B)\#G$-module with the action given by*

$$(x \otimes y)(a \otimes b\#g) := (x.(a\#g)) \otimes (y.(b\#g))$$

for $x \in M, y \in N, a \in A, b \in B, g \in G$.

Let A be a unital associative algebra. Clearly, the tensor product algebra $A \otimes A$ (also denoted by $A^{\otimes 2}$) admits a natural right $\mathbb{S}_2$-action given by $(a \otimes b)(2,1) := b \otimes a$. So we obtain a skew group algebra $(A \otimes A)\#\mathbb{S}_2$ (also denoted by $A^{\otimes 2}\#\mathbb{S}_2$). Let M be a left A- right $A^{\otimes 2}\#\mathbb{S}_2$-bimodule. Equivalently, M is both a right $\Bbbk\mathbb{S}_2$-module and a left A- right $(A \otimes A)$-bimodule satisfying

$$a(\mu(2,1)) = (a\mu)(2,1), \tag{E4.1.1}$$

$$(\mu.(a \otimes b))(2,1) = (\mu(2,1)).(b \otimes a), \tag{E4.1.2}$$

for all $a, b \in A$, $\mu \in M$.

REMARK 4.2. Observe that if M is both a right $\Bbbk\mathbb{S}_2$-module and a right A-module with the action $M \otimes A \to M, (\mu, a) \mapsto \mu \underset{1}{\cdot} a$, then M admits another right A-module action given by

$$\mu \underset{2}{\cdot} a = ((\mu.(2,1)) \underset{1}{\cdot} a).(2,1),$$

which is called the *congruence action*. Therefore, a right $A^{\otimes 2}\#\mathbb{S}_2$-module action on M is equivalent to a right $\Bbbk\mathbb{S}_2$-action together with a right A-module action satisfying

$$(\mu \underset{1}{\cdot} a) \underset{2}{\cdot} b = (\mu \underset{2}{\cdot} b) \underset{1}{\cdot} a \tag{E4.2.1}$$

for all $\mu \in M, a \in A$.

4.1. Tridents. In this subsection we introduce a new algebraic system.

Let $A = \Bbbk\mathbb{1}_1 \oplus \bar{A}$ be an augmented algebra with augmentation ideal $\bar{A}$. Obviously, the right regular $A^{\otimes 2}$-module $A \otimes A$ with an action of $\mathbb{S}_2$ given by $(a \otimes b)(2,1) = b \otimes a$ admits a right $A^{\otimes 2}\#\mathbb{S}_2$-module structure. Furthermore, its quotient module $(A \otimes A)/(\bar{A} \otimes \bar{A})$ admits an $(A, A^{\otimes 2}\#\mathbb{S}_2)$-bimodule structure, where the left A-action is given by

$$a \cdot [1_A \otimes 1_A] := [a \otimes 1_A] + [1_A \otimes a],$$

$$a \cdot [b \otimes 1_A] := [(ab) \otimes 1_A], \quad a \cdot [1_A \otimes b] := [1_A \otimes (ab)]$$

for $a, b \in \bar{A}$, where $[x \otimes y]$ denotes the element $x \otimes y + \bar{A} \otimes \bar{A} \in (A \otimes A)/(\bar{A} \otimes \bar{A})$ for $x \otimes y \in A \otimes A$.

In fact, $(A \otimes A)/(\bar{A} \otimes \bar{A})$ is isomorphic to $\Bbbk(1_A \otimes 1_A) \oplus (\bar{A} \otimes \Bbbk 1_A) \oplus (\Bbbk 1_A \otimes \bar{A})$ as a vector space.

Let A be an augmented algebra with the augmented ideal $\bar{A}$, and M be an $(A, A^{\otimes 2}\#\mathbb{S}_2)$-bimodule. Suppose that E is an extension of M by $(A \otimes A)/(\bar{A} \otimes \bar{A})$ as an $(A, A^{\otimes 2}\#\mathbb{S}_2)$-bimodule. Then the triple (A, M, E) is called a *trident*.

Let (A, M, E) and (A', M', E') be two tridents. Suppose that $\alpha: A \to A'$ is a homomorphism of augmented algebras, and $\beta: M \to M'$ is a homomorphism of $(A, A^{\otimes 2}\#\mathbb{S}_2)$-bimodules with the $(A, A^{\otimes 2}\#\mathbb{S}_2)$-action on M' given by the algebra homomorphisms $\alpha: A \to A'$ and $\alpha \otimes \alpha: A \otimes A \to A' \otimes A'$. Clearly, $[\alpha \otimes \alpha]: (A \otimes$

$A)/(\bar{A}\otimes\bar{A}) \to (A'\otimes A')/(\bar{A}'\otimes\bar{A}')$ is a homomorphism of $(A, A^{\otimes 2}\#\mathbb{S}_2)$-bimodules. Then one can obtain a homomorphism of extensions of $(A, A^{\otimes 2}\#\mathbb{S}_2)$-bimodules

$$\begin{array}{ccccccccc} 0 & \longrightarrow & M & \longrightarrow & E & \longrightarrow & (A\otimes A)/(\bar{A}\otimes\bar{A}) & \longrightarrow & 0 \\ & & \downarrow{\scriptstyle\beta} & & \downarrow{\scriptstyle\gamma} & & \downarrow{\scriptstyle[\alpha\otimes\alpha]} & & \\ 0 & \longrightarrow & M' & \longrightarrow & E' & \longrightarrow & (A'\otimes A')/(\bar{A}'\otimes\bar{A}') & \longrightarrow & 0 \end{array}$$

which is also called a homomorphism of tridents. Consequently, we obtain a category $\mathcal{T}$, called the *trident category*. The following lemma follows from Lemma 4.1.

LEMMA 4.3. *Let A, B be associative algebras over $\Bbbk$. Suppose that M and N are $(A, A^{\otimes 2}\#\mathbb{S}_2)$- and $(B, B^{\otimes 2}\#\mathbb{S}_2)$-bimodules, respectively. Then $M\otimes N$ is an $(A\otimes B, (A\otimes B)^{\otimes 2}\#\mathbb{S}_2)$-bimodule.*

4.2. Trident systems. There is another way of introducing a trident. Denote by mod-$\mathbb{S}_2$ the category of finite dimensional right $\Bbbk\mathbb{S}_2$-modules. It is well known that $V\otimes W\in$ mod-$\mathbb{S}_2$ with the diagonal action

$$(v\otimes w)*(2,1) = (v*(2,1))\otimes(w*(2,1))$$

for $V, W\in$ mod-$\mathbb{S}_2$. Let B be an associative algebra in the category mod-$\mathbb{S}_2$. Such an algebra is also called a $\Bbbk\mathbb{S}_2$-module algebra (in Hopf algebra language). Sometimes it is called an algebra with involution (but $\Bbbk$ may not be the complex field). A right (*resp.* left) module V over a $\Bbbk\mathbb{S}_2$-module algebra B means $V\in$ mod-$\mathbb{S}_2$ and the action $V\otimes B\to V$ (*resp.* $B\otimes V\to V$) is a homomorphism of $\Bbbk\mathbb{S}_2$-modules.

DEFINITION 4.4. A pair (A, M) with morphisms f, g, or equivalently, a quadruple (A, M, f, g), is called a *trident system* if

(10) $A=\Bbbk 1\!\!1_1\oplus\bar{A}$ is an augmented algebra with the augmentation ideal $\bar{A}$,
(20) M is an $(A, A\otimes A)$-bimodule in mod-$\mathbb{S}_2$,
(30) $f\colon \bar{A}\to M$ is a $\Bbbk$-linear map in mod-$\mathbb{S}_2$ where the $\mathbb{S}_2$-action on $\bar{A}$ is trivial,
(40) $g\colon \bar{A}\otimes\bar{A}\to M$ is a homomorphism of right $A\otimes A$-modules in mod-$\mathbb{S}_2$,

such that the following identities hold

(E4.4.1) $$f(ab) = af(b) + f(a)\cdot(b\otimes 1_A) + f(a)\cdot(1_A\otimes b) + g(a,b) + g(b,a),$$

(E4.4.2) $$f(a)\cdot(b,c) = ag(b,c) - g(ab,c) - g(b,ac),$$

for all $a,b,c\in\bar{A}$.

We define morphisms between trident systems as follows. Let (A, M, f, g) and (A', M', f', g') be two trident systems. A morphism $(\alpha,\beta)\colon (A, M, f, g)\to (A', M', f', g')$ is given by an algebra homomorphism $\alpha\colon A\to A'$ and a trident A-module homomorphism $\beta\colon M\to M'$ such that the following diagrams commute

$$\begin{array}{ccc} \bar{A} & \xrightarrow{f} & M \\ {\scriptstyle\alpha}\downarrow & & \downarrow{\scriptstyle\beta} \\ \bar{A}' & \xrightarrow{f'} & M' \end{array} \qquad\text{and}\qquad \begin{array}{ccc} \bar{A}\otimes\bar{A} & \xrightarrow{g} & M \\ {\scriptstyle\alpha\otimes\alpha}\downarrow & & \downarrow{\scriptstyle\beta} \\ \bar{A}'\otimes\bar{A}' & \xrightarrow{g'} & M' \end{array}$$

where the $(A, A\otimes A)$-bimodule structure on M' is determined by $(A', A'\otimes A')$-bimodule action and the algebra homomorphisms $\alpha\colon A\to A'$ and $\alpha\otimes\alpha\colon A\otimes A\to A'\otimes A'$.

One can define a category $\mathcal{C}$ consisting of all trident systems and morphisms defined above.

PROPOSITION 4.5. *Retain the above notation. The trident category is isomorphic to the category of trident systems.*

PROOF. Let (A, M, f, g) be a trident system. Recall that $A^{\otimes 2}$ is a subring of $A^{\otimes 2}\#\mathbb{S}_2$. Then f and g satisfy

(1) $f(a) = f(a) * (2,1)$,
(2) $g(a,b) = g(b,a) * (2,1)$,
(3) $f(ab) = a\cdot f(b) + f(a)\cdot(b\otimes 1_A\#(1)) + f(a)\cdot(1_A\otimes b\#(1)) + g(a,b) + g(b,a)$,
(4) $f(a)\cdot(b\otimes c\#(1)) = a\cdot g(b,c) - g(ab,c) - g(b,ac)$,

for all $a,b,c\in \bar{A}$. Using these equations, one can define an extension E of M by $(A\otimes A)/(\bar{A}\otimes\bar{A})$. To be precise, $E = M\oplus(A\otimes A)/(\bar{A}\otimes\bar{A})$ as a right $\mathbb{S}_2$-module with the $(A, A\otimes A)$-bimodule action given by

$$\begin{aligned} a\cdot(x, \lambda[1_A\otimes 1_A] + [b\otimes 1_A] + [1_A\otimes c]) &\\ = (ax + \lambda f(a) + f(a)\cdot(b\otimes 1_A) + f(a)\cdot(1_A\otimes c) + g(b,a) + g(a,c),&\\ \lambda[a\otimes 1_A] + \lambda[1_A\otimes a] + [ab\otimes 1_A] + [1_A\otimes ac]),& \end{aligned}$$

$$\begin{aligned} (x, \lambda[1_A\otimes 1_A] + [b\otimes 1_A] + [1_A\otimes c])\cdot(a\otimes 1_A) &\\ = (x\cdot(a\otimes 1_A) + g(a,c), \lambda[a\otimes 1_A] + [ba\otimes 1_A]),& \end{aligned}$$

$$\begin{aligned} (x, \lambda[1_A\otimes 1_A] + [b\otimes 1_A] + [1_A\otimes c])\cdot(1_A\otimes a) &\\ = (x\cdot(1_A\otimes a) + g(b,a), \lambda[1_A\otimes a] + [1\otimes ca]),& \end{aligned}$$

$$\begin{aligned} (x, \lambda[1_A\otimes 1_A] + [b\otimes 1_A] + [1_A\otimes c])\cdot(a\otimes a') &\\ = (x\cdot(a\otimes a') + \lambda g(a,a') + g(ba,a') + g(a,ca'), 0)& \end{aligned}$$

for all $\lambda\in\Bbbk$, $a,b,c\in\bar{A}$ and $x\in M$. It is easy to verify that (A, M, E) is a trident.

Conversely, given a trident (A, M, E), we construct a trident system as follows. Suppose that

$$0\to M\to E\xrightarrow{\pi}(A\otimes A)/(\bar{A}\otimes\bar{A})\to 0$$

is the corresponding short exact sequence of $(A, A^{\otimes 2}\#\mathbb{S}_2)$-bimodules. Without loss of generality, we assume M is a sub-bimodule of E. Fix an element $\mathbb{1}_2\in E$ with $\pi(\mathbb{1}_2) = [1_A\otimes 1_A]\in(A\otimes A)/(\bar{A}\otimes\bar{A})$. For all $a,b\in\bar{A}$, we define

$$\begin{aligned} f(a)\colon &= a\cdot\mathbb{1}_2 - \mathbb{1}_2\cdot(a\otimes 1_A\#(1)) - \mathbb{1}_2\cdot(1_A\otimes a\#(1)),\\ g(a,b)\colon &= \mathbb{1}_2\cdot(a\otimes b\#(1)) \end{aligned}$$

in E. Clearly, in $(A\otimes A)/(\bar{A}\otimes\bar{A})$, we have

$$\begin{aligned} \pi(f(a)) &= a\cdot[1_A\otimes 1_A] - [a\otimes 1_A] - [1_A\otimes a] = 0,\\ \pi(g(a,b)) &= [1_A\otimes 1_A]\cdot(a\otimes b) = 0 \end{aligned}$$

Therefore, we obtain two $\Bbbk$-linear maps

$$f\colon \bar{A}\to M, \quad\text{and}\quad g\colon \bar{A}\otimes\bar{A}\to M.$$

It can be directly checked that (A, M, f, g) is a trident system.

Since both constructions above are canonical, these defines two functors that are inverse to each other. $\square$

DEFINITION 4.6. A *trident algebra* means either a trident (A, M, E) or a trident system (A, M, f, g).

4.3. Examples. We give some examples of trident algebras.

EXAMPLE 4.7. Let A be an augmented algebra and let M be an $(A, A^{\otimes 2}\#\mathbb{S}_2)$-bimodule. Consider the trivial extension of M by $(A \otimes A)/(\bar{A} \otimes \bar{A})$. Equivalently, $f : \bar{A} \to M$ and $g : \bar{A}^{\otimes 2} \to M$ are zero maps in the corresponding trident system. In this case, $(A, M, 0, 0)$ is called a *trivial* trident algebra.

In particular, in case $A = \Bbbk$, any trident algebra is isomorphic to $(\Bbbk, M, 0, 0)$ for some right $\mathbb{S}_2$-module M and hence trivial. Also see Corollary 6.1.

Next we show some examples of nontrivial and natural trident algebras.

EXAMPLE 4.8. Let A be an augmented algebra and let $E = A \otimes A \otimes A$ be the $(A, A^{\otimes 2}\#\mathbb{S}_2)$-bimodule with the actions given by

$$a \cdot (x \otimes y \otimes z) \cdot (b \otimes c\#(2,1)) \colon = ax \otimes zb \otimes yc$$

for all $a \in A, x \otimes y \otimes z \in E, b \otimes c\#(2,1) \in A^{\otimes 2}\#\mathbb{S}_2$. Consider the homomorphism $\pi\colon E \to (A \otimes A)/(\bar{A} \otimes \bar{A})$ of $(A, A^{\otimes 2}\#\mathbb{S}_2)$-bimodules induced by

$$\pi(1_A \otimes 1_A \otimes 1_A) = [1_A \otimes 1_A].$$

Clearly, π is an epimorphism. Take $M = \operatorname{Ker} \pi$, and we obtain a trident algebra (A, M, E). Equivalently, we have a trident system $(A, \operatorname{Ker} \pi, f, g)$ with

$$f(a) = a \otimes 1_A \otimes 1_A - 1_A \otimes a \otimes 1_A - 1_A \otimes 1_A \otimes a,$$

$$g(b \otimes c) = 1_A \otimes b \otimes c,$$

for all $a \in \bar{A}, b \otimes c \in \bar{A} \otimes \bar{A}$.

We construct another related trident algebra. Let J be the sub-bimodule of E generated by

$$\{a \otimes 1_A \otimes 1_A - 1_A \otimes a \otimes 1_A - 1_A \otimes 1_A \otimes a \mid a \in \bar{A}\}.$$

Clearly, $J \subset \operatorname{Ker} \pi$. Set $E' = E/J$. Then we have an epimorphism

$$\pi'\colon E' \to (A \otimes A)/(\bar{A} \otimes \bar{A})$$

of $(A, A^{\otimes 2}\#\mathbb{S}_2)$-bimodules, and hence a trident algebra $(A, \operatorname{Ker} \pi', E')$. This trident algebra corresponds to the trident system $(A, \operatorname{Ker} \pi', f', g')$ with

$$f' = 0, \quad g'(b \otimes c) = \overline{1_A \otimes b \otimes c}$$

for all $b \otimes c \in \bar{A} \otimes \bar{A}$.

EXAMPLE 4.9. This is the trident algebra corresponding to $\mathcal{D}_A \underset{\mathrm{H}}{\otimes} \mathcal{D}_B$, where $\mathcal{D}_A$ and $\mathcal{D}_B$ are the 2-unitary operad defined in [**BYZ**, Example 2.4].

Let A and B be augmented algebras. Clearly, $A \otimes B$ is also an augmented algebra with $\overline{A \otimes B} := \bar{A} \otimes \Bbbk 1_B + \Bbbk 1_A \otimes \bar{B} + \bar{A} \otimes \bar{B}$. From Subsection 4.1, we know that $(A \otimes A)/(\bar{A} \otimes \bar{A})$ and $(B \otimes B)/(\bar{B} \otimes \bar{B})$ are $(A, A^{\otimes 2}\#\mathbb{S}_2)$-bimodule and $(B, B^{\otimes 2}\#\mathbb{S}_2)$-bimodule, respectively. By Lemma 4.1, we obtain an $(A \otimes B, (A \otimes B)^{\otimes 2}\#\mathbb{S}_2)$-bimodule

$$E = \{(A \otimes A)/(\bar{A} \otimes \bar{A})\} \otimes \{(B \otimes B)/(\bar{B} \otimes \bar{B})\}.$$

Observe that,

$$E = (\Bbbk[1_A \otimes 1_A] \oplus (\bar{A} \otimes \Bbbk 1_A) \oplus (\Bbbk 1_A \otimes \bar{A}))\otimes \\ (\Bbbk[1_B \otimes 1_B] \oplus (\bar{B} \otimes \Bbbk 1_B) \oplus (\Bbbk 1_B \otimes \bar{B}))$$

and

$$M := \{(\bar{A} \otimes \Bbbk 1_A) \otimes (\Bbbk 1_B \otimes \bar{B})\} \oplus \{(\Bbbk 1_A \otimes \bar{A}) \otimes (\bar{B} \otimes \Bbbk 1_B)\}$$

is a sub-bimodule of E. By an easy computation, we have

$$\begin{aligned} E/M \cong & \Bbbk[1_{A\otimes B} \otimes 1_{A\otimes B}] \oplus [(\Bbbk 1_A \otimes \bar{B}) \otimes \Bbbk 1_{A\otimes B}] \oplus [(\Bbbk 1_{A\otimes B} \otimes (\Bbbk 1_A \otimes \bar{B}))] \\ & \oplus [(\bar{A} \otimes \Bbbk 1_B) \otimes \Bbbk 1_{A\otimes B}] \oplus [(\bar{A} \otimes \bar{B}) \otimes \Bbbk 1_{A\otimes B}] \\ & \oplus [\Bbbk 1_{A\otimes B} \otimes (\bar{A} \otimes \Bbbk 1_B)] \oplus [\Bbbk 1_{A\otimes B} \otimes (\bar{A} \otimes \bar{B})] \\ = & \Bbbk[1_{A\otimes B} \otimes 1_{A\otimes B}] \oplus (\overline{A \otimes B} \otimes \Bbbk 1_{A\otimes B}) \oplus (\Bbbk 1_{A\otimes B} \otimes \overline{A \otimes B}) \\ = & [(A \otimes B) \otimes (A \otimes B)]/[(\overline{A \otimes B}) \otimes (\overline{A \otimes B})] \end{aligned}$$

Therefore, we obtain a trident $(A \otimes B, M, E)$ which is denoted by $A \odot B$. Using the language of trident system, we have

(1) $f\colon \overline{A \otimes B} \to M$ is determined by

$$\begin{aligned} f(a \otimes 1_B) &= 0, \\ f(1_A \otimes b) &= 0, \\ f(a \otimes b) &= (a \otimes 1_A) \otimes (1_B \otimes b) + (1_A \otimes a) \otimes (b \otimes 1_B) \end{aligned}$$

for all $a \in \bar{A}$ and $b \in \bar{B}$.

(2) $g\colon \overline{A \otimes B} \otimes \overline{A \otimes B} \to M$ is determined by

$$\begin{aligned} g((a \otimes 1_B) \otimes (a' \otimes 1_B)) &= 0, \\ g((a \otimes 1_B) \otimes (1_A \otimes b')) &= (a \otimes 1_A) \otimes (1_B \otimes b'), \\ g((a \otimes 1_B) \otimes (a' \otimes b')) &= 0, \\ g((1_A \otimes b) \otimes (a' \otimes 1_B)) &= (1_A \otimes a') \otimes (b \otimes 1_B), \\ g((1_A \otimes b) \otimes (1_A \otimes b')) &= 0, \\ g((1_A \otimes b) \otimes (a' \otimes b')) &= 0, \\ g((a \otimes b) \otimes (a' \otimes 1_B)) &= 0, \\ g((a \otimes b) \otimes (1_A \otimes b')) &= 0, \\ g((a \otimes b) \otimes (a' \otimes b')) &= 0 \end{aligned}$$

for all $a, a' \in \bar{A}$ and $b, b' \in \bar{B}$.

5. Classification of 2-unitary operads of GK-dimension 3

5.1. An operad constructed from a trident algebra. In this subsection, we construct a 2-unitary operad $\mathcal{P}$ by (A, M, E), with $\mathcal{P}(0) = \Bbbk \mathbb{1}_0$, $\mathcal{P}(1) = A$ and $\mathcal{P}(2) = E$, where the composition $\mathcal{P}(1) \underset{1}{\circ} \mathcal{P}(1) \to \mathcal{P}(1)$ is given by the multiplication of A, the compositions $\mathcal{P}(1) \underset{1}{\circ} \mathcal{P}(2) \to \mathcal{P}(2)$, $\mathcal{P}(2) \circ (\mathcal{P}(1), \mathcal{P}(1)) \to \mathcal{P}(2)$ are given by the corresponding actions of A on E.

Precisely, given a trident system (A, M, f, g) (where $A = \Bbbk \oplus \bar{A}$), consider the operad generated by the $\Bbbk\mathbb{S}$-module $(\Bbbk\mathbb{1}_0, A, \Bbbk\mathbb{1}_2 \oplus M, 0, 0, \cdots)$ and subject to the following relations

$$\begin{gathered}
a \circ \mathbb{1}_0 = 0, \text{ for all } a \in \bar{A}, \\
\mu \underset{i}{\circ} \mathbb{1}_0 = 0, \text{ for all } \mu \in M, \\
\mathbb{1}_2 \underset{i}{\circ} \mathbb{1}_0 = \mathbb{1}_1, \text{ for } i = 1, 2, \\
a \circ b = ab, \text{ for all } a, b \in \bar{A}, \\
a \circ \mu = a \cdot \mu, \text{ for all } a \in \bar{A}, \mu \in M, \\
a \circ \mathbb{1}_2 = \mathbb{1}_2 \underset{1}{\circ} a + \mathbb{1}_2 \underset{2}{\circ} a + f(a), \text{ for all } a \in \bar{A}, \\
\mathbb{1}_2 \circ (a, b) = g(a, b), \text{ for all } a, b \in \bar{A}, \\
\mu \circ (a, b) = \mu \cdot (a \otimes b), \text{ for all } \mu \in M, a, b \in \bar{A}, \\
\mathbb{1}_2 \underset{1}{\circ} \mathbb{1}_2 = \mathbb{1}_2 \underset{2}{\circ} \mathbb{1}_2, \\
\mu \underset{1}{\circ} \mathbb{1}_2 = \mathbb{1}_2 \underset{2}{\circ} \mu + (\mathbb{1}_2 \underset{2}{\circ} \mu) * (2, 1, 3), \text{ for all } \mu \in M, \\
\mu \underset{2}{\circ} \mathbb{1}_2 = \mathbb{1}_2 \underset{1}{\circ} \mu + (\mathbb{1}_2 \underset{1}{\circ} \mu) * (1, 3, 2), \text{ for all } \mu \in M, \\
\mathbb{1}_2 \circ (\mu, a) = 0, \text{ for all } \mu \in M, a \in \bar{A}, \\
\mu \underset{i}{\circ} \mu' = 0, \text{ for all } \mu, \mu' \in M, i = 1, 2,
\end{gathered}$$

where $\mathbb{1}_2 * (2, 1) = \mathbb{1}_2$, and $ab, a \cdot \mu, \mu \cdot (a \otimes b)$ are given by the multiplication of A, the left module action of A on M, and the right module action of $A \otimes A$ on M, respectively.

Next we give an explicit description of the above operad. Let $\mathcal{P} = F(A, M, f, g)$ be the following $\Bbbk\mathbb{S}$-module with partial compositions.

(C1) The vector space $\mathcal{P}(n)$:
(C1.1) $\mathcal{P}(0) = \Bbbk\mathbb{1}_0$.
(C1.2) $\mathcal{P}(1) = A = \Bbbk\mathbb{1}_1 \oplus \bar{A}$.
(C1.3) for each $n \geq 2$,

$$\mathcal{P}(n) = \Bbbk\mathbb{1}_n \oplus \bigoplus_{k=1}^{n} \bar{A}_k^{(n)} \oplus \bigoplus_{1 \leq i < j \leq n} M_{ij}^{(n)},$$

where $\Bbbk\mathbb{1}_n$ is a 1-dimensional vector space with the basis $\mathbb{1}_n$, $\bar{A}_k^{(n)}$ is a vector space isomorphic to $\bar{A}$ for $1 \leq k \leq n$, $n \geq 2$, and $M_{ij}^{(n)}$ is a vector space isomorphic to M for $1 \leq i < j \leq n$, $n \geq 2$. By convention $M_{12}^{(2)} = M$. So $\mathcal{P}(2) = \Bbbk\mathbb{1}_2 \oplus \bar{A}_1^{(1)} \oplus \bar{A}_2^{(1)} \oplus M$. In order to write elements in $\bar{A}_k^{(n)}$ and $M_{ij}^{(n)}$, we fix two families of $\Bbbk$-linear isomorphisms

$$\varphi_k^n \colon \bar{A} \to \bar{A}_k^{(n)} \quad \text{and} \quad \psi_{ij}^n \colon M \to M_{ij}^{(n)},$$

for $1 \leq k \leq n, 1 \leq i < j \leq n$ and $n \geq 2$. In fact, we will see that $\bar{A}_k^{(n)} = \{\mathbb{1}_n \underset{k}{\circ} a \mid a \in \bar{A}\}$ and $M_{ij}^{(n)} = \{(\mathbb{1}_{n-1} \underset{1}{\circ} \mu) * c_{ij} \mid \mu \in M\}$, where $c_{ij} = (i, j, 1, \cdots, i-1, \hat{i}, i+1, \cdots, j-1, \hat{j}, j+1, \cdots, n)$.

(C2) The right action of $\Bbbk\mathbb{S}_n$ on $\mathcal{P}(n)$: for each $\sigma \in \mathbb{S}_n$,
(C2.1) $\mathbb{1}_n * \sigma = \mathbb{1}_n$,

(C2.2) $\varphi_i^{(n)}(a) * \sigma = \varphi_{\sigma^{-1}(i)}^{(n)}(a)$,

(C2.3) $\psi_{ij}^{(n)}(\mu) * \sigma = \begin{cases} \psi_{\sigma^{-1}(i),\sigma^{-1}(j)}^{(n)}(\mu), & \sigma^{-1}(i) < \sigma^{-1}(j), \\ \psi_{\sigma^{-1}(j),\sigma^{-1}(i)}^{(n)}(\mu * (2,1)), & \sigma^{-1}(i) > \sigma^{-1}(j). \end{cases}$

(C3) The partial composition $\mathcal{P}(m) \underset{s}{\circ} \mathcal{P}(n) \to \mathcal{P}(m+n-1)$:

(C3.1) $\mathbb{1}_m \underset{s}{\circ} \mathbb{1}_n = \mathbb{1}_{m+n-1}$.

(C3.2) $\mathbb{1}_m \underset{s}{\circ} \varphi_i^{(n)}(a) = \varphi_{s+i-1}^{(m+n-1)}(a)$.

(C3.3) $\mathbb{1}_m \underset{s}{\circ} \psi_{i_1,i_2}^{(n)}(\mu) = \psi_{s+i_1-1,s+i_2-1}^{(m+n-1)}(\mu)$.

(C3.4)

$$\varphi_i^{(m)}(a) \underset{s}{\circ} \mathbb{1}_n = \begin{cases} \varphi_i^{(m+n-1)}(a), & i < s, \\ \sum\limits_{k=i}^{i+n-1} \varphi_k^{(m+n-1)}(a) + \sum\limits_{i \le k_1 < k_2 \le i+n-1} \psi_{k_1 k_2}^{(m+n-1)}(f(a)), & i = s, \\ \varphi_{i+n-1}^{(m+n-1)}(a), & i > s. \end{cases}$$

(C3.5)

$$\varphi_i^{(m)}(a) \underset{s}{\circ} \varphi_j^{(n)}(b) = \begin{cases} \psi_{i,s+j-1}^{(m+n-1)}(g(a,b)), & i < s, \\ \sum\limits_{k=i}^{i+j-2} \psi_{k,i+j-1}^{(m+n-1)}(g(a,b) + f(a) \underset{2}{\cdot} b) + \varphi_{i+j-1}^{(m+n-1)}(ab) & \\ \quad + \sum\limits_{k=i+j}^{i+n-1} \psi_{i+j-1,k}^{(m+n-1)}(g(b,a) + f(a) \underset{1}{\cdot} b), & i = s, \\ \psi_{s+j-1,i+n-1}^{(m+n-1)}(g(b,a)), & i > s. \end{cases}$$

(C3.6) $\varphi_i^{(m)}(a) \underset{s}{\circ} \psi_{j_1 j_2}^{(n)}(\mu) = \begin{cases} 0, & i \neq s, \\ \psi_{i+j_1-1,i+j_2-1}^{(m+n-1)}(a\mu), & i = s. \end{cases}$

(C3.7) $\psi_{i_1 i_2}^{(m)}(\mu) \underset{s}{\circ} \mathbb{1}_n = \begin{cases} \psi_{i_1+n-1,i_2+n-1}^{(m+n-1)}(\mu), & 1 \le s < i_1, \\ \sum\limits_{k=i_1}^{i_1+n-1} \psi_{k,i_2+n-1}^{(m+n-1)}(\mu), & s = i_1, \\ \psi_{i_1,i_2+n-1}^{(m+n-1)}(\mu), & i_1 < s < i_2, \\ \sum\limits_{k=i_2}^{i_2+n-1} \psi_{i_1,k}^{(m+n-1)}(\mu), & s = i_2, \\ \psi_{i_1 i_2}^{(m+n-1)}(\mu), & i_2 < s \le m. \end{cases}$

(C3.8) $\psi_{i_1 i_2}^{(m)}(\mu) \underset{s}{\circ} \varphi_j^{(n)}(b) = \begin{cases} 0, & s \neq i_1, i_2, \\ \psi_{i_1+j-1,i_2+n-1}^{(m+n-1)}(\mu \underset{1}{\cdot} b), & s = i_1, \\ \psi_{i_1,i_2+j-1}^{(m+n-1)}(\mu \underset{2}{\cdot} b), & s = i_2. \end{cases}$

(C3.9) $\psi_{i_1 i_2}^{(m)}(\mu) \underset{s}{\circ} \psi_{j_1 j_2}^{(n)}(\nu) = 0$.

THEOREM 5.1. *Retain the above notation. Let (A, M, f, g) be a trident algebra. Suppose $M \neq 0$. Then $F(A, M, f, g)$ is a 2-unitary Com-augmented operad of GK-dimension ≤ 3. Further,* $\operatorname{GKdim} F(A, M, f, g) \le 2$ *if and only if $M = 0$.*

PROOF. We firstly check that $\mathcal{P} := F(A, M, f, g)$ is an operad. Set $\mathbb{A} = \{\mathbb{1}_n\}_{n=0}^{\infty}$, $\Phi = \{\varphi_i^{(n)}(a) \mid a \in \bar{A}\}_{i,n}$, and $\Psi = \{\psi_{ij}^{(n)}(y) \mid y \in M\}_{i,j,n}$. It suffices to show that (OP1), (OP2) and (OP3) hold for these elements.

Verification of (OP1): By (C3.1), (C3.2) and (C3.3), $\mathbb{1} \underset{1}{\circ} \theta = \theta$ for all $\theta \in \mathcal{P}$. By (C3.1), (C3.4) and (C3.7), we have $\theta \underset{i}{\circ} \mathbb{1} = \theta$ for all $\theta \in \mathcal{P}$ and $1 \leq i \leq \mathrm{Ar}(\theta)$. Therefore (OP1) holds.

Verification of (OP2): We make a discussion on λ, μ, ν. Clearly we have the following types, Ψ^3, $\Phi\Psi^2, \mathbb{A}\Psi^2, \Phi^2\Psi, \mathbb{A}\Phi\Psi, \mathbb{A}^2\Psi, \Phi^3, \mathbb{A}\Phi^2, \mathbb{A}^2\Phi, \mathbb{A}^3$. Here for instance, by the type $\Phi\Psi^2$ we mean that one of λ, μ, ν is taken from Φ and the other two are from Ψ. Other types are similar.

For each type there are several subcases, for instance, the case of type $\Phi\Psi^2$ has 3 subcases, say $\lambda \in \Phi, \mu, \nu \in \Psi$, or $\mu \in \Phi, \lambda, \nu \in \Psi$, or $\nu \in \Phi, \lambda, \mu \in \Psi$. Altogether we have 27 subcases.

By (C3.7), (C3.8) and (C3.9), if at least two of λ, μ, ν are from Ψ, say in the types Ψ^3, $\Phi\Psi^2$, $\mathbb{A}\Psi^2$, then both sides of (E2.1.1) and (E2.1.2) are obviously 0 and hence (OP2) holds automatically. Thus there are 20 subcases left. The proof is routine but tedious and we only check a special case, namely, (E2.1.1) holds for $\lambda \in \Psi$ and $\mu, \nu \in \Phi$, and other cases can be checked similarly.

Assume that $\lambda = \psi^{(l)}_{k_1k_2}(y)$, $\mu = \varphi^{(m)}_s(a)$, and $\nu = \varphi^{(n)}_t(b)$. Then

$$\text{LHS of (E2.1.1)} = (\psi^{(l)}_{k_1k_2}(y) \underset{i}{\circ} \varphi^{(m)}_s(a)) \underset{i-1+j}{\circ} \varphi^{(n)}_t(b)$$

$$\underset{\text{by (C3.8)}}{=} \begin{cases} 0 & i \neq k_1, k_2, \\ \psi^{(l+m-1)}_{k_1+s-1,k_2+m-1}(y \underset{1}{\cdot} a) & i = k_1, \\ \psi^{(l+m-1)}_{k_1,k_2+s-1}(y \underset{2}{\cdot} a) & i = k_2, \end{cases} \quad \underset{i-1+j}{\circ} \varphi^{(n)}_t(b)$$

$$\underset{\text{by (C3.8)}}{=} \begin{cases} 0 & i \neq k_1, k_2, i+j-1 \neq k_1+s-1, k_2+m-1, k_1, k_2+s-1, \\ \psi^{(l+m+n-2)}_{k_1+s+t-2,k_2+m+n-2}((y \underset{1}{\cdot} a) \underset{1}{\cdot} b) & i = k_1, i+j-1 = k_1+s-1, \\ \psi^{(l+m+n-2)}_{k_1+s-1,k_2+m+t-2}((y \underset{1}{\cdot} a) \underset{2}{\cdot} b) & i = k_1, i+j-1 = k_2+m-1, \ (\textit{impossible}) \\ \psi^{(l+m+n-2)}_{k_1+t-1,k_2+s+n-2}((y \underset{2}{\cdot} a) \underset{1}{\cdot} b) & i = k_2, i+j-1 = k_1, \ (\textit{impossible}) \\ \psi^{(l+m+n-2)}_{k_1,k_2+s+t-2}((y \underset{2}{\cdot} a) \underset{2}{\cdot} b) & i = k_2, i+j-1 = k_2+s-1, \end{cases}$$

$$= \begin{cases} \psi^{(l+m+n-2)}_{k_1+s+t-2,k_2+m+n-2}((y \underset{1}{\cdot} a) \underset{1}{\cdot} b) & i = k_1, j = s, \\ \psi^{(l+m+n-2)}_{k_1,k_2+s+t-2}((y \underset{2}{\cdot} a) \underset{2}{\cdot} b) & i = k_2, j = s, \\ 0 & \text{otherwise}, \end{cases}$$

and

$$\text{RHS of (E2.1.1)} = \psi^{(l)}_{k_1k_2}(y) \underset{i}{\circ} (\varphi^{(m)}_s(a) \underset{j}{\circ} \varphi^{(n)}_t(b))$$

$$\underset{\text{by (C3.5)}}{=} \psi^{(l)}_{k_1k_2}(y) \underset{i}{\circ} \begin{cases} \psi^{(m+n-1)}_{s,j+t-1}(g(a,b)), & s < j, \\ \sum\limits_{k=s}^{s+t-2} \psi^{(m+n-1)}_{k,s+t-1}(g(a,b) + f(a) \underset{2}{\cdot} b) + \varphi^{(m+n-1)}_{s+t-1}(ab) & \\ \quad + \sum\limits_{k=s+t}^{s+n-1} \psi^{(m+n-1)}_{s+t-1,k}(g(b,a) + f(a) \underset{1}{\cdot} b), & s = j, \\ \psi^{(m+n-1)}_{j+t-1,s+n-1}(g(b,a)), & s > j. \end{cases}$$

$$\underset{\text{by (C3.9)}}{=} \begin{cases} 0 & s < j, \\ \psi^{(l)}_{k_1k_2}(y) \underset{i}{\circ} \varphi^{(m+n-1)}_{s+t-1}(ab) & s = j, \\ 0 & s > j, \end{cases}$$

$$\underset{\text{by (C3.6)}}{=} \begin{cases} 0 & s < j, \\ 0 & s = j, i \neq k_1, k_2, \\ \psi^{(l+m+n-2)}_{k_1+s+t-2,k_2+m+n-2}(y \underset{1}{\cdot} ab) & s = j, i = k_1 \\ \psi^{(l+m+n-2)}_{k_1,k_2+s+t-2}(y \underset{2}{\cdot} ab) & s = j, i = k_2 \\ 0 & s > j, \end{cases}$$

$$= \begin{cases} \psi^{(l+m+n-2)}_{k_1+s+t-2,k_2+m+n-2}((y \underset{1}{\cdot} a) \underset{1}{\cdot} b) & i = k_1, j = s, \\ \psi^{(l+m+n-2)}_{k_1,k_2+s+t-2}((y \underset{2}{\cdot} a) \underset{2}{\cdot} b) & i = k_2, j = s, \\ 0 & \text{otherwise}, \end{cases}$$

which implies that (E2.1.1) holds.

Verification of (OP3): Since each of μ and ν has 3 choices, say, belongs to $\mathbb{A}$, Φ or Ψ, we have 9 cases. (OP3) for the case $\mu, \nu \in \mathbb{A}$ is trivial and we have 8 cases left. We only check a special case, namely, (E2.1.3) for $\mu \in \Phi$ and $\nu \in \mathbb{A}$, and $\sigma \in \mathbb{S}_n$, and other cases can be checked similarly. Recall that $\sigma' = 1_m \underset{i}{\circ} \sigma$. Write

$$\sigma = (k_1,\ k_2,\ \cdots,\ k_n) \tag{E5.1.1}$$

where by convention $k_w = \sigma^{-1}(w)$ for all w. Then, by definition,

$$\sigma' = (1, \cdots, i-1, k_1+i-1, k_2+i-1, \cdots, k_n+i-1, i+n, \cdots, n+m-1). \tag{E5.1.2}$$

By (E5.1.2), $(\sigma')^{-1}(s) = s$ if $s < i$ and $s \geq i+n$ and $(\sigma')^{-1}(k+i-1) = \sigma^{-1}(k)+i-1$ for $1 \leq k \leq n$. We refer to [**BYZ**, Section 8] for more details concerning σ'.

Write $\mu = \varphi_k^{(m)}(a)$ and $\nu = \mathbb{1}_n$. Then

$$\text{LHS of (E2.1.3)} = \varphi_k^{(m)}(a) \underset{i}{\circ} (\mathbb{1}_n * \sigma) \underset{\text{by (C3.3)}}{=} \varphi_k^{(m)}(a) \underset{i}{\circ} \mathbb{1}_n$$

$$\underset{\text{by (C3.4)}}{=} \begin{cases} \varphi_k^{(m+n-1)}(a), & k < i, \\ \sum\limits_{w=k}^{k+n-1} \varphi_k^{(m+n-1)}(a) + \sum\limits_{k \leq k_1 < k_2 \leq k+n-1} \psi_{k_1 k_2}^{(m+n-1)}(f(a)), & k = i, \\ \varphi_{k+n-1}^{(m+n-1)}(a), & k > i. \end{cases}$$

and in the following computation we use the fact that $f(a)*(2,1) = f(a)$ [Definition 4.4(3)] and notation $(k_1', k_2') = ((\sigma')^{-1}(k_1), (\sigma')^{-1}(k_2))$ or $((\sigma')^{-1}(k_2), (\sigma')^{-1}(k_1))$,

$$\text{RHS of (E2.1.3)} = (\varphi_k^{(m)}(a) \underset{i}{\circ} \mathbb{1}_n) * \sigma'$$

$$\underset{\text{by (C3.4)}}{=} \begin{cases} \varphi_k^{(m+n-1)}(a), & k < i, \\ \sum\limits_{w=k}^{k+n-1} \varphi_k^{(m+n-1)}(a) + \sum\limits_{k \leq k_1 < k_2 \leq k+n-1} \psi_{k_1 k_2}^{(m+n-1)}(f(a)), & k = i, \\ \varphi_{k+n-1}^{(m+n-1)}(a), & k > i. \end{cases} * \sigma'$$

$$\underset{\text{by (C2.2) and (C2.3)}}{=} \begin{cases} \varphi_{(\sigma')^{-1}(k)}^{(m+n-1)}(a), & k < i, \\ \sum\limits_{w=k}^{k+n-1} \varphi_{(\sigma')^{-1}(k)}^{(m+n-1)}(a) + \sum\limits_{k \leq k_1' < k_2' \leq k+n-1} \psi_{k_1' k_2'}^{(m+n-1)}(f(a)), & k = i, \\ \varphi_{(\sigma')^{-1}(k+n-1)}^{(m+n-1)}(a), & k > i. \end{cases}$$

$$= \begin{cases} \varphi_k^{(m+n-1)}(a), & k < i, \\ \sum\limits_{w=k}^{k+n-1} \varphi_k^{(m+n-1)}(a) + \sum\limits_{k \leq k_1 < k_2 \leq k+n-1} \psi_{k_1 k_2}^{(m+n-1)}(f(a)), & k = i, \\ \varphi_{k+n-1}^{(m+n-1)}(a), & k > i. \end{cases}$$

Hence (E2.1.3) holds. It follows that $\mathcal{P}$ is an operad.

By (C3.1), we know that $\mathbb{1}_2$ is a 2-unit and $\mathcal{P}$ is 2-unitary operad. We define a morphism $u_{\mathcal{P}} : \mathcal{Com} \to \mathcal{P}$ by sending $\mathbb{1}_m \to \mathbb{1}_m$ for all $m \geq 0$. Note that $u_{\mathcal{P}}$ is an operadic morphism by (C2.1) and (C3.1). It follows that $\mathcal{P}$ is $\mathcal{Com}$-augmented.

Let $\{a_j\}_{j=1}^d$ be a basis of $\bar{A}$ and $\{\mu_k\}_{k=1}^m$ a basis of M where d is the dimension of $\bar{A}$ and m is the dimension of M. By construction,

$$\{\mathbb{1}_n,\ \ \varphi_i^{(n)}(a_j) := \mathbb{1}_n \underset{i}{\circ} a_j,\ \ \psi_{i_1 i_2}^{(n)}(\mu_k) := (\mathbb{1}_{n-1} \underset{1}{\circ} \mu_k) * c_{i_1 i_2}$$
$$\mid i \in [n], j \in [d], k \in [m], 1 \leq i_1 < i_2 \leq n\}$$

is a $\Bbbk$-basis of $\mathcal{P}(n)$. As a consequence, the generating function of $\mathcal{P}$ is

$$G_{\mathcal{P}}(t) = \sum_{n=0}^{\infty}(1+dn+m\frac{n(n-1)}{2})t^n = \frac{1}{1-t} + d\frac{t}{(1-t)^2} + m\frac{t^2}{(1-t)^3}.$$

Therefore $\mathcal{P}$ has GK-dimension ≤ 3. It is clear that $\operatorname{GKdim}\mathcal{P} \leq 2$ if and only if $m = 0$ if and only if $M = 0$. □

EXAMPLE 5.2. Let A be an augmented algebra with the augmentation ideal $\bar{A}$, and let $\mathcal{P}$ be the corresponding operad of the trident algebra $(A, \operatorname{Ker}\pi, E)$ as in Example 4.8. Then an algebra Λ over $\mathcal{P}$ is a commutative associative algebra endowed with operators $\delta\colon \Lambda \to \Lambda$ (for all $\delta \in \bar{A}$) satisfying

$$\delta(xy)\delta'(z) = x\delta(y)\delta'(z) + y\delta(x)\delta'(z),$$

and

$$\delta(x)\delta'(y)\delta''(z) = 0$$

for all $\delta, \delta', \delta'' \in \bar{A}$ and $x, y, z \in \Lambda$.

Let $\mathcal{P}'$ be the operad corresponding to the second trident algebra $(A, \operatorname{Ker}\pi', E')$ as in Example 4.8. Then an algebra Λ over $\mathcal{P}'$ is a commutative associative algebra endowed with derivations $\delta\colon \Lambda \to \Lambda$ (for all $\delta \in \bar{A}$) satisfying

$$\delta(x)\delta'(y) + \delta'(x)\delta(y) = 0$$

and

$$\delta(x)\delta'(y)\delta''(z) = 0$$

for all $\delta, \delta', \delta'' \in \bar{A}$ and $x, y, z \in \Lambda$.

5.2. Classification of 2-unitary operads of GK-dimension 3. Now we are ready to prove the main theorem.

THEOREM 5.3. *The category $\mathcal{C}$ of trident algebras (A, M, f, g) is equivalent to the category $\mathcal{D}$ of $\mathcal{C}om$-augmented operads of GK-dimension ≤ 3.*

PROOF. Define a functor $\mathcal{F} : \mathcal{C} \longrightarrow \mathcal{D}$ as follows:

(i) For any trident algebra (A, M, f, g),

$$\mathcal{F}(A, M, f, g) := F(A, M, f, g)$$

where $F(A, M, f, g)$ is given in Theorem 5.1, namely,

$$\mathcal{F}(A, M, f, g) = \{\mathcal{P}(n)\}_{n\geq 0} = \{\Bbbk\mathbb{1}_n \oplus \bigoplus_{k=1}^{n} \bar{A}_k^{(n)} \oplus \bigoplus_{1\leq i<j\leq n} M_{ij}^{(n)}\}_{n\geq 0}.$$

By the main theorem [Theorem 5.1], $\mathcal{F}(A, M, f, g)$ is a $\mathcal{C}om$-augmented operad of GK-dimension ≤ 3.

(ii) For a morphism $(\alpha, \beta)\colon (A, M, f, g) \to (A', M', f', g')$, we define an operadic morphism

$$\Phi = \mathcal{F}(\alpha, \beta)\colon \{\mathcal{P}(n)\} \to \{\mathcal{P}'(n)\}$$

as follows:

$$\begin{aligned} \Phi_n(\mathbb{1}_n)\colon &= \mathbb{1}'_n;\\ \Phi_n(\varphi_k^n(a))\colon &= \varphi'^n_k(\alpha(a)), \quad \text{for } a \in \bar{A};\\ \Phi_n(\psi_{ij}^n(\mu))\colon &= \psi'^n_{ij}(\beta(\mu)), \quad \text{for } \mu \in M. \end{aligned}$$

By a direct calculation, it follows easily from (C2.1)–(C2.3) and (C3.1)–(C3.9) that Φ is a morphism of operads since (α, β) is a morphism in the category $\mathcal{C}$.

Conversely, we define a functor $\mathcal{G} : \mathcal{D} \longrightarrow \mathcal{C}$ as follows: for an object $\mathcal{P}$ in category $\mathcal{D}$, we define

$$\mathcal{G}(\mathcal{P}) = (A, M, f, g)$$

where $A = \mathcal{P}(1)$, $M = {}^2\Upsilon_{\mathcal{P}}(2)$, and

$$f: \quad {}^1\Upsilon(1) \to {}^2\Upsilon(2), \qquad a \mapsto a \circ \mathbb{1}_2 - \mathbb{1}_2 \underset{1}{\circ} a - \mathbb{1}_2 \underset{2}{\circ} a;$$

$$g: \quad {}^1\Upsilon(1) \otimes {}^1\Upsilon(1) \to {}^2\Upsilon(2), \qquad (a, b) \mapsto (\mathbb{1}_2 \underset{1}{\circ} a) \underset{2}{\circ} b.$$

We show next that (A, M, f, g) is a trident algebra. By definitions, $A := \mathcal{P}(1)$ is an associative $\Bbbk$-algebra with identity $\mathbb{1}_1$. Considering the map $\pi^{\varnothing} : \mathcal{P}(1) \to \mathcal{P}(0) = \Bbbk\mathbb{1}_0, \theta \mapsto \theta \circ \mathbb{1}_0$, we know that A is an augmented algebra with the augmentation ideal $\operatorname{Ker} \pi^{\varnothing} = {}^1\Upsilon_{\mathcal{P}}(1)$. By the definition of truncation ideals, $M := {}^2\Upsilon_{\mathcal{P}}(2)$ is a $\Bbbk\mathbb{S}_2$-submodule of $\mathcal{P}(2)$, and is an $(A, A \otimes A)$-bimodule with the module actions given by the related composition map. Since $\mathcal{P}$ is a $\mathcal{C}om$-augmented (hence 2-unitary) operad of GK-dimension ≤ 3, we have

$$(a \circ \mathbb{1}_2 - \mathbb{1}_2 \underset{1}{\circ} a - \mathbb{1}_2 \underset{2}{\circ} a) \underset{i}{\circ} \mathbb{1}_0 = 0,$$

for all $a \in \bar{A}$ and $i = 1, 2$, and

$$((\mathbb{1}_2 \underset{1}{\circ} a) \underset{2}{\circ} b) \underset{i}{\circ} \mathbb{1}_0 = 0,$$

for all $a, b \in \bar{A}$ and $i = 1, 2$. Therefore $a \circ \mathbb{1}_2 - \mathbb{1}_2 \underset{1}{\circ} a - \mathbb{1}_2 \underset{2}{\circ} a$ and $(\mathbb{1}_2 \underset{1}{\circ} a) \underset{2}{\circ} b$ are in M. Therefore f maps from $\bar{A} \to M$ and g maps from $\bar{A}^{\otimes 2} \to M$.

For $a, b \in A$ and $\mu \in M$, let $a \cdot \mu$ be $a \circ \mu$ and $\mu \cdot (a \otimes b)$ be $\mu \circ (a, b)$ both of which are in M. Then, for all $a \in A$, $\mu \in M$,

$$a \cdot (\mu * (2,1)) = a \underset{1}{\circ} (\mu * (2,1)) = (a \underset{1}{\circ} \mu) * (\mathbb{1}_1 \underset{1}{\circ} (2,1)) = (a \cdot \mu) * (2,1)$$

which shows that (E4.1.1) holds. For $a, b \in A$, we have

$$\begin{aligned}(\mu * (2,1)) \cdot (a \otimes b) =& (\mu * (2,1)) \underset{1}{\circ} a \underset{2}{\circ} b = ((\mu \underset{2}{\circ} a) * ((2,1) \underset{1}{\circ} \mathbb{1}_1)) \underset{2}{\circ} b \\ =& ((\mu \underset{2}{\circ} a) \underset{1}{\circ} b) * ((2,1) \underset{2}{\circ} \mathbb{1}_1)) = (\mu \cdot (b \otimes a)) * (2,1),\end{aligned}$$

which shows that (E4.1.2) holds. Hence M is a trident A-module.

For $\bar{a}, \bar{b} \in \bar{A}$,

$$\begin{aligned} f(\bar{a}\bar{b}) =& (\bar{a}\bar{b}) \circ \mathbb{1}_2 - \mathbb{1}_2 \underset{1}{\circ} (\bar{a}\bar{b}) - \mathbb{1}_2 \underset{2}{\circ} (\bar{a}\bar{b}) \\ =& \bar{a} \circ (\bar{b} \circ \mathbb{1}_2 - \mathbb{1}_2 \underset{1}{\circ} \bar{b} - \mathbb{1}_2 \underset{2}{\circ} \bar{b}) + (\bar{a} \circ \mathbb{1}_2 - \mathbb{1}_2 \underset{1}{\circ} \bar{a} - \mathbb{1}_2 \underset{2}{\circ} \bar{a}) \underset{1}{\circ} \bar{b} \\ &+ (\bar{a} \circ \mathbb{1}_2 - \mathbb{1}_2 \underset{1}{\circ} \bar{a} - \mathbb{1}_2 \underset{2}{\circ} \bar{a}) \underset{2}{\circ} \bar{b} + (\mathbb{1}_2 \underset{1}{\circ} \bar{a}) \underset{2}{\circ} \bar{b} + (\mathbb{1}_2 \underset{1}{\circ} \bar{b}) \underset{2}{\circ} \bar{a} \\ =& \bar{a} f(\bar{b}) + f(\bar{a}) \underset{1}{\circ} \bar{b} + f(\bar{a}) \underset{2}{\circ} \bar{b} + g(\bar{a}, \bar{b}) + g(\bar{b}, \bar{a}). \end{aligned}$$

Hence (E4.4.1) holds. For $\bar{a}, \bar{b}, \bar{c} \in \bar{A}$,

$$\begin{aligned} f(\bar{a}) \cdot (\bar{b}, \bar{c}) =& ((\bar{a} \circ \mathbb{1}_2 - \mathbb{1}_2 \underset{1}{\circ} \bar{a} - \mathbb{1}_2 \underset{2}{\circ} \bar{a}) \underset{1}{\circ} \bar{b}) \underset{2}{\circ} \bar{c} \\ =& \bar{a}((\mathbb{1}_2 \underset{1}{\circ} \bar{b}) \underset{2}{\circ} \bar{c}) - (\mathbb{1}_2 \underset{1}{\circ} (\bar{a}\bar{b})) \underset{2}{\circ} \bar{c} - (\mathbb{1}_2 \underset{1}{\circ} \bar{b}) \underset{2}{\circ} (\bar{a}\bar{c}) \\ =& \bar{a} g(\bar{b}, \bar{c}) - g(\bar{a}\bar{b}, \bar{c}) - g(\bar{b}, \bar{a}\bar{c}). \end{aligned}$$

Hence (E4.4.2) holds. Therefore (A, M, f, g) is a trident algebra. It follows that $\mathcal{G}(\mathcal{P})$ is an object in the category $\mathcal{C}$.

For the operad morphism $\Psi\colon \mathcal{P} \to \mathcal{P}'$, we define the morphism

$$(\alpha, \beta) = \mathcal{G}(\Psi)\colon (\mathcal{P}(1), {}^2\Upsilon_{\mathcal{P}}(2), f, g) \to (\mathcal{P}'(1), {}^2\Upsilon'_{\mathcal{P}'}(2), f', g')$$

as follows:

$$\mathcal{G}(\Psi) := (\Psi(1), \Psi(2)|_{{}^2\Upsilon(2)}).$$

It is routine to check that $\mathcal{G}$ is a functor from $\mathcal{D} \to \mathcal{C}$.

Finally, it is clear from the definition that $\mathcal{GF}$ is the identity functor and it follows from Lemma 3.4 that $\mathcal{FG}$ is naturally isomorphic to the identity functor. The assertion follows. □

6. Connected 2-unitary operads of GK-dimension 3

In this short section, we focus on connected 2-unitary operads of GK-dimension 3. Recall that a connected operad means a unitary operad with $\mathcal{P}(1) = \Bbbk\mathbb{1}$ [Definition 2.5(1)]. Let $\mathcal{P}$ be a connected 2-unitary operad. By Theorems 5.1 and 5.3, it is easily seen that the corresponding trident algebra of $\mathcal{P}$ is of the form $(\Bbbk, M, 0, 0)$, where M is a right $\mathbb{S}_2$-module. It follows that there is one-to-one correspondence betweeen connected 2-unitary operads $\mathcal{P}$ of GK-dimension 3 and nonzero right $\mathbb{S}_2$-modules M.

For the rest of this section we assume that char $\Bbbk \neq 2$. Suppose that

$$M = \Bbbk\tau_1 \oplus \cdots \oplus \Bbbk\tau_p \oplus \Bbbk\tau_{p+1} \oplus \cdots \Bbbk\tau_{p+q}$$

is an irreducible decomposition of the right $\mathbb{S}_2$-module M, where $\tau_i * (2,1) = -\tau_i$ for $1 \leq i \leq p$, $\tau_j * (2,1) = \tau_j$ for $p+1 \leq j \leq p+q$. Denote by $\mathcal{T}_{p,q}$ the connected 2-unitary operad corresponding to M. Then we have

$$\mathcal{T}_{p,q}(0) = \Bbbk\mathbb{1}_0, \quad \mathcal{T}_{p,q}(1) = \Bbbk\mathbb{1}_1,$$

$$\mathcal{T}_{p,q}(2) = \Bbbk\mathbb{1}_2 \oplus \Bbbk\tau_1 \oplus \cdots \oplus \Bbbk\tau_p \oplus \Bbbk\tau_{p+1} \oplus \cdots \oplus \Bbbk\tau_{p+q},$$

and for $n \geq 3$

$$\mathcal{T}_{p,q}(n) = \Bbbk\mathbb{1}_n \oplus \bigoplus_{1\leq i_1<i_2\leq n, 1\leq s\leq p+q} \Bbbk\tau^n_{(i_1,i_2),s}$$

as a $\Bbbk$-vector space.

The right action of $\mathbb{S}_n$ on $\mathcal{T}_{p,q}(n)$ is given by

(1) $\mathbb{1}_n * \sigma = \mathbb{1}_n$;

(2) $$\tau^{(n)}_{(i_1,i_2),s} * \sigma = \begin{cases} \tau^n_{(\sigma^{-1}(i_1),\sigma^{-1}(i_2)),s}, & \sigma^{-1}(i_1) < \sigma^{-1}(i_2), 1 \leq s \leq p+q, \\ -\tau^n_{(\sigma^{-1}(i_2),\sigma^{-1}(i_1)),s}, & \sigma^{-1}(i_1) > \sigma^{-1}(i_2), 1 \leq s \leq p, \\ \tau^n_{(\sigma^{-1}(i_2),\sigma^{-1}(i_1)),s}, & \sigma^{-1}(i_1) > \sigma^{-1}(i_2), p+1 \leq s \leq p+q. \end{cases}$$

for all $\sigma \in \mathbb{S}_n$.

The partial composition

$$\mathcal{T}_{p,q}(m) \underset{i}{\circ} \mathcal{T}_{p,q}(n) \to \mathcal{T}_{p,q}(m+n-1)$$

is given by

(i) $\mathbb{1}_m \underset{i}{\circ} \mathbb{1}_n = \mathbb{1}_{m+n-1}$ for all $1 \leq i \leq m$.

(ii) $\mathbb{1}_m \underset{i}{\circ} \tau^n_{(i_1 i_2),s} = \tau^{m+n-1}_{(i+i_1-1,i+i_2-1),s}$.

(iii) $\tau^m_{(i_1 i_2),s} \underset{i}{\circ} \mathbb{1}_n = \begin{cases} \sum\limits_{k=i_1}^{i_1+n-1} \tau^{m+n-1}_{(k,i_2+n-1),s}, & i = i_1 < i_2, \\ \sum\limits_{k=i_2}^{i_2+n-1} \tau^{m+n-1}_{(i_1,k),s}, & i_1 < i = i_2, \\ \tau^{m+n-1}_{(i_1,i_2+n-1),s}, & i_1 < i < i_2, \\ \tau^{m+n-1}_{(i_1 i_2),s}, & i_1 < i_2 < i, \\ \tau^{m+n-1}_{(i_1+n-1,i_2+n-1),s}, & i < i_1 < i_2. \end{cases}$

(iv) $\tau^m_{(i_1 i_2),s_1} \underset{i}{\circ} \tau^n_{(j_1 j_2),s_2} = 0$ for all $1 \le i \le m$.

As a consequence of Theorem 1.2 we have the following.

COROLLARY 6.1. *The category of connected 2-unitary operads of GK-dimension ≤ 3 is equivalent to the category of right $\mathbb{S}_2$-modules.*

EXAMPLE 6.2. Let $M = \Bbbk\tau$ be a 1-dimensional vector space with right $\mathbb{S}_2$-action $\tau * (2,1) = -\tau$. Consider the trivial trident algebra $(\Bbbk, \Bbbk\tau, 0, 0)$. The correspond operad $\mathcal{P} = F(\Bbbk, \Bbbk\tau, 0, 0)$ in Theorem 5.1 is just $\mathcal{T}_{1,0}$, which is isomorphic to the quotient operad $\mathcal{P}ois/(\tau \underset{1}{\circ} \tau = 0)$. An algebra Λ over $\mathcal{P}$ is just a Poisson algebra with $\{\{x,y\},z\} = 0$ for any $x, y, z \in \Lambda$.

On the other hand, the quotient operad $\mathcal{A}ss/{}^3\Upsilon_{\mathcal{A}ss}$ is also isomorphic to $\mathcal{T}_{1,0}$. Therefore we obtain an isomorphism of operads $\mathcal{P}ois/(\tau \underset{1}{\circ} \tau) \cong \mathcal{A}ss/{}^3\Upsilon_{\mathcal{A}ss}$.

In fact, an algebra Λ over $\mathcal{A}ss/{}^3\Upsilon_{\mathcal{A}ss}$ is an associative algebra satisfying $z(xy - yx) = (xy - yx)z$, or equivalently, $[[x,y],z] = 0$ for all $x, y, z \in \Lambda$, where $[x,y] = xy - yx$. Then we can define a new product $\star$ by $x \star y = \frac{xy+yx}{2}$ for all $x, y \in \Lambda$. It is direct to check that $(\Lambda, \star, [-,-])$ forms a Poisson algebra, and hence an algebra over $\mathcal{P}ois/(\tau \underset{1}{\circ} \tau)$ since $[[x,y],z] = 0$ for all $x, y, z \in \Lambda$.

Conversely, let Λ be a Poisson algebra with $\{\{x,y\},z\} = 0$ for any $x, y, z \in \Lambda$. We may define a new product $\circ$ on Λ by $x \circ y = \frac{xy + \{x,y\}}{2}$. Then $\{x,y\} = x \circ y - y \circ x$ for any $x, y \in \Lambda$ and $\circ$ satisfies the associativity law, which means that $(\Lambda, \circ)$ is an algebra over $\mathcal{A}ss/{}^3\Upsilon_{\mathcal{A}ss}$.

It is also easy to see that $\mathcal{P}ois/{}^3\Upsilon_{\mathcal{P}ois} \cong \mathcal{A}ss/{}^3\Upsilon_{\mathcal{A}ss}$.

Similar to Example 6.2, for any 2-unitary operad $\mathcal{P}$, $\mathcal{P}/{}^3\Upsilon_{\mathcal{P}}$ has GK-dimension ≤ 3. If further $\mathcal{P}$ is connected, then $\mathcal{P}/{}^3\Upsilon_{\mathcal{P}}$ is a connected 2-unitary operad with GK-dimension ≤ 3.

EXAMPLE 6.3. Let $M = \Bbbk\nu$ be a 1-dimensional vector space with right $\mathbb{S}_2$-action given by $\nu * (2,1) = \nu$. Then the corresponding operad $\mathcal{P} = F(\Bbbk, \Bbbk\nu, 0, 0)$ is just $\mathcal{T}_{0,1}$. An algebra over $\mathcal{P}$ is a commutative associative algebra Λ equipped with a binary operation $\nu\colon \Lambda \otimes \Lambda \to \Lambda$ satisfying

$$\begin{aligned} \nu(a,b) &= \nu(b,a), \\ \nu(ab,c) &= a\nu(b,c) + \nu(a,c)b, \\ \nu(\nu(a,b),c) &= 0, \end{aligned}$$

for all $a, b, c \in \Lambda$. By Corollary 6.1, $\mathcal{T}_{0,1} \not\cong \mathcal{T}_{1,0}$.

7. Comments, examples, and remarks

We refer to [**LV**] for the definition of Hadamard product $-\underset{\mathrm{H}}{\otimes}-$ and a Hopf operad. Recall that the *Hadamard product* $\mathcal{P}\underset{\mathrm{H}}{\otimes}\mathcal{Q}$ of the operads $\mathcal{P}$ and $\mathcal{Q}$ is defined to be

$$(\mathcal{P}\underset{\mathrm{H}}{\otimes}\mathcal{Q})(n)=\mathcal{P}(n)\otimes\mathcal{Q}(n),$$

for all $n\geq 0$ with the partial composition

$$(\mu_1\otimes\nu_1)\underset{i}{\circ}(\mu_2\otimes\nu_2)=(\mu_1\underset{i}{\circ}\mu_2)\otimes(\nu_1\underset{i}{\circ}\nu_2),$$

for $\mu_1\otimes\nu_1\in(\mathcal{P}\underset{\mathrm{H}}{\otimes}\mathcal{Q})(m)$, $\mu_2\otimes\nu_2\in(\mathcal{P}\underset{\mathrm{H}}{\otimes}\mathcal{Q})(n)$, and $m\geq 1, n\geq 0, 1\leq i\leq m$. Clearly, the operad $\mathcal{C}om$ is obviously a unit for Hadamard product.

A *Hopf operad* is a symmetric operad $\mathcal{P}$ with a morphism of operads $\Delta\colon\mathcal{P}\to\mathcal{P}\underset{\mathrm{H}}{\otimes}\mathcal{P}$ called the coproduct of $\mathcal{P}$ and a morphism of operads $\epsilon_{\mathcal{P}}\colon\mathcal{P}\to\mathcal{C}om$ called the counit, which is supposed to be coassociative and counital.

Definition 7.1. Let $\mathcal{P}$ be a Hopf operad. We say that $\mathcal{P}$ is a *$\mathcal{C}om$-augmented Hopf operad* if

(1) $\mathcal{P}$ is $\mathcal{C}om$-augmented and the composition

$$\mathcal{C}om\xrightarrow{u_{\mathcal{P}}}\mathcal{P}\xrightarrow{\epsilon}\mathcal{C}om$$

is the identity map, and

(2) the following diagram is commutative

$$\begin{array}{ccc}\mathcal{C}om & \xrightarrow{\cong} & \mathcal{C}om\underset{\mathrm{H}}{\otimes}\mathcal{C}om\\ {\scriptstyle u_{\mathcal{P}}}\downarrow & & \downarrow{\scriptstyle u_{\mathcal{P}}\underset{\mathrm{H}}{\otimes}u_{\mathcal{P}}}\\ \mathcal{P} & \xrightarrow[\Delta]{} & \mathcal{P}\underset{\mathrm{H}}{\otimes}\mathcal{P}.\end{array}$$

Remark 7.2. A Hopf operad satisfying the condition (1) in Definition 7.1 is also called a unital augmented connected Hopf operad in [**Kh**, Definition 2.5].

Proposition 7.3. *Let $\mathcal{P}$ be a $\mathcal{C}om$-augmented Hopf operad of GK-dimension ≤ 2. Then $\mathcal{P}=\mathcal{C}om$.*

Proof. Note that $\Delta:\mathcal{P}\to\mathcal{P}\underset{\mathrm{H}}{\otimes}\mathcal{P}$ is an operadic morphism. Write $\mathcal{P}$ as $F(A,0,0,0)$ as given by Theorem 1.2 (and as in Theorem 1.1).

Suppose $\bar{A}\neq 0$. Then $\operatorname{GKdim}\mathcal{P}=2$ and $\operatorname{GKdim}\mathcal{P}\underset{\mathrm{H}}{\otimes}\mathcal{P}=3$. Since $\mathcal{P}\underset{\mathrm{H}}{\otimes}\mathcal{P}$ has GK-dimension 3, it is of the form $F(A\odot A)$ where $A\odot A$ is given in Example 4.9. By Theorem 5.3, Δ induces a morphism of trident algebras $(A,0,0,0)\to A\odot A:=(A\otimes A,M,f,g)$. Then $f\Delta\mid_{\mathcal{P}(1)}\colon A\to A\otimes A\to M$ is zero. We claim that $A=\Bbbk$. If not, let $0\neq a\in\bar{A}$ and write $\Delta\mid_{\mathcal{P}(1)}(a)=1\otimes a+a\otimes 1+\sum a_{(1)}\otimes a_{(2)}$ where $a_{(1)},a_{(2)}\in\bar{A}$. Then, by three equations in Example 4.9(1),

$$\begin{aligned}0=f\Delta(a)&=f(1\otimes a+a\otimes 1+\sum a_{(1)}\otimes a_{(2)})\\&=\sum((a_{(1)}\otimes 1_A)\otimes(1_A\otimes a_{(2)})+(1_A\otimes a_{(1)})\otimes(a_{(2)}\otimes 1_A)).\end{aligned}$$

Therefore $\sum a_{(1)}\otimes a_{(2)}=0$, and consequently, a is a primitive element. By Definition 7.1(2), $\mathbb{1}_n$ is group-like, i.e., $\Delta(\mathbb{1}_n)=\mathbb{1}_n\otimes\mathbb{1}_n$ for all n. Since each a is primitive, it follows from (C3.2) that each $\varphi_i^{(n)}(a)$ is primitive, i.e., $\Delta(\varphi_i^{(n)}(a)) = \varphi_i^{(n)}(a) \otimes \mathbb{1}_n + \mathbb{1}_n \otimes \varphi_i^{(n)}(a)$ for all i, n. Since $\mathcal{P}$ has GK-dimension ≤ 2, $\varphi_1^{(2)}(a) \underset{1}{\circ} \varphi_1^{(3)}(a) = 0$ by (C3.5). But

$$\begin{aligned}
&\Delta(\varphi_1^{(2)}(a) \underset{1}{\circ} \varphi_1^{(3)}(a)) \\
&\quad =(\varphi_1^{(2)}(a) \otimes \mathbb{1}_2 + \mathbb{1}_2 \otimes \varphi_1^{(2)}(a)) \underset{1}{\circ} (\varphi_1^{(3)}(a) \otimes \mathbb{1}_3 + \mathbb{1}_3 \otimes \varphi_1^{(3)}(a)) \\
&\quad =(\varphi_1^{(2)}(a) \otimes \mathbb{1}_2) \underset{1}{\circ} (\mathbb{1}_3 \otimes \varphi_1^{(3)}(a)) + (\mathbb{1}_2 \otimes \varphi_1^{(2)}(a)) \underset{1}{\circ} (\varphi_1^{(3)}(a) \otimes \mathbb{1}_3) \\
&\quad =(\sum_{k=1}^{2} \varphi_k^{(4)}(a)) \otimes \varphi_1^{(4)}(a) + \varphi_1^{(4)}(a) \otimes (\sum_{k=1}^{2} \varphi_k^{(4)}(a)) \\
&\quad =\varphi_2^{(4)}(a) \otimes \varphi_1^{(4)}(a) + \varphi_1^{(4)}(a) \otimes \varphi_2^{(4)}(a) + 2\varphi_1^{(4)}(a) \otimes \varphi_1^{(4)}(a) \\
&\quad \neq 0,
\end{aligned}$$

yielding a contradiction. Therefore $\bar{A} = 0$ and $\mathcal{P} = \mathcal{C}om$. □

The proof of the above proposition uses both Theorems 1.1 and 1.2, as well as the structure maps of trident algebras. We do know an easy proof without using trident algebras.

Unlike in the GK-dimension 2 case, the 2-unit of an operad needs not be unique.

EXAMPLE 7.4. Let $\mathcal{P} = F(A, M, f, g)$ where $M \neq 0$. Let $\mathbb{1}_2$ be the canonical 2-unit of $\mathcal{P}$ given in the construction of $F(A, M, f, g)$. Let $\mathbb{1}_2' = \mathbb{1}_2 + \psi_{12}^{(2)}(m)$. It is easy to check that $\mathbb{1}_2'$ is a $2a$-unit [Lemma 3.2]. Suppose that $m * (2,1) \neq m$. Then $(\mathcal{P}, \mathbb{1}_0, \mathbb{1}_1, \mathbb{1}_2')$ is a 2-unitary operad, but not $\mathcal{C}om$-augmented. As a consequence, we can not replace "$\mathcal{C}om$-augmented" by "2-unitary" in Theorem 1.2.

REMARK 7.5. For non-2-unitary operads, we have the following remarks.

(1) By [**QXZZ**, Construction 7.1] there are a lot of symmetric operads of GK-dimension 3 that are not 2-unitary.
(2) In [**QXZZ**], an analog of Bergman's gap theorem of nonsymmetric operads is proved, namely, no finitely generated locally finite nonsymmetric operad has GK-dimension strictly between 1 and 2. In [**LQXZZZ**] the authors proved that there is no finitely generated locally finite symmetric operad with GK-dimension strictly between 1 and 2.
(3) It is an open question if there are finitely generated locally finite symmetric operads with GK-dimension strictly between 2 and 3, see [**QXZZ**, Question 0.8].
(4) For every $r \in \{0\} \cup \{1\} \cup [2, \infty)$ or $r = \infty$, the authors in [**QXZZ**] constructed an explicit non-symmetric operad of GK-dimension r.

The following lemma was proved in [**BYZ**, Theorem 6.5].

LEMMA 7.6. *Let $\mathcal{P}$ be a 2-unitary operad of finite GK-dimension ≥ 3. Then $\mathcal{P}$ is not semiprime.*

PROOF. If $\mathcal{P}$ is semiprime, by the proof of [**BYZ**, Theorem 6.5], ${}^2\Upsilon = 0$. So $\operatorname{GKdim} \mathcal{P} \leq 2$, yielding a contradiction. □

The following example shows that $\underset{\mathrm{H}}{\otimes}$ does not preserve primeness.

EXAMPLE 7.7. Let $A = \Bbbk \oplus M_2(\Bbbk)$, considered as an augmented algebra. Then $\mathcal{P} := F(A, 0, 0, 0)$ is prime of GK-dimension 2 which can be verified by using structure Theorem 1.1. Since $\mathcal{P} \underset{\mathrm{H}}{\otimes} \mathcal{P}$ has GK-dimension 3, it is not semiprime by Lemma 7.6.

Acknowledgment

We would like to thank the referee for his/her careful reading and many useful suggestions.

References

[BV] J. M. Boardman and R. M. Vogt, *Homotopy invariant algebraic structures on topological spaces*, Lecture Notes in Mathematics, Vol. 347, Springer-Verlag, Berlin-New York, 1973. MR420609

[BYZ] Y.-H. Bao, Y. Ye, and J. J. Zhang, *Truncation of unitary operads*, Adv. Math. **372** (2020), 107290, 75, DOI 10.1016/j.aim.2020.107290. MR4125516

[Dot] V. Dotsenko, *A Quillen adjunction between algebras and operads, Koszul duality, and the Lagrange inversion formula*, Int. Math. Res. Not. IMRN **14** (2019), 4281–4301, DOI 10.1093/imrn/rnx257. MR4023760

[DK] V. Dotsenko and A. Khoroshkin, *Gröbner bases for operads*, Duke Math. J. **153** (2010), no. 2, 363–396, DOI 10.1215/00127094-2010-026. MR2667136

[DMR] V. Dotsenko, M. Markl, and E. Remm, *Veronese powers of operads and pure homotopy algebras*, Eur. J. Math. **6** (2020), no. 3, 829–863, DOI 10.1007/s40879-019-00351-6. MR4151721

[DT] V. Dotsenko and P. Tamaroff *Tangent complexes and the Diamond Lemma*, Preprint, `arXiv:2010.14792`, (2020).

[Fi] D. R. Finston, *Polynomial algebras have polynomial growth*, Trans. Amer. Math. Soc. **300** (1987), no. 2, 535–556, DOI 10.2307/2000356. MR876465

[Fr1] B. Fresse, *Homotopy of operads and Grothendieck-Teichmüller groups. Part 1*, Mathematical Surveys and Monographs, vol. 217, American Mathematical Society, Providence, RI, 2017. The algebraic theory and its topological background, DOI 10.1090/surv/217.1. MR3643404

[Fr2] B. Fresse, *Homotopy of operads and Grothendieck-Teichmüller groups. Part 2*, Mathematical Surveys and Monographs, vol. 217, American Mathematical Society, Providence, RI, 2017. The applications of (rational) homotopy theory methods, DOI 10.1090/surv/217.2. MR3616816

[Kh] A. Khoroshkin, *Quadratic algebras arising from Hopf operads generated by a single element*, Lett. Math. Phys. **110** (2020), no. 8, 2053–2082, DOI 10.1007/s11005-020-01283-z. MR4126873

[KP] A. Khoroshkin and D. Piontkovski, *On generating series of finitely presented operads*, J. Algebra **426** (2015), 377–429, DOI 10.1016/j.jalgebra.2014.12.012. MR3301915

[KL] G. R. Krause and T. H. Lenagan, *Growth of algebras and Gelfand-Kirillov dimension*, Revised edition, Graduate Studies in Mathematics, vol. 22, American Mathematical Society, Providence, RI, 2000, DOI 10.1090/gsm/022. MR1721834

[LQXZZZ] Y. Li, Z.-H. Qi, Y.-J. Xu, J. J. Zhang, Z.-R. Zhang, and X.-G. Zhao, *Operads of Gelfand–Kirillov dimension one*, in preparation.

[LV] J.-L. Loday and B. Vallette, *Algebraic operads*, Grundlehren der mathematischen Wissenschaften [Fundamental Principles of Mathematical Sciences], vol. 346, Springer, Heidelberg, 2012, DOI 10.1007/978-3-642-30362-3. MR2954392

[Lo] J.-L. Loday, *Generalized bialgebras and triples of operads* (English, with English and French summaries), Astérisque **320** (2008), x+116. MR2504663

[MSS] M. Markl, S. Shnider, and J. Stasheff, *Operads in algebra, topology and physics*, Mathematical Surveys and Monographs, vol. 96, American Mathematical Society, Providence, RI, 2002, DOI 10.1090/surv/096. MR1898414

[Ma] J. P. May, *The geometry of iterated loop spaces*, Lecture Notes in Mathematics, Vol. 271, Springer-Verlag, Berlin-New York, 1972. MR420610

[QXZZ] Z. Qi, Y. Xu, J. J. Zhang, and X. Zhao, *Growth of nonsymmetric operads*, Indiana Univ. Math. J. **72** (2023), no. 2, 651–697. MR4596188

School of Mathematical Sciences, Anhui University, Hefei 230601, People's Republic of China; and Anhui University Center for Pure Mathematics, Hefei 230601, People's Republic of China

Email address: baoyh@ahu.edu.cn, yhbao@ustc.edu.cn

School of Mathematical Sciences, Anhui University, Hefei 230601, People's Republic of China

Email address: fudongxing920811@126.com

School of Mathematical Sciences, University of Science and Technology of China, Hefei 230001, People's Republic of China

Email address: yeyu@ustc.edu.cn

Department of Mathematics, Box 354350, University of Washington, Seattle, Washington 98195

Email address: zhang@math.washington.edu

Contemporary Mathematics
Volume **814**, 2025
https://doi.org/10.1090/conm/814/16301

On a necessary condition for unitary categorification of fusion rings

Pavel Etingof, Dmitri Nikshych, and Victor Ostrik

To the memory of Earl J. Taft

ABSTRACT. In Liu, Palcoux, and Wu's article [Adv. Math. **390** (2021), Paper No. 107905, 63], they proved a remarkable necessary condition for a fusion ring to admit a unitary categorification, by constructing invariants of the fusion ring that have to be positive if it is unitarily categorifiable. The main goal of this note is to provide a somewhat more direct proof of this result. In the last subsection we discuss integrality properties of the Liu–Palcoux–Wu invariants.

1. Introduction

In [**LPW**] Liu, Palcoux, and Wu proved a remarkable necessary condition for a fusion ring to be unitarily categorifiable, which can be used to disqualify many new fusion rings from having such a categorification. They did so by constructing invariants of the fusion ring that have to be positive if it is the Grothendieck ring of a unitary fusion category. The first goal of this article is to provide a more direct proof of this criterion using the notion of a formal codegree and the Drinfeld center of the underlying category, which is done in Section 4, after discussing preliminaries in Sections 2 and 3. Thus Section 4 does not contain any essentially new results, it is just our attempt to give an exposition of the result of [**LPW**] in a somewhat more algebraic language.[1]

Our second goal is to discuss integrality properties of the Liu–Palcoux–Wu invariants and their relation to the generalized Kaplansky 6th conjecture stating that in any spherical fusion category, dimensions of objects divide the dimension of the category. This is implemented in Section 5.

2. Fusion rings

Let A be a fusion ring with a finite basis $\{b_i\}$ (where $b_0 = 1$) and let $\tau : A \to \mathbb{Z}$ be the trace function given by $\tau(\sum_i \lambda_i b_i) := \lambda_0$. There is an inner product $(a, b) := \tau(ab^*)$, for $a, b \in A$, with respect to which the basis b_i is orthonormal. Let $N_{ijm} :=$

The first author's work was partially supported by the NSF grant DMS - 1916120.

The second author was partially supported by the NSF grant DMS-1800198 and DMS-2302267.

The third author was partially supported by the NSF grant DMS-1702251.

[1]We note that the paper [**LPW**] contains a nice diagrammatic proof of this result, see the proof of Proposition 7.7 there.

(b_ib_j, b_m) be the fusion coefficients and let d_j be the Frobenius–Perron dimensions of the basis elements b_j. Recall that $N_{ijm} = N_{jm^*i^*} = N_{m^*ij^*} = N_{j^*i^*m^*}$ and $d_j = d_{j^*}$. For more details on fusion rings and Frobenius–Perron dimensions we refer the reader to [**EGNO**, Chapter 3].

We now give a shorter proof of [**LPW**, Proposition 8.1]. (This proposition will not be used below.)

PROPOSITION 2.1 ([**LPW**, Proposition 8.1]).
(i) $\sum_m N_{ijm}^2 \le \min(d_i^2, d_j^2)$.
(ii) $N_{ijm} \le d_id_j/d_m$; *hence* $N_{ijm} \le d_m(d_i/d_j)^{\frac{2-t}{t}}$ *for any* $t \ge 1$.
(iii) $N_{ijm} \le \min(d_i, d_j, d_m)$.
(iv) $\sum_m N_{i_1i_2m}N_{i_3i_4m} \le d_{i_p}d_{i_q}$ *for any distinct* $1 \le p, q \le 4$.

PROOF.
(i) $\sum_m N_{ijm}^2 = \sum_m (b_ib_j, b_m)^2 = (b_ib_j, b_ib_j) = (b_{i^*}b_ib_j, b_j) \le \mathrm{FPdim}(b_{i^*}b_i) = d_i^2$. Similarly, $\sum_m N_{ijm}^2 \le d_j^2$.
(ii) The first statement follows from the equality $\sum_m N_{ijm}d_m = d_id_j$. Using the symmetry properties of N_{ijm}, this implies that $N_{ijm} \le d_md_i/d_j$ and $N_{ijm} \le d_md_j/d_i$. This implies the second statement, since for $t \ge 1$ the number $\frac{2-t}{t}$ varies between 1 and -1.
(iii) N_{ijm} are matrix elements of the matrix of multiplication by b_i in the orthonormal basis b_j, so they do not exceed the norm d_i of this matrix. The rest follows from the symmetry properties of N_{ijm}.
(iv) $\sum_m N_{i_1i_2m}N_{i_3i_4m} = \sum_m N_{i_1i_2m}N_{i_3^*mi_4} = (b_{i_3^*}b_{i_1}b_{i_2}, b_{i_4}) = (b_{i_1}b_{i_2}b_{i_4^*}, b_{i_3})$. This is a matrix element of each of the following operators on A: $L(b_{i_3^*})L(b_{i_1})$; $R(b_{i_1})R(b_{i_2})$; $L(b_{i_3^*})R(b_{i_2})$; $L(b_{i_1})L(b_{i_2})$; $R(b_{i_2})R(b_{i_4^*})$; $L(b_{i_1})R(b_{i_4^*})$, where L denotes left multiplication and R right multiplication. Thus it is at most the norms of these operators, as desired. □

3. Formal codegrees and the Drinfeld center

Recall that $A_{\mathbb{C}} = A \otimes_{\mathbb{Z}} \mathbb{C}$ is a $*$-algebra. Consider an irreducible representation $(V, \rho : A_{\mathbb{C}} \to \mathrm{End}V)$ of $A_{\mathbb{C}}$ (which we will often shortly denote just by ρ). The **formal codegree** α_ρ of ρ is the eigenvalue of the central element $z_\rho := \sum_i \mathrm{Tr}(\rho(b_i))b_i^*$ on ρ, an algebraic integer (the eigenvalues of this element on all the other irreducible representations are 0), see, e.g., [**O1**, Lemma 2.3]. Thus $\tau(z_\rho) = \dim\rho = \alpha_\rho^{-1}\mathrm{Tr}\,\rho(z_\rho)$. Hence

$$\tau(a) = \sum_\rho \alpha_\rho^{-1}\,\mathrm{Tr}\,\rho(a),\ a \in A.$$

In particular, $\alpha_\rho > 0$ for all ρ, see also [**O2**, Remark 2.12].

Let $v_1, v_2 \in V$, $f_1, f_2 \in V^*$.

LEMMA 3.1. *We have*

$$\sum_i f_1(\rho(b_i)v_1)f_2(\rho(b_i^*)v_2) = \alpha_\rho f_2(v_1)f_1(v_2).$$

PROOF. It suffices to show that for any $a_1, a_2 \in A$ we have

$$\sum_i \mathrm{Tr}\,\rho(a_1b_i)\,\mathrm{Tr}\,\rho(b_i^*a_2) = \alpha_\rho\,\mathrm{Tr}\,\rho(a_1a_2);$$

then we can take a_1, a_2 such that $\rho(a_1) = v_1 \otimes f_1^*$, $\rho(a_2) = v_2 \otimes f_2^*$ in End$V$, which gives the desired statement. But this is equivalent to the obvious relation

$$\sum_i \operatorname{Tr} \rho(a_1 b_i)(a_2, b_i) = \operatorname{Tr} \rho(a_1 a_2), \qquad a_1, a_2 \in A. \qquad \square$$

Suppose now that $\mathcal{C}$ is a spherical fusion category that categorifies a fusion ring A. Recall that irreducible (unitary) representations V of $A_{\mathbb{C}}$ have the form $V_Z = \operatorname{Hom}_{\mathcal{C}}(\mathbb{1}, Z)$ where Z is a simple object of the Drinfeld center $\mathcal{Z}(\mathcal{C})$ of $\mathcal{C}$ whose image in $\mathcal{C}$ (which we, abusing notation, will also denote by Z) contains $\mathbb{1}$ (see [**O2**, Theorem 2.13] and [**S**, Theorem 5.9]). Namely, the map ρ_Z is constructed as follows. Note that $V_Z^* = \operatorname{Hom}_{\mathcal{C}}(Z, \mathbb{1})$ using the composition pairing. Given $X \in \mathcal{C}$, $v \in V_Z$, $f \in V_Z^*$, we define $\rho_Z(X)$ by

$$f(\rho_Z(X)v) = \operatorname{Tr}((1_X \otimes f) \circ c_{X,Z} \circ (v \otimes 1_X)).$$

where $c_{X,Z} : Z \otimes X \to X \otimes Z$ is the central structure of Z. Using the identity

$$c_{Z,X\otimes Y} = (1_X \otimes c_{Z,Y}) \circ (c_{Z,X} \otimes 1_Y),$$

it is not hard to show that ρ_Z is a representation. Note that $\rho_Z(X^*) = \rho_Z(X)^\dagger$ (the adjoint operator) and $\rho_{Z^*} \cong \overline{\rho}_Z$ (the same action on the complex conjugate space).

THEOREM 3.2 ([**O2**, Theorem 2.13]). *If $\rho = \rho_Z$ then one has $\alpha_\rho = \frac{\dim \mathcal{C}}{\dim Z}$. In particular, if $[Z : \mathbb{1}] > 0$ then $\dim Z > 0$.*

COROLLARY 3.3. *Under the assumptions of Lemma* 3.1

$$\sum_i f_1(\rho(b_i)v_1) f_2(\rho(b_i^*)v_2) = \frac{\dim \mathcal{C}}{\dim Z} f_2(v_1) f_1(v_2).$$

4. Positivity results

Recall that a **unitary fusion category** is a fusion category with a $*$-structure ([**T**]; see also [**G**, Subsection 2.1] for a full definition). A **unitary categorification** of a fusion ring is a realization of this ring as the Grothendieck ring of a unitary fusion category.

Let $(V_s, \rho_s : A \to \operatorname{End} V_s)$ be a collection of irreducible (unitary) representations of $A_{\mathbb{C}}$ and $v_s \in V_s$, $s = 1, \ldots, n$.

THEOREM 4.1 ([**LPW**, Proposition 8.3]).

(i) *If A admits a unitary categorification then we have*

$$\sum_i \frac{1}{d_i^{n-2}} (\rho_1(b_i)v_1, v_1) \ldots (\rho_n(b_i)v_n, v_n) \geq 0.$$

(ii) *([**LPW**, Corollary 8.5]) If in (i) A is commutative then*

$$\sum_i \frac{1}{d_i^{n-2}} \rho_1(b_i) \ldots \rho_n(b_i) \geq 0.$$

EXAMPLE 4.2.

(1) Let $n = 1$. Then Theorem 4.1 says that $(v, \rho(\sum_i d_i b_i)v) \geq 0$. This holds (regardless of A being unitarily categorifiable) since $R := \sum_i d_i b_i$ is the regular element of A, hence $Ra = \operatorname{FPdim}(a)R$ for any $a \in A$.

(2) Let $n = 2$. Then Theorem 4.1 says that $\sum_i (\rho_1(b_i)v_1, v_1)(\rho_2(b_i)v_2, v_2) \geq 0$. This follows (again regardless of A being unitarily categorifiable) from Lemma 3.1. Indeed, since $\rho(b_i^*) = \rho(b_i)^\dagger$, Lemma 3.1 implies that this sum is zero unless $\rho_1 \cong \overline{\rho}_2$, and

$$\sum_i (\rho(b_i)v_1, v_1)(\overline{\rho}(b_i)v_2, v_2) = \alpha_\rho |(v_1, v_2)|^2.$$

However, for $n \geq 3$, as shown in [**LPW**], unitary categorifiability of A is essential.

PROOF. It suffices to prove (i) for $n \geq 3$. Let $\mathcal{C}$ be a unitary fusion category categorifying A. Recall ([**EGNO**, Section 9.5]) that it has a canonical spherical structure in which the dimensions of simple objects X_i are d_i; so let us endow $\mathcal{C}$ with this structure. Let $Z_1, \dots, Z_n \in \mathcal{Z}(\mathcal{C})$, and consider the vector space $\mathrm{Hom}_{\mathcal{Z}(\mathcal{C})}(\mathbb{1}, Z_1 \otimes \cdots \otimes Z_n)$. Since $\mathcal{C}$ is unitary, this space has a positive definite Hermitian inner product given by $(v, w) = w^\dagger \circ v$. Now let $Z_1, \dots, Z_n \in \mathcal{Z}(\mathcal{C})$ be objects containing $\mathbb{1}$ as objects of $\mathcal{C}$. Let (V_i, ρ_i) be the corresponding representations of $A_{\mathbb{C}}$, $i = 1, \dots, n$. We have a natural map

$$\phi : V_1 \otimes \cdots \otimes V_n \to \mathrm{Hom}_{\mathcal{Z}(\mathcal{C})}(\mathbb{1}, Z_1 \otimes \cdots \otimes Z_n)$$

given by the orthogonal projection of $v_1 \otimes \cdots \otimes v_n \in \mathrm{Hom}_{\mathcal{C}}(\mathbb{1}, Z_1 \otimes \cdots \otimes Z_n)$ to the space $\mathrm{Hom}_{\mathcal{Z}(\mathcal{C})}(\mathbb{1}, Z_1 \otimes \cdots \otimes Z_n)$, where $v_i \in \mathrm{Hom}(\mathbb{1}, Z_i)$. In other words, we may view $v_1 \otimes \cdots \otimes v_n$ as an element of $\mathrm{Hom}_{\mathcal{Z}(\mathcal{C})}(Z_1 \otimes \cdots \otimes Z_n, \mathbb{1})^*$ by taking composition, and $\phi(v_1 \otimes \cdots \otimes v_n)$ is the corresponding element of $\mathrm{Hom}_{\mathcal{Z}(\mathcal{C})}(\mathbb{1}, Z_1 \otimes \cdots \otimes Z_n)$. This element may be viewed as a $\mathcal{Z}(\mathcal{C})$-morphism $Z_1^* \to Z_2 \otimes \cdots \otimes Z_n$. Now, we have

$$\begin{aligned}\mathrm{Hom}_{\mathcal{C}}(\mathbb{1}, Z_2 \otimes \cdots \otimes Z_n) &= \oplus_{Z \in \mathrm{Irr}\, \mathcal{Z}(\mathcal{C})} \mathrm{Hom}_{\mathcal{C}}(\mathbb{1}, Z^*) \otimes \mathrm{Hom}_{\mathcal{Z}(\mathcal{C})}(Z^*, Z_2 \otimes \cdots \otimes Z_n) \\ &= \oplus_{Z \in \mathrm{Irr}\, \mathcal{Z}(\mathcal{C})} \mathrm{Hom}_{\mathcal{C}}(Z, \mathbb{1}) \otimes \mathrm{Hom}_{\mathcal{Z}(\mathcal{C})}(Z^*, Z_2 \otimes \cdots \otimes Z_n).\end{aligned}$$

Let $\{e_{Z,j}\}$ is an orthonormal basis of V_Z. Then we get

$$v_2 \otimes \cdots \otimes v_n = \sum_{Z \in \mathrm{Irr}\, \mathcal{Z}(\mathcal{C})} \sum_{j=1}^{\dim V_Z} e_{Z,j}^\dagger \otimes \phi(e_{Z,j} \otimes v_2 \otimes \cdots \otimes v_n). \tag{4.1}$$

Recall that

$$c_{Z \otimes Z', X} = (c_{Z,X} \otimes 1_{Z'}) \circ (1_Z \otimes c_{Z',X}).$$

Therefore, using (4.1), it is easy to see that

$$\begin{aligned}&\frac{1}{d_i^{n-2}} (\rho_2(X_i)v_2, v_2) \dots (\rho_n(X_i)v_n, v_n) \\ &= \sum_{Z \in \mathrm{Irr}\, \mathcal{Z}(\mathcal{C})} \dim Z \sum_{j,k=1}^{\dim V_Z} (\overline{\rho}_Z(X_i)e_{Z,j}, e_{Z,k})(\phi(e_{Z,j} \otimes v_2 \otimes \cdots \otimes v_n), \\ &\qquad\qquad \phi(e_{Z,k} \otimes v_2 \otimes \cdots \otimes v_n)) \\ &= \sum_{Z \in \mathrm{Irr}\, \mathcal{Z}(\mathcal{C})} \dim Z \sum_{j=1}^{\dim V_Z} (\phi(e_{Z,j} \otimes v_2 \otimes \cdots \otimes v_n), \\ &\qquad\qquad \phi(\rho_{Z^*}(X_i)e_{Z,j} \otimes v_2 \otimes \cdots \otimes v_n)).\end{aligned}$$

Let $B_Z : V_Z \to V_Z$ be an operator whose matrix elements in the basis $e_{Z,j}$ are

$$B_{Z,jk} = (\phi(e_{Z,j} \otimes v_2 \otimes \cdots \otimes v_n), \phi(e_{Z,k} \otimes v_2 \otimes \cdots \otimes v_n)).$$

Then

$$\frac{1}{d_i^{n-2}}(\rho_2(X_i)v_2, v_2)\dots(\rho_n(X_i)v_n, v_n)$$
$$= \sum_{Z\in \mathrm{Irr}\,\mathcal{Z}(\mathcal{C})} \dim Z \sum_{j=1}^{\dim V_Z} (B_Z e_{Z,j}, \rho_{Z^*}(X_i)e_{Z,j}).$$

Thus,

$$\sum_i \frac{1}{d_i^{n-2}}(\rho_1(X_i)v_1, v_1)(\rho_2(X_i)v_2, v_2)\dots(\rho_n(X_i)v_n, v_n)$$
$$= \sum_i (\rho_1(X_i)v_1, v_1) \sum_{Z\in \mathrm{Irr}\,\mathcal{Z}(\mathcal{C})} \dim Z \sum_{j=1}^{\dim V_Z} (B_Z e_{Z,j}, \rho_{Z^*}(X_i)e_{Z,j})$$
$$= \sum_i \dim Z_1(\rho_{Z_1}(X_i)v_1, v_1) \sum_{j=1}^{\dim V_{Z_1}} (B_{Z_1}e_{Z_1,j}, \rho_{Z_1^*}(X_i)e_{Z_1,j})$$
$$= \sum_{j=1}^{\dim V_{Z_1}} \dim Z_1(B_{Z_1}e_{Z_1,j}, v_1)(v_1, \rho_{Z_1^*}(X_i)e_{Z_1,j})$$
$$= (B_{Z_1}v_1, v_1) = \dim \mathcal{C}\cdot \|\phi(v_1\otimes\cdots\otimes v_n)\|^2,$$

where for the last equality we used Lemma 3.1 and Theorem 3.2. Since $\mathcal{C}$ is unitary, so is its Drinfeld center, so this squared norm is ≥ 0, as desired. □

The formula at the end of the proof of Theorem 4.1 can, in fact, be generalized to the situation when the fusion category $\mathcal{C}$ is spherical but not assumed unitary or even Hermitian. Namely, let $v_i \in V_i$, $f_i \in V_i^*$, and let

$$\psi : V_1^*\otimes\cdots\otimes V_n^* \to \mathrm{Hom}(\mathbb{1}, Z_1\otimes\cdots\otimes Z_n)^* \cong \mathrm{Hom}(Z_1\otimes\cdots\otimes Z_n, \mathbb{1})$$

be the natural map.

PROPOSITION 4.3. *We have*

$$\sum_i \frac{1}{(\dim X_i)^{n-2}} f_1(\rho_1(X_i)v_1)\dots f_n(\rho_n(X_i)v_n)$$
$$= \dim(\mathcal{C})(\phi(v_1\otimes\cdots\otimes v_n), \psi(f_1\otimes\cdots\otimes f_n)).$$

where dim *denotes the categorical dimensions.*

Consider the operator $\psi^*\phi \in \mathrm{End}(V_1\otimes\cdots\otimes V_n)$. Proposition 4.3 immediately implies the following.

COROLLARY 4.4. *Let $a_1, \dots, a_n \in A$. Then*

$$\sum_i \frac{1}{(\dim X_i)^{n-2}} \mathrm{Tr}\,\rho_1(a_1X_i)\dots \mathrm{Tr}\,\rho_n(a_nX_i)$$
$$= \dim \mathcal{C}\cdot \mathrm{Tr}\left(\psi^*\phi\circ(\rho_1(a_1)\otimes\cdots\otimes\rho_n(a_n))\right).$$

In particular, we have the following corollary.

COROLLARY 4.5. *If V_i are 1-dimensional then*

$$I_n(\rho_1,\dots,\rho_n) := \sum_i \frac{1}{(\dim X_i)^{n-2}}\rho_1(X_i)\dots\rho_n(X_i) = \dim \mathcal{C}\cdot(\phi,\psi).$$

Here we treat ϕ,ψ as vectors since the space $V_1\otimes\cdots\otimes V_n$ is 1-dimensional.

REMARK 4.6. It is easy to see that the invariants I_n satisfy the recursion

$$I_n(\rho_1,\dots,\rho_n) = \sum_\rho \alpha_\rho^{-1} I_{n-1}(\rho_1,\dots,\rho_{n-2},\rho)I_3(\overline{\rho},\rho_{n-1},\rho_n),\ n\geq 3.$$

(see [**LPW**]). This implies that the multiplication law on the complexified Grothendieck group of the category of $A_\mathbb{C}$-modules given by

$$\rho_1 * \rho_2 = \sum_\rho \alpha_\rho^{-1} I_3(\rho_1,\rho_2,\overline{\rho})\rho$$

is commutative and associative, giving this group the structure of a commutative Frobenius algebra. This is nothing but the dual fusion ring considered in [**LPW**].

5. Integrality properties of spherical fusion categories

In this section we explore integrality properties for spherical fusion categories over $\mathbb{C}$. For simplicity we will restrict ourselves to the case of commutative fusion rings.

5.1. Isaacs criterion and Frobenius type. Let $\mathcal{C}$ be a fusion category with fusion ring A. Let $s\geq 0$ be a rational number.

DEFINITION 5.1. We say that $\mathcal{C}$ is s-**Isaacs** if for any character $\rho: A\to\mathbb{C}$ and any simple object $X\in\mathcal{C}$, the number

$$\lambda_s(\rho,X) := (\dim\mathcal{C})^s(\dim Z_\rho)^{1-s}\frac{\rho(X)}{\dim X}$$

is an algebraic integer.

Since $\dim Z_\rho$ divides $\dim\mathcal{C}$, if $\mathcal{C}$ is s-Isaacs then it is t-Isaacs for any $t>s$.

The 0-Isaacs property will simply be called the **Isaacs property**. It was introduced in [**LPR1, LPR2**]. This definition was motivated by the following proposition.

PROPOSITION 5.2. *Any ribbon fusion category is Isaacs.*

PROOF. In a ribbon category $\mathcal{C}$ we have $\rho(X) = \frac{\mathbf{s}_{Z_\rho X}}{\dim Z_\rho}$. where $(\mathbf{s}_{ij})$ is the S-matrix of the Drinfeld center $\mathcal{Z}(\mathcal{C})$. Thus

$$\lambda_0(\rho,X) = \frac{\mathbf{s}_{Z_\rho X}}{\dim X},$$

which is an algebraic integer (an eigenvalue of an integer matrix) by the Verlinde formula, see [**EGNO**, Corollary 8.14.4]. □

In [**LPR1**] Liu, Palcoux, and Ren conjectured that any spherical pseudo-unitary fusion category is Isaacs ([**LPR1**, Conjecture 2.5]). However, this conjecture was disproved in [**BP**], where it is shown that the extended Haagerup fusion category is not Isaacs. In fact, they show that this category is s-Isaacs if and only if $s\geq 1$.

Recall that $\mathcal{C}$ is said to be **Frobenius type** if dimensions of its simple objects divide $\dim \mathcal{C}$. The Kaplansky 6th conjecture for fusion categories states that any spherical fusion category is Frobenius type. Generalizing, let us say that a category $\mathcal{C}$ is s-**Frobenius type** for a rational number $s \geq 1$ if the dimensions of its simple objects divide $(\dim \mathcal{C})^s$; so 1-Frobenius type is Frobenius type in the usual sense. Clearly, s-Frobenius type implies t-Frobenius type for any $t > s$. Finally, it is clear that any Frobenius type category $\mathcal{C}$ is 1-Isaacs.

This motivates the following weakening of Kaplansky 6-th conjecture for fusion categories.

CONJECTURE 5.3. *Any spherical fusion category is 1-Isaacs.*

The following proposition generalizes [**LPR1**, Proposition 2.6].

PROPOSITION 5.4. *If $\mathcal{C}$ is s-Isaacs for $s \geq \frac{1}{2}$ then it is $(s + \frac{1}{2})$-Frobenius type. In particular, if $\mathcal{C}$ is $\frac{1}{2}$-Isaacs then it is Frobenius type.*

PROOF. If $\mathcal{C}$ is s-Isaacs with $s \geq \frac{1}{2}$ then the number

$$\sum_\rho \lambda_s(\rho, X)\lambda_s(\overline{\rho}, X)(\dim Z_\rho)^{2s-1} = \sum_\rho (\dim \mathcal{C})^{2s} \dim Z_\rho \frac{\rho(X)\overline{\rho}(X)}{(\dim X)^2} = \frac{(\dim \mathcal{C})^{2s+1}}{(\dim X)^2}$$

is an algebraic integer, hence $\mathcal{C}$ is $(s + \frac{1}{2})$-Frobenius type, as claimed. □

5.2. Integrality properties of I_n. For any characters $\rho_i : A \to \mathbb{C}$, $i = 1, \dots, n$, and a rational number $s \geq 0$ define the number

$$J_{n,s}(\rho_1, \dots, \rho_n) := (\dim \mathcal{C})^{(n-2)s}(\dim Z_{\rho_1} \dots \dim Z_{\rho_n})^{1-s} I_n(\rho_1, \dots, \rho_n).$$

For example,

$$J_{2,s}(\rho, \eta) = \dim \mathcal{C}(\dim Z_\rho)^{1-2s}\delta_{\eta,\overline{\rho}}.$$

THEOREM 5.5. *$\mathcal{C}$ is s-Isaacs if and only if for any $n \geq 2$ and ρ_i, $i = 1, \dots, n$, the number $\frac{J_{n,s}(\rho_1,\dots,\rho_n)}{(\dim Z_{\rho_1} \dim Z_{\rho_2})^{1-s}}$ is an algebraic integer.*

PROOF. We have

$$\frac{J_{n,s}(\rho_1, \dots, \rho_n)}{(\dim Z_{\rho_1} \dim Z_{\rho_2})^{1-s}} = \sum_X \rho_1(X)\rho_2(X)\lambda_s(\rho_3, X) \dots \lambda_s(\rho_n, X),$$

which proves the "only if" part. For the "if" part, consider the sum

$$\sum_\eta (\dim Z_\eta)^s \frac{J_{n,s}(\rho, \dots, \rho, \eta)}{(\dim Z_\rho)^{2(1-s)}} \overline{\eta}(Y) =$$

$$= (\dim \mathcal{C})^{(n-2)s+1}(\dim Z_\rho)^{(n-3)(1-s)} \frac{\rho(Y)^{n-1}}{\dim(Y)^{n-2}} =$$

$$\dim \mathcal{C}(\dim Z_\rho)^{s-1}\rho(Y)\lambda_s(\rho, Y)^{n-2}.$$

This is an algebraic integer for all n, hence $\lambda_s(\rho, Y)$ is an algebraic integer, as claimed. □

DEFINITION 5.6. Let us say that $\mathcal{C}$ is **strongly Isaacs** if the number

$$\frac{J_{n,0}(\rho_1, \dots, \rho_n)}{\dim \mathcal{C}\sqrt{\dim Z_{\rho_1} \dim Z_{\rho_2}}}$$

is an algebraic integer.

It is easy to see that if $\mathcal{C}$ is strongly Isaacs then it is (0-)Isaacs.

THEOREM 5.7.
(i) *The category* $\mathcal{C} = \mathrm{Rep}(G)$ *for a finite group* G *is strongly Isaacs.*
(ii) *Any modular category is strongly Isaacs.*

PROOF.
(i) The representations ρ_i correspond to the conjugacy classes C_i in G, and $Z_{\rho_i} = \mathrm{Fun}(C_i)$, so $\dim Z_{\rho_i} = |C_i|$. Also it is well known that in this case

$$\frac{J_{n,0}(\rho_1, \ldots, \rho_n)}{\dim \mathcal{C}} = \frac{\dim Z_{\rho_1} \ldots \dim Z_{\rho_n}}{\dim \mathcal{C}} I_n(\rho_1, \ldots, \rho_n) = |S|, \tag{5.1}$$

where S is the set of tuples $(g_1, \ldots, g_n) \in G^n$ such that $g_1 \ldots g_n = 1$ and $g_i \in C_i$. The set S carries an action of G by conjugation, and the stabilizer of every element is contained in the centralizer G_i of the class C_i for all i, so the size of each orbit is divisible by $|C_i|$. This implies that (5.1) is an integer and divisible by $\dim Z_{\rho_i}$ for all i, hence by $\sqrt{\dim Z_{\rho_1} \dim Z_{\rho_2}}$, which yields the desired statement.

(ii) Each ρ_j corresponds to a simple object X_j of $\mathcal{C}$. A direct calculation using the Verlinde formula yields

$$\frac{J_{n,0}(\rho_1, \ldots, \rho_n)}{\dim \mathcal{C}} = \dim X_1 \ldots \dim X_n \dim \mathrm{Hom}(\mathbb{1}, X_1 \otimes \cdots \otimes X_n).$$

Note also that $Z_{\rho_j} = X_j \boxtimes X_j^*$, so $\dim Z_{\rho_j} = (\dim X_j)^2$. This implies the statement. □

In the original version of this article we asked the following question.

QUESTION 5.8. Is any spherical fusion category strongly Isaacs? Is it true at least for ribbon categories?[2]

As mentioned above, the first part of this question was answered negatively in [**BP**], a counterexample being the extended Haagerup category. However, the second part of the question (regarding ribbon categories) remains open.

Acknowledgment

We are very grateful to Sebastien Palcoux for useful discussions.

References

[BP] S. Burciu and S. Palcoux, *Structure constants, Isaacs property, and Extended-Haagerup fusion categories*, Preprint, arXiv:2210.13936, (2023).

[EGNO] P. Etingof, S. Gelaki, D. Nikshych, and V. Ostrik, *Tensor categories*, Mathematical Surveys and Monographs, vol. 205, American Mathematical Society, Providence, RI, 2015, DOI 10.1090/surv/205. MR3242743

[G] C. Galindo, *On braided and ribbon unitary fusion categories*, Canad. Math. Bull. **57** (2014), no. 3, 506–510, DOI 10.4153/CMB-2013-017-5. MR3239112

[LPW] Z. Liu, S. Palcoux, and J. Wu, *Fusion bialgebras and Fourier analysis: analytic obstructions for unitary categorification*, Adv. Math. **390** (2021), Paper No. 107905, 63, DOI 10.1016/j.aim.2021.107905. MR4292959

[2] S. Palcoux has informed us that he checked many examples of commutative integral fusion rings which are perfect (i.e., have no invertible basis elements other than the unit) and pass the extended cyclotomic criterion (see [**LPR2**, 5.4]), and all the Isaacs ones turned out to be strongly Isaacs.

[LPR1] Z. Liu, S. Palcoux, and Y. Ren, *Classification of Grothendieck rings of complex fusion categories of multiplicity one up to rank six*, Lett. Math. Phys. **112** (2022), no. 3, Paper No. 54, 37, DOI 10.1007/s11005-022-01542-1. MR4434141

[LPR2] Z. Liu, S. Palcoux, and Y. Ren, *Interpolated family of non-group-like simple integral fusion rings of Lie type*, Internat. J. Math. **34** (2023), no. 6, Paper No. 2350030, 51, DOI 10.1142/S0129167X23500301. MR4591939

[O1] V. Ostrik, *On formal codegrees of fusion categories*, Math. Res. Lett. **16** (2009), no. 5, 895–901, DOI 10.4310/MRL.2009.v16.n5.a11. MR2576705

[O2] V. Ostrik, *Pivotal fusion categories of rank 3* (English, with English and Russian summaries), Mosc. Math. J. **15** (2015), no. 2, 373–396, 405, DOI 10.17323/1609-4514-2015-15-2-373-396. MR3427429

[S] K. Shimizu, *The monoidal center and the character algebra*, J. Pure Appl. Algebra **221** (2017), no. 9, 2338–2371, DOI 10.1016/j.jpaa.2016.12.037. MR3631720

[T] V. G. Turaev, *Quantum invariants of knots and 3-manifolds*, De Gruyter Studies in Mathematics, vol. 18, De Gruyter, Berlin, 2016. Third edition [of MR1292673], DOI 10.1515/9783110435221. MR3617439

Department of Mathematics, Massachusetts Institute of Technology, Cambridge, Massachussetts 02139

Email address: `etingof@math.mit.edu`

Department of Mathematics and Statistics, University of New Hampshire, Durham, New Hampshire 03824

Email address: `dmitri.nikshych@unh.edu`

Department of Mathematics, University of Oregon, Eugene, Oregon 97403; and Laboratory of Algebraic Geometry, National Research University Higher School of Economics, Moscow, Russia

Email address: `vostrik@uoregon.edu`

Contemporary Mathematics
Volume **814**, 2025
https://doi.org/10.1090/conm/814/16302

On finite group scheme-theoretical categories, I

Shlomo Gelaki and Guillermo Sanmarco

In memory of Earl J. Taft

ABSTRACT. Let $\mathscr{C}(G, H, \psi)$ be a finite group scheme-theoretical category over an algebraically closed field of characteristic $p \geq 0$ (Gelaki, Quantum Topol. **6** (2015), no. 1, 1–37). For any indecomposable exact $\mathscr{C}(G, H, \psi)$-module category, we classify its simple objects and provide an expression for their projective covers, in terms of double cosets and projective representations of certain closed subgroup schemes of G, which upgrades a result by Ostrik (Int. Math. Res. Not. **27** (2003), 1507–1520) for group-theoretical fusion categories. As a byproduct, we describe the simples and indecomposable projectives of $\mathscr{C}(G, H, \psi)$, and parametrize the Brauer–Picard group of $\mathrm{Coh}(G)$ for any finite connected group scheme G. Finally, we apply our results to describe the blocks of the center of $\mathrm{Coh}(G)$.

1. Introduction

The main purpose of this paper is to study the structure of indecomposable exact module categories over finite group scheme-theoretical categories [**11**] witha focus on classifying their simple objects and describing their projective covers. Our study in particular sheds some light on the structure of group scheme-theoretical categories themselves, and applies to the representation category of the twisted double of a finite group scheme, for which a more precise description is obtained. Throughout, we work over an algebraically closed field k of characteristic $p \geq 0$. All schemes in this paper are assumed to be finite over k, unless otherwise stated.

Group-theoretical fusion categories [**24**] play a fundamental role in the theory of fusion categories in characteristic 0. These categories can be explicitly constructed from finite group data, which makes them treatable enough to serve as a common testing ground for several questions within the theory [**8, 12, 21**]. At the same time, this class of fusion categories encompasses enough to play a part in general classification efforts as explained, for example, in [**10**, §9.13].

From an abstract point of view, group-theoretical fusion categories are characterized as those that are dual to a pointed fusion category with respect to an indecomposable module category. For a concrete construction, one starts with a finite group G and a normalized 3-cocycle $\omega \in Z^3(G, k^\times)$, and considers the pointed fusion category Vec_G^ω of finite dimensional G-graded vector spaces with associativity constraint given by ω. By [**21, 24**], indecomposable module categories over

Key words and phrases. finite group schemes, equivariant sheaves, finite tensor categories, exact module categories.

Vec_G^ω are classified, up to equivalence, by certain conjugacy classes of pairs (H, ψ), where $H \subset G$ is a subgroup and $\psi \in C^2(H, k^\times)$ is a normalized 2-cochain such that $d\psi = \omega_{|H}$. Given such a pair (H, ψ), the twisted group algebra $k^\psi[H]$ becomes an algebra in Vec_G^ω, and the group-theoretical category $\mathscr{C}(G, \omega, H, \psi)$ associated to this data is defined as the category of $k^\psi[H]$-bimodules in Vec_G^ω.

By [**24**] (see also [**19**]), equivalence classes of indecomposable module categories over $\mathscr{C}(G, \omega, H, \psi)$ are classified by certain conjugacy classes of pairs (K, η), where $K \subset G$ is a subgroup and $\eta \in C^2(K, k^\times)$ such that $d\eta = \omega_{|K}$. Ostrik also provided a formula that computes the rank of the module category associated to (K, η) in terms of (H, K)-double cosets in G and certain subgroups built from them. Ostrik developed this theory mainly to classify indecomposable module categories, particularly fiber functors, over the group-theoretical modular category $\mathscr{Z}(\mathrm{Vec}_G^\omega)$.

The problem of classifying indecomposable exact module categories over the finite tensor category $\mathrm{Rep}_k(G)$ of finite dimensional representations over k of a finite group G had been previously solved by Ostrik [**23**] when $p = 0$, and then extended by Etingof and Ostrik [**9**] to $p > 0$. However, the dual of $\mathrm{Rep}_k(G)$ with respect to the module category given by the standard forgetful functor $\mathrm{Rep}_k(G) \to \mathrm{Vec}$ coincides with the fusion category Vec_G of finite dimensional G-graded k-vector spaces. Thus, one can alternatively solve this problem using the classification of module categories over Vec_G.

In a pursuit of a classification of indecomposable exact module categories over $\mathrm{Rep}_k(G)$ for an affine k-group scheme G, in [**11**] the first author generalized Ostrik's construction [**24**] to affine group schemes in any characteristic, but in this paper we focus on the *finite* case. Given a finite group scheme G and a normalized 3-cocycle $\omega \in Z^3(G, \mathbb{G}_m)$ (equivalently, a *Drinfeld associator* $\omega \in \mathscr{O}(G)^{\otimes 3}$ for the coordinate algebra), one first defines the category $\mathrm{Coh}(G, \omega)$ of sheaves on G with tensor product given by convolution, and associativity by the action of ω. Now, for any closed subgroup scheme $H \subset G$ and normalized 2-cochain $\psi \in C^2(H, \mathbb{G}_m)$ such that $d\psi = \iota_H^{\sharp\otimes 3}(\omega)$, there is a notion of a (H, ψ)-equivariant sheaf on (G, ω). Moreover, the category $\mathscr{M}(H, \psi)$ they form has the structure of an indecomposable exact module category over $\mathrm{Coh}(G, \omega)$ under convolution of sheaves.

THEOREM 1.1 ([**11**, Theorem 5.3]). *The assignment $(H, \psi) \mapsto \mathscr{M}(H, \psi)$ determines a bijection between conjugacy classes of pairs (H, ψ), where $H \subset G$ is a closed subgroup scheme and $\psi \in C^2(H, \mathbb{G}_m)$ is normalized such that $d\psi = \iota_H^{\sharp\otimes 3}(\omega)$, and equivalence classes of (left) indecomposable exact module categories over* $\mathrm{Coh}(G, \omega)$. □

By [**9**, Theorem 3.31], since $\mathrm{Coh}(G)$ is dual to $\mathrm{Rep}_k(G)$ with respect to the standard forgetful functor, [**11**, Theorem 5.3] also classifies indecomposable exact module categories over $\mathrm{Rep}_k(G)$: as in the étale case [**9**, Prop. 4.1], these are all of the form $\mathrm{Rep}_k(H, \psi)$ for a closed subgroup scheme $H \subset G$, and a normalized 2-cocycle $\psi \in Z^2(H, \mathbb{G}_m)$.

Going back to the data above, one defines the *group scheme-theoretical* category $\mathscr{C}(G, \omega, H, \psi)$ as the dual of $\mathrm{Coh}(G, \omega)$ with respect to $\mathscr{M}(H, \psi)$. The first author also classified indecomposable exact module categories over the finite tensor category $\mathscr{C}(G, \omega, H, \psi)$.

THEOREM 1.2 ([**11**, Theorem 5.7]). *The assignment*

$$\mathscr{M}(K,\eta) \mapsto \mathrm{Fun}_{\mathrm{Coh}(G,\omega)}\left(\mathscr{M}(H,\psi),\mathscr{M}(K,\eta)\right)$$

determines an equivalence between the 2-category of indecomposable exact module categories over $\mathrm{Coh}(G,\omega)$ *and the 2-category of indecomposable exact module categories over* $\mathscr{C}(G,\omega,H,\psi)$. □

In contrast with the fusion case in characteristic 0, group scheme-theoretical categories and their module categories are rarely semisimple if $p>0$, and a plethora of questions arises. In this paper we focus on the case $\omega=1$, and delegate to future work the treatment in the presence of a general 3-cocycle. So, fix a finite group scheme G, a closed subgroup scheme $H\subset G$, and a normalized $\psi\in Z^2(H,\mathbb{G}_m)$ (equivalently, a *Drinfeld twist* $\psi\in\mathscr{O}(H)^{\otimes 2}$).

We begin our work in Section 2 with preliminaries about finite group schemes and module categories over their categories of sheaves. Our main result here is Theorem 2.11, where we adapt [**20**, Theorem 1(B), p.112] to show that the abelian category $\mathscr{M}(H,\psi)$ of (H,ψ)-equivariant sheaves on G is equivalent to the category $\mathrm{Coh}(G/H)$ of sheaves over the finite quotient scheme G/H.

In Section 3 we consider an embedding $\partial\colon A\xrightarrow{1:1}B$ of finite group schemes, a twist $\Psi\in Z^2(B,\mathbb{G}_m)$, and its restriction $\xi:=\partial^{\sharp\otimes 2}(\Psi)$ to A.

THEOREM 3.7. *There are equivalences of abelian categories*

$$\begin{array}{c} \xrightarrow{\qquad\qquad\qquad\mathrm{Ind}^{(B,\Psi)}_{(A,\xi^{-1})}\qquad\qquad\qquad} \\ \mathrm{Rep}_k(A,\xi^{-1}) \longleftrightarrow \mathrm{Coh}^{(A\times B,\xi^{-1}\times\Psi)}(B) \longleftrightarrow \mathrm{Coh}^{(B,\Psi)}(A\backslash B). \\ \xleftarrow[\qquad\qquad\qquad\mathrm{Res}^{(B,\Psi)}_{(A,\xi^{-1})}\qquad\qquad\qquad]{} \end{array}$$

Moreover, $\mathrm{Ind}^{(B,\Psi)}_{(A,\xi^{-1})}$ *and* $\mathrm{Res}^{(B,\Psi)}_{(A,\xi^{-1})}$ *are mutually inverse, and can be described explicitly.* □

Using this, we get yet another abelian equivalence that will be used later on for a different purpose, see Section 3 for precise statements.

THEOREM 3.11. *We have an equivalence of abelian categories*

$$\mathrm{F}\colon \mathrm{Rep}_k(A,\xi^{-1}) \xrightarrow{\cong} \mathrm{Coh}^{(B,\Psi)}(A\backslash B),\quad V\mapsto \mathscr{O}(A\backslash B)\otimes_k V.$$

In Section 4 we discuss double cosets in finite group schemes, and biequivariant sheaves.

In Section 5 we apply Theorems 3.7 and 3.11 (using Section 4) to describe indecomposable exact module categories over a group scheme-theoretical category $\mathscr{C}:=\mathscr{C}(G,H,\psi)$, which are classified by [**11**, Theorem 5.7] as recalled above. As a first step, we show in Lemma 5.2 that there is an equivalence of abelian categories

$$\mathscr{M}:=\mathrm{Coh}^{\left(H\times K,\psi^{-1}\times\eta\right)}(G)\cong\mathrm{Fun}_{\mathrm{Coh}(G)}\left(\mathscr{M}(H,\psi),\mathscr{M}(K,\eta)\right),$$

where $H\times K$ acts on G on the right via $(g,a,b)\mapsto a^{-1}gb$. Fix a closed point $Z\in Y(k)$ of the quotient finite scheme $Y:=G/(H\times K)$ (see Section 2.3), and let $\mathscr{M}_Z\subset\mathscr{M}$ denote the full abelian subcategory consisting of all objects annihilated by the defining ideal $\mathscr{I}(Z)\subset\mathscr{O}(G)$ of Z. Our next result classifies simples in $\mathscr{M}$; we interpret it as a generalization of [**24**, Proposition 3.2, Remark 3.3(i)].

THEOREM 5.3. *Let* $\mathscr{M} = \mathrm{Coh}^{(H\times K,\psi^{-1}\times\eta)}(G)$ *be as above.*

(1) *For any closed point* $Z \in Y(k)$ *with representative closed point* $g \in Z(k)$, *we have an equivalence of abelian categories*

$$\mathbf{Ind}_Z : \mathrm{Rep}_k(L^g, \xi_g^{-1}) \xrightarrow{\cong} \mathscr{M}_Z,$$

where $L^g = H \cap gKg^{-1}$ *and* $\xi_g \in Z^2(L^g, \mathbb{G}_m)$ *is defined in* (4.7).

(2) *There is a bijection between equivalence classes of pairs* (Z, V), *where* $Z \in Y(k)$ *with representative closed point* $g \in Z(k)$, *and* $V \in \mathrm{Rep}_k(L^g, \xi_g^{-1})$ *is simple, and simple objects of* $\mathscr{M}$, *assigning* (Z, V) *to* $\mathbf{Ind}_Z(V)$.

(3) *We have a direct sum decomposition of abelian categories*

$$\mathscr{M} = \bigoplus_{Z\in Y(k)} \overline{\mathscr{M}_Z},$$

where $\overline{\mathscr{M}_Z} \subset \mathscr{M}$ *denotes the Serre closure of* $\mathscr{M}_Z$ *inside* $\mathscr{M}$.

This result relies on the equivalence from Theorem 3.7. On the other hand, Theorem 3.11 is a better tool to compute projective covers.

THEOREM 5.5. *Let* $\mathscr{M} = \mathrm{Coh}^{(H\times K,\psi^{-1}\times\eta)}(G)$ *be as above.*

(1) *For any closed point* $Z \in Y(k)$ *with representative closed point* $g \in Z(k)$, *we have an equivalence of abelian categories*

$$\mathbf{F}_Z : \mathrm{Rep}_k(L^g, \xi_g^{-1}) \xrightarrow{\cong} \mathscr{M}_Z.$$

(2) *For any simple* $V \in \mathrm{Rep}_k(L^g, \xi_g^{-1})$, *we have*

$$P_{\mathscr{M}}(\mathbf{F}_Z(V)) \cong \mathscr{O}(G^\circ) \otimes \mathscr{O}(Z(k)) \otimes_k P_{(L^g,\xi_g^{-1})}(V).$$

We conclude Section 5 with Corollary 5.8, which provides a classification of fiber functors on $\mathscr{C}(G, H, \psi)$.

Section 6 uses the previous results to describe the abelian structure of $\mathscr{C}(G, H, \psi)$. In Theorem 6.1 we classify simple objects, compute their Frobenius–Perron dimensions, describe how they relate under dualization and provide a graded decomposition for $\mathscr{C}(G, H, \psi)$. Then in Theorem 6.2 we give an alternative parametrization of the simples, use it to compute their projective covers and deduce that $\mathscr{C}(G, H, \psi)$ is unimodular if so is $\mathrm{Rep}_k(H)$. The description provided for $\mathscr{C}(G, H, \psi)$ is nicer when G is either étale or connected, or H is normal (see Sections 6.1–6.3).

In Section 7, we use Sections 6.2–6.3 to classify invertible bimodule categories over $\mathrm{Coh}(G)$ in the connected case. When G is also abelian, we provide in Theorem 7.4 an expression for the multiplication in the Brauer–Picard group of $\mathrm{Coh}(G)$.

We conclude the paper with Section 8, in which we focus on the center $\mathscr{Z}(G) := \mathscr{Z}(\mathrm{Coh(G)})$. Since $\mathscr{Z}(G)$ is a group scheme-theoretical category, we can apply the results from 6 to provide in Theorem 8.3 a description of its simple and projective objects. In particular, we obtain a decomposition of abelian categories

$$\mathscr{Z}(G) = \bigoplus_{C\in \mathrm{C}(k)} \overline{\mathscr{Z}(G)_C},$$

where C denotes the finite scheme of conjugacy orbits in G, and for each subcategory $\mathscr{Z}(G)_C$, we have an explicit abelian equivalence

$$\mathbf{F}_C : \mathrm{Rep}_k(G_C) \xrightarrow{\cong} \mathscr{Z}(G)_C,\ \ V \mapsto \mathscr{O}(C) \otimes_k V,$$

where G_C is the stabilizer of $g \in C(k)$.

On the other hand, by [**15**, Theorem 3.5], if we set $\mathscr{D} := \mathrm{Coh}(G)$ and $\mathscr{D}^\circ := \mathrm{Coh}(G^\circ)$, there is an equivalence of tensor categories

$$F : \mathscr{Z}(G) \xrightarrow{\cong} \mathscr{Z}_{\mathscr{D}^\circ}(\mathscr{D})^{G(k)} = \left(\bigoplus_{a \in G(k)} \mathscr{Z}_{\mathscr{D}^\circ}(\mathscr{D}^\circ \boxtimes a) \right)^{G(k)}.$$

THEOREM 8.6. *For any $C \in \mathrm{C}(k)$, the functor F restricts to an equivalence of abelian categories*

$$F_C : \overline{\mathscr{Z}(G)_C} \xrightarrow{\cong} \bigoplus_{a \in C(k)} \mathscr{Z}_{\mathscr{D}^\circ}(\mathscr{D}^\circ \boxtimes a)^{G(k)}.$$

In particular, F restricts to an equivalence of tensor categories

$$F_1 : \overline{\mathrm{Rep}_k(G)} \xrightarrow{\cong} \mathscr{Z}(G^\circ)^{G(k)}.$$

2. Preliminaries

In this section we fix notation and recall some basic results about finite (group) schemes to be used throughout the paper (see [**16, 27**] for more details). We assume familiarity with the theory of finite tensor categories and their modules categories, and refer to [**10**] for any unexplained notion. Recall that we work over an algebraically closed field k of characteristic $p \geq 0$. All schemes considered in this paper are assumed to be *finite* over k, unless otherwise stated.

2.1. Sheaves on finite schemes. Let X be a finite scheme, i.e., $\mathscr{O}(X)$ is a finite dimensional commutative algebra. Let $\mathrm{Coh}(X)$ be the abelian category of sheaves on X, i.e., the category of finite dimensional representations of $\mathscr{O}(X)$. For each closed point $x \in X(k)$, let $\mathrm{Coh}(X)_x \subset \mathrm{Coh}(X)$ denote the abelian subcategory of sheaves on X supported on x. Each simple object of $\mathrm{Coh}(X)$ is contained in a unique $\mathrm{Coh}(X)_x$, which contains only one simple (up to isomorphism), so we denote that simple object by δ_x. It is well known that there are no extensions of δ_x by any other nonisomorphic simple object, i.e., we have a decomposition

$$\mathrm{Coh}(X) = \bigoplus_{x \in X(k)} \mathrm{Coh}(X)_x$$

of abelian categories. The projective cover $P_x := P(\delta_x)$ lies in $\mathrm{Coh}(X)_x$.

Let Y be a finite scheme, $\varphi : Y \to X$ a scheme morphism, and $\varphi^\sharp : \mathscr{O}(X) \to \mathscr{O}(Y)$ the corresponding algebra homomorphism. Recall that φ induces a pair (φ^*, φ_*) of adjoint functors of abelian categories

$$\varphi^* : \mathrm{Coh}(X) \to \mathrm{Coh}(Y), \ S \mapsto S \otimes_{\mathscr{O}(X)} \mathscr{O}(Y), \tag{2.1}$$

where $\mathscr{O}(X)$ acts on $\mathscr{O}(Y)$ via $\varphi^\sharp$, and

$$\varphi_* : \mathrm{Coh}(Y) \to \mathrm{Coh}(X), \ T \mapsto T_{|\mathscr{O}(X)}, \tag{2.2}$$

where $\mathscr{O}(X)$ acts on T via $\varphi^\sharp$.

2.2. Finite group schemes. A finite group scheme G is a finite scheme whose coordinate algebra $\mathcal{O}(G)$ is a (finite dimensional commutative) Hopf algebra (see, e.g., [**16**, **27**]).

A finite group scheme G is called *étale* if $G = G(k)$, where $G(k)$ is the group of closed points of G. Since k is algebraically closed, there is a categorical equivalence

$$\{\text{Étale group schemes}\} \longleftrightarrow \{\text{Finite abstract groups}\}$$

that assigns to an étale group scheme G the group $G(k)$, and to a finite group Γ the étale group scheme represented by $\mathcal{O}(\Gamma)$ (see, e.g., [**27**, Section 6.4] for more details).

A finite group scheme G is *connected* if $G(k) = 1$. Assume that $p > 0$, and let $F\colon G \to G^{(1)}$ denote the Frobenius map. Since G is finite and connected, it is annihilated by some Frobenius power $F^n\colon G \to G^{(n)}$, and a filtration by normal closed subgroup schemes

$$1 \triangleleft \operatorname{Ker}(F) \triangleleft \cdots \triangleleft \operatorname{Ker}(F^n) = G,$$

where each composition factor $\operatorname{Ker}(F^k)/\operatorname{Ker}(F^{k-1})$ is a group scheme of *height one*, i.e. with vanishing Frobenius.

EXAMPLE 2.1. Consider a finite dimensional p-Lie algebra $\mathfrak{g}$ over k, and let $u(\mathfrak{g})$ denote its p-restricted universal enveloping algebra, which is a finite dimensional cocommutative Hopf algebra; see, e.g., [**26**]. Then the dual commutative Hopf algebra $u(\mathfrak{g})^*$ is local, and satisfies $x^p = 0$ for all x in the augmentation ideal.

By [**5**, Section 2.7], there is a categorical equivalence

$$\{\text{Height one finite group schemes}\} \longleftrightarrow \{\text{Finite dim } p\text{-Lie algebras}\}$$

that assigns to a height one group scheme G its Lie algebra $\operatorname{Lie}(G)$ with p-operation given by the differential of $x \mapsto x^p$. Conversely, the height one group scheme corresponding to a finite dimensional p-Lie algebra $\mathfrak{g}$ is represented by $u(\mathfrak{g})^*$.

By a theorem of Cartier (see [**27**, Section 11.4]), if $p = 0$ then every finite group scheme is étale. On the other hand, if $p > 0$ then any finite group scheme G fits into a split exact sequence

$$1 \to G^\circ \xrightarrow{i} G \underset{q}{\overset{\pi}{\rightleftarrows}} G(k) \to 1, \tag{2.3}$$

where G° is connected and $G(k)$ is étale, i.e., any finite group scheme is an extension of a connected group scheme by an étale one. In particular, $G(k)$ acts on $\operatorname{Coh}(G)_1 \cong \operatorname{Coh}(G^\circ)$, $a \mapsto T_a$, and

$$\operatorname{Coh}(G) \cong \operatorname{Coh}(G^\circ) \rtimes G(k) \tag{2.4}$$

is a crossed product category. Namely, $\operatorname{Coh}(G) = \operatorname{Coh}(G^\circ) \boxtimes \operatorname{Coh}(G(k))$ as abelian categories, and for any $X_1, X_2 \in \operatorname{Coh}(G^\circ)$ and $a_1, a_2 \in G(k)$, we have

$$(X_1 \boxtimes a_1) \otimes (X_2 \boxtimes a_2) = (X_1 \otimes T_{a_1}(X_2)) \boxtimes a_1 a_2.$$

Thus, the classes of connected and étale group schemes play an important role in this paper. Restricting to these classes allows us, on the one hand, to illustrate our general results, and on the other hand to give a more specialized description of our objects of interest.

2.3. Quotients and free actions. Let H be a finite group scheme, and X a finite scheme equipped with a *right* H-action

$$\mu := \mu_{X\times H} : X \times H \to X. \tag{2.5}$$

Equivalently, the algebra homomorphism

$$\mu^\sharp : \mathscr{O}(X) \to \mathscr{O}(X) \otimes \mathscr{O}(H) \tag{2.6}$$

endows the algebra $\mathscr{O}(X)$ with a structure of a right $\mathscr{O}(H)$-comodule algebra (see, e.g., [**1**]). Since H is finite, there exists a geometrical quotient finite scheme

$$\pi : X \twoheadrightarrow X/H, \tag{2.7}$$

with coordinate algebra

$$\mathscr{O}(X/H) = \mathscr{O}(X)^H := \{f \in \mathscr{O}(X) \mid \mu^\sharp(f) = f \otimes 1\}. \tag{2.8}$$

Recall that the action of H on X (2.5) is called *free* if the morphism

$$(p_1, \mu) : X \times H \to X \times X$$

is a closed immersion. Recall [**20**, Theorem 1(B), p.112] that in this case, the morphism (p_1, μ) induces a scheme isomorphism

$$X \times H \xrightarrow{\cong} X \times_{X/H} X.$$

Equivalently, (p_1, μ) induces an algebra isomorphism

$$\mathscr{O}(X) \otimes_{\mathscr{O}(X/H)} \mathscr{O}(X) \xrightarrow{\cong} \mathscr{O}(X) \otimes_k \mathscr{O}(H),$$

given by $f \otimes_{\mathscr{O}(X/H)} \tilde{f} \mapsto (f \otimes 1)\mu^\sharp(\tilde{f})$; $f, \tilde{f} \in \mathscr{O}(X)$.

2.4. Equivariant sheaves. Retain the notation of Section 2.3. Let m be the multiplication map of H, and set

$$\nu := \mu(\mathrm{id}_X \times \mathrm{m}) = \mu(\mu \times \mathrm{id}_H) : X \times H \times H \to X. \tag{2.9}$$

Consider the obvious projections

$$p_1 : X \times H \twoheadrightarrow X,\ \ \mathrm{p}_1 : X \times H \times H \twoheadrightarrow X,$$
$$\text{and } p_{12} : X \times H \times H \twoheadrightarrow X \times H.$$

Clearly, $p_1 \circ p_{12} = \mathrm{p}_1$.

Let $\psi \in Z^2(H, \mathbb{G}_m)$ be a normalized 2-cocycle, i.e., a Drinfeld twist for $\mathscr{O}(H)$. Namely, $\psi \in \mathscr{O}(H)^{\otimes 2}$ is an invertible element that satisfies the equations

$$(\Delta \otimes \mathrm{id})(\psi)(\psi \otimes 1) = (\mathrm{id} \otimes \Delta)(\psi)(1 \otimes \psi),$$

$$(\varepsilon \otimes \mathrm{id})(\psi) = (\mathrm{id} \otimes \varepsilon)(\psi) = 1.$$

Note that multiplication by ψ defines an automorphism of any sheaf on $H \times H$, which we still denote by ψ. Note also that $Z^2(H, \mathbb{G}_m)$ is a group since $\mathscr{O}(H)$ is commutative.

DEFINITION 2.2. Let (H, ψ) and X be as above.

(1) An (H, ψ)-equivariant sheaf on X consists of a pair (S, ρ), where $S \in \mathrm{Coh}(X)$ and ρ is an isomorphism $\rho : p_1^* S \xrightarrow{\cong} \mu^* S$ of sheaves on $X \times H$ such that the diagram

$$\begin{array}{ccc} p_1^* S & \xrightarrow{p_{12}^*(\rho)} & (\mu \circ p_{12})^* S \\ {\scriptstyle (\mathrm{id} \times \mathrm{m})^*(\rho)} \downarrow & & \downarrow {\scriptstyle (\mu \times \mathrm{id})^*(\rho)} \\ \nu^* S & \xrightarrow[\mathrm{id} \boxtimes \psi]{} & \nu^* S \end{array}$$

of morphisms of sheaves on $X \times H \times H$ is commutative.

(2) Let (S, ρ), (T, τ) be two (H, ψ)-equivariant sheaves on X. A morphism $\phi : S \to T$ in $\mathrm{Coh}(X)$ is (H, ψ)-equivariant if the diagram

$$\begin{array}{ccc} p_1^* S & \xrightarrow{p_1^*(\phi)} & p_1^* T \\ {\scriptstyle \rho} \downarrow & & \downarrow {\scriptstyle \tau} \\ \mu^* S & \xrightarrow[\mu^*(\phi)]{} & \mu^* T \end{array}$$

of morphisms of sheaves on $X \times H$ commutes.

(3) Let $\mathrm{Coh}^{(H,\psi)}(X)$ denote the category of (H, ψ)-equivariant sheaves on X, with (H, ψ)-equivariant morphisms.

REMARK 2.3. Morally, a (H, ψ)-equivariant sheaf on X is one equipped with a *lift* of the H-action on X to a H-action on the sheaf, which is consistent with the automorphism of sheaves on $H \times H$ induced by ψ.

EXAMPLE 2.4. Consider $\mathscr{O}(X) \otimes \mathscr{O}(H)$ as an $\mathscr{O}(X)$-module via the algebra map $\mu^\sharp$ (2.6). Then $\mu^\sharp$ determines a H-equivariant structure on $\mathscr{O}(X)$ via the adjunction

$$\begin{aligned} \mathrm{Hom}_{\mathscr{O}(X)}(\mathscr{O}(X), \mathscr{O}(X) \otimes \mathscr{O}(H)) &= \mathrm{Hom}_{\mathscr{O}(X)}(\mathscr{O}(X), \mu_* p_1^* \mathscr{O}(X)) \\ &= \mathrm{Hom}_{\mathscr{O}(X \times H)}(\mu^* \mathscr{O}(X), p_1^* \mathscr{O}(X)). \end{aligned}$$

However, for a nontrivial ψ, it is not always the case that $\mathscr{O}(X)$ admits a canonical (H, ψ)-equivariant structure (see Example 2.5).

EXAMPLE 2.5. Let H_ψ denote the finite group scheme central extension of H by $\mathbb{G}_m$ associated to the 2-cocycle ψ. Let $\mathrm{Rep}_k(H, \psi)$ denote the category of finite dimensional representations of H_ψ on which $\mathbb{G}_m$ acts trivially. Then there is an equivalence of abelian categories

$$\mathrm{Rep}_k(H, \psi) \cong \mathrm{Coh}^{(H,\psi)}(\mathrm{pt}).$$

2.5. Equivariant sheaves and comodules. Given a finite group scheme H and a twist $\psi \in \mathscr{O}(H)^{\otimes 2}$, we let $\mathscr{O}(H)_\psi$ be the coalgebra with underlying vector space $\mathscr{O}(H)$ and comultiplication Δ_ψ given by $\Delta_\psi(f) := \Delta(f)\psi$, where Δ is the standard comultiplication of $\mathscr{O}(H)$. It is clear that $\mathscr{O}(H)_\psi$ is a H-coalgebra, which is isomorphic to the regular representation of H as an H-module.

Note that the category $\mathrm{Rep}_k(H, \psi)$ (see Example 2.5) is equivalent to the category $\mathrm{Comod}(\mathscr{O}(H)_\psi)_k$ of finite dimensional *right* k-comodules over $\mathscr{O}(H)_\psi$.

DEFINITION 2.6. Let X be as in Section 2.4. Let $\mathrm{Comod}(\mathscr{O}(H)_\psi)_{\mathrm{Coh}(X)}$ be the abelian category of *right* $\mathscr{O}(H)_\psi$-comodules in $\mathrm{Coh}(X)$. Namely, the objects of this category are pairs (S,ρ), where $S \in \mathrm{Coh}(X)$ and $\rho : S \to S \otimes \mathscr{O}(H)_\psi$, such that

$$\rho(f \cdot s) = \mu^\sharp(f) \cdot \rho(s);\ \ f \in \mathscr{O}(X),\ s \in S$$

(see (2.6)), and $(\rho \otimes \mathrm{id})\rho = (\mathrm{id} \otimes \Delta_\psi)\rho$. The morphisms in this category are ones that preserve the actions and coactions. □

PROPOSITION 2.7. *The following hold:*

(1) *There is a k-linear equivalence of categories*

$$\mathrm{Coh}^{(H,\psi)}(X) \cong \mathrm{Comod}(\mathscr{O}(H)_\psi)_{\mathrm{Coh}(X)}.$$

In particular, $\mathrm{Coh}^{(H,\psi)}(X)$ *is an abelian category.*

(2) *If* $\mathscr{I} \subset \mathscr{O}(X)$ *is a* H*-stable ideal, then for any* $S \in \mathrm{Coh}^{(H,\psi)}(X)$, $\mathscr{I}S \subset S$ *is a subobject of* S *in* $\mathrm{Coh}^{(H,\psi)}(X)$.

PROOF.

(1) The proof is similar to [**11**, Proposition 3.7(3)].

(2) It follows from (1) since the equivariant structure of S restricts to $\mathscr{I}S$ via the H-equivariant inclusion morphism $\mathscr{I}S \hookrightarrow S$. □

2.6. Equivariant morphisms. Assume X, Y are finite schemes on which a finite group scheme H acts from the right via $\mu_X := \mu_{X\times H}$, $\mu_Y := \mu_{Y\times H}$, respectively. Let $\psi \in Z^2(H, \mathbb{G}_m)$.

PROPOSITION 2.8. *For any* H*-equivariant morphism* $\varphi : X \to Y$*, the functors* φ^* (2.1) *and* φ_* (2.2) *lift to a pair* (φ^*, φ_*) *of adjoint functors of abelian categories*

$$\varphi^* : \mathrm{Coh}^{(H,\psi)}(Y) \to \mathrm{Coh}^{(H,\psi)}(X), \quad \varphi_* : \mathrm{Coh}^{(H,\psi)}(X) \to \mathrm{Coh}^{(H,\psi)}(Y).$$

PROOF. Let $(S,\rho) \in \mathrm{Coh}^{(H,\psi)}(Y)$, i.e., $S \in \mathrm{Coh}(Y)$ and ρ is an $\mathscr{O}(Y)$-morphism endowing S with a structure $\rho : S \to S \otimes \mathscr{O}(H)_\psi$ of a right $\mathscr{O}(H)_\psi$-comodule. It is straightforward to verify that

$$\rho^* := \mu_X^\sharp \bar{\otimes} \rho : \mathscr{O}(X) \otimes_{\mathscr{O}(Y)} S \to \mathscr{O}(X) \otimes_{\mathscr{O}(Y)} S \otimes \mathscr{O}(H)_\psi$$

endows $\varphi^* S$ with a structure of an object $(\varphi^* S, \rho^*)$ in $\mathrm{Coh}^{(H,\psi)}(X)$.

Let $(T,\tau) \in \mathrm{Coh}^{(H,\psi)}(X)$, i.e., $T \in \mathrm{Coh}(X)$ and τ is an $\mathscr{O}(X)$-morphism endowing T with a structure $\tau : T \to T \otimes \mathscr{O}(H)_\psi$ of a right $\mathscr{O}(H)_\psi$-comodule. It is easy to verify that $\tau_* := \tau$ endows $\varphi_* T$ with a structure of an object $(\varphi_* T, \tau_*)$ in $\mathrm{Coh}^{(H,\psi)}(Y)$.

It is straightforward to verify that the adjunction isomorphism

$$\mathrm{Hom}_{\mathrm{Coh}(X)}(\varphi^* S, T) \cong \mathrm{Hom}_{\mathrm{Coh}(Y)}(S, \varphi_* T)$$

preserves (H,ψ)-equivariant maps (see Definition 2.2(2)), hence lifts to an isomorphism

$$\mathrm{Hom}_{\mathrm{Coh}^{(H,\psi)}(X)}((\varphi^* S, \rho^*), (T,\tau)) \cong \mathrm{Hom}_{\mathrm{Coh}^{(H,\psi)}(Y)}((S,\rho), (\varphi_* T, \tau_*)),$$

as claimed. □

EXAMPLE 2.9. Let $\mathrm{u} : X \to \mathrm{Spec}(k) = \{\mathrm{pt}\}$ be the scheme structure morphism. Since u is H-equivariant, we have a functor

$$\mathrm{u}^* : \mathrm{Rep}_k(H, \psi) \to \mathrm{Coh}^{(H,\psi)}(X)$$

(see Example 2.5).

2.7. Principal homogeneous spaces. In this section we assume that

$$\iota : H \xrightarrow{1:1} G$$

is an embedding of finite group schemes. Consider the free right action of H on G, given by

$$\mu_{G\times H} : G \times H \to G, \ \ (g, h) \mapsto gh,^{1} \tag{2.10}$$

and (see Section 2.3) the corresponding quotient morphism

$$\pi : G \twoheadrightarrow G/H. \tag{2.11}$$

Recall that $\mathcal{O}(G/H) \subset \mathcal{O}(G)$ is a *right* $\mathcal{O}(H)$-Hopf Galois cleft extension. Namely, $\mathcal{O}(G)$ is a right $\mathcal{O}(H)$-comodule algebra via $\mu^\sharp_{G\times H}$, $\mathcal{O}(G/H) \subset \mathcal{O}(G)$ is the *left* coideal subalgebra of coinvariants, the map

$$\mathcal{O}(G) \otimes_{\mathcal{O}(G/H)} \mathcal{O}(G) \to \mathcal{O}(G) \otimes_k \mathcal{O}(H), \ \ \mathrm{f} \otimes \tilde{\mathrm{f}} \mapsto (\mathrm{f} \otimes 1)\mu^\sharp_{G\times H}(\tilde{\mathrm{f}}),$$

is bijective, and there exists a (unique up to multiplication by a convolution invertible element in $\mathrm{Hom}_k(\mathcal{O}(H), \mathcal{O}(H\backslash G))$) unitary convolution invertible right $\mathcal{O}(H)$-colinear map, called the *cleaving map*,

$$\mathfrak{c} : \mathcal{O}(H) \xrightarrow{1:1} \mathcal{O}(G). \tag{2.12}$$

A *choice* of $\mathfrak{c}$ determines a convolution invertible 2-cocycle

$$\sigma : \mathcal{O}(H)^{\otimes 2} \to \mathcal{O}(G/H), \ \ f \otimes \tilde{f} \mapsto \mathfrak{c}(f_1)\mathfrak{c}(\tilde{f}_1)\mathfrak{c}^{-1}(f_1\tilde{f}_2), \tag{2.13}$$

from which one can define the crossed product algebra $\mathcal{O}(G/H)\#_\sigma\mathcal{O}(H)$ (with the trivial action of $\mathcal{O}(H)$ on $\mathcal{O}(G/H)$), which is naturally a left $\mathcal{O}(G/H)$-module and right $\mathcal{O}(H)$-comodule algebra, where $\mathcal{O}(H)$ coacts via $a\#f \mapsto a\#f_1 \otimes f_2$. Moreover, the map

$$\phi : \mathcal{O}(G/H)\#_\sigma\mathcal{O}(H) \xrightarrow{\cong} \mathcal{O}(G), \ \ a\#f \mapsto a\mathfrak{c}(f), \tag{2.14}$$

is an algebra isomorphism, $\mathcal{O}(G/H)$-linear and right $\mathcal{O}(H)$-colinear. The inverse of ϕ is given by

$$\phi^{-1} : \mathcal{O}(G) \xrightarrow{\cong} \mathcal{O}(G/H)\#_\sigma\mathcal{O}(H), \ \ \mathrm{f} \mapsto \mathrm{f}_1\mathfrak{c}^{-1}(\iota^\sharp(\mathrm{f}_2))\#\iota^\sharp(\mathrm{f}_3). \tag{2.15}$$

We also have the $\mathcal{O}(G/H)$-linear map

$$\alpha := (\mathrm{id}\,\bar{\otimes}\varepsilon)\phi^{-1} : \mathcal{O}(G) \twoheadrightarrow \mathcal{O}(G/H), \ \ \mathrm{f} \mapsto \mathrm{f}_1\mathfrak{c}^{-1}(\iota^\sharp(\mathrm{f}_2)). \tag{2.16}$$

(For details, see [**18**, Section 3] and references therein.)

Since σ (2.13) is a 2-cocycle, it follows that for any $S \in \mathrm{Coh}(G/H)$, the vector space $S \otimes_k \mathcal{O}(H)$ is an $\mathcal{O}(G/H)\#_\sigma\mathcal{O}(H)$-module via

$$(a\#f) \cdot \left(s \otimes \tilde{f}\right) = a\sigma(f_1, \tilde{f}_1) \cdot s \otimes f_2\tilde{f}_2.$$

Consequently, we have the following lemma.

[1]Whenever there is no confusion, we will write gh instead of $g\iota(h)$.

LEMMA 2.10. *For any $U \in \mathrm{Coh}(G/H)$, the following hold:*

(1) *The vector space $U \otimes_k \mathscr{O}(H)$ is an $\mathscr{O}(G)$-module via ϕ^{-1} (2.15):*

$$\mathrm{f} \cdot \left(u \otimes \tilde{f}\right) := \phi^{-1}(\mathrm{f}) \cdot \left(u \otimes \tilde{f}\right); \ \mathrm{f} \in \mathscr{O}(G),\ u \otimes \tilde{f} \in U \otimes \mathscr{O}(H).$$

(2) *We have an $\mathscr{O}(G)$-linear isomorphism*

$$F_U : U \otimes_{\mathscr{O}(G/H)} \mathscr{O}(G) \xrightarrow{\cong} U \otimes_k \mathscr{O}(H),\ \ u \otimes_{\mathscr{O}(G/H)} \mathrm{f} \mapsto \phi^{-1}(\mathrm{f}) \cdot (u \otimes 1),$$

whose inverse is given by

$$F_U^{-1} : U \otimes_k \mathscr{O}(H) \xrightarrow{\cong} U \otimes_{\mathscr{O}(G/H)} \mathscr{O}(G),\ \ u \otimes_k f \mapsto u \otimes_{\mathscr{O}(G/H)} \phi(1 \otimes_k f).$$

PROOF. Follow from the preceding remarks. □

For any $U \in \mathrm{Coh}(G/H)$, define

$$\rho_U^{\psi} := (F_U^{-1} \otimes \mathrm{id})(\mathrm{id} \otimes \Delta_{\psi}) F_U. \tag{2.17}$$

For any $(S, \rho) \in \mathrm{Coh}^{(H,\psi)}(G)$, define the subspace of coinvariants

$$S^{(H,\psi)} := \{s \in S \mid \rho(s) = \mathfrak{c}(\psi^1) \cdot s \otimes \psi^2\} \subset S. \tag{2.18}$$

THEOREM 2.11. *The following hold:*

(1) *There is an equivalence of abelian categories*

$$\mathrm{Coh}^{(H,1)}(G) \xrightarrow{\cong} \mathrm{Coh}^{(H,\psi)}(G),\ (S, \rho) \mapsto \left(S^{(H,1)} \otimes_k \mathscr{O}(H), \mathrm{id} \otimes \Delta_{\psi}\right)$$

(see (2.17)), whose inverse is given by

$$\mathrm{Coh}^{(H,\psi)}(G) \xrightarrow{\cong} \mathrm{Coh}^{(H,1)}(G),\ (S, \rho) \mapsto \left(S^{(H,\psi)} \otimes_k \mathscr{O}(H), \mathrm{id} \otimes \Delta\right).$$

(2) *The quotient morphism $\pi : G \twoheadrightarrow G/H$ (2.11) induces an equivalence of abelian categories*

$$\pi^* : \mathrm{Coh}(G/H) \xrightarrow{\cong} \mathrm{Coh}^{(H,\psi)}(G),\ \ U \mapsto \left(U \otimes_{\mathscr{O}(G/H)} \mathscr{O}(G), \rho_U^{\psi}\right),$$

whose inverse is given by

$$\pi_*^{(H,\psi)} : \mathrm{Coh}^{(H,\psi)}(G) \xrightarrow{\cong} \mathrm{Coh}(G/H),\ \ (S, \rho) \mapsto \pi_*^{(H,\psi)} S.$$

PROOF.

(1) Follows from Lemma 2.10 in a straightforward manner.
(2) For $\psi = 1$, this is [**20**, p. 112], so the claim follows from (1). □

Now for any closed point $\bar{g} := gH \in (G/H)(k)$, let $\delta_{\bar{g}}$ denote the corresponding simple object of $\mathrm{Coh}(G/H)$ (see Section 2.1). Let

$$S_{\bar{g}} := \pi^* \delta_{\bar{g}} \cong \delta_{\bar{g}} \otimes_k \mathscr{O}(H) \in \mathrm{Coh}^{(H,\psi)}(G) \tag{2.19}$$

(see Theorem 2.11), and let

$$P_{\bar{g}} := P(\delta_{\bar{g}}) \in \mathrm{Coh}(G/H),\ \text{ and }\ P(S_{\bar{g}}) \in \mathrm{Coh}^{(H,\psi)}(G) \tag{2.20}$$

be the projective covers of $\delta_{\bar{g}}$ and $S_{\bar{g}}$, respectively.

COROLLARY 2.12. *The following hold:*

(1) *For any $\bar{g} \in (G/H)(k)$, there are $\mathrm{Coh}^{(H,\psi)}(G)$-isomorphisms*

$$S_{\bar{g}} \cong \delta_g \otimes S_{\bar{1}},\ \textit{and}\ P_{\bar{g}} \cong \delta_g \otimes P_{\bar{1}},$$

where $\mathscr{O}(G)$ acts diagonally and $\mathscr{O}(H)_{\psi}$ coacts on the right.

(2) *The assignment* $\bar{g} \mapsto S_{\bar{g}}$ *is a bijection between* $(G/H)(k)$ *and the set of isomorphism classes of simples in* $\mathrm{Coh}^{(H,\psi)}(G)$.

(3) *We have* $P(S_{\bar{g}}) \cong \pi^* P_{\bar{g}} \cong P_{gH} := \bigoplus_{h\in H(k)} P_{gh}$.

PROOF. Follow immediately from Lemma 2.10 and Theorem 2.11. □

LEMMA 2.13. *Let* $\Psi \in \mathscr{O}(G)^{\otimes 2}$ *be a twist, and set* $\psi := \left(\iota^\sharp \otimes \iota^\sharp\right)(\Psi)$. *Then* $\psi \in \mathscr{O}(H)^{\otimes 2}$ *is a twist, and the linear map*

$$\rho_\Psi : \mathscr{O}(G) \to \mathscr{O}(G) \otimes \mathscr{O}(H)_\psi,\ \ f \mapsto f_1\Psi^1 \otimes \iota^\sharp\left(f_2\Psi^2\right),$$

endows $\mathscr{O}(G)$ *with a structure of a right* $\mathscr{O}(H)_\psi$*-comodule in* $\mathrm{Coh}(G)$, *with respect to the right action* $\mu_{G\times H}$ (2.10).

PROOF. Follows since $\iota^\sharp : \mathscr{O}(G)_\Psi \twoheadrightarrow \mathscr{O}(H)_\psi$ is a coalgebra map. □

Consider next the free *left* action of H on G given by

$$\widetilde{\mu_{H\times G}} : H \times G \to G,\ \ (h, g) \mapsto hg, \tag{2.21}$$

and the corresponding quotient morphism

$$\widetilde{\pi} : G \twoheadrightarrow H\backslash G. \tag{2.22}$$

Similarly to the above, we can *choose* a (unique up to multiplication by an invertible element in $\mathrm{Hom}_k(\mathscr{O}(H), \mathscr{O}(H\backslash G))$) cleaving map

$$\widetilde{\mathfrak{c}} : \mathscr{O}(H) \xrightarrow{1:1} \mathscr{O}(G), \tag{2.23}$$

which then determines a 2-cocycle $\widetilde{\sigma}$ and an $\mathscr{O}(H\backslash G)$-linear and left $\mathscr{O}(H)$-colinear isomorphism of algebras

$$\widetilde{\phi} : \mathscr{O}(H\backslash G)\#_{\widetilde{\sigma}}\mathscr{O}(H) \xrightarrow{\cong} \mathscr{O}(G),\ \ a\#f \mapsto a\widetilde{\mathfrak{c}}(f), \tag{2.24}$$

whose inverse is given by

$$\widetilde{\phi}^{-1} : \mathscr{O}(G) \xrightarrow{\cong} \mathscr{O}(H\backslash G)\#_{\widetilde{\sigma}}\mathscr{O}(H),\ \ \mathrm{f} \mapsto \mathrm{f}_1\widetilde{\mathfrak{c}}^{-1}(\iota^\sharp(\mathrm{f}_2))\#\iota^\sharp(\mathrm{f}_3), \tag{2.25}$$

and an $\mathscr{O}(H\backslash G)$-linear map

$$\widetilde{\alpha} := (\mathrm{id}\,\bar{\otimes}\varepsilon)\widetilde{\phi}^{-1} : \mathscr{O}(G) \twoheadrightarrow \mathscr{O}(H\backslash G),\ \ \mathrm{f} \mapsto \mathrm{f}_1\widetilde{\mathfrak{c}}^{-1}(\iota^\sharp(\mathrm{f}_2)). \tag{2.26}$$

2.8. Twisting by the inverse 2-cocycle. Retain the notation of Section 2.7. We note an explicit relation between the twisted coalgebras associated to ψ and ψ^{-1}.

Set $Q_\psi := \sum \mathrm{S}(\psi^1)\psi^2$, where S is the antipode map of $\mathscr{O}(H)$. It is well known (see, e.g., [**2**, **17**]) that

$$Q_\psi^{-1} = \sum \psi^{-1}\mathrm{S}(\psi^{-2}) \text{ and } \Delta(Q_\psi) = (\mathrm{S}\otimes\mathrm{S})(\psi_{21}^{-1})(Q_\psi \otimes Q_\psi)\psi^{-1}.$$

In particular, $\Delta(Q_\psi^{-1}\mathrm{S}(Q_\psi)) = Q_\psi^{-1}\mathrm{S}(Q_\psi) \otimes Q_\psi^{-1}\mathrm{S}(Q_\psi)$, i.e., $Q_\psi^{-1}\mathrm{S}(Q_\psi)$ is a grouplike element of $\mathscr{O}(H)$. Thus, we have a coalgebra isomorphism

$$\mathscr{O}(H)_\psi^{\mathrm{cop}} \xrightarrow{\cong} \mathscr{O}(H)_{\psi^{-1}},\ \ f \mapsto \mathrm{S}(fQ_\psi^{-1}),$$

which induces the following equivalence of abelian categories

$$\mathrm{Corep}(H, \psi^{-1})_k \xrightarrow{\cong} \mathrm{Corep}_k(H, \psi),\ \ (V, r) \mapsto (V, \widetilde{r}), \tag{2.27}$$

where $\widetilde{r} : V \to \mathscr{O}(H)_\psi \otimes V$, $v \mapsto \mathrm{S}(v)Q_\psi \otimes v^0$.

LEMMA 2.14. *Let $\Psi \in \mathscr{O}(G)^{\otimes 2}$ be a twist, and set $\psi := \left(\iota^\sharp \otimes \iota^\sharp\right)(\Psi)$. Then $\psi \in \mathscr{O}(H)^{\otimes 2}$ is a twist, and the linear map*

$$\widetilde{\rho_\Psi} : \mathscr{O}(G) \to \mathscr{O}(G) \otimes \mathscr{O}(H)_\psi, \ \ f \mapsto \Psi^{-2} f_2 \otimes \iota^\sharp \mathrm{S}\left(f_1 \Psi^{-1}\right) Q_\psi, \tag{2.28}$$

endows $\mathscr{O}(G)$ with a structure of a right $\mathscr{O}(H)_\psi$-comodule in $\mathrm{Coh}(G)$, *with respect to the right action $G \times H \to G$, $(g,h) \mapsto h^{-1}g$.*

PROOF. It follows from the above lemma and the fact that $\iota^\sharp : \mathscr{O}(G)_\Psi \twoheadrightarrow \mathscr{O}(H)_\psi$ is a coalgebra map. Indeed, set $\rho := \widetilde{\rho_\Psi}$ and $Q := Q_\psi$. Then on the one hand, we have

$$\begin{aligned}
&(\rho \otimes \mathrm{id})\rho(f) = \rho(\Psi^{-2} f_2) \otimes \iota^\sharp \mathrm{S}\left(f_1 \Psi^{-1}\right) Q \\
&= \Omega^{-2}(\Psi^{-2} f_2)_2 \otimes \iota^\sharp \mathrm{S}\left((\Psi^{-2} f_2)_1 \Omega^{-1}\right) Q \otimes \iota^\sharp \mathrm{S}\left(f_1 \Psi^{-1}\right) Q \\
&= \Omega^{-2} \Psi_2^{-2} f_3 \otimes \iota^\sharp \mathrm{S}\left(\Psi_1^{-2} f_2 \Omega^{-1}\right) Q \otimes \iota^\sharp \mathrm{S}\left(f_1 \Psi^{-1}\right) Q \\
&= \left(\Omega^{-2} \Psi_2^{-2} \otimes \iota^\sharp \mathrm{S}\left(\Psi_1^{-2} \Omega^{-1}\right) Q \otimes \iota^\sharp \mathrm{S}\left(\Psi^{-1}\right) Q\right) \widetilde{\mu}(f),
\end{aligned}$$

and on the other hand, we have

$$\begin{aligned}
&(\mathrm{id} \otimes \Delta_\psi)\rho(f) = \Psi^{-2} f_2 \otimes \Delta_\psi\left(\iota^\sharp \mathrm{S}\left(f_1 \Psi^{-1}\right) Q\right)) \\
&= \Psi^{-2} f_2 \otimes \psi\left(\left(\iota^\sharp \mathrm{S}\left(f_1 \Psi^{-1}\right)_1 \otimes \iota^\sharp \mathrm{S}\left(f_1 \Psi^{-1}\right)_2\right) \Delta(Q)\right) \\
&= \Psi^{-2} f_3 \otimes \psi\left(\left(\iota^\sharp \mathrm{S}\left(f_2 \Psi_2^{-1}\right) \otimes \iota^\sharp \mathrm{S}\left(f_1 \Psi_1^{-1}\right)\right) \Delta(Q)\right) \\
&= \left(\Psi^{-2} \otimes \left(\iota^\sharp \mathrm{S}\left(\Psi_2^{-1}\right) \otimes \iota^\sharp \mathrm{S}\left(\Psi_1^{-1}\right)\right) \Delta(Q)\psi\right) \widetilde{\mu}(f).
\end{aligned}$$

Thus, we have

$$\begin{aligned}
&\Omega^{-2} \Psi_2^{-2} \otimes \iota^\sharp \mathrm{S}\left(\Psi_1^{-2} \Omega^{-1}\right) Q \otimes \iota^\sharp \mathrm{S}\left(\Psi^{-1}\right) Q \\
&= \Psi^{-2} \otimes \iota^\sharp \mathrm{S}\left(\Psi_2^{-1} \Omega^{-2}\right) Q \otimes \iota^\sharp \mathrm{S}\left(\Psi_1^{-1} \Omega^{-1}\right) Q \\
&= \Psi^{-2} \otimes \left(\iota^\sharp \mathrm{S}\left(\Psi_2^{-1}\right) \otimes \iota^\sharp \mathrm{S}\left(\Psi_1^{-1}\right)\right) (\mathrm{S} \otimes \mathrm{S})\left(\psi_{21}^{-1}\right)(Q \otimes Q) \\
&= \Psi^{-2} \otimes \left(\iota^\sharp \mathrm{S}\left(\Psi_2^{-1}\right) \otimes \iota^\sharp \mathrm{S}\left(\Psi_1^{-1}\right)\right) \Delta(Q)\psi,
\end{aligned}$$

which implies the claim. □

2.9. Module categories over $\mathrm{Coh}(G)$**.** Fix a finite group scheme G. Let $H \subset G$ be a closed subgroup scheme, and let $\psi \in Z^2(H, \mathbb{G}_m)$ be a normalized 2-cocycle. Recall [**11**] the abelian category

$$\mathscr{M}(H,\psi) := \mathrm{Coh}^{(H,\psi)}(G)$$

with respect to the right action (2.10) of H on G. Recall [**11**] that $\mathscr{M} := \mathscr{M}(H,\psi)$ admits a canonical structure of an indecomposable exact left module category over $\mathrm{Coh}(G)$ given by convolution of sheaves. Namely, the action of $X \in \mathrm{Coh}(G)$ on $(S,\rho) \in \mathscr{M}$ is given by

$$X \otimes^{\mathscr{M}} (S,\rho) = (X \otimes S, \mathrm{id} \otimes \rho).$$

EXAMPLE 2.15. The following hold:

(1) $\mathscr{M}(1,1) \cong \mathrm{Coh}(G)$ is the regular $\mathrm{Coh}(G)$-module.
(2) $\mathscr{M}(G,1) \cong \mathrm{Vec}$ is the usual fiber functor on $\mathrm{Coh}(G)$.

For any closed point $g \in G(k)$, set

$$\psi^g := \mathrm{Ad}g(\psi) \in Z^2(g^{-1}Hg, \mathbb{G}_m). \tag{2.29}$$

DEFINITION 2.16. Two pairs (H,ψ), (K,η) as above are called equivalent if there is a closed point $g\in G(k)$ such that $K=g^{-1}Hg$ and $[\eta]=[\psi^g]$ in $H^2(K,\mathbb{G}_m)$.

THEOREM 2.17 ([**11**]). *The following hold:*

(1) *There is a bijection between equivalence classes of pairs (H,ψ) (in the sense of Definition 2.16) and equivalence classes of indecomposable exact left module categories over* Coh(G), *assigning (H,ψ) to $\mathscr{M}(H,\psi)$.*
(2) *There exists an equivalence of* Coh(G)*-module categories*

$$\mathscr{M}(H,\psi)\cong \mathrm{Comod}(\mathscr{O}(H)_\psi)_{\mathrm{Coh}(G)}.$$

REMARK 2.18. If G is connected then Theorem 2.17 says that equivalence classes of indecomposable exact module categories over Coh(G) correspond bijectively to pairs (H,ψ). □

REMARK 2.19. Let $\mathscr{M}:=\mathscr{M}(H,\psi)$. Corollary 2.12(1) states that

$$S_{\bar g}\cong \delta_g\otimes^{\mathscr{M}} S_{\bar 1},\ \text{ and } P_{\bar g}\cong \delta_g\otimes^{\mathscr{M}} P_{\bar 1}$$

for any simple $S_{\bar g}\in\mathscr{M}$. Also, using Corollary 2.12(3), we see that

$$\begin{aligned} P_1\otimes^{\mathscr{M}} S_{\bar 1} &\cong P_1\otimes^{\mathscr{M}} \mathscr{O}(H)\cong \pi_*^{(H,\psi)}\pi^*\left(P_1\otimes\mathscr{O}(H)\right)\\ &\cong \pi_*^{(H,\psi)}P_H\cong |H^\circ|P_{\bar 1}, \end{aligned}$$

which in particular demonstrates that $\mathscr{M}$ is exact (see [**11**]). □

REMARK 2.20. For any pair (H,ψ) as above, one can consider also the abelian category

$$\mathscr{N}(H,\psi):=\mathrm{Comod}_{\mathrm{Coh}(G)}(\mathscr{O}(H)_\psi)$$

of **left** $\mathscr{O}(H)_\psi$-comodules in Coh(G) with respect to the left H-action

$$\mu_{H\times G}:H\times G\to G$$

by left translation. Then similarly, the assignment $(H,\psi)\mapsto\mathscr{N}(H,\psi)$ classifies exact indecomposable **right** module categories over Coh(G), up to equivalence.

Recall [**11**] that the group scheme-theoretical category

$$\mathscr{C}(G,H,\psi):=\mathrm{Coh}(G)^*_{\mathscr{M}(H,\psi)}$$

associated to this data is the dual category of Coh(G) with respect to $\mathscr{M}(H,\psi)$. In other words,

$$\mathscr{C}(G,H,\psi)=\mathrm{Fun}_{\mathrm{Coh}(G)}\left(\mathscr{M}(H,\psi),\mathscr{M}(H,\psi)\right) \tag{2.30}$$

is the category of Coh(G)-module endofunctors of $\mathscr{M}(H,\psi)$. Recall that $\mathscr{C}(G,H,\psi)$ is a finite tensor category with tensor product given by composition of module functors [**9**].

Recall [**11**] that convolution of sheaves on G lifts to endow the category

$$\mathscr{M}\left(H\times H,\psi^{-1}\times\psi\right)$$

with the structure of a finite tensor category.

Finally, let Bicomod$_{\mathrm{Coh}(G)}(\mathscr{O}(H)_\psi)$ be the finite tensor category with tensor product $\otimes^{\mathscr{O}(H)_\psi}$ and unit object $\mathscr{O}(H)_\psi$. Recall that objects in this category are triples (V,λ,ρ) where (V,λ) is a left $\mathscr{O}(H)_\psi$-comodule in Coh(G) with respect to the left action $\mu_{H\times G}$, and (V,ρ) is a right $\mathscr{O}(H)_\psi$-comodule in Coh(G) with respect to the right action $\mu_{G\times H}$.

LEMMA 2.21. *There are equivalences of tensor categories*

$$\mathscr{C}(G,H,\psi)\cong \mathrm{Bicomod}_{\mathrm{Coh}(G)}(\mathscr{O}(H)_{\psi})^{\mathrm{rev}},$$

and

$$\mathscr{C}(G,H,\psi)\cong \mathscr{M}(H\times H,\psi^{-1}\times\psi).$$

PROOF. All equivalences are standard. For the first one see, e.g., [10], and for the second one see, e.g., Lemma 5.2. □

EXAMPLE 2.22. The following hold:

(1) $\mathscr{C}(G,1,1)\cong \mathrm{Coh}(G)$.
(2) $\mathscr{C}(G,G,1)\cong \mathrm{Rep}(G)$.

3. Equivariant sheaves on group schemes

Let $\partial: A\xrightarrow{1:1} B$ be an embedding of finite group schemes. Let $\Psi\in Z^2(B,\mathbb{G}_m)$, and $\xi:=\Psi_{|A}$ be its restriction to A via ∂, i.e.,

$$\xi:=\left(\partial^\sharp\otimes\partial^\sharp\right)(\Psi)\in Z^2(A,\mathbb{G}_m). \tag{3.1}$$

Let $\mathrm{Coh}^{(A\times B,\xi^{-1}\times\Psi)}(B)$ be the category of $(A\times B,\xi^{-1}\times\Psi)$-equivariant sheaves on B with respect to the right action

$$\mu_{B\times(A\times B)}: B\times(A\times B)\to B,\ \ (b_1,a,b_2)\mapsto \partial(a)^{-1}b_1b_2,$$

and $\mathrm{Coh}^{(B,\Psi)}(A\backslash B)$ the category of (B,Ψ)-equivariant sheaves on $A\backslash B$ with respect to the right action

$$\mu_{A\backslash B\times B}: A\backslash B\times B\to A\backslash B,\ \ (\overline{b_1},b_2)\mapsto\overline{b_1b_2},$$

where $A\backslash B$ is the quotient scheme with respect to $\widetilde{\mu_{A\times B}}$ (2.21). Our goal is to construct explicit equivalences of abelian categories

$$\begin{array}{c}
\xrightarrow{\qquad\qquad\qquad \mathrm{Ind}^{(B,\Psi)}_{(A,\xi^{-1})}\qquad\qquad\qquad}\\
\mathrm{Rep}_k(A,\xi^{-1})\ \underset{\mathrm{t}_*^{(B,\Psi)}}{\overset{\mathrm{t}^*}{\rightleftarrows}}\ \mathrm{Coh}^{(A\times B,\xi^{-1}\times\Psi)}(B)\ \underset{\mathrm{p}^*}{\overset{\mathrm{p}_*^{(A,\xi^{-1})}}{\rightleftarrows}}\ \mathrm{Coh}^{(B,\Psi)}(A\backslash B).\\
\xleftarrow{\qquad\qquad\qquad \mathrm{Res}^{(B,\Psi)}_{(A,\xi^{-1})}\qquad\qquad\qquad}
\end{array}$$

3.1. The equivalence $\mathrm{Rep}_k(A,\xi^{-1})\cong\mathrm{Coh}^{(A\times B,\xi^{-1}\times\Psi)}(B)$.

For any

$$(S,\rho)\in\mathrm{Coh}^{(A\times B,\xi^{-1}\times\Psi)}(B),$$

define the maps

$$\begin{aligned}\rho_2&:=(\mathrm{id}\otimes\varepsilon\otimes\mathrm{id})\rho: S\to S\otimes\mathscr{O}(B)_\Psi,\\ \rho_3&:=(\mathrm{id}\otimes\mathrm{id}\otimes\varepsilon)\rho: S\to S\otimes\mathscr{O}(A)_{\xi^{-1}},\end{aligned}\tag{3.2}$$

and the coinvariants subsheaf

$$S^{(B,\Psi)}:=\{s\in S\mid \rho_2(s)=\Psi^1\cdot s\otimes\Psi^2\}\subset S. \tag{3.3}$$

LEMMA 3.1. *For any $(S,\rho)\in\mathrm{Coh}^{(A\times B,\xi^{-1}\times\Psi)}(B)$, the following hold:*

(1) $(S,\rho_2)\in\mathrm{Coh}^{(B,\Psi)}(B)$.
(2) $\mathrm{Coh}^{(B,\Psi)}(B)\cong\mathrm{Vec}$, *with the unique simple object being* $(\mathscr{O}(B),\Delta_\Psi)$.

(3) *There is a* $\mathrm{Coh}^{(B,\Psi)}(B)$*-isomorphism*

$$(S,\rho_2)\cong(\mathscr{O}(B),\Delta_\Psi)\otimes_k S^{(B,\Psi)}.$$

(4) $(S,\rho_3)\in\mathrm{Rep}_k(A,\xi^{-1})$, *and* $(S^{(B,\Psi)},\rho_3)\subset(S,\rho_3)$ *in* $\mathrm{Rep}_k(A,\xi^{-1})$.

PROOF. (1) is clear, (2) follows from [**11**], and (3)–(4) from (1)–(2). □

Consider the trivial morphism $\mathrm{t}:B\twoheadrightarrow 1$.

THEOREM 3.2. *The*[2] *functor* $\mathrm{t}^*:\mathrm{Vec}\to\mathrm{Coh}(B)$ *lifts to an equivalence of abelian categories*

$$\mathrm{t}^*:\mathrm{Rep}_k(A,\xi^{-1})\xrightarrow{\cong}\mathrm{Coh}^{(A\times B,\xi^{-1}\times\Psi)}(B),\ \ V\mapsto\left(\mathscr{O}(B)\otimes_k V,\rho^V\right)$$

(see (3.4)*), whose inverse is given by*

$$\mathrm{t}_*^{(B,\Psi)}:\mathrm{Coh}^{(A\times B,\xi^{-1}\times\Psi)}(B)\xrightarrow{\cong}\mathrm{Rep}_k(A,\xi^{-1}),\ \ (S,\rho)\mapsto\left(S^{(B,\Psi)},\rho_3\right).$$

PROOF. Let $(V,r)\in\mathrm{Rep}_k(A,\xi^{-1})=\mathrm{Coh}^{(A,\xi^{-1})}(1)$ be any object, and write $r(v)=\sum v^0\otimes v^1\in V\otimes\mathscr{O}(A)_{\xi^{-1}}$. Consider the free $\mathscr{O}(B)$-module $\mathrm{t}^*V=\mathscr{O}(B)\otimes_k V$, and map

$$\rho^V:=((\widetilde{\rho_1}\otimes\mathrm{id})\Delta_\Psi)\,\bar{\otimes}r$$

(see (2.28)). Namely, we have

$$\rho^V:\mathscr{O}(B)\otimes_k V\to\mathscr{O}(B)\otimes_k V\otimes\mathscr{O}(A)_{\xi^{-1}}\otimes\mathscr{O}(B)_\Psi,$$
$$f\otimes v\mapsto\Psi_2^1f_2\otimes v^0\otimes\partial^\sharp\mathrm{S}\left(\Psi_1^1f_1\right)v^1\otimes\Psi^2f_3. \tag{3.4}$$

It is straightforward to verify that

$$\left(\mathscr{O}(B)\otimes_k V,\rho^V\right)\in\mathrm{Coh}^{(A\times B,\xi^{-1}\times\Psi)}(B).$$

Thus, we have defined a functor

$$\mathrm{Rep}_k(A,\xi^{-1})\to\mathrm{Coh}^{(A\times B,\xi^{-1}\times\Psi)}(B),\ \ V\mapsto\left(\mathscr{O}(B)\otimes_k V,\rho^V\right).$$

Conversely, take any $(S,\rho)\in\mathrm{Coh}^{(A\times B,\xi^{-1}\times\Psi)}(B)$, and consider the sheaf t_*S (that is, the underlying vector space of S). Then by Lemma 3.1(4),

$$\left(S^{(B,\Psi)},\rho_3\right)\in\mathrm{Rep}_k(A,\xi^{-1}).$$

Thus, we have a functor

$$\mathrm{Coh}^{(A\times B,\xi^{-1}\times\Psi)}(B)\to\mathrm{Rep}_k(A,\xi^{-1}),\ \ (S,\rho)\mapsto\left(S^{(B,\Psi)},\rho_3\right).$$

Finally, it is straightforward to verify that the two functors constructed above are inverse to each other. □

REMARK 3.3. If B is any affine group scheme, then Theorem 3.2 and its proof hold after replacing Coh by Coh_f (see [**11**]).

---

[2] The assumption $\xi=\Psi_{|A}$ (3.1) is not needed here.

3.2. The equivalence $\mathrm{Coh}^{(A\times B,\xi^{-1}\times\Psi)}(B)\cong\mathrm{Coh}^{(B,\Psi)}(A\backslash B)$. Consider the free left action $A\times B\to B$, $(a,b)\mapsto\partial(a)b$, and let

$$\mathrm{p}:B\twoheadrightarrow A\backslash B \tag{3.5}$$

be the corresponding quotient morphism.

For any $(S,\rho)\in\mathrm{Coh}^{(A\times B,\xi^{-1}\times\Psi)}(B)$, define the coinvariants subsheaf

$$S^{(A,\xi^{-1})}:=\{s\in S\mid \rho_3(s)=\Psi^2\cdot s\otimes\partial^{\sharp}\mathrm{S}(\Psi^1)Q_{\xi^{-1}}\}\subset S \tag{3.6}$$

(see (3.2)).

LEMMA 3.4. *For any $(S,\rho)\in\mathrm{Coh}^{(A\times B,\xi^{-1}\times\Psi)}(B)$, $(S,\rho_3)\in\mathrm{Coh}^{(A,\xi^{-1})}(B)$.*

PROOF. Similar to the proof of Lemma 3.1. □

THEOREM 3.5. *The functor $\mathrm{p}^*:\mathrm{Coh}(A\backslash B)\to\mathrm{Coh}(B)$ (3.5) lifts to an equivalence of abelian categories*

$$\mathrm{p}^*:\mathrm{Coh}^{(B,\Psi)}(A\backslash B)\xrightarrow{\cong}\mathrm{Coh}^{(A\times B,\xi^{-1}\times\Psi)}(B),$$
$$(S,\rho)\mapsto\left(\mathscr{O}(B)\otimes_{\mathscr{O}(A\backslash B)}S,\rho^*\right)$$

(see (3.7)), whose inverse is given by

$$\mathrm{p}_*^{(A,\xi^{-1})}:\mathrm{Coh}^{(A\times B,\xi^{-1}\times\Psi)}(B)\xrightarrow{\cong}\mathrm{Coh}^{(B,\Psi)}(A\backslash B),$$
$$(S,\rho)\mapsto\left(\mathrm{p}_*^{(A,\xi^{-1})}S,\rho_2\right).$$

PROOF. Let (S,ρ) be any object in $\mathrm{Coh}^{(B,\Psi)}(A\backslash B)$. Consider the sheaf $\mathrm{p}^*S=\mathscr{O}(B)\otimes_{\mathscr{O}(A\backslash B)}S$, and map

$$\rho^*:=\left(\widetilde{(\rho_{\Psi^{-1}}}\otimes\mathrm{id})\Delta\right)\bar{\otimes}\rho$$

(see (2.28)). Namely, we have

$$\begin{aligned}&\rho^*:\mathrm{p}^*S\to\mathrm{p}^*S\otimes\mathscr{O}(A)_{\xi^{-1}}\otimes\mathscr{O}(B)_{\Psi},\\&f\otimes_{\mathscr{O}(A\backslash B)}s\mapsto\Psi^2f_2\otimes_{\mathscr{O}(A\backslash B)}s^0\otimes\partial^{\sharp}\mathrm{S}\left(\Psi^1f_1\right)Q_{\xi^{-1}}\otimes f_3s^1.\end{aligned} \tag{3.7}$$

It is straightforward to verify that $(\mathrm{p}^*S,\rho^*)\in\mathrm{Coh}^{(A\times B,\xi^{-1}\times\Psi)}(B)$, so we have a functor

$$\mathrm{p}^*:\mathrm{Coh}^{(B,\Psi)}(A\backslash B)\to\mathrm{Coh}^{(A\times B,\xi^{-1}\times\Psi)}(B),\ \ (S,\rho)\mapsto(\mathrm{p}^*S,\rho^*).$$

Conversely, let $(S,\rho)\in\mathrm{Coh}^{(A\times B,\xi^{-1}\times\Psi)}(B)$. Then by Lemma 3.4, we have a functor

$$\mathrm{p}_*^{(A,\xi^{-1})}:\mathrm{Coh}^{(A\times B,\xi^{-1}\times\Psi)}(B)\to\mathrm{Coh}^{(B,\Psi)}(A\backslash B),\ (S,\rho)\mapsto\left(\mathrm{p}_*^{(A,\xi^{-1})}S,\rho_2\right).$$

Finally, it is straightforward to verify that the two functors constructed above are inverse to each other. □

REMARK 3.6. If B is any affine group scheme such that the quotient scheme $A\backslash B$ is affine, then Theorem 3.5 and its proof hold after replacing Coh by $\mathrm{Coh}_{\mathrm{f}}$ (see [**11**]).

3.3. The first equivalence $\operatorname{Rep}_k(A,\xi^{-1}) \cong \operatorname{Coh}^{(B,\Psi)}(A\backslash B)$**.** Set

$$\operatorname{Ind}^{(B,\Psi)}_{(A,\xi^{-1})} := \mathrm{p}^{(A,\xi^{-1})}_* \mathrm{t}^*, \text{ and } \operatorname{Res}^{(B,\Psi)}_{(A,\xi^{-1})} := \mathrm{t}^{(B,\Psi)}_* \mathrm{p}^*. \tag{3.8}$$

THEOREM 3.7. *The equivalences*

$$\operatorname{Ind}^{(B,\Psi)}_{(A,\xi^{-1})} : \operatorname{Rep}_k(A,\xi^{-1}) \xrightarrow{\cong} \operatorname{Coh}^{(B,\Psi)}(A\backslash B),$$
$$V \mapsto \left(\mathrm{p}^{(A,\xi^{-1})}_* (\mathscr{O}(B)\otimes_k V), \rho^V_2\right),$$

and

$$\operatorname{Res}^{(B,\Psi)}_{(A,\xi^{-1})} : \operatorname{Coh}^{(B,\Psi)}(A\backslash B) \xrightarrow{\cong} \operatorname{Rep}_k(A,\xi^{-1}),$$
$$(S,\rho) \mapsto \left(\left(\mathscr{O}(B)\otimes_{\mathscr{O}(A\backslash B)} S\right)^{(B,\Psi)}, \rho^*_3\right)$$

(see (3.4), (3.7)*), are inverse to each other.*

PROOF. Follows from Theorems 3.2 and 3.5. □

Recall (2.27) that for any *right* $\mathscr{O}(A)_{\xi^{-1}}$-comodule (V,r), the map

$$\widetilde{r} : V \to \mathscr{O}(A)_\xi \otimes V, \quad v \mapsto \mathrm{S}(v^1)Q_\xi \otimes v^0,$$

defines a *left* $\mathscr{O}(A)_\xi$-comodule structure on V.

View $\mathscr{O}(B)$ as a right $\mathscr{O}(A)_\xi$-comodule via $\widetilde{\rho_\Psi}$ (2.28) given by

$$\widetilde{\rho_\Psi}(f) = \Psi^{-2}f_2 \otimes \partial^\sharp \mathrm{S}\left(\Psi^{-1}f_1\right)Q_\xi; \;\; f \in \mathscr{O}(B).$$

Recall that $\mathscr{O}(B) \otimes^{\mathscr{O}(A)_\xi} V = \operatorname{Ker}(\widetilde{\rho_\Psi}\otimes \mathrm{id} - \mathrm{id}\otimes \widetilde{r})$.

PROPOSITION 3.8. *For any right* $\mathscr{O}(A)_{\xi^{-1}}$*-comodule* (V,r)*, we have*

$$\mathscr{O}(B) \otimes^{\mathscr{O}(A)_\xi} V = \left(\mathscr{O}(B)\otimes_k V\right)^{(A,\xi^{-1})}.$$

PROOF. By (3.4), $\mathscr{O}(A)_{\xi^{-1}}$ coacts on $\mathscr{O}(B)\otimes_k V$ via

$$f\otimes v \mapsto \Psi^1_2 f_2 \otimes v^0 \otimes \partial^\sharp \mathrm{S}\left(\Psi^1_1 f_1\right)v^1\varepsilon(\Psi^2 f_3) = f_2\otimes v^0 \otimes \partial^\sharp\mathrm{S}(f_1)v^1.$$

Thus, by definition, $\sum_i f_i\otimes v_i \in (\mathscr{O}(B)\otimes_k V)^{(A,\xi^{-1})}$ if and only if

$$\sum_i f_{i2}\otimes v_i^0 \otimes \partial^\sharp\mathrm{S}(f_{i1})v_i^1 = \sum_i \Psi^2 f_i \otimes v_i \otimes \partial^\sharp\mathrm{S}(\Psi^1)Q_{\xi^{-1}}. \tag{3.9}$$

Now assume that $\sum_i f_i\otimes v_i$ belongs to $(\mathscr{O}(B)\otimes_k V)^{(A,\xi^{-1})}$. We have to verify that $\sum_i f_i\otimes v_i$ lies in $\mathscr{O}(B)\otimes^{\mathscr{O}(A)_\xi} V$, i.e., that

$$\sum_i \widetilde{\rho_\Psi}(f_i)\otimes v_i = \sum_i f_i \otimes \widetilde{r}(v_i).$$

Thus, we have to show that

$$\sum_i \Psi^{-2}f_{i2}\otimes \partial^\sharp\mathrm{S}(\Psi^{-1}f_{i1})Q_\xi \otimes v_i = \sum_i f_i\otimes \mathrm{S}(v_i^1)Q_\xi\otimes v_i^0. \tag{3.10}$$

To this end, apply $(\mathrm{id}\otimes\widetilde{r}\otimes\mathrm{id})$ to (3.9) to obtain

$$\sum_i f_{i2}\otimes \mathrm{S}(v_i^1)Q_\xi \otimes v_i^0 \otimes \partial^\sharp\mathrm{S}(f_{i1})v_i^2 = \sum_i \Psi^2 f_i \otimes \mathrm{S}(v_i^1)Q_\xi\otimes v_i^0\otimes \partial^\sharp\mathrm{S}(\Psi^1)Q_{\xi^{-1}}.$$

Thus, we have

$$\sum_i f_{i2}\otimes \mathrm{S}(v_i^1)\otimes v_i^0\otimes \partial^\sharp \mathrm{S}(f_{i1})v_i^2=\sum_i \Psi^2 f_i\otimes \mathrm{S}(v_i^1)\otimes v_i^0\otimes \partial^\sharp \mathrm{S}(\Psi^1)Q_{\xi^{-1}}.$$

Multiplying the second and fourth factors, yields

$$\sum_i f_{i2}\otimes \partial^\sharp \mathrm{S}(f_{i1})\mathrm{S}(v_i^1)v_i^2\otimes v_i^0=\sum_i \Psi^2 f_i\otimes \mathrm{S}(v_i^1)\partial^\sharp \mathrm{S}(\Psi^1)Q_{\xi^{-1}}\otimes v_i^0,$$

or equivalently (since $\mathrm{S}(v_i^1)v_i^2=\varepsilon(v_i^1)Q_{\xi^{-1}}$),

$$\sum_i f_{i2}\otimes \partial^\sharp \mathrm{S}(f_{i1})\varepsilon(v_i^1)Q_{\xi^{-1}}\otimes v_i^0=\sum_i \Psi^2 f_i\otimes \mathrm{S}(v_i^1)\partial^\sharp \mathrm{S}(\Psi^1)Q_{\xi^{-1}}\otimes v_i^0.$$

Thus, we have

$$\sum_i f_{i2}\otimes \partial^\sharp \mathrm{S}(f_{i1})\otimes v_i=\sum_i \Psi^2 f_i\otimes \mathrm{S}(v_i^1)\partial^\sharp \mathrm{S}(\Psi^1)\otimes v_i^0,$$

which is equivalent to (3.10).

Similarly, if $\sum_i f_i\otimes v_i\in \mathscr{O}(B)\otimes^{\mathscr{O}(A)_\xi} V$, then $\sum_i f_i\otimes v_i$ lies in $(\mathscr{O}(B)\otimes_k V)^{(A,\xi^{-1})}$. □

THEOREM 3.9. *The equivalences*

$$\mathrm{Ind}_{(A,\xi^{-1})}^{(B,\Psi)}:\mathrm{Rep}_k(A,\xi^{-1})\xrightarrow{\cong}\mathrm{Coh}^{(B,\Psi)}(A\backslash B),$$
$$V\mapsto \left(\mathscr{O}(B)\otimes^{\mathscr{O}(A)_\xi} V,\rho_2^V\right),$$

and

$$\mathrm{Res}_{(A,\xi^{-1})}^{(B,\Psi)}:\mathrm{Coh}^{(B,\Psi)}(A\backslash B)\xrightarrow{\cong}\mathrm{Rep}_k(A,\xi^{-1}),$$
$$(S,\rho)\mapsto \left(\left(\mathscr{O}(B)\otimes_{\mathscr{O}(A\backslash B)} S\right)^{(B,\Psi)},\rho_3^*\right),$$

are inverse to each other.

PROOF. Follows from Theorem 3.7 and Proposition 3.8. □

EXAMPLE 3.10. There is a canonical isomorphism

$$\mathrm{Ind}_{(A,\xi^{-1})}^{(B,\Psi)}\left(\mathscr{O}(A),\Delta_{\xi^{-1}}\right)\cong\left(\mathscr{O}(B),\Delta_\Psi\right)$$

in $\mathrm{Coh}^{(B,\Psi)}(A\backslash B)$. □

3.4. The second equivalence $\mathrm{Rep}_k(A,\xi^{-1})\cong\mathrm{Coh}^{(B,\Psi)}(A\backslash B)$. *Choose* a cleaving map (2.23) $\widetilde{\mathfrak{c}}:\mathscr{O}(A)\xrightarrow{1:1}\mathscr{O}(B)$. Recall (2.26), and let

$$\widetilde{\alpha}:\mathscr{O}(B)\twoheadrightarrow\mathscr{O}(A\backslash B),\quad f\mapsto f_1\widetilde{\mathfrak{c}}^{-1}(\partial^\sharp(f_2)).$$

Also, for any $V\in\mathrm{Rep}_k(A,\xi^{-1})$, consider the k-linear isomorphism

$$\mathrm{f}_V:=\widetilde{\alpha}\otimes\mathrm{id}:\mathscr{O}(B)\otimes^{\mathscr{O}(A)_\xi}V\xrightarrow{\cong}\mathscr{O}(A\backslash B)\otimes_k V,$$
$$f\otimes v\mapsto f_1\widetilde{\mathfrak{c}}^{-1}(\partial^\sharp(f_2))\otimes v,$$

whose inverse is given by

$$\mathrm{f}_V^{-1}:\mathscr{O}(A\backslash B)\otimes_k V\xrightarrow{\cong}\mathscr{O}(B)\otimes^{\mathscr{O}(A)_\xi}V,$$
$$f\otimes v\mapsto f\widetilde{\mathfrak{c}}(v^1)\otimes v^0,$$

and the map

$$\rho_V := (\mathrm{f}_V \otimes \mathrm{id})(23)(\Delta_\Psi \otimes \mathrm{id})\mathrm{f}_V^{-1}.$$

Note that we have

$$\begin{aligned} &\rho_V : \mathscr{O}(A\backslash B) \otimes_k V \to \mathscr{O}(A\backslash B) \otimes_k V \otimes \mathscr{O}(B)_\Psi, \\ &f \otimes v \mapsto \Psi_1^1 f_1 \widetilde{\mathfrak{c}}(v^1)_1 \widetilde{\mathfrak{c}}^{-1} \partial^\sharp \left(\Psi_2^1 f_2 \widetilde{\mathfrak{c}}(v^1)_2\right) \otimes v^0 \otimes \Psi^2 f_3 \widetilde{\mathfrak{c}}(v^1)_3. \end{aligned} \tag{3.11}$$

THEOREM 3.11. *We have an equivalence of abelian categories*

$$\mathrm{f} : \mathrm{Rep}_k(A, \xi^{-1}) \xrightarrow{\cong} \mathrm{Coh}^{(B,\Psi)}(A\backslash B), \;\; V \mapsto \left(\mathscr{O}(A\backslash B) \otimes_k V, \rho_V\right).$$

PROOF. Follows from Theorem 3.7 and the preceding remarks. □

4. Double cosets and biequivariant sheaves

Recall that double cosets in finite abstract groups play a fundamental role in the theory of group-theoretical fusion categories (see, e.g, [**24**]).

Let $\iota_H : H \xrightarrow{1:1} G$ and $\iota_K : K \xrightarrow{1:1} G$ be two embeddings of finite group schemes, and consider the right action of $H \times K$ on G given by

$$\mu_{G\times(H\times K)} : G \times (H \times K) \to G, \;\; (g, h, k) \mapsto h^{-1}gk. \tag{4.1}$$

The algebra map $\mu^\sharp_{G\times(H\times K)} : \mathscr{O}(G) \to \mathscr{O}(G) \otimes \mathscr{O}(H \times K)$ is given by

$$\mu^\sharp_{G\times(H\times K)}(f) = f_2 \otimes \iota_H^\sharp \mathrm{S}(f_1) \otimes \iota_K^\sharp(f_3); \;\; f \in \mathscr{O}(G). \tag{4.2}$$

Recall that since $H \times K$ is a finite group scheme, there exists a finite quotient scheme $Y := G/(H \times K)$ (see Section 2.3).

For any closed point $g \in G(k)$, let $Z_g := HgK \subset G$ denote the orbit of g under the action (4.1)—it is a closed subset of G. It is clear that $Z_g \in Y(k)$, and all closed points of Y are such. Letting $\mathrm{Coh}(Y)_{Z_g} \subset \mathrm{Coh}(Y)$ denote the abelian subcategory of sheaves on Y supported on Z_g (see Section 2.1), we have a direct sum decomposition of abelian categories

$$\mathrm{Coh}(Y) = \bigoplus_{Z_g \in Y(k)} \mathrm{Coh}(Y)_{Z_g}. \tag{4.3}$$

For the remaining of this section, we fix $Z \in Y(k)$ with representative $g \in Z(k)$. Let $L^g := H \cap gKg^{-1}$ denote the (group scheme-theoretical) intersection of H and gKg^{-1}, i.e., $L^g = H \times_G gKg^{-1}$. Clearly, L^g is a closed subgroup scheme of both H and gKg^{-1}, and

$$\mathscr{O}(L^g) = \mathscr{O}(H) \otimes_{\mathscr{O}(G)} \mathscr{O}(gKg^{-1}) = \mathscr{O}(G)/\left(\mathscr{I}(H) + \mathscr{I}(gKg^{-1})\right).$$

Let $\iota_g : L^g \hookrightarrow G$ be the inclusion morphism, and consider the group scheme embedding of L^g in $H \times K$ given by

$$\partial_g := \left(\iota_g, (\mathrm{Ad}g^{-1}) \circ \iota_g\right) : L^g \xrightarrow{1:1} H \times K, \;\; \ell \mapsto (\ell, g^{-1}\ell g). \tag{4.4}$$

It is clear that the subgroup scheme $\partial_g(L^g) \subset H \times K$ is the stabilizer of g for the action (4.1). Namely, Z is a quotient scheme for the free *left* action of L^g on $H \times K$ given by

$$\mu_{L^g\times(H\times K)} : L^g \times (H \times K) \to H \times K, \;\; (\ell, h, k) \mapsto (\ell h, g^{-1}\ell gk). \tag{4.5}$$

Thus, we have scheme isomorphisms

$$(4.6)\qquad \begin{aligned} &\mathrm{j}_g : L^g\backslash(H\times K)\xrightarrow{\cong} Z,\ \ \overline{(h,k)}\mapsto h^{-1}gk,\ \text{ and}\\ &\mathrm{j}_g^{-1} : Z\xrightarrow{\cong} L^g\backslash(H\times K),\ \ hgk\mapsto \overline{(h^{-1},k)}. \end{aligned}$$

Assume that $\psi\in Z^2(H,\mathbb{G}_m)$ and $\eta\in Z^2(K,\mathbb{G}_m)$. Set

$$(4.7)\qquad \xi_g := \left(\partial_g^\sharp\otimes\partial_g^\sharp\right)\left(\psi^{-1}\times\eta\right) = \psi^{-1}\eta^{g^{-1}}\in Z^2(L^g,\mathbb{G}_m).$$

Next *choose* a cleaving map (2.23) $\widetilde{\mathfrak{c}_g} : \mathscr{O}(L^g)\xrightarrow{1:1}\mathscr{O}(H\times K)$, and let

$$(4.8)\qquad \widetilde{\alpha_g} : \mathscr{O}(H\times K)\twoheadrightarrow\mathscr{O}(L^g\backslash(H\times K)),\ \ \mathrm{f}\mapsto \mathrm{f}_1\widetilde{\mathfrak{c}_g}^{-1}\left(\partial_g^\sharp(\mathrm{f}_2)\right)$$

(see (2.26)). Set (see Theorems 3.9 and 3.11)

$$(4.9)\qquad \mathrm{Ind}_Z := \mathrm{Ind}_{\left(L^g,\xi_g^{-1}\right)}^{\left(H\times K,\psi^{-1}\times\eta\right)},\ \text{ and } \mathrm{f}_Z := \mathrm{f}.$$

COROLLARY 4.1. *For any closed point $Z\in Y(k)$ with representative $g\in Z(k)$, we have the following equivalences of abelian categories:*

(1) $\mathrm{Ind}_Z : \mathrm{Rep}_k(L^g,\xi_g^{-1})\xrightarrow{\cong}\mathrm{Coh}^{\left(H\times K,\psi^{-1}\times\eta\right)}(L^g\backslash(H\times K))$,
$V\mapsto\left(\mathscr{O}(H\times K)\otimes^{\mathscr{O}(L^g)_{\xi_g}} V,\rho_2^V\right)$.

(2) $\mathrm{f}_Z : \mathrm{Rep}_k(L^g,\xi_g^{-1})\xrightarrow{\cong}\mathrm{Coh}^{\left(H\times K,\psi^{-1}\times\eta\right)}(L^g\backslash(H\times K))$,
$V\mapsto(\mathscr{O}(L^g\backslash(H\times K))\otimes_k V,\rho_V)$.

(3) $\mathrm{j}_{g*} : \mathrm{Coh}^{\left(H\times K,\psi^{-1}\times\eta\right)}(L^g\backslash(H\times K))\xrightarrow{\cong}\mathrm{Coh}^{\left(H\times K,\psi^{-1}\times\eta\right)}(Z)$.

PROOF. Follow from Theorems 3.9 and 3.11 for $A := L^g$, $B := H\times K$, $\partial := \partial_g$, $\Psi := \psi^{-1}\times\eta$, and $\xi := \xi_g$. $\square$

5. Module categories over $\mathscr{C}(G,H,\psi)$

Fix a finite group scheme-theoretical category $\mathscr{C} := \mathscr{C}(G,H,\psi)$ as in Section 2.9. In this section we study the abelian structure of indecomposable exact $\mathscr{C}$-module categories, which can be classified as categories of $\mathrm{Coh}(G)$-module functors [**11**].

THEOREM 5.1 ([**11**, Theorem 5.7]). *The assignment*

$$\mathscr{M}(K,\eta)\mapsto \mathrm{Fun}_{\mathrm{Coh}(G)}\left(\mathscr{M}(H,\psi),\mathscr{M}(K,\eta)\right)$$

determines an equivalence between the 2-category of indecomposable exact module categories over Coh(G) *and the 2-category of indecomposable exact module categories over $\mathscr{C}(G,H,\psi)$.* $\square$

5.1. Three descriptions of module categories.

For H,K,ψ,η as above, consider the category

$$(5.1)\qquad \mathscr{M}\left(H\times K,\psi^{-1}\times\eta\right) = \mathrm{Coh}^{\left(H\times K,\psi^{-1}\times\eta\right)}(G)$$

of $\left(H\times K,\psi^{-1}\times\eta\right)$-equivariant sheaves on G with respect to the right action $\mu_{G\times(H\times K)}$ from (4.1). Note that there is an equivalences of abelian categories

$$\mathscr{M}\left(H\times K,\psi^{-1}\times\eta\right)\cong \mathrm{Bicomod}_{\mathrm{Coh}(G)}\left(\mathscr{O}(H)_\psi,\mathscr{O}(K)_\eta\right).$$

LEMMA 5.2. *There is an equivalence of abelian categories*

$$\mathrm{Fun}_{\mathrm{Coh}(G)}\left(\mathscr{M}(H,\psi),\mathscr{M}(K,\eta)\right)\cong\mathscr{M}\left(H\times K,\psi^{-1}\times\eta\right).$$

PROOF. A functor $\mathscr{M}(H,\psi) \to \mathscr{M}(K,\eta)$ is determined by a (K,η)-equivariant sheaf S on G (its value on $\mathscr{O}(H)_\psi$), and the fact that the functor is a $\mathrm{Coh}(G)$-module (hence, $\mathrm{Coh}(H)$-module) functor gives S a commuting H-equivariant structure for the right action $\widetilde{\mu_{G\times H}}$ of H on G, i.e., $S \in \mathscr{M}\left(H \times K, \psi^{-1} \times \eta\right)$.

Conversely, it is clear that any $S \in \mathscr{M}\left(H \times K, \psi^{-1} \times \eta\right)$, viewed as an object in $\mathrm{Bicomod}_{\mathrm{Coh}(G)}\left(\mathscr{O}(H)_\psi, \mathscr{O}(K)_\eta\right)$, defines a $\mathrm{Coh}(G)$-module functor

$$\mathscr{M}(H,\psi) \to \mathscr{M}(K,\eta), \ \ T \mapsto T \otimes^{\mathscr{O}(H)_\psi} S,$$

where $T \otimes^{\mathscr{O}(H)_\psi} S$ is the cotensor product. □

5.2. The simple objects of $\mathscr{M}\left(H \times K, \psi^{-1} \times \eta\right)$. Fix an indecomposable exact left $\mathscr{C}$-module category $\mathscr{M} := \mathscr{M}\left(H \times K, \psi^{-1} \times \eta\right)$ (5.1).

Fix a closed point $Z \in Y(k)$, and let

$$\mathscr{M}_Z := \mathscr{M}_Z\left(H \times K, \psi^{-1} \times \eta\right) \subset \mathscr{M} \tag{5.2}$$

denote the full abelian subcategory of $\mathscr{M}$ consisting of all objects annihilated by the defining ideal $\mathscr{I}(Z) \subset \mathscr{O}(G)$ of Z. Namely, let

$$\iota_Z : Z \hookrightarrow G$$

denote the inclusion morphism (it is $H \times K$-equivariant). Then $\mathscr{M}_Z$ is the image of the injective functor

$$\iota_{Z*} : \mathrm{Coh}^{\left(H\times K,\psi^{-1}\times\eta\right)}(Z) \xrightarrow{1:1} \mathrm{Coh}^{\left(H\times K,\psi^{-1}\times\eta\right)}(G). \tag{5.3}$$

Let $\overline{\mathscr{M}_Z} \subset \mathscr{M}$ denote the Serre closure of $\mathscr{M}_Z$ inside $\mathscr{M}$, i.e., $\overline{\mathscr{M}_Z}$ is the full abelian subcategory of $\mathscr{M}$ consisting of all objects whose composition factors lie in $\mathscr{M}_Z$.

Fix a representative $g \in Z(k)$, and define the functor

$$\mathbf{Ind}_Z := \iota_{Z*} \circ \mathrm{j}_{g*} \circ \mathrm{Ind}_Z. \tag{5.4}$$

THEOREM 5.3. *Let $\mathscr{M} = \mathscr{M}\left(H \times K, \psi^{-1} \times \eta\right)$ be as above.*

(1) *For any closed point $Z \in Y(k)$ with representative $g \in Z(k)$, we have an equivalence of abelian categories*

$$\mathbf{Ind}_Z : \mathrm{Rep}_k(L^g, \xi_g^{-1}) \xrightarrow{\cong} \mathscr{M}_Z,$$
$$V \mapsto (\iota_Z \mathrm{j}_g)_* \left(\mathscr{O}(H \times K) \otimes^{\mathscr{O}(L^g)_{\xi_g}} V, \rho_2^V\right).$$

(2) *There is a bijection between equivalence classes of pairs (Z,V), where $Z \in Y(k)$ is a closed point with representative $g \in Z(k)$, and*

$$V \in \mathrm{Rep}_k(L^g, \xi_g^{-1})$$

is simple, and simple objects of $\mathscr{M}$, assigning (Z,V) to $\mathbf{Ind}_Z(V)$.

(3) *We have a direct sum decomposition of abelian categories*

$$\mathscr{M} = \bigoplus_{Z \in Y(k)} \overline{\mathscr{M}_Z}.$$

PROOF.

(1) Follows from Corollary 4.1.

(2) Assume $S \in \mathscr{M}$ is simple. Since for any $Z \in Y(k)$, $\mathscr{I}(Z)$ is a $H \times K$-stable ideal, it follows from Proposition 2.7(3) that either $\mathscr{I}(Z)S = 0$ or $\mathscr{I}(Z)S = S$. If $\mathscr{I}(Z)S = S$ for every $Z \in Y(k)$, then $(\Pi_Z \mathscr{I}(Z))S = S$. But, $\Pi_Z \mathscr{I}(Z)$ is a nilpotent ideal of $\mathscr{O}(G)$ (being contained in the radical of $\mathscr{O}(G)$), so $S = 0$, a contradiction.

Thus, there exists $Z \in Y(k)$ such that $\mathscr{I}(Z)S = 0$. Assume that $\mathscr{I}(Z')S = 0$ for some $Z' \in Y(k)$, $Z' \neq Z$. Then $\mathscr{O}(G)S = 0$ (since $\mathscr{O}(G) = \mathscr{I}(Z) + \mathscr{I}(Z')$), a contradiction. It follows that there exists a unique $Z \in Y(k)$ such that $\mathscr{I}(Z)S = 0$, i.e., $S \in \mathscr{M}_Z$, so the claim follows from (1).

(3) Follows from (2), and the fact that there are no nontrivial cross extensions of $\mathscr{O}(G)$-modules which are supported on distinct closed points of G (see Section 2.1). □

5.3. Projectives in $\mathscr{M}\left(H \times K, \psi^{-1} \times \eta\right)$. Let $\mathscr{M}$ be as in Section 5.2.

For any $Z \in Y(k)$ with representative $g \in Z(k)$, recall the functor f_Z from Corollary 4.1, and define the functor

$$\mathbf{F}_Z := \iota_{Z*} \circ \mathrm{j}_{g*} \circ \mathrm{f}_Z. \tag{5.5}$$

Also, set $Z^\circ := H^\circ g K^\circ$,

$$|Z^\circ| := \frac{|H^\circ||K^\circ|}{|(L^g)^\circ|} \in \mathbb{Z}^{\geq 1}, \text{ and } |Z(k)| := \frac{|H(k)||K(k)|}{|(L^g)(k)|} \in \mathbb{Z}^{\geq 1}. \tag{5.6}$$

Note that (2.3) induces a split exact sequence of schemes

$$1 \to Z^\circ \xrightarrow{i_{Z^\circ}} Z \underset{q_Z}{\overset{\pi_Z}{\rightleftarrows}} Z(k) \to 1.$$

LEMMA 5.4. *For any $Z \in Y(k)$, with representative closed point $g \in Z(k)$, the following hold:*

(1) *The surjective $\mathscr{O}(G)$-linear map*
$$\theta_Z := (\mathrm{id} \otimes q_Z^\sharp) : \mathscr{O}(G^\circ) \otimes \mathscr{O}(Z) \twoheadrightarrow \mathscr{O}(G^\circ) \otimes \mathscr{O}(Z(k))$$
splits via the map
$$\lambda_Z := (\mathrm{id} \otimes \pi_Z^\sharp) : \mathscr{O}(G^\circ) \otimes \mathscr{O}(Z(k)) \xrightarrow{1:1} \mathscr{O}(G^\circ) \otimes \mathscr{O}(Z).$$
Thus, $\mathscr{O}(G^\circ) \otimes \mathscr{O}(Z(k))$ is a direct summand of $\mathscr{O}(G^\circ) \otimes \mathscr{O}(Z)$ as an $\mathscr{O}(G)$-module.

(2) *For any simple $V \in \mathrm{Rep}_k(L^g, \xi_g^{-1})$, we have an $\mathscr{M}$-isomorphism*
$$\mathscr{O}(G^\circ) \otimes \mathbf{F}_Z\left(P_{(L^g,\xi_g^{-1})}(V)\right) \cong \mathscr{O}(Z^\circ) \otimes_k P_{\mathscr{M}}\left(\mathbf{F}_Z(V)\right).$$
Here, $\mathscr{O}(G)$ acts diagonally on $\mathscr{O}(G^\circ) \otimes \mathbf{F}_Z\left(P_{(L^g,\xi_g^{-1})}(V)\right)$, and
$$\mathscr{O}(H \times K)_{\psi^{-1} \times \eta}$$
coacts on the second factor, and the right hand side is a direct sum of $|Z^\circ|$ copies of $P_{\mathscr{M}}\left(\mathbf{F}_Z(V)\right)$.

PROOF.

(1) Follows from the preceding remarks.

(2) By Example 3.10, $\mathbf{F}_Z\left(\mathscr{O}(L^g)_{\xi_g^{-1}}\right) \cong \mathscr{O}(H \times K)$ in $\mathscr{M}$, so

$$\mathscr{O}(G) \otimes \mathbf{F}_Z\left(\mathscr{O}(L^g)_{\xi_g^{-1}}\right) \cong \mathscr{O}(G) \otimes \mathscr{O}(H \times K)$$

is projective in $\mathscr{M}$ (where $\mathscr{O}(G)$ acts diagonally, and $\mathscr{O}(H \times K)_{\psi^{-1}\times\eta}$ coacts on the second factor via $\Delta_{\psi^{-1}\times\eta}$). Since $\mathscr{O}(G^\circ)$ is a direct summand of $\mathscr{O}(G)$, and $\mathbf{F}_Z\left(P_{(L^g,\xi_g^{-1})}(V)\right)$ is a direct summand of $\mathbf{F}_Z\left(\mathscr{O}(L^g)_{\xi_g^{-1}}\right)$, it follows that the object $\mathscr{O}(G^\circ) \otimes \mathbf{F}_Z\left(P_{(L^g,\xi_g^{-1})}(V)\right)$ is a direct summand of $\mathscr{O}(G) \otimes \mathscr{O}(H \times K)_{\psi^{-1}\times\eta}$, hence projective in $\mathscr{M}$. Thus, $\mathscr{O}(G^\circ) \otimes \mathbf{F}_Z\left(P_{(L^g,\xi_g^{-1})}(V)\right)$ is projective in $\mathscr{M}$.

Now, on the one hand, we have

$$\begin{aligned}
&\operatorname{Hom}_{\mathscr{M}}\left(\mathscr{O}(G^\circ) \otimes \mathbf{F}_Z\left(\mathscr{O}(L^g)_{\xi_g^{-1}}\right), \mathbf{F}_Z(V)\right)\\
&\quad= \operatorname{Hom}_{\mathscr{M}}\left(\mathscr{O}(G^\circ) \otimes \mathscr{O}(H \times K)_{\psi^{-1}\times\eta}, \mathbf{F}_Z(V)\right)\\
&\quad= \operatorname{Hom}_{\operatorname{Coh}(G)}\left(\mathscr{O}(G^\circ), \mathscr{O}(Z)\right) \otimes_k V.
\end{aligned}$$

On the other hand, observe that for any simple $W \in \operatorname{Rep}_k(L^g, \xi_g^{-1})$, the objects $\mathscr{O}(G^\circ) \otimes \mathbf{F}_Z\left(P_{(L^g,\xi_g^{-1})}(W)\right)$ and $\mathbf{F}_Z\left(P_{(L^g,\xi_g^{-1})}(W)\right)$ have the same composition factors as $\mathscr{O}(G)$-modules. Hence, if V and W are nonisomorphic simples in $\operatorname{Rep}_k(L^g, \xi_g^{-1})$, then

$$\operatorname{Hom}_{\mathscr{M}}\left(\mathscr{O}(G^\circ) \otimes \mathbf{F}_Z\left(P_{(L^g,\xi_g^{-1})}(W)\right), \mathbf{F}_Z(V)\right) = 0.$$

This implies that

$$\begin{aligned}
&\operatorname{Hom}_{\mathscr{M}}\left(\mathscr{O}(G^\circ) \otimes \mathbf{F}_Z\left(\mathscr{O}(L^g)_{\xi_g^{-1}}\right), \mathbf{F}_Z(V)\right)\\
&= \bigoplus_{W \in \operatorname{Irr}\left(\mathscr{O}(L^g)_{\xi_g^{-1}}\right)} \operatorname{Hom}_{\mathscr{M}}\left(\mathscr{O}(G^\circ) \otimes \mathbf{F}_Z\left(P_{(L^g,\xi_g^{-1})}(W)\right), \mathbf{F}_Z(V)\right) \otimes_k W\\
&= \operatorname{Hom}_{\mathscr{M}}\left(\mathscr{O}(G^\circ) \otimes \mathbf{F}_Z\left(P_{(L^g,\xi_g^{-1})}(V)\right), \mathbf{F}_Z(V)\right) \otimes_k V.
\end{aligned}$$

Thus, it follows from the above that

$$\begin{aligned}
&\dim \operatorname{Hom}_{\mathscr{M}}\left(\mathscr{O}(G^\circ) \otimes \mathbf{F}_Z\left(P_{(L^g,\xi_g^{-1})}(V)\right), \mathbf{F}_Z(V)\right)\\
&\quad= \dim \operatorname{Hom}_{\operatorname{Coh}(G)}\left(\mathscr{O}(G^\circ), \mathscr{O}(Z)\right) = |Z^\circ|,
\end{aligned}$$

which implies the statement. $\square$

THEOREM 5.5. *Let $\mathscr{M} = \mathscr{M}\left(H \times K, \psi^{-1} \times \eta\right)$ be as above.*

(1) *For any $Z \in Y(k)$ with representative $g \in Z(k)$, we have an equivalence of abelian categories*

$$\mathbf{F}_Z : \operatorname{Rep}_k(L^g, \xi_g^{-1}) \xrightarrow{\cong} \mathscr{M}_Z, \quad V \mapsto \iota_{Z*}\left(\mathscr{O}(Z) \otimes_k V, \rho_V^g\right),$$

where, $\rho_V^g := \left(\left(\mathrm{j}_g^\sharp\right)^{-1} \otimes \mathrm{id}^{\otimes 3}\right) \rho_V \left(\mathrm{j}_g^\sharp \otimes \mathrm{id}\right)$ (3.11).

(2) *For any simple* $V \in \operatorname{Rep}_k(L^g, \xi_g^{-1})$, *we have*

$$P_{\mathscr{M}}(\mathbf{F}_Z(V)) \cong \left(\mathscr{O}(G^\circ) \otimes \mathscr{O}(Z(k)) \otimes_k P_{(L^g,\xi_g^{-1})}(V), R_V^g\right),$$

where $R_V^g := (\theta_Z \otimes \mathrm{id}^{\otimes 3}) \left(\mathrm{id} \otimes \rho^g_{P_{(L^g,\xi_g^{-1})}(V)}\right) (\lambda_Z \otimes \mathrm{id})$.

PROOF.

(1) Follows from Theorems 3.11 and 5.3.

(2) We have $\mathscr{M}$-isomorphisms

$$\begin{aligned}\mathscr{O}(G^\circ) \otimes \mathbf{F}_Z\left(P_{(L^g,\xi_g^{-1})}(V)\right) &= \mathscr{O}(G^\circ) \otimes \mathscr{O}(Z) \otimes_k P_{(L^g,\xi_g^{-1})}(V)\\ &\cong \mathscr{O}(G^\circ) \otimes \mathscr{O}(Z(k)) \otimes_k \mathscr{O}(Z^\circ) \otimes_k P_{(L^g,\xi_g^{-1})}(V),\end{aligned}$$

thus by Proposition 5.4, $\mathscr{O}(G^\circ) \otimes \mathscr{O}(Z(k)) \otimes_k P_{(L^g,\xi_g^{-1})}(V)$ is a direct summand in a projective object of $\mathscr{M}$, hence is projective. Moreover, by Proposition 5.4 again, we have

$$\dim \operatorname{Hom}_{\mathscr{M}}\left(\mathscr{O}(G^\circ) \otimes \mathscr{O}(Z(k)) \otimes_k P_{(L^g,\xi_g^{-1})}(V), \mathbf{F}_Z(V)\right) = 1.$$

Hence, $\mathscr{O}(G^\circ) \otimes \mathscr{O}(Z(k)) \otimes_k P_{(L^g,\xi_g^{-1})}(V)$ is the projective cover of $\mathbf{F}_Z(V)$ in $\mathscr{M}$, as claimed. □

EXAMPLE 5.6. Take $K = 1$. Then for any $Z \in Y(k)$ with representative $g \in G(k)$, we have $Z = Hg$, $L^g = 1$, and the functor

$$\mathbf{F}_{Hg} : \mathrm{Vec} \xrightarrow{\cong} \mathrm{Coh}^{(H,\psi^{-1})}(Hg),\ \ V \mapsto \mathscr{O}(Hg) \otimes_k V,$$

is an equivalence of abelian categories (where $\mathscr{O}(H)$ acts on $\mathscr{O}(Hg)$ via multiplication by $\rho_{g^{-1}}(\cdot)$ and $\mathscr{O}(H)_{\psi^{-1}}$-coacts via $\Delta_{\psi^{-1}}$).

Now by Theorem 5.5, there is a bijection between the set of closed points $Hg \in Y(k)$ (i.e., isomorphism classes of simples of $\mathrm{Coh}(H\backslash G)$) and simple objects of $\mathscr{M}$, assigning Hg to $\mathscr{O}(Hg) \otimes_k k = \mathscr{O}(Hg)$, and we have $\overline{\mathscr{M}_{Hg}} = \langle \mathscr{O}(Hg) \rangle$.

Moreover, by Theorem 5.5 again, for any simple $\mathscr{O}(Hg)$ in $\mathscr{M}$,

$$P_{\mathscr{M}}(\mathscr{O}(Hg)) \cong P_{Hg} \otimes_k k = P_{Hg}$$

as $\mathscr{O}(G)$-modules, where $\mathscr{O}(H)_{\psi^{-1}}$ coacts on $P_{Hg} \otimes_k k = P_{Hg}$ via the map $\rho_k = \Delta_{\psi^{-1}}$. (Compare with Remark 2.19.) □

EXAMPLE 5.7. Theorems 5.3 and 5.5 apply to the representation categories of two-sided twisted function (Hopf) algebras. Namely, let $I, J \in (\mathscr{O}(G)^{\otimes 2})^*$ be two Hopf 2-cocycles for G, and let ${}_I\mathscr{O}(G)_J$ be the corresponding two-sided twisted function algebra (if $I = J$, then ${}_J\mathscr{O}(G)_J$ is a Hopf algebra). By [**11**, Corollary 4.8], I, J correspond to pairs (H, ψ), (K, η), where $H, K \subset G$ are closed subgroup schemes and $\psi \in Z^2(H, \mathbb{G}_m)$, $\eta \in Z^2(K, \mathbb{G}_m)$ are *nondegenerate* 2-cocycles (i.e., $\operatorname{Rep}_k(H, \psi) \cong \mathrm{Vec}$ and $\operatorname{Rep}_k(K, \eta) \cong \mathrm{Vec}$). Then it is straightforward to verify that we have an equivalence of abelian categories

$$\operatorname{Rep}_k({}_I\mathscr{O}(G)_J) \cong \mathscr{M}\left(H \times K, \psi^{-1} \times \eta\right).$$

In particular, we have an equivalence of tensor categories

$$\operatorname{Rep}_k({}_J\mathscr{O}(G)_J) \cong \mathscr{C}(G, H, \psi).$$

(In the fusion case it was proved in [**2**].) □

5.4. Fiber functors on $\mathscr{C}(G,H,\psi)$. Recall that a fiber functor on a finite tensor category $\mathscr{C}$ is the same as a $\mathscr{C}$-module category of rank 1.

The following corollary is known in the fusion case [**24**].

COROLLARY 5.8. *Let $\mathscr{C} := \mathscr{C}(G,H,\psi)$ be a group scheme-theoretical category. There is a bijection between equivalence classes of fiber functors on $\mathscr{C}$ and equivalence classes of pairs (K,η), where $K \subset G$ is a closed subgroup scheme and $\eta \in Z^2(K,\mathbb{G}_m)$, such that $HK = G$ and the 2-cocycle $\xi^{-1} := \xi_1^{-1} \in Z^2(H\cap K,\mathbb{G}_m)$ is nondegenerate.*

PROOF. Let $\mathscr{M} := \mathscr{M}\left(H\times K,\psi^{-1}\times\eta\right)$ be an indecomposable exact module category over $\mathscr{C}$. By Theorem 5.3, $\mathscr{M} \cong \mathrm{Vec}$ if and only if $\mathscr{M} = \mathscr{M}_{HK}$ and $\mathrm{Rep}(H\cap K,\xi^{-1}) \cong \mathrm{Vec}$. Thus, the statement follows from the fact that $\mathscr{M} = \mathscr{M}_{HK}$ if and only if ι_{HK*} is an equivalence, i.e., if and only if $G = HK$. □

EXAMPLE 5.9. Any exact factorization $G = HK$ of finite group schemes [**3**] determines a fiber functor on $\mathscr{C}(G,H,\psi)$ given by $\mathscr{M}(K,1)$.

6. The structure of $\mathscr{C}(G,H,\psi)$

Fix a group scheme-theoretical category $\mathscr{C} := \mathscr{C}(G,H,\psi)$ Section 2.9. For any closed point $g\in G(k)$, let $H^g := H\cap gHg^{-1}$. Note that $\xi_1 = 1$.

THEOREM 6.1. *The following hold:*

(1) *For any closed point $Z\in Y(k)$ with representative $g\in Z(k)$, we have an equivalence of abelian categories*

$$\mathbf{Ind}_Z : \mathrm{Rep}_k(H^g,\xi_g^{-1}) \xrightarrow{\cong} \mathscr{C}_Z,$$
$$V \mapsto (\iota_Z \jmath_g)_*\left(\mathscr{O}(H\times H)\otimes^{\mathscr{O}(H^g)_{\xi_g}} V,\rho_2^V\right).$$

In particular,

$$\mathbf{Ind}_H : \mathrm{Rep}_k(H) \xrightarrow{\cong} \mathscr{C}_H,$$
$$V \mapsto (\iota_H \jmath_1)_*\left(\mathscr{O}(H\times H)\otimes^{\mathscr{O}(H)} V,\rho_2^V\right),$$

is an equivalence of tensor categories.

(2) *There is a bijection between equivalence classes of pairs (Z,V), where $Z\in Y(k)$ is a closed point with representative $g\in Z(k)$, and*

$$V\in \mathrm{Rep}_k(H^g,\xi_g^{-1})$$

is simple, and simple objects of $\mathscr{C}$, assigning (Z,V) to $\mathbf{Ind}_Z(V)$. Moreover, we have a direct sum decomposition of abelian categories

$$\mathscr{C} = \bigoplus_{Z\in Y(k)} \overline{\mathscr{C}_Z},$$

and $\overline{\mathscr{C}_H}\subset\mathscr{C}$ is a tensor subcategory.

(3) *For any $V\in \mathrm{Rep}_k(H^g,\xi_g^{-1})$, we have*

$$\mathrm{FPdim}\left(\mathbf{Ind}_Z(V)\right) = \frac{|H|}{|H^g|}\dim(V).$$

(4) *For any $V\in\mathrm{Rep}_k(H^g,\xi_g^{-1})$, we have $\mathbf{Ind}_Z(V)^*\cong\mathbf{Ind}_{Z^{-1}}(V^*)$, where $Z^{-1}\in Y(k)$ such that $g^{-1}\in Z^{-1}(k)$.*

PROOF. It follows from Theorem 5.3. □

THEOREM 6.2. *The following hold:*

(1) *For any closed point $Z \in Y(k)$ with representative $g \in Z(k)$, we have an equivalence of abelian categories*
$$\mathbf{F}_Z : \operatorname{Rep}_k(H^g, \xi_g^{-1}) \xrightarrow{\cong} \mathscr{C}_Z, \quad V \mapsto \iota_{Z*}\left(\mathscr{O}(Z) \otimes_k V, \rho_V^g\right).$$
In particular, we have an equivalence of tensor categories
$$\mathbf{F}_H : \operatorname{Rep}_k(H) \xrightarrow{\cong} \mathscr{C}_H, \quad V \mapsto \iota_{H*}\left(\mathscr{O}(H) \otimes_k V, \rho_V^1\right).$$

(2) *For any $V \in \operatorname{Rep}_k(H^g, \xi_g^{-1})$, we have $\mathbf{F}_Z(V)^* \cong \mathbf{F}_{Z^{-1}}(V^*)$.*

(3) *For any simple $V \in \operatorname{Rep}_k(H^g, \xi_g^{-1})$, we have*
$$P_{\mathscr{C}}\left(\mathbf{F}_Z(V)\right) \cong \left(\mathscr{O}(G^\circ) \otimes \mathscr{O}(Z(k)) \otimes_k P_{(H^g, \xi_g^{-1})}(V), R_V^g\right).$$
In particular,
$$\operatorname{FPdim}\left(P_{\mathscr{C}}\left(\mathbf{F}_Z\left(V\right)\right)\right) = \frac{|G^\circ||H(k)|}{|H^\circ||H^g(k)|} \dim\left(P_{(H^g, \xi_g^{-1})}\left(V\right)\right).$$

(4) *For any closed point $Z \in Y(k)$, $\operatorname{FPdim}\left(\overline{\mathscr{C}_Z}\right) = |G^\circ||Z(k)|$.*

(5) *If $\operatorname{Rep}_k(H)$ is unimodular, so is $\mathscr{C}$ (but not necessarily vice versa).*

PROOF. (1)–(4) follows from Theorem 5.5 in a straightforward manner.

(5) recalls that $\operatorname{Rep}_k(H)$ is unimodular if and only if $P(\mathbf{1}) \cong P(\mathbf{1})^*$, where $P(\mathbf{1})$ is the projective cover of $\mathbf{1}$ in $\operatorname{Rep}_k(H)$. Thus, if $\operatorname{Rep}_k(H)$ is unimodular then
$$\begin{aligned}
|H^\circ| P_{\mathscr{C}}(\mathbf{1}) &\cong \mathscr{O}(G^\circ) \otimes \mathbf{F}\left(P(\mathbf{1})\right) \\
&\cong \mathscr{O}(G^\circ) \otimes \mathbf{F}\left(P(\mathbf{1})^*\right) \cong \mathscr{O}(G^\circ) \otimes \mathbf{F}\left(P(\mathbf{1})\right)^* \\
&\cong \mathscr{O}(G^\circ)^* \otimes \mathbf{F}\left(P(\mathbf{1})\right)^* \cong \left(\mathscr{O}(G^\circ) \otimes \mathbf{F}\left(P(\mathbf{1})\right)\right)^* \\
&\cong |H^\circ| P_{\mathscr{C}}(\mathbf{1})^*,
\end{aligned}$$
so $P_{\mathscr{C}}(\mathbf{1}) \cong P_{\mathscr{C}}(\mathbf{1})^*$. □

REMARK 6.3. Theorem 6.2(5) for étale groups follows from [**28**].

6.1. The étale case. Assume G is *étale*, i.e, $G = G(k)$. The following result is known in characteristic 0 [**14, 24**].

COROLLARY 6.4. *The following hold:*

(1) *For any (H, H)-double coset Z with representative $g \in G$, and simple $V \in \operatorname{Rep}_k(H^g, \xi_g^{-1})$, we have*
$$P_{\mathscr{C}}\left(\mathbf{F}_Z(V)\right) = \mathbf{F}_Z\left(P_{(H^g, \xi_g^{-1})}\left(V\right)\right) = \left(\mathscr{O}\left(Z\right) \otimes_k P_{(H^g, \xi_g^{-1})}\left(V\right), R_V^g\right).$$

(2) *We have a direct sum decomposition of abelian categories*
$$\mathscr{C} \cong \bigoplus_{Z \in Y} \operatorname{Rep}_k(H^g, \xi_g^{-1}),$$
i.e., $\overline{\mathscr{C}_Z} = \mathscr{C}_Z$ for every Z.

(3) *$\mathscr{C}$ is fusion if and only if p does not divide $|H|$.*

PROOF. Follows immediately from Theorem 6.2. □

6.2. The normal case. Assume H is normal in G. Then $G/H = H\backslash G$ is a finite group scheme, and the quotient morphism $\pi : G \to G/H$ (2.11) is a group scheme morphism.

Recall (2.17) the maps ρ_U^{ψ}, $U \in \mathrm{Coh}(G/H)$.

THEOREM 6.5. *The following hold:*

(1) *The injective tensor functor* $\pi^* : \mathrm{Coh}(G/H) \xrightarrow{1:1} \mathrm{Coh}(G)$ *lifts to an injective tensor functor*

$$\pi^* : \mathrm{Coh}(G/H) \xrightarrow{1:1} \mathscr{C}, \;\; U \mapsto \left(\pi^* U, \left(\rho_U^{\psi^{-1}} \otimes \mathrm{id}\right) \rho_U^{\psi}\right).$$

(2) *Set* $\mathbf{F} := \mathbf{F}_H$. *We have an equivalence of abelian categories*

$$\mathrm{Coh}(G/H) \boxtimes \mathrm{Rep}_k(H) \xrightarrow{\cong} \mathscr{C}, \;\; U \boxtimes V \mapsto \pi^* U \otimes \mathbf{F}(V).$$

PROOF.

(1) Follows from Theorem 2.11 (since H is normal).

(2) Since for any $\bar{g} \in (G/H)(k)$, the object $\pi^* \delta_{\bar{g}} \in \mathscr{C}_{gH}$ is invertible, we have an equivalence of abelian categories

$$\pi^* \delta_{\bar{g}} \otimes - : \overline{\mathscr{C}_H} \xrightarrow{\cong} \overline{\mathscr{C}_{gH}}, \;\; S \mapsto \pi^* \delta_{\bar{g}} \otimes S.$$

Now since we have an equivalence

$$\mathrm{Coh}(G/H)_{\bar{1}} \boxtimes \mathrm{Rep}_k(H) \xrightarrow{\cong} \overline{\mathscr{C}_H}, \;\; U \boxtimes V \mapsto \pi^* U \otimes \mathbf{F}(V),$$

it follows that for any $\bar{g} \in (G/H)(k)$, the functor

$$\mathrm{Coh}(G/H)_{\bar{g}} \boxtimes \mathrm{Rep}_k(H) \xrightarrow{\cong} \overline{\mathscr{C}_{gH}}, \;\; U \boxtimes V \mapsto \pi^* U \otimes \mathbf{F}(V),$$

is an equivalence of abelian categories, which implies the statement.[3] □

6.3. The connected case. Assume that $G = G^{\circ}$. Recall the group scheme embedding $\partial := \partial_1 : H \xrightarrow{1:1} H \times H$, $h \mapsto (h, h)$ (4.4). It is clear that $\partial^{\sharp} : \mathscr{O}(H)^{\otimes 2} \to \mathscr{O}(H)$ is the multiplication map of $\mathscr{O}(H)$.

Note that the map

$$\widetilde{\mathfrak{c}} : \mathscr{O}(H) \to \mathscr{O}(H \times H), \;\; f \mapsto 1 \otimes f,$$

is a cleaving map (2.23) with convolution inverse

$$\widetilde{\mathfrak{c}}^{-1} : \mathscr{O}(H) \to \mathscr{O}(H \times H), \;\; f \mapsto 1 \otimes \mathrm{S}(f).$$

Since the induced 2-cocycle $\widetilde{\sigma}$ is trivial, it follows that

$$\widetilde{\phi} : \mathscr{O}\left(H \backslash (H \times H)\right) \otimes \mathscr{O}(H) \xrightarrow{\cong} \mathscr{O}(H \times H), \;\; \mathrm{f} \otimes f \mapsto f'(1 \otimes f),$$

$$\widetilde{\phi}^{-1} : \mathscr{O}(H \times H) \xrightarrow{\cong} \mathscr{O}\left(H \backslash (H \times H)\right) \otimes \mathscr{O}(H), \;\; \mathrm{f} \mapsto \mathrm{f}_1 \widetilde{\mathfrak{c}}^{-1}(\partial^{\sharp}(\mathrm{f}_2)) \otimes \partial^{\sharp}(\mathrm{f}_3),$$

and $\widetilde{\alpha} : \mathscr{O}(H \times H) \twoheadrightarrow \mathscr{O}\left(H \backslash (H \times H)\right)$, $\mathrm{f} \mapsto \mathrm{f}_1 \widetilde{\mathfrak{c}}^{-1}(\partial^{\sharp}(\mathrm{f}_2))$ (see Section 2.7).

For any $V \in \mathrm{Rep}_k(H)$, let $\mathrm{f}_V := \widetilde{\alpha} \otimes \mathrm{id}$ (see Section 3.4). We have

$$\mathrm{f}_V : \mathscr{O}(H \times H) \otimes^{\mathscr{O}(H)} V \xrightarrow{\cong} \mathscr{O}(H \backslash (H \times H)) \otimes_k V,$$
$$\mathrm{f} \otimes v \mapsto \mathrm{f}_1 \widetilde{\mathfrak{c}}^{-1}(\partial^{\sharp}(\mathrm{f}_2)) \otimes v,$$

---

[3] See [**3**, Example 5.8] for a different proof.

and

$$\mathrm{f}_V^{-1} : \mathscr{O}(H\backslash(H\times H))\otimes_k V \xrightarrow{\cong} \mathscr{O}(H\times H)\otimes^{\mathscr{O}(H)} V,$$
$$\mathrm{f}\otimes v\mapsto \mathrm{f}\widetilde{\mathfrak{c}}(v^1)\otimes v^0.$$

Recall also (3.11) the map

$$\rho_V : \mathscr{O}(H\backslash(H\times H))\otimes_k V\to \mathscr{O}(H\backslash(H\times H))\otimes_k V\otimes \mathscr{O}(H\times H)_{\psi^{-1}\times\psi}.$$

LEMMA 6.6. *For any* $\mathrm{f}\otimes v\in\mathscr{O}(H\backslash(H\times H))\otimes_k V$, *we have*

$$\rho_V(\mathrm{f}\otimes v)=\mathrm{f}_1\left(\psi_1^{-1}\otimes \mathrm{S}(\psi_2^{-1}\partial^\sharp(\mathrm{f}_2))\right)\otimes v^0\otimes(\psi^{-2}\otimes v^1)\mathrm{f}_3.$$

PROOF. Set $\Psi:=\psi^{-1}\times\psi$. By (3.11), we have

$$\begin{aligned}
\rho_V(\mathrm{f}\otimes v) &= \Psi_1^1\mathrm{f}_1\widetilde{\mathfrak{c}}(v^1)_1\widetilde{\mathfrak{c}}^{-1}\partial^\sharp\left(\Psi_2^1\mathrm{f}_2\widetilde{\mathfrak{c}}(v^1)_2\right)\otimes v^0\otimes\Psi^2\mathrm{f}_3\widetilde{\mathfrak{c}}(v^1)_3\\
&= \Psi_1^1\mathrm{f}_1(1\otimes v_1^1)\left(1\otimes \mathrm{S}\partial^\sharp\left(\Psi_2^1\mathrm{f}_2(1\otimes v_2^1)\right)\right)\otimes v^0\otimes\Psi^2\mathrm{f}_3(1\otimes v_3^1)\\
&= \Psi_1^1\mathrm{f}_1\left(1\otimes v_1^1\mathrm{S}(v_2^1)\right)\left(1\otimes \mathrm{S}\partial^\sharp\left(\Psi_2^1\mathrm{f}_2\right)\right)\otimes v^0\otimes\Psi^2\mathrm{f}_3(1\otimes v_3^1)\\
&= \Psi_1^1\mathrm{f}_1\left(1\otimes \mathrm{S}\partial^\sharp\left(\Psi_2^1\mathrm{f}_2\right)\right)\otimes v^0\otimes\Psi^2\mathrm{f}_3(1\otimes v^1)\\
&= (\psi_1^{-1}\otimes\psi_1^1)\mathrm{f}_1\left(1\otimes \mathrm{S}\partial^\sharp\left((\psi_2^{-1}\otimes\psi_2^1)\mathrm{f}_2\right)\right)\otimes v^0\otimes(\psi^{-2}\otimes\psi^2)\mathrm{f}_3(1\otimes v^1)\\
&= (\psi_1^{-1}\otimes\psi_1^1)\mathrm{f}_1\left(1\otimes \mathrm{S}(\psi_2^{-1}\psi_2^1)\mathrm{S}\partial^\sharp(\mathrm{f}_2)\right)\otimes v^0\otimes(\psi^{-2}\otimes\psi^2)\mathrm{f}_3(1\otimes v^1)\\
&= \mathrm{f}_1\left(\psi_1^{-1}\otimes\psi_1^1\mathrm{S}(\psi_2^{-1}\psi_2^1)\mathrm{S}\partial^\sharp(\mathrm{f}_2)\right)\otimes v^0\otimes(\psi^{-2}\otimes\psi^2)\mathrm{f}_3(1\otimes v^1)\\
&= \mathrm{f}_1\left(\psi_1^{-1}\otimes \mathrm{S}(\psi_2^{-1})\mathrm{S}\partial^\sharp(\mathrm{f}_2)\right)\otimes v^0\otimes(\psi^{-2}\otimes 1)\mathrm{f}_3(1\otimes v^1)\\
&= \mathrm{f}_1\left(\psi_1^{-1}\otimes \mathrm{S}(\psi_2^{-1}\partial^\sharp(\mathrm{f}_2))\right)\otimes v^0\otimes(\psi^{-2}\otimes v^1)\mathrm{f}_3,
\end{aligned}$$

for any $\mathrm{f}\otimes v\in\mathscr{O}(H\backslash H\times H)\otimes V$, as claimed. □

Now recall the scheme isomorphism $\mathrm{j}:=\mathrm{j}_1$ (4.6). We have

$$\mathrm{j}^\sharp:\mathscr{O}(H)\xrightarrow{\cong}\mathscr{O}\left(H\backslash\left(H\times H\right)\right),\quad f\mapsto \mathrm{S}(f_1)\otimes f_2,$$

and

$$(\mathrm{j}^\sharp)^{-1}:\mathscr{O}\left(H\backslash\left(H\times H\right)\right)\xrightarrow{\cong}\mathscr{O}(H),\quad \mathrm{f}\mapsto(\varepsilon\bar{\otimes}\,\mathrm{id})(\mathrm{f}).$$

THEOREM 6.7. *The following hold:*

(1) *We have* $\mathscr{C}=\overline{\mathscr{C}_H}$.

(2) *We have an equivalence of tensor categories*

$$\mathbf{F}:=\mathbf{F}_H:\mathrm{Rep}_k(H)\xrightarrow{\cong}\mathscr{C}_H,\quad V\mapsto\iota_*\left(\mathscr{O}(H)\otimes_k V,\rho_V^1\right),$$

where for every $f\otimes v\in\mathscr{O}(H)\otimes V$,

$$\rho_V^1(f\otimes v)=\mathrm{S}(\psi^{-1})f_2\otimes v^0\otimes\psi^{-2}\mathrm{S}(f_1)\otimes v^1f_3.$$

(3) *For any simple* $V\in\mathrm{Rep}_k(H)$, *we have*

$$P_{\mathscr{C}}(\mathbf{F}(V))\cong\left(\mathscr{O}(G)\otimes_k P_H(V),R_V^1\right),$$

where for every $\mathrm{f}\otimes x\in\mathscr{O}(G)\otimes P_H(V)$,

$$R_V^1(\mathrm{f}\otimes x)=\mathrm{f}\otimes x^0\otimes 1\otimes x^1.$$

(4) *For any* $V\in\mathrm{Rep}_k(H)$, *we have*

$$\mathrm{FPdim}\left(P_{\mathscr{C}}\left(\mathbf{F}\left(V\right)\right)\right)=\frac{|G|}{|H|}\dim\left(P_H\left(V\right)\right).$$

PROOF. Follow from Theorem 6.2, except for the formulas of ρ_V^1 and R_V^1.

(1) By Lemma 6.6, for every $f\otimes v\in \mathscr{O}(H)\otimes_k V$, we have

$$\begin{aligned}
&\rho_V\left(\mathrm{j}_1^\sharp\otimes\mathrm{id}\right)(f\otimes v)\\
&=\rho_V\left(\mathrm{S}(f_1)\otimes f_2\otimes v\right)\\
&=(\mathrm{S}(f_1)\otimes f_2)_1\left(\psi_1^{-1}\otimes \mathrm{S}(\psi_2^{-1}\partial^\sharp\left((\mathrm{S}(f_1)\otimes f_2)_2\right))\right)\otimes v^0\otimes(\psi^{-2}\otimes v^1)\,(\mathrm{S}(f_1)\otimes f_2)_3\\
&=(\mathrm{S}(f_3)\otimes f_4)\left(\psi_1^{-1}\otimes \mathrm{S}(\psi_2^{-1}\partial^\sharp\left((\mathrm{S}(f_2)\otimes f_5)\right))\right)\otimes v^0\otimes(\psi^{-2}\otimes v^1)\,(\mathrm{S}(f_1)\otimes f_6)\\
&=(\mathrm{S}(f_3)\otimes f_4)\left(\psi_1^{-1}\otimes \mathrm{S}(\psi_2^{-1}\,\mathrm{S}(f_2)f_5)\right)\otimes v^0\otimes(\psi^{-2}\otimes v^1)\,(\mathrm{S}(f_1)\otimes f_6)\\
&=(\mathrm{S}(f_3)\otimes f_4)\left(\psi_1^{-1}\otimes \mathrm{S}(\psi_2^{-1})f_2\,\mathrm{S}(f_5)\right)\otimes v^0\otimes(\psi^{-2}\otimes v^1)\,(\mathrm{S}(f_1)\otimes f_6)\\
&=\left(\mathrm{S}(f_3)\psi_1^{-1}\otimes \mathrm{S}(\psi_2^{-1})f_2f_4\,\mathrm{S}(f_5)\right)\otimes v^0\otimes(\psi^{-2}\otimes v^1)\,(\mathrm{S}(f_1)\otimes f_6)\\
&=\left(\mathrm{S}(f_3)\psi_1^{-1}\otimes \mathrm{S}(\psi_2^{-1})f_2\varepsilon(f_4)\right)\otimes v^0\otimes(\psi^{-2}\otimes v^1)\,(\mathrm{S}(f_1)\otimes f_5)\\
&=\left(\mathrm{S}(f_3)\psi_1^{-1}\otimes \mathrm{S}(\psi_2^{-1})f_2\right)\otimes v^0\otimes(\psi^{-2}\otimes v^1)\,(\mathrm{S}(f_1)\otimes f_4)\\
&=\mathrm{S}(f_3)\psi_1^{-1}\otimes \mathrm{S}(\psi_2^{-1})f_2\otimes v^0\otimes\psi^{-2}\,\mathrm{S}(f_1)\otimes v^1f_4.
\end{aligned}$$

Thus, by Theorem 5.5, we have

$$\begin{aligned}
\rho_V^1(f\otimes v)&=\left(\left(\mathrm{j}_1^\sharp\right)^{-1}\otimes\mathrm{id}^{\otimes 3}\right)\rho_V\left(\mathrm{j}_1^\sharp\otimes\mathrm{id}\right)(f\otimes v)\\
&=\left(\left(\mathrm{j}_1^\sharp\right)^{-1}\otimes\mathrm{id}^{\otimes 3}\right)\left(\mathrm{S}(f_3)\psi_1^{-1}\otimes \mathrm{S}(\psi_2^{-1})f_2\otimes v^0\otimes\psi^{-2}\,\mathrm{S}(f_1)\otimes v^1f_4\right)\\
&=\left(\mathrm{j}_1^\sharp\right)^{-1}\left(\mathrm{S}(f_3)\psi_1^{-1}\otimes \mathrm{S}(\psi_2^{-1})f_2\right)\otimes v^0\otimes\psi^{-2}\,\mathrm{S}(f_1)\otimes v^1f_4\\
&=\varepsilon\left(\mathrm{S}(f_3)\psi_1^{-1}\right)\mathrm{S}(\psi_2^{-1})f_2\otimes v^0\otimes\psi^{-2}\,\mathrm{S}(f_1)\otimes v^1f_4\\
&=\varepsilon\left(\mathrm{S}(f_3)\right)\mathrm{S}(\psi^{-1})f_2\otimes v^0\otimes\psi^{-2}\,\mathrm{S}(f_1)\otimes v^1f_4\\
&=\mathrm{S}(\psi^{-1})f_2\otimes v^0\otimes\psi^{-2}\,\mathrm{S}(f_1)\otimes v^1f_3,
\end{aligned}$$

for every $f\otimes v\in\mathscr{O}(H)\otimes_k V$, as claimed.

(2) Set $\theta:=\theta_H$ and $\lambda:=\lambda_H$ (see Lemma 5.4). We have

$$\theta:\mathscr{O}(G)\otimes\mathscr{O}(H)\twoheadrightarrow\mathscr{O}(G),\ \ \mathrm{f}\otimes f\mapsto\varepsilon(f)\mathrm{f},$$

and

$$\lambda:\mathscr{O}(G)\xrightarrow{1:1}\mathscr{O}(G)\otimes\mathscr{O}(H),\ \ \mathrm{f}\mapsto\mathrm{f}\otimes 1.$$

Also, by (1), we have

$$\rho^1_{P_H(V)}(1\otimes x)=\mathrm{S}(\psi^{-1})\otimes x^0\otimes\psi^{-2}\otimes x^1.$$

Thus, by Theorem 5.5, we have

$$\begin{aligned}
R_V^1(\mathrm{f}\otimes x)&=\left(\theta\otimes\mathrm{id}^{\otimes 3}\right)\left(\mathrm{id}\otimes\rho^1_{P_H(V)}\right)(\lambda\otimes\mathrm{id})(\mathrm{f}\otimes x)\\
&=\left(\theta\otimes\mathrm{id}^{\otimes 3}\right)\left(\mathrm{id}\otimes\rho^1_{P_H(V)}\right)(\mathrm{f}\otimes 1\otimes x)\\
&=\left(\theta\otimes\mathrm{id}^{\otimes 3}\right)\left(\mathrm{f}\otimes\mathrm{S}(\psi^{-1})\otimes x^0\otimes\psi^{-2}\otimes x^1\right)\\
&=\mathrm{f}\varepsilon(\mathrm{S}(\psi^{-1}))\otimes x^0\otimes\psi^{-2}\otimes x^1\\
&=\mathrm{f}\otimes x^0\otimes 1\otimes x^1,
\end{aligned}$$

for any $\mathrm{f}\otimes x\in\mathscr{O}(G)\otimes P_H(V)$, as claimed. □

EXAMPLE 6.8. Let $\mathfrak{g}$ be a finite dimensional restricted p-Lie algebra, and let $\mathfrak{h} \subset \mathfrak{g}$ be a restricted p-Lie subalgebra. Let G, H be the associated finite group schemes (see Example 2.1). Consider the finite tensor category

$$\mathscr{C}(\mathfrak{g}, \mathfrak{h}) := \mathscr{C}(G, H, 1).$$

By Corollary 6.7, we have $\mathscr{C}(\mathfrak{g}, \mathfrak{h}) = \overline{\mathrm{Rep}(\mathfrak{h})}$.

For example, if $\mathfrak{h}$ is unipotent then $\mathscr{C} := \mathscr{C}(\mathfrak{g}, \mathfrak{h})$ is a unipotent tensor category, with unique simple object $\mathbf{1} := \mathbf{F}(k) = \mathscr{O}(H)$, and

$$P_{\mathscr{C}}(\mathbf{1}) \cong \mathscr{O}(G) \otimes_k P_H(k) \cong \mathscr{O}(G) \otimes_k \mathscr{O}(H)$$

is the free $\mathscr{O}(G)$-module of rank $|H|$, with trivial left $\mathscr{O}(H)$-cocation and right $\mathscr{O}(H)$-cocation $\mathrm{id} \otimes \Delta$. So, $\mathscr{C} \cong \mathrm{Coh}(G)$ as abelian categories, but not necessarily as tensor categories (see Theorem 7.3). □

7. Invertible bimodule categories over Coh(G)

Let $\mathscr{A}$ be a finite tensor category, and let $\mathscr{A}^{\mathrm{rev}}$ be the category $\mathscr{A}$ with reversed tensor product. Let $\mathscr{M}$ be a left $\mathscr{A}$-module category. Recall that the opposite category $\mathscr{M}^{\mathrm{op}}$ is a right $\mathscr{A}$-module category, with the $\mathscr{A}$-action given by

$$M \otimes^{\mathscr{M}^{\mathrm{op}}} A := A^* \otimes^{\mathscr{M}} M; \;\; A \in \mathscr{A},\, M \in \mathscr{M}.$$

Recall that an exact $\mathscr{A}$-bimodule category $\mathscr{B}$ is called *invertible* if there exists an $\mathscr{A}$-bimodule equivalence

$$\mathscr{B}^{\mathrm{op}} \boxtimes_{\mathscr{A}} \mathscr{B} \cong \mathscr{A},$$

or equivalently, if the functor

$$R_{\mathscr{B}} : \mathscr{A}^{\mathrm{rev}} \to \mathscr{A}^*_{\mathscr{B}} = \mathrm{Fun}_{\mathscr{A}}(\mathscr{B}, \mathscr{B}), \;\; A \mapsto - \otimes^{\mathscr{B}} A, \tag{7.1}$$

is an equivalence of tensor categories. Here, $\mathscr{A}^*_{\mathscr{B}}$ is with respect to the left $\mathscr{A}$-module structure on $\mathscr{B}$, while $- \otimes^{\mathscr{B}} A$ is with respect to the right $\mathscr{A}$-module structure on $\mathscr{B}$. Recall also that any invertible exact bimodule category over $\mathscr{A}$ is indecomposable both as left and right module category over $\mathscr{A}$.

Recall [**6**] that $\mathrm{BrPic}(\mathscr{A})$ denotes the group of invertible exact $\mathscr{A}$-bimodule categories, $\mathrm{Pic}(\mathscr{Z}(\mathscr{A}))$ denotes the group of one-sided exact $\mathscr{Z}(\mathscr{A})$-bimodule categories, and there exist group isomorphisms

$$\mathrm{Pic}(\mathscr{Z}(\mathscr{A})) \cong \mathrm{Aut}^{\mathrm{br}}(\mathscr{Z}(\mathscr{A})) \cong \mathrm{BrPic}(\mathscr{A}). \tag{7.2}$$

Recall also that for any $\mathscr{A}$-bimodule categories $\mathscr{B}_1$ and $\mathscr{B}_2$, there exists an equivalence of $\mathscr{A}$-bimodule categories

$$\mathscr{B}_1 \boxtimes_{\mathscr{A}} \mathscr{B}_2 \cong \mathrm{Fun}_{\mathscr{A}}(\mathscr{B}_1^{\mathrm{op}}, \mathscr{B}_2).$$

In the following we describe the group $\mathrm{BrPic}(\mathrm{Coh}(G))$, which we will denote by $\mathrm{BrPic}(G)$.

THEOREM 7.1. *There is a one to one correspondence between equivalence classes of pairs (H, ψ) and equivalence classes of indecomposable exact* Coh(G)*-bimodule categories, $(H, \psi) \mapsto \mathscr{B}(H, \psi)$, where $H \subset G$ is an abelian normal closed subgroup scheme and $\psi \in Z^2(H, \mathbb{G}_m)$.*

PROOF. Assume (H,ψ) is as stated, and consider the abelian category $\mathscr{B} := \mathrm{Coh}(G/H)$. Recall that (H,ψ) determines a structure of a left indecomposable exact $\mathrm{Coh}(G)$-module category on $\mathscr{B}$, which we denote by $\mathscr{M}(H,\psi)$ (see Section 2.9). Similarly, since $H \subset G^{\mathrm{op}}$ is a closed subgroup scheme (as H is abelian), (H,ψ) determines a structure of a left indecomposable exact $\mathrm{Coh}(G^{\mathrm{op}})$-module category on the category $\mathrm{Coh}(G^{\mathrm{op}}/H) = \mathrm{Coh}(H\backslash G) = \mathscr{B}$ (as H is normal). Thus, since $\mathrm{Coh}(G^{\mathrm{op}})^{\mathrm{rev}} = \mathrm{Coh}(G)$, (H,ψ) determines a structure of a right indecomposable exact $\mathrm{Coh}(G)$-module category on $\mathscr{B}$, which we denote by $\mathscr{N}(H,\psi)$ (see Section 2.9). Thus, (H,ψ) determines a canonical structure of an indecomposable exact $\mathrm{Coh}(G)$-bimodule category on $\mathscr{B}$, which we will denote by $\mathscr{B}(H,\psi)$.

Conversely, let $\mathscr{B}$ be an indecomposable exact $\mathrm{Coh}(G)$-bimodule category. Then by [**11**], $\mathscr{B} = \mathscr{M}(H,\psi)$ as *left* $\mathrm{Coh}(G)$-module categories, and $\mathscr{B} = \mathscr{N}(L,\eta)$ as *right* $\mathrm{Coh}(G)$-module categories. (Note that by Theorem 2.17, the pairs (H,ψ), (L,η) are unique since G is connected.) In particular, since $(\mathscr{O}(H),\Delta_\psi)$ is the unique simple object of $\mathscr{M}(H,\psi)$, and $(\mathscr{O}(L),\Delta_\eta)$ is the unique simple object of $\mathscr{N}(L,\eta)$ [**11**], it follows that $(H,\psi) = (L,\eta)$ (since G is connected). Thus, $\mathrm{Coh}(G/H) = \mathrm{Coh}(H\backslash G)$, so H is normal. Moreover, since $\mathscr{N}(H,\eta)$ is a left $\mathrm{Coh}(G^{\mathrm{op}})$-module category, it follows that $H \subset G^{\mathrm{op}}$ is a closed subgroup scheme, so H is abelian.

Finally, it is straightforward to verify that the above two assignments are inverse to each other. □

Recall that for a finite abelian group scheme H, H^D denotes the Cartier dual of H (i.e., $\mathscr{O}(H^D) = k[H]$).

DEFINITION 7.2. Let $\mathrm{IB}(G)$ denote the set of equivalence classes of pairs (H,ψ) such that

(1) $H \subset G$ is an abelian normal closed subgroup scheme,
(2) $\psi \in Z^2(H,\mathbb{G}_m)$ is a G-invariant nondegenerate twist, and
(3) $G^{\mathrm{op}} = H^D \rtimes K$, where $K := G/H$.

Here, (H_1,ψ_1) is equivalent to (H_2,ψ_2) if $H_1 = H_2 = H$ and $[\psi_1] = [\psi_2]$ in $H^2(H,\mathbb{G}_m)$ (see Definition 2.16, Remark 2.18). Also, H^D is viewed as an abelian normal closed subgroup scheme of G via the G-equivariant group scheme isomorphism $f : H^D \xrightarrow{\cong} H$ induced by $\psi\psi_{21}^{-1}$. □

THEOREM 7.3. *The assignment* $(H,\psi) \mapsto \mathscr{B}(H,\psi)$ *determines a one to one correspondence between* $\mathrm{IB}(G)$ *and* $\mathrm{BrPic}(G)$. *Moreover, if* $\mathscr{B}(H,\psi)$ *is invertible then* $\mathscr{B}(H,\psi)^{-1} = \mathscr{B}(H,\psi^{-1})$.

PROOF. Assume $(H,\psi) \in \mathrm{IB}(G)$. We have an embedding

$$f_* : \mathrm{Rep}_k(H) = \mathrm{Coh}(H^D) \xrightarrow{\cong} \mathrm{Coh}(H) \hookrightarrow \mathrm{Coh}(G^{\mathrm{op}}) \tag{7.3}$$

of tensor categories, whose tensor structure is given by ψ.

Let $\mathscr{B} := \mathscr{B}(H,\psi)$, and set $\mathscr{C} := \mathscr{C}(G,1,H,\psi)$. Recall that

$$\mathscr{C} \cong \mathrm{Fun}_{\mathrm{Coh}(G)}\left(\mathscr{M}\left(H,\psi\right),\mathscr{M}\left(H,\psi\right)\right)$$

as tensor categories. We have to show that the tensor functor

$$R := R_{\mathscr{B}} : \mathrm{Coh}(G^{\mathrm{op}}) \to \mathscr{C},\ \ X \mapsto -\otimes^{\mathscr{N}(H,\psi)} X$$

is an equivalence (see (7.1)).

Since $\mathrm{Coh}(G^{\mathrm{op}})$ and $\mathscr{C}$ have the same FP dimension, it suffices to show that R is surjective. To this end, observe that the restriction of R to $\mathrm{Rep}_k(H)$ (via (7.3)) coincides with $\mathbf{F}$, and the restriction of R to $\mathrm{Coh}(G/H)$ coincides with $\mathbf{I}$, so the claim follows from Corollary 6.5.

Conversely, assume that $\mathscr{B} \in \mathrm{BrPic}(G)$. Then by Theorem 7.1, $\mathscr{B} = \mathscr{B}(H, \psi)$, where $H \subset G$ is an abelian normal closed subgroup scheme and $\psi \in Z^2(H, \mathbb{G}_m)$. Set $\mathscr{C} := \mathscr{C}(G, 1, H, \psi)$, so

$$\mathscr{C} \cong \mathrm{Fun}_{\mathrm{Coh}(G)}\left(\mathscr{M}(H,\psi), \mathscr{M}(H,\psi)\right)$$

as tensor categories. By assumption (see (7.1)), we have a tensor equivalence

$$R := R_{\mathscr{B}} : \mathrm{Coh}(G^{\mathrm{op}}) \xrightarrow{\cong} \mathscr{C}, \quad X \mapsto - \otimes^{\mathscr{N}(H,\psi)} X. \tag{7.4}$$

Note that for any X in $\mathrm{Coh}(H) = \mathrm{Coh}(H^{\mathrm{op}})$, we have

$$R(X) \in \mathrm{Fun}_{\mathrm{Coh}(H)}\left(\mathscr{M}(H,\psi), \mathscr{M}(H,\psi)\right) = \mathscr{C}_H.$$

Since $\mathscr{C}_H$ has Frobenious-Perron dimension $|H|$, we see that R restricts to a tensor equivalence

$$R\iota_* : \mathrm{Coh}(H) \xrightarrow{\cong} \mathscr{C}_H, \quad X \mapsto - \otimes^{\mathscr{N}(H,\psi)} X.$$

Thus, by Corollary 6.7, we obtain a tensor equivalence

$$T : \mathrm{Coh}(H) \xrightarrow{R\iota_*} \mathscr{C}_H \xrightarrow{\mathbf{F}^{-1}} \mathrm{Rep}(H) = \mathrm{Coh}(H^D). \tag{7.5}$$

Moreover, since the tensor structure on T is given by ψ, T is induced from the skew-symmetric bicharacter $\psi\psi_{21}^{-1}$ on H. Hence, ψ is G-invariant and nondegenerate, i.e., it defines a G-equivariant group scheme isomorphism $H^D \xrightarrow{\cong} H$, which allows us to view H^D as an abelian normal closed subgroup scheme of G.

Finally, by (7.4) and Corollary 6.5, there is an injective tensor functor

$$R^{-1}\mathbf{I} : \mathrm{Coh}(G/H) \xrightarrow{1:1} \mathrm{Coh}(G^{\mathrm{op}}).$$

But since G is connected, this implies that $K := G/H$ is a closed subgroup scheme of G^{op}. Also, since

$$\left(R^{-1}\mathbf{I}\right)\left(\mathrm{Coh}(G/H)\right) \cap \left(R^{-1}\mathbf{F}\right)\left(\mathrm{Coh}(H^D)\right) = \mathrm{Vec},$$

it follows that $H^D \cap K = 1$. This implies that $G^{\mathrm{op}} = H^D \rtimes K$. Thus, $(H, \psi) \in \mathrm{IB}(G)$, as claimed. □

Let $\mathscr{B}(H_1, \psi_1)$ and $\mathscr{B}(H_2, \psi_2)$ be two invertible $\mathrm{Coh}(G)$-bimodule categories. Let $L := \Delta(H_1 \cap H_2)$, $A := (H_1 \times H_2)/L$, $\pi : H_1 \times H_2 \twoheadrightarrow A$ be the canonical surjective group scheme morphism, and $H := \pi(L^\perp)$ the image of the orthogonal complement of L under $\psi_1 \times \psi_2$. The latter means that $L^\perp \subset H_1 \times H_2$ is the largest closed subgroup scheme such that the Hopf algebra map

$$\mathscr{O}(L)^* \xrightarrow{1:1} \mathscr{O}(H_1 \times H_2)^* \xrightarrow{(\psi_1\times\psi_2)(\psi_1\times\psi_2)_{21}^{-1}} \mathscr{O}(H_1 \times H_2) \twoheadrightarrow \mathscr{O}(L^\perp)$$

is the trivial map $x \mapsto \varepsilon(x)1$.

Now let $j : L^\perp \xrightarrow{1:1} H_1 \times H_2$ be the inclusion morphism. We have

$$\tau := j^{\sharp\otimes 2}(\psi_1 \times \psi_2) \in Z^2\left(L^\perp / \left(L \cap L^\perp\right), \mathbb{G}_m\right). \tag{7.6}$$

Since π induces a group scheme isomorphism

$$\pi : L^\perp/(L \cap L^\perp) \xrightarrow{\cong} H, \tag{7.7}$$

it follows that

$$\psi := \left(\pi^{\sharp\otimes 2}\right)^{-1}(\tau) \in Z^2(H, \mathbb{G}_m) \tag{7.8}$$

is nondegenerate.

Assume from now on that G is *abelian.* Then $\mathscr{B}(H,\psi) = \mathscr{M}(H,\psi)$, and $\mathscr{M}(H,\psi)^{\mathrm{op}} = \mathscr{M}(H,\psi^{-1})$ for any pair (H,ψ).

Also, in this case, we have an injective group scheme morphism

$$\iota : A = (H_1 \times H_2)/L \xrightarrow{1:1} G, \; \overline{(h_1,h_2)} \mapsto h_1^{-1}h_2, \tag{7.9}$$

so we can view A (hence, H) as a closed subgroup scheme of G. Thus, it follows from the above and Theorem 7.1 that $\mathscr{M}(H,\psi)$ is an exact indecomposable $\mathrm{Coh}(G)$-bimodule category.

THEOREM 7.4. *Assume G is abelian. We have*

$$\mathscr{M}(H_1,\psi_1) \boxtimes_{\mathrm{Coh}(G)} \mathscr{M}(H_2,\psi_2) \cong \mathscr{M}(H,\psi)$$

as $\mathrm{Coh}(G)$*-module categories. In particular, $\mathscr{M}(H,\psi)$ is invertible.*

PROOF. Recall that there is a $\mathrm{Coh}(G)$-module equivalence

$$\mathscr{M}(H_1,\psi_1) \boxtimes_{\mathrm{Coh}(G)} \mathscr{M}(H_2,\psi_2) \cong \mathrm{Fun}_{\mathrm{Coh}(G)}(\mathscr{M}(H_1,\psi_1^{-1}), \mathscr{M}(H_2,\psi_2)).$$

Thus, by Theorem 5.3, there are $\mathrm{Coh}(G)$-module equivalences

$$\begin{aligned}\mathscr{M}(H_1,\psi_1) \boxtimes_{\mathrm{Coh}(G)} \mathscr{M}(H_2,\psi_2) &\cong \mathrm{Coh}^{\left((H_1,\psi_1^{-1}),(H_2,\psi_2)\right)}(G)\\ &\cong \mathrm{Coh}^{(H_1\times H_2,\psi_1\times\psi_2)}(G).\end{aligned}$$

Hence, our job is to show that $\mathrm{Coh}^{(H_1\times H_2,\psi_1\times\psi_2)}(G) \cong \mathrm{Coh}^{(H,\psi)}(G)$ as $\mathrm{Coh}(G)$-module categories. To this end, consider the functor

$$F : \mathrm{Coh}^{(H_1\times H_2,\psi_1\times\psi_2)}(G) \to \mathrm{Coh}^{(H,\psi)}(G), \; (S,\rho) \mapsto (T,\lambda),$$

where

$$T := S^{\left(L^{\perp}/(L\cap L^{\perp}),\tau\right)} \subset S, \quad \lambda := \left(\mathrm{id} \otimes \pi^{\sharp}\right)^{-1} \circ (\rho_{|T}).$$

It is clear that F is a functor of $\mathrm{Coh}(G)$-module categories.

Now by Theorem 5.3, the unique simple object of $\mathrm{Coh}^{(H_1\times H_2,\psi_1\times\psi_2)}(G)$ is $\delta := \mathscr{O}(A) \otimes_k V \cong \mathscr{O}(H_1 \times H_2) \otimes^{\mathscr{O}(L)_{\psi_1\psi_2}} V$, where V is the unique simple object of $\mathrm{Corep}_k(L, \psi_1^{-1}\psi_2^{-1})$. Since $F(\delta) = \delta' := \mathscr{O}(H)$ is the unique simple object of $\mathrm{Coh}^{(H,\psi)}(G)$, and every object of $\mathrm{Coh}^{(H,\psi)}(G)$ has the form $X \otimes \delta'$ for some $X \in \mathrm{Coh}(G)$ [**11**], it follows that F is an equivalence of $\mathrm{Coh}(G)$-module categories. □

8. The center of $\mathrm{Coh}(G)$

Fix a finite group scheme G. Let $\Delta\colon G \to G \times G$ be the diagonal morphism, $\mathbb{G} := G \times G$, and $\mathbb{H} := \Delta(G)$.

Let $\mathscr{Z}(G) := \mathscr{Z}(\mathrm{Coh}(G))$ be the center of $\mathrm{Coh}(G)$. Recall that objects of $\mathscr{Z}(G)$ are pairs (X,γ), where $X \in \mathrm{Coh}(G)$ and

$$\gamma : (- \otimes X) \xrightarrow{\cong} (X \otimes -)$$

is a natural isomorphism satisfying a certain property, usually known as a half-braiding (see, e.g., [**7**, Section 7.13]). The center $\mathscr{Z}(G)$ is a finite nondegenerate braided tensor category (see, e.g., [**7**, Section 8.6.3]).

Recall that there is a canonical equivalence of tensor categories

$$\mathscr{Z}(G) \xrightarrow{\cong} \left(\mathrm{Coh}(G) \boxtimes \mathrm{Coh}\,(G)^{\mathrm{rev}}\right)^*_{\mathrm{Coh}(G)} \tag{8.1}$$

assigning to a pair (X, γ) the functor $X \otimes -\colon \mathrm{Coh}(G) \to \mathrm{Coh}(G)$, equipped with a module structure coming from γ (see, e.g., [**7**, Proposition 7.13.8]).

Since there is a canonical equivalence of tensor categories

$$\mathrm{Coh}(\mathbb{G})^*_{\mathscr{M}(\mathbb{H},1)} \xrightarrow{\cong} \left(\mathrm{Coh}(G) \boxtimes \mathrm{Coh}\,(G)^{\mathrm{rev}}\right)^*_{\mathrm{Coh}(G)}, \tag{8.2}$$

it follows from (8.1)–(8.2) that there is a canonical equivalence of tensor categories

$$\mathscr{C}\,(\mathbb{G}, \mathbb{H}, 1) \cong \mathscr{Z}(G). \tag{8.3}$$

Finally, recall that there is a canonical tensor equivalence

$$\mathrm{Coh}^{(G)}(G) \cong \mathscr{Z}(G), \tag{8.4}$$

where $\mathrm{Coh}^{(G)}(G)$ is the category of G-equivariant sheaves on G with respect to right conjugation. Thus, (8.3)–(8.4) yield a canonical equivalence of tensor categories

$$\mathscr{C}\,(\mathbb{G}, \mathbb{H}, 1) \cong \mathrm{Coh}^{(G)}(G). \tag{8.5}$$

In this section, we study the tensor category $\mathscr{Z}(G)$ from two perspectives. First, we use Theorem 6.2, and the descriptions of $\mathscr{Z}(G)$ mentioned above, to study the abelian structure of $\mathscr{Z}(G)$. We then use (2.4) and [**15**] to describe $\mathscr{Z}(G)$ as a $G(k)$-equivariantization. Each of these descriptions provides a certain direct sum decomposition of $\mathscr{Z}(G)$, and we end the discussion by establishing a relation between the components coming from the two decompositions.

8.1. The structure of $\mathscr{C}\,(\mathbb{G}, \mathbb{H}, 1)$. Let $\mathrm{Y} := \mathbb{G}/(\mathbb{H} \times \mathbb{H})$ with respect to the right action $\mu_{\mathbb{G}\times(\mathbb{H}\times\mathbb{H})}$ (4.1).

THEOREM 8.1. *Let $\mathscr{C} := \mathscr{C}\,(\mathbb{G}, \mathbb{H}, 1)$. The following hold:*

(1) *For any $\mathrm{Z} \in \mathrm{Y}(k)$ with representative $\mathrm{g} \in \mathrm{Z}(k)$, we have an equivalence of abelian categories*

$$\mathbf{F}_{\mathrm{Z}} : \mathrm{Rep}_k(\mathbb{H}^{\mathrm{g}}) \xrightarrow{\cong} \mathscr{C}_{\mathrm{Z}}, \ \ V \mapsto \iota_{\mathrm{Z}*}\left(\mathscr{O}\,(\mathrm{Z}) \otimes_k V, \rho_V^{\mathrm{g}}\right).$$

In particular, we have a tensor equivalence

$$\mathbf{F}_{\mathbb{H}} : \mathrm{Rep}_k(\mathbb{H}) \xrightarrow{\cong} \mathscr{C}_{\mathbb{H}}, \ \ V \mapsto \iota_{\mathbb{H}*}\left(\mathscr{O}\,(\mathbb{H}) \otimes_k V, \rho_V^{1}\right).$$

(2) *There is a bijection between equivalence classes of pairs (Z, V), where $\mathrm{Z} \in \mathrm{Y}(k)$ is a closed point with representative $\mathrm{g} \in \mathrm{Z}(k)$, and $V \in \mathrm{Rep}_k(\mathbb{H})$ is simple, and simple objects of $\mathscr{C}$, assigning (Z, V) to $\mathbf{F}_{\mathrm{Z}}(V)$. Moreover, we have a direct sum decomposition of abelian categories*

$$\mathscr{C} = \bigoplus_{\mathrm{Z}\in\mathrm{Y}(k)} \overline{\mathscr{C}_{\mathrm{Z}}},$$

and $\overline{\mathscr{C}_{\mathbb{H}}} \subset \mathscr{C}$ is a tensor subcategory.

(3) *For any $V \in \mathrm{Rep}_k(\mathbb{H})$, we have $\mathbf{F}_{\mathrm{Z}}(V)^* \cong \mathbf{F}_{\mathrm{Z}^{-1}}(V^*)$.*

(4) *For any $\mathrm{Z} \in \mathrm{Y}(k)$ with representative $\mathrm{g} \in \mathrm{Z}(k)$, and V in $\mathrm{Rep}_k(\mathbb{H})$, we have*

$$\mathrm{FPdim}\,(\mathbf{F}_{\mathrm{Z}}(V)) = \frac{|\mathbb{H}|}{|\mathbb{H}^{\mathrm{g}}|}\,\mathrm{dim}(V).$$

(5) *For any* $\mathrm{Z} \in \mathrm{Y}(k)$ *with representative* $\mathrm{g} \in \mathrm{Z}(k)$*, and simple* $V \in \operatorname{Rep}_k(\mathbb{H})$*, we have*

$$P_{\mathscr{C}}\left(\mathbf{F}_{\mathrm{Z}}(V)\right) \cong \left(\mathscr{O}(\mathbb{G}^\circ) \otimes \mathscr{O}(\mathrm{Z}(k)) \otimes_k P_{\mathbb{H}^{\mathrm{g}}}(V), R_V^{\mathrm{g}}\right), \quad \textit{and}$$

$$\operatorname{FPdim}\left(P_{\mathscr{C}}\left(\mathbf{F}_{\mathrm{Z}}(V)\right)\right) = \frac{|\mathbb{G}^\circ||\mathbb{H}(k)|}{|\mathbb{H}^\circ||\mathbb{H}^{\mathrm{g}}(k)|} \dim\left(P_{\mathbb{H}^{\mathrm{g}}}(V)\right).$$

(6) *For any* $\mathrm{Z} \in \mathrm{Y}(k)$*, we have* $\operatorname{FPdim}\left(\overline{\mathscr{C}_{\mathrm{Z}}}\right) = |\mathbb{G}^\circ||\,\mathrm{Z}(k)|$.

PROOF. Follow immediately from Theorem 6.2. □

COROLLARY 8.2. *Equivalence classes of fiber functors on* $\mathscr{C}(\mathbb{G}, \mathbb{H}, 1)$ *are classified by equivalence classes of pairs* $(\mathbb{K}, \eta)$*, where* $\mathbb{K} \subset \mathbb{G}$ *is a closed subgroup scheme,* $\eta \in Z^2(\mathbb{K}, \mathbb{G}_m)$*,* $\mathbb{K}\mathbb{H} = \mathbb{G}$*, and* $\xi_1^{-1} \in Z^2(\mathbb{K} \cap \mathbb{H}, \mathbb{G}_m)$ *is nondegenerate.*

PROOF. Follows from Corollary 5.8. □

8.2. The structure of $\operatorname{Coh}^{(G)}(G)$**.** Let C be the finite scheme of conjugacy orbits in G. Then for any closed point $C \in \mathrm{C}(k)$, $C \subset G$ is closed and $C(k) \subset G(k)$ is a conjugacy class. Fix a representative $g = g_C \in C(k)$, and let G_C denote the centralizer of g in G (so $G_C(k)$ is the centralizer of g in $G(k)$).

Note that the map

$$\mathrm{C}(k) \to \mathrm{Y}(k), \ \ C_g \mapsto \mathrm{Z}_{(g,1)},$$

is bijective with inverse given by

$$\mathrm{Y}(k) \to \mathrm{C}(k), \ \ \mathrm{Z}_{(g_1,g_2)} \mapsto C_{g_1 g_2^{-1}}.$$

Also, for any $C \in \mathrm{C}(k)$ with representative $g \in C(k)$, we have

$$\mathbb{H}^{(g,1)} = \mathbb{H} \cap (g,1)\mathbb{H}(g^{-1},1) = \Delta(G_C).$$

For any $C \in \mathrm{C}(k)$ with representative $g \in C(k)$, let $\iota_g : G_C \hookrightarrow G$ be the inclusion morphism.

Now *choose* a cleaving map (2.23) $\widetilde{\mathfrak{c}_g} : \mathscr{O}(G_C) \xrightarrow{1:1} \mathscr{O}(G)$, and let

$$\widetilde{\alpha_g} : \mathscr{O}(G) \twoheadrightarrow \mathscr{O}(G_C \backslash G), \ \ f \mapsto f_1 \widetilde{\mathfrak{c}_g}^{-1}\left(\iota_g^\sharp(f_2)\right) \tag{8.6}$$

(see (2.26)). Consider also the split exact sequence of schemes

$$1 \to C^\circ \xrightarrow{i_{C^\circ}} C \underset{q_C}{\overset{\pi_C}{\rightleftarrows}} C(k) \to 1$$

induced from (2.3), and define the $\mathscr{O}(G)$-linear algebra maps

$$\theta_C := \operatorname{id} \otimes q_C^\sharp : \mathscr{O}(G^\circ) \otimes \mathscr{O}(C) \twoheadrightarrow \mathscr{O}(G^\circ) \otimes \mathscr{O}(C(k)), \ \ \text{and}$$

$$\lambda_C := \operatorname{id} \otimes \pi_C^\sharp : \mathscr{O}(G^\circ) \otimes \mathscr{O}(C(k)) \xrightarrow{1:1} \mathscr{O}(G^\circ) \otimes \mathscr{O}(C).$$

THEOREM 8.3. *Set* $\mathscr{Z} := \operatorname{Coh}^{(G)}(G)$*. The following hold:*

(1) *For any* $C \in \mathrm{C}(k)$ *with representative* $g \in C(k)$*, we have an equivalence of abelian categories*

$$\mathbf{F}_C : \operatorname{Rep}_k(G_C) \xrightarrow{\cong} \mathscr{Z}_C, \ \ V \mapsto \iota_{C*}\left(\mathscr{O}(C) \otimes_k V, \rho_V^g\right),$$

where $\rho_V^g : \mathscr{O}(C) \otimes_k V \to \mathscr{O}(C) \otimes_k V \otimes \mathscr{O}(G)$ *is given by*

$$\rho_V^g(f \otimes v) = \left(\mathfrak{j}_g^{-1}\right)^\sharp \widetilde{\alpha_g}\left(\mathfrak{j}_g^\sharp(f)_1 \widetilde{\mathfrak{c}_g}\left(v^1\right)_1\right) \otimes v^0 \otimes \mathfrak{j}_g^\sharp(f)_2 \widetilde{\mathfrak{c}_g}\left(v^1\right)_2.$$

(Here, $\mathfrak{j}_g : G_C \backslash G \xrightarrow{\cong} C$ *is the canonical scheme isomorphism.)*

In particular, $\mathbf{F}_1 : \mathrm{Rep}_k(G) \xrightarrow{\cong} \mathscr{Z}_1 \hookrightarrow \mathscr{Z}$ *coincides with the canonical embedding of braided tensor categories.*

(2) *There is a bijection between equivalence classes of pairs* (C,V)*, where* $C \in \mathrm{C}(k)$ *with representative* $g \in C(k)$*, and* $V \in \mathrm{Rep}_k(G_C)$ *is simple, and simple objects of* $\mathscr{Z}$*, assigning* (C,V) *to* $\mathbf{F}_C(V)$*. Moreover, we have a decomposition of abelian categories*

$$\mathscr{Z} = \bigoplus_{C\in \mathrm{C}(k)} \overline{\mathscr{Z}_C},$$

and $\overline{\mathscr{Z}_1} \subset \mathscr{Z}$ *is a tensor subcategory.*

(3) *For any* $V \in \mathrm{Rep}_k(G_C)$*, we have* $\mathbf{F}_C(V)^* \cong \mathbf{F}_{C^{-1}}(V^*)$*.*

(4) *For any* $V \in \mathrm{Rep}_k(G_C)$*, we have*

$$\mathrm{FPdim}\left(\mathbf{F}_C(V)\right) = \frac{|G|}{|G_C|}\dim(V) = |C|\dim(V).$$

(5) *For any simple* $V \in \mathrm{Rep}_k(G_C)$*, we have*

$$P_{\mathscr{Z}}\left(\mathbf{F}_C(V)\right) \cong \left(\mathscr{O}(G^\circ)\otimes\mathscr{O}(C(k))\otimes_k P_{G_C}(V), R_V^g\right),$$

where $\mathscr{O}(G)$ *acts diagonally, and*

$$R_V^g := \left(\theta_C\otimes \mathrm{id}^{\otimes 2}\right)\left(\mathrm{id}_{\mathscr{O}(G^\circ)}\otimes \rho^g_{P_{G_C}(V)}\right)\left(\lambda_C\otimes \mathrm{id}\right).$$

In particular,

$$\mathrm{FPdim}\left(P_{\mathscr{Z}}\left(\mathbf{F}_C(V)\right)\right) = \frac{|G|}{|G_C(k)|}\dim\left(P_{G_C}(V)\right).$$

(6) *For any* $C \in \mathrm{C}(k)$*, we have* $\mathrm{FPdim}\left(\overline{\mathscr{Z}_C}\right) = \frac{|G|^2}{|G_C(k)|}$*.* □

PROOF. Using the preceding remarks, it is straightforward to verify that Theorem 8.1 translates to the theorem via the equivalence (8.5). □

EXAMPLE 8.4. Assume G is connected (e.g., G is the finite group scheme associated to a finite dimensional restricted p-Lie algebra $\mathfrak{g}$ as in Example 2.1), and consider the finite braided tensor category $\mathscr{Z} := \mathrm{Coh}^{(G)}(G)$. By Theorem 8.3, we have $\mathscr{Z} = \overline{\mathrm{Rep}_k(G)}$, and for any $(V,r) \in \mathrm{Rep}_k(G)$, we have $\rho_V^1 = r : V \to V\otimes\mathscr{O}(G)$. Moreover, for any simple $V \in \mathrm{Rep}_k(G)$, we have

$$P_{\mathscr{Z}}\left(\mathbf{F}(V)\right) \cong \left(\mathscr{O}(G)\otimes_k P_G(V), R_V^1\right),$$

where $\mathscr{O}(G)$ acts on the first factor, and

$$R_V^1 := \left(\theta_1\otimes \mathrm{id}^{\otimes 2}\right)\left(\mathrm{id}_{\mathscr{O}(G)}\otimes \rho^1_{P_G(V)}\right)\left(\lambda_1\otimes \mathrm{id}\right).$$

Now since

$$\theta_1 : \mathscr{O}(G) \xrightarrow{1:1} \mathscr{O}(G)\otimes\mathscr{O}(G),\ \ f\mapsto f\otimes 1,$$
$$\lambda_1 : \mathscr{O}(G)\otimes\mathscr{O}(G) \twoheadrightarrow \mathscr{O}(G),\ \ f\otimes f'\mapsto f\varepsilon(f'),$$
$$\widetilde{\mathfrak{c}_1} = \mathrm{id},\ \text{ and }\ \widetilde{\alpha_1} = \varepsilon,$$

we see that $R_V^1 = \mathrm{id}_{\mathscr{O}(G)}\otimes \rho_{P_G(V)}$, so

$$P_{\mathscr{Z}}\left(\mathbf{F}(V)\right) \cong \left(\mathscr{O}(G)\otimes_k P_G(V), \mathrm{id}\otimes\rho_{P_G(V)}\right).$$

8.3. Short exact sequence of centers. Recall (2.3) the split exact sequence of group schemes

$$1 \to G^\circ \xrightarrow{i} G \underset{q}{\overset{\pi}{\rightleftarrows}} G(k) \to 1.$$

By Proposition 2.8, it induces the tensor functors

$$i_* : \mathrm{Coh}(G^\circ) \xrightarrow{1:1} \mathrm{Coh}(G), \;\; i^* : \mathrm{Coh}(G) \twoheadrightarrow \mathrm{Coh}(G^\circ),$$

$$\pi_* : \mathrm{Coh}(G) \twoheadrightarrow \mathrm{Coh}(G(k)), \;\; q_* : \mathrm{Coh}(G(k)) \xrightarrow{1:1} \mathrm{Coh}(G).$$

THEOREM 8.5. *The following hold:*

(1) *The functor* π_* *lifts (using* q*) to a surjective tensor functor*

$$\pi_* : \mathrm{Coh}^{(G)}(G) \twoheadrightarrow \mathrm{Coh}^{(G(k))}(G(k)).$$

(2) *The functor* q_* *lifts (using* π*) to an injective tensor functor*

$$q_* : \mathrm{Coh}^{(G(k))}(G(k)) \xrightarrow{1:1} \mathrm{Coh}^{(G)}(G).$$

(3) $\pi_* q_* = \mathrm{id}$ *(as abelian functors).*

(4) *The functor* i^* *lifts to a surjective tensor functor*

$$i^* : \mathrm{Coh}^{(G)}(G) \twoheadrightarrow \mathrm{Coh}^{(G^\circ)}(G^\circ).$$

(5) $i^* q_* : \mathrm{Coh}^{(G(k))}(G(k)) \to \mathrm{Vec} = \langle \mathbf{1} \rangle \subset \mathrm{Coh}^{(G^\circ)}(G^\circ)$ *is the forgetful functor (as abelian functors).*

(6) *The identity functor* $\mathrm{Coh}(G) \to \mathrm{Coh}(G)$ *lifts (using* i*) to a surjective tensor functor*

$$\mathrm{Coh}^{(G)}(G) \twoheadrightarrow \mathrm{Coh}^{(G^\circ)}(G).$$

PROOF.

(1) Take (V, r) in $\mathrm{Coh}^{(G)}(G)$. Namely, V is an $\mathscr{O}(G)$-module and

$$r : V \to V \otimes \mathscr{O}(G)$$

is an $\mathscr{O}(G)$-coaction with respect to right conjugation. By definition, $\pi_* V = V$ is an $\mathscr{O}(G(k))$-module via $\pi^\sharp : \mathscr{O}(G(k)) \xrightarrow{1:1} \mathscr{O}(G)$. Thus, the $\mathscr{O}(G(k))$-coaction

$$(\mathrm{id} \otimes q^\sharp) r : \pi_* V \to \pi_* V \otimes \mathscr{O}(G(k))$$

endows $\pi_* V$ with a structure of an object in $\mathrm{Coh}^{(G(k))}(G(k))$.

(2) Take (V, r) in $\mathrm{Coh}^{(G(k))}(G(k))$. Namely, V is an $\mathscr{O}(G(k))$-module and $r : V \to V \otimes \mathscr{O}(G(k))$ is an $\mathscr{O}(G(k))$-coaction with respect to right conjugation. Now by definition, $q_* V = V$ is an $\mathscr{O}(G)$-module via $q^\sharp : \mathscr{O}(G) \twoheadrightarrow \mathscr{O}(G(k))$. Thus, the $\mathscr{O}(G)$-coaction

$$(\mathrm{id} \otimes \pi^\sharp) r : q_* V \to q_* V \otimes \mathscr{O}(G)$$

endows $q_* V$ with a structure of an object in $\mathrm{Coh}^{(G)}(G)$.

(3) Follows from $\pi_* q_* = (\pi q)_* = \mathrm{id}$.

(4) Take (V, r) in $\mathrm{Coh}^{(G)}(G)$. By Proposition 2.8, the $\mathscr{O}(G^\circ)$-module $i^* V = \mathscr{O}(G^\circ) \otimes_{\mathscr{O}(G)} V$ is equipped with the $\mathscr{O}(G)$-coaction given by $\mu^\sharp_{G^\circ \times G} \bar{\otimes} r$ (where $\mu_{G^\circ \times G} : G^\circ \times G \to G^\circ$ is the right conjugation action of G on G°). Thus, the map

$$(\mathrm{id} \otimes \mathrm{id} \otimes i^\sharp) \left(\mu^\sharp_{G^\circ \times G} \bar{\otimes} r \right),$$

endows $i^* V$ with a structure of an object in $\mathrm{Coh}^{(G^\circ)}(G^\circ)$.

(5) Take (V,r) in $\mathrm{Coh}^{(G(k))}(G(k))$. By (2),

$$q_*(V,r) = (q_*V, (\mathrm{id}\otimes\pi^\sharp)r) \in \mathrm{Coh}^{(G)}(G),$$

where $q_*V = V$ is an $\mathscr{O}(G)$-module via $q^\sharp$. Thus by (4),

$$\begin{aligned} i^*q_*(V,r) &= i^*\left(q_*V, \left(\mathrm{id}\otimes\pi^\sharp\right)r\right) \\ &= \left(\mathscr{O}\left(G^\circ\right)\otimes_{\mathscr{O}(G)} q_*V, \left(\mathrm{id}\otimes\mathrm{id}\otimes i^\sharp\right)\left(\mu^\sharp_{G^\circ\times G}\bar{\otimes}\left(\mathrm{id}\otimes\pi^\sharp\right)r\right)\right) \\ &= \left(\mathscr{O}\left(G^\circ\right)\otimes_{\mathscr{O}(G)} q_*V, \mu^\sharp_{G^\circ\times G^\circ}\bar{\otimes}V_{\mathrm{tr}}\right) \end{aligned}$$

(where $\mu_{G^\circ\times G^\circ} : G^\circ\times G^\circ\to G^\circ$ is the right conjugation action of G° on itself). In other words, i^*q_* sends (V,r) to the direct sum of $\dim(V)$ copies of the identity object $\left(\mathscr{O}(G^\circ), \mu^\sharp_{G^\circ\times G^\circ}\right)\in\mathrm{Coh}^{(G^\circ)}(G^\circ)$.

(6) Similar. □

8.4. $\mathscr{Z}(G)$ as $G(k)$-equivariantization. Set $\mathscr{D} := \mathrm{Coh}(G)$, and let

$$\mathscr{D}^\circ := \mathrm{Coh}(G^\circ).$$

By (2.4), we have

$$\mathscr{D}\cong\mathscr{D}^\circ\rtimes G(k)\cong\bigoplus_{a\in G(k)}\mathscr{D}^\circ\boxtimes a$$

as tensor categories, where the associativity constraint on the right hand side category is given by ω in the obvious way.

Let $\mathscr{M}$ be any $\mathscr{D}^\circ$-bimodule category. Recall [**15**] that the relative center $\mathscr{Z}_{\mathscr{D}^\circ}(\mathscr{M})$ is the abelian category whose objects are pairs (M,c), where $M\in\mathscr{M}$ and

$$\mathrm{c} = \{\mathrm{c}_X : X\otimes M\xrightarrow{\cong} M\otimes X \mid X\in\mathscr{D}^\circ\} \tag{8.7}$$

is a natural family of isomorphisms satisfying some compatibility conditions. In particular, the relative center $\mathscr{Z}_{\mathscr{D}^\circ}(\mathscr{D})$ is a finite $G(k)$-crossed braided tensor category [**15**, Theorem 3.3]. The $G(k)$-grading on $\mathscr{Z}_{\mathscr{D}^\circ}(\mathscr{D})$ is given by

$$\mathscr{Z}_{\mathscr{D}^\circ}(\mathscr{D}) = \bigoplus_{a\in G(k)}\mathscr{Z}_{\mathscr{D}^\circ}(\mathscr{D}^\circ\boxtimes a),$$

and the action of $G(k)$ on $\mathscr{Z}_{\mathscr{D}^\circ}(\mathscr{D})$, $h\mapsto\widetilde{T}_h$, is induced from the action of $G(k)$ on $\mathscr{D}^\circ$, $h\mapsto T_h$, in the following way. For any $h\in G(k)$, $X\in\mathscr{D}^\circ$, and

$$(Y\boxtimes a,\mathrm{c})\in\mathscr{Z}_{\mathscr{D}^\circ}(\mathscr{D}^\circ\boxtimes a),$$

we have an isomorphism

$$\widetilde{\mathrm{c}}_X := (T_h\otimes T_h)\mathrm{c}_{T_{h^{-1}}(X)} : X\otimes T_h(Y)\xrightarrow{\cong} T_h(Y)\otimes T_{hah^{-1}}(X). \tag{8.8}$$

Set

$$\widetilde{T}_h(Y\boxtimes a,\mathrm{c}) := (T_h(Y)\boxtimes hah^{-1}, \widetilde{\mathrm{c}}).$$

Then $\widetilde{T}_h$ maps $\mathscr{Z}_{\mathscr{D}^\circ}(\mathscr{D}^\circ\boxtimes a)$ to $\mathscr{Z}_{\mathscr{D}^\circ}(\mathscr{D}^\circ\boxtimes hah^{-1})$.

Note that $\mathscr{Z}_{\mathscr{D}^\circ}(\mathscr{D})\cong\mathrm{Coh}^{(G^\circ)}(G)$ as tensor categories, and the obvious forgetful tensor functor

$$\mathscr{Z}(G) = \mathscr{Z}_{\mathscr{D}}(\mathscr{D})\to\mathscr{Z}_{\mathscr{D}^\circ}(\mathscr{D}),\ \ (X,\mathrm{c})\mapsto(X,\mathrm{c}_{|\mathscr{D}^\circ}) \tag{8.9}$$

coincides with the surjective tensor functor given in Theorem 8.5(6).

By [**15**, Theorem 3.5], there is an equivalence of tensor categories

$$F : \mathscr{Z}(G) \xrightarrow{\cong} \mathscr{Z}_{\mathscr{D}^\circ}(\mathscr{D})^{G(k)} = \left(\bigoplus_{a \in G(k)} \mathscr{Z}_{\mathscr{D}^\circ}(\mathscr{D}^\circ \boxtimes a) \right)^{G(k)}. \tag{8.10}$$

For any $C \in \mathrm{C}(k)$, set

$$\mathscr{E}_C := \bigoplus_{a \in C(k)} \mathscr{Z}_{\mathscr{D}^\circ}(\mathscr{D}^\circ \boxtimes a)^{G(k)} \subset \mathscr{Z}_{\mathscr{D}^\circ}(\mathscr{D})^{G(k)}.$$

By (8.10), $\mathscr{E}_C$ is a Serre subcategory of $\mathscr{Z}_{\mathscr{D}^\circ}(\mathscr{D})^{G(k)}$, and we have a tensor equivalence

$$F : \mathscr{Z}(G) \xrightarrow{\cong} \bigoplus_{C \in \mathrm{C}(k)} \mathscr{E}_C. \tag{8.11}$$

THEOREM 8.6. *For any $C \in \mathrm{C}(k)$, the functor F* (8.11) *restricts to an equivalence of abelian categories*

$$F_C : \overline{\mathscr{Z}(G)_C} \xrightarrow{\cong} \mathscr{E}_C = \bigoplus_{a \in C(k)} \mathscr{Z}_{\mathscr{D}^\circ}(\mathscr{D}^\circ \boxtimes a)^{G(k)}.$$

In particular, F restricts to an equivalence of tensor categories

$$F_1 : \overline{\mathrm{Rep}(G)} \xrightarrow{\cong} \mathscr{Z}(G^\circ)^{G(k)}.$$

PROOF. Fix $C \in \mathrm{C}(k)$ with representative $g \in C(k)$. To see that

$$F\left(\overline{\mathscr{Z}(G)_C}\right) \subset \mathscr{E}_C$$

it is enough to show that $F(\mathscr{Z}(G)_C) \subset \mathscr{E}_C$ (by the discussion above). To this end, it is enough to show that for any simple $\mathrm{V} \in \mathscr{Z}(G)_C$, the simple $G(k)$-equivariant object $F(\mathrm{V}) \in \mathscr{Z}_{\mathscr{D}^\circ}(\mathscr{D})$ is supported on C.

So, let $\mathrm{V} \in \mathscr{Z}(G)_C$ be simple. By Theorem 8.3, there exists a unique simple V in $\mathrm{Rep}(G_C)$ such that $\mathrm{V} = \mathbf{F}_C(V) = \mathscr{O}(C) \otimes_k V$. It follows that the forgetful image of $F(\mathrm{V})$ in $\mathscr{Z}_{\mathscr{D}^\circ}(\mathscr{D})$ (8.9) lies in $\bigoplus_{a \in C(k)} \mathscr{Z}_{\mathscr{D}^\circ}(\mathscr{D}^\circ \boxtimes a)$, so by the proof of [**15**, Theorem 3.5], $F(\mathrm{V})$ is supported on C, as claimed. □

References

[1] E. Abe, *Hopf algebras*, Cambridge Tracts in Mathematics, vol. 74, Cambridge University Press, Cambridge-New York, 1980. Translated from the Japanese by Hisae Kinoshita and Hiroko Tanaka. MR594432

[2] E. Aljadeff, P. Etingof, S. Gelaki, and D. Nikshych, *On twisting of finite-dimensional Hopf algebras*, J. Algebra **256** (2002), no. 2, 484–501, DOI 10.1016/S0021-8693(02)00092-3. MR1939116

[3] T. Basak and S. Gelaki, *Exact factorizations and extensions of finite tensor categories*, J. Algebra **631** (2023), 46–71, DOI 10.1016/j.jalgebra.2023.04.022. MR4589637

[4] P. Deligne, *Catégories tannakiennes* (French), The Grothendieck Festschrift, Vol. II, Progr. Math., vol. 87, Birkhäuser Boston, Boston, MA, 1990, pp. 111–195. MR1106898

[5] M. Demazure and P. Gabriel, *Groupes algébriques. Tome I: Géométrie algébrique, généralités, groupes commutatifs* (French), Masson & Cie, Éditeurs, Paris; North-Holland Publishing Co., Amsterdam, 1970. Avec un appendice *Corps de classes local* par Michiel Hazewinkel. MR302656

[6] A. Davydov and D. Nikshych, *The Picard crossed module of a braided tensor category*, Algebra Number Theory **7** (2013), no. 6, 1365–1403, DOI 10.2140/ant.2013.7.1365. MR3107567

[7] V. Drinfeld, S. Gelaki, D. Nikshych, and V. Ostrik, *On braided fusion categories. I*, Selecta Math. (N.S.) **16** (2010), no. 1, 1–119, DOI 10.1007/s00029-010-0017-z. MR2609644

[8] P. Etingof, R. Kinser, and C. Walton, *Tensor algebras in finite tensor categories*, Int. Math. Res. Not. IMRN **24** (2021), 18529–18572, DOI 10.1093/imrn/rnz332. MR4365996

[9] P. Etingof and V. Ostrik, *Finite tensor categories* (English, with English and Russian summaries), Mosc. Math. J. **4** (2004), no. 3, 627–654, 782–783, DOI 10.17323/1609-4514-2004-4-3-627-654. MR2119143

[10] P. Etingof, S. Gelaki, D. Nikshych, and V. Ostrik, *Tensor categories*, Mathematical Surveys and Monographs, vol. 205, American Mathematical Society, Providence, RI, 2015, DOI 10.1090/surv/205. MR3242743

[11] S. Gelaki, *Module categories over affine group schemes*, Quantum Topol. **6** (2015), no. 1, 1–37, DOI 10.4171/QT/59. MR3335005

[12] S. Gelaki, *Exact factorizations and extensions of fusion categories*, J. Algebra **480** (2017), 505–518, DOI 10.1016/j.jalgebra.2017.02.034. MR3633318

[13] S. Gelaki, *Minimal extensions of Tannakian categories in positive characteristic*, J. Algebra **603** (2022), 213–234, DOI 10.1016/j.jalgebra.2022.03.025. MR4412201

[14] S. Gelaki and D. Naidu, *Some properties of group-theoretical categories*, J. Algebra **322** (2009), no. 8, 2631–2641, DOI 10.1016/j.jalgebra.2009.05.047. MR2559854

[15] S. Gelaki, D. Naidu, and D. Nikshych, *Centers of graded fusion categories*, Algebra Number Theory **3** (2009), no. 8, 959–990, DOI 10.2140/ant.2009.3.959. MR2587410

[16] J. C. Jantzen, *Representations of algebraic groups*, 2nd ed., Mathematical Surveys and Monographs, vol. 107, American Mathematical Society, Providence, RI, 2003. MR2015057

[17] S. Majid, *Quantum double for quasi-Hopf algebras*, Lett. Math. Phys. **45** (1998), no. 1, 1–9, DOI 10.1023/A:1007450123281. MR1631648

[18] S. Montgomery, *Hopf Galois theory: a survey*, New topological contexts for Galois theory and algebraic geometry (BIRS 2008), Geom. Topol. Monogr., vol. 16, Geom. Topol. Publ., Coventry, 2009, pp. 367–400, DOI 10.2140/gtm.2009.16.367. MR2544394

[19] Y. Morales, M. Müller, J. Plavnik, A. Ros Camacho, A. Tabiri, and C. Walton, *Algebraic structures in group-theoretical fusion categories*, Algebr. Represent. Theory **26** (2023), no. 6, 2399–2431, DOI 10.1007/s10468-022-10186-7. MR4681321

[20] D. Mumford, *Abelian varieties*, Tata Institute of Fundamental Research Studies in Mathematics, vol. 5, Published for the Tata Institute of Fundamental Research, Bombay; by Oxford University Press, London, 1970. MR282985

[21] S. Natale, *Frobenius-Schur indicators for a class of fusion categories*, Pacific J. Math. **221** (2005), no. 2, 353–377, DOI 10.2140/pjm.2005.221.353. MR2196640

[22] S. Natale, *On the equivalence of module categories over a group-theoretical fusion category*, SIGMA Symmetry Integrability Geom. Methods Appl. **13** (2017), Paper No. 042, 9, DOI 10.3842/SIGMA.2017.042. MR3663587

[23] V. Ostrik, *Module categories, weak Hopf algebras and modular invariants*, Transform. Groups **8** (2003), no. 2, 177–206, DOI 10.1007/s00031-003-0515-6. MR1976459

[24] V. Ostrik, *Module categories over the Drinfeld double of a finite group*, Int. Math. Res. Not. **27** (2003), 1507–1520, DOI 10.1155/S1073792803205079. MR1976233

[25] H.-J. Schneider, *Normal basis and transitivity of crossed products for Hopf algebras*, J. Algebra **152** (1992), no. 2, 289–312, DOI 10.1016/0021-8693(92)90034-J. MR1194305

[26] H. Strade and R. Farnsteiner, *Modular Lie algebras and their representations*, Monographs and Textbooks in Pure and Applied Mathematics, vol. 116, Marcel Dekker, Inc., New York, 1988. MR929682

[27] W. C. Waterhouse, *Introduction to affine group schemes*, Graduate Texts in Mathematics, vol. 66, Springer-Verlag, New York-Berlin, 1979. MR547117

[28] H. Yadav, *On unimodular module categories*, Adv. Math. **432** (2023), Paper No. 109264, 41, DOI 10.1016/j.aim.2023.109264. MR4632574

Department of Mathematics, Iowa State University, Ames, Iowa 50100
Email address: gelaki@iastate.edu

Department of Mathematics, Iowa State University, Ames, Iowa 50100
Email address: sanmarco@uw.edu

Contemporary Mathematics
Volume **814**, 2025
https://doi.org/10.1090/conm/814/16303

Left coideal subalgebras of Nichols algebras

Istvan Heckenberger and Katharina Schäfer

ABSTRACT. We deal with the problem of the existence and the construction of not-primitively generated left coideal subalgebras of Nichols algebras of Yetter–Drinfeld modules. In our main result we look at finite-dimensional Yetter–Drinfeld modules over groups with at most one-dimensional homogeneous components, and classify those where all left coideal subalgebras in the category of $\mathbb{N}_0$-graded comodules of the associated Nichols algebra are generated in degree one. We present a strategy to extend left coideal subalgebras by adding a suitable generator in degree two, three or four to a smaller left coideal subalgebra. We also discuss some methods for the construction of left coideal subalgebras of a Nichols algebra in the category of $\mathbb{N}_0$-graded H-comodules, where H is a Hopf algebra, that is not necessarily a group algebra.

Introduction

Nichols algebras are important objects in the lifting method for the classification of pointed Hopf algebras introduced in [**AS98**]. Although Nichols algebras appeared much earlier in [**Nic78**], their appearance in the lifting method has increased interest in better understanding their structure. An overall problem is the classification of finite-dimensional Nichols algebras of Yetter-Drinfeld modules over groups. The theory of Weyl groupoids was introduced in [**Hec06**] for diagonal braidings and it enabled to classify finite-dimensional Nichols algebras of diagonal braidings [**Hec09**]. One-sided coideal subalgebras play a role in a generalised theory of Weyl groupoids which is described in [**AHS10**], [**HS10**] and [**HS13**] and which is powerful for the classification of Nichols algebras of direct sums of several irreducible Yetter–Drinfeld modules. In comparison, it is still little known about the general structure of (finite-dimensional) Nichols algebras of irreducible Yetter–Drinfeld modules over non-abelian groups and there is a great need for new approaches. In order to better understand this class of Nichols algebras, we take the approach of looking at nontrivial classes of left coideal subalgebras.

In a Nichols algebra $\mathcal{B}(V)$ of an irreducible Yetter–Drinfeld module V over a Hopf algebra H with bijective antipode the only nonzero left coideal subalgebras in the category of Yetter–Drinfeld modules over H are $\Bbbk 1$ and $\mathcal{B}(V)$ (see [**HS20**, Lemma 14.1.1]). However, typically there are more left coideal subalgebras of $\mathcal{B}(V)$ in the category of $\mathbb{N}_0$-graded H-comodules. Left coideal subalgebras of $\mathcal{B}(V)$ which are $\mathbb{N}_0$-graded H-comodules are in bijection with $\mathbb{N}_0$-graded left coideal subalgebras

This work was supported by the Deutsche Forschungsgemeinschaft under grant number HE 5375/9-1.

of the bosonization $\mathcal{B}(V)\#H$ intersecting H trivially (by [**HS20**, Lemma 12.4.4] and since the objects are $\mathbb{N}_0$-graded). One-sided coideal subalgebras of pointed braided Hopf algebras are studied in several recent papers focusing on different aspects (e.g., [**BLM16**], [**HS13**], [**Kha05**], [**Kha08**], [**ST16**], [**HS22**]).

We determine all Nichols algebras of finite-dimensional Yetter–Drinfeld modules over groups G under the restriction that each G-homogeneous component is at most one-dimensional such that all its left coideal subalgebras in the category of $\mathbb{N}_0$-graded $\Bbbk G$-comodules are generated in degree one as an algebra. In [**BLM16**], subalgebras of Fomin-Kirillov algebras are studied which are generated in degree one by $\mathbb{S}_n$-homogeneous elements. These subalgebras are by construction left coideal subalgebras in the category of $\mathbb{N}_0$-graded $\Bbbk\mathbb{S}_n$-comodules.

To decide whether a given Nichols algebra contains a left coideal subalgebra that is not generated in degree one we describe and study the braided vector space of the underlying Yetter–Drinfeld module in terms of racks and two-cocycles. It turns out that there are only six indecomposable racks of conjugacy classes and in each case only a few two-cocycles such that all such left coideal subalgebras of the corresponding Nichols algebra are generated in degree one (see Theorem 8.8). In Section 2 we discuss left coideal subalgebras of Nichols algebras over a Hopf algebra H, that is not necessarily a group algebra. There we give some methods for the construction of left coideal subalgebras of $\mathcal{B}(V)$ in the category of $\mathbb{N}_0$-graded H-comodules. In particular, Theorem 2.12 provides a useful condition which implies that all such left coideal subalgebras of $\mathcal{B}(V)$ are generated in degree one. In Section 3 we study the description of braided vector spaces coming from Yetter–Drinfeld modules over group algebras in terms of racks and two-cocycles. We provide arguments which show that it suffices to answer the question of the existence of not primitively generated left coideal subalgebras for indecomposable racks of conjugacy classes with two-cocycles and their Yetter–Drinfeld realization over the enveloping group of the rack. In Section 4 we present a strategy to extend left coideal subalgebras by adding a suitable degree 2, 3 or 4 generator to a smaller left coideal subalgebra. In Sections 5–7 we discuss racks of transpositions, the rack associated to the vertices of a tetrahedron and the rack associated to the faces of a cube in detail. We determine all two-cocycles on these racks for which all left coideal subalgebras of the corresponding Nichols algebra are generated in degree one. In Section 8 we prove that these (and the rack with one element) are the only indecomposable racks of conjugacy classes that admit two-cocycles such that all left coideal subalgebras are generated in degree one.

1. Preliminaries

We recall some definitions and facts about Yetter–Drinfeld modules and Nichols algebras. Our main reference is the book [**HS20**].

Let $\Bbbk$ be an algebraically closed field with $\operatorname{char}\Bbbk = 0$. Let H be a Hopf algebra over $\Bbbk$ with invertible antipode S. We use Sweedler notation for the comultiplication, i.e., writing $\Delta(h) = h_{(1)} \otimes h_{(2)}$ for all $h \in H$. For the left coaction of a left H-comodule V we use the notation ${}^H\delta(v) = v_{(-1)} \otimes v_{(0)}$ for all $v \in V$ and we abbreviate ${}^H\delta = \delta$ if no confusion is possible in the context. A **Yetter–Drinfeld module** over H is a left H-module V that is also a left H-comodule with structure map $\delta : V \to H \otimes V$ satisfying the compatibility condition

$$\delta(h \cdot v) = h_{(1)}v_{(-1)}S(h_{(3)}) \otimes h_{(2)} \cdot v_{(0)}$$

for all $h \in H$, $v \in V$. We denote the category of Yetter–Drinfeld modules over H by ${}^H_H\mathcal{YD}$. Morphisms in ${}^H_H\mathcal{YD}$ are morphisms of H-modules and H-comodules. The category of Yetter–Drinfeld modules over H is a braided monoidal category (see, e.g., [**HS20**, Theorem 3.4.13]). Let V, W be Yetter–Drinfeld modules over H. The monoidal structure is given by diagonal action and diagonal coaction on $V \otimes W$. The braiding

$$c_{V,W} : V \otimes W \to W \otimes V$$

is defined by $c_{V,W}(v \otimes w) = v_{(-1)} \cdot w \otimes v_{(0)}$ for all $v \in V$ and $w \in W$. In particular, V is a braided vector space with braiding $c_V = c_{V,V}$, i.e., c_V is a linear automorphism of $V^{\otimes 2}$ satisfying the so-called Yang-Baxter equation

$$(c_V \otimes \mathrm{id})(\mathrm{id} \otimes c_V)(c_V \otimes \mathrm{id}) = (\mathrm{id} \otimes c_V)(c_V \otimes \mathrm{id})(\mathrm{id} \otimes c_V).$$

The **Artin braid group** $\mathbb{B}_n$ is the group generated by elements $\sigma_1, \dots, \sigma_{n-1}$ with the **braid relations**

$$\sigma_i\sigma_{i+1}\sigma_i = \sigma_{i+1}\sigma_i\sigma_{i+1} \qquad \text{for all } 1 \le i \le n-2, \tag{1.1}$$

$$\sigma_i\sigma_j = \sigma_j\sigma_i \qquad \text{for all } 1 \le i, j \le n-1, |i-j| > 1. \tag{1.2}$$

Let $\{s_1, \dots, s_{n-1}\}$ be the standard generators $s_i = (i\; i+1)$ of the symmetric group $\mathbb{S}_n$. Then $\nu : \mathbb{S}_n \to \mathbb{B}_n$ defined by mapping an element $s = s_{i_1}s_{i_2}\cdots s_{i_m} \in \mathbb{S}_n$ onto $\sigma = \sigma_{i_1}\sigma_{i_2}\cdots\sigma_{i_m}$, where m is the length of the permutation s, is a well-defined map, the so-called Matsumoto section. Let V be a braided vector space with braiding c. For $1 \le i \le n-1$ define $c_i \in \mathrm{Aut}(V^{\otimes n})$ by applying c at the i-th position. Then

$$\tau : \mathbb{B}_n \to \mathrm{Aut}(V^{\otimes n}), \quad \sigma_i \mapsto c_i$$

is a group homomorphism since the automorphisms c_i satisfy the braid relations. For $n \ge 2$ and $1 \le i \le n-1$ the **braided symmetrizer** map $S_n : V^{\otimes n} \to V^{\otimes n}$ is defined by

$$S_n = \sum_{s \in \mathbb{S}_n} \tau(\nu(s)).$$

The **Nichols algebra** of a braided vector space (V, c) is

$$\mathcal{B}(V) = \bigoplus_{n \in \mathbb{N}_0} \mathcal{B}(V)(n) = \Bbbk \oplus V \oplus \bigoplus_{n \ge 2} T(V)(n)/\ker(S_n),$$

where $T(V)$ is the tensor algebra of (V, c). We denote the multiplication in $\mathcal{B}(V)$ by μ and the subalgebra (with one) generated by a subset $X \subseteq \mathcal{B}(V)$ by

$$\langle X \rangle.$$

If H is a Hopf algebra with bijective antipode and if V is a Yetter–Drinfeld module over H with Yetter–Drinfeld braiding c_V, then the Nichols algebra $\mathcal{B}(V)$ becomes a Hopf algebra in the category ${}^H_H\mathcal{YD}$. For details see, e.g., [**HS20**, Section 7.1]. But note that the Nichols algebra as an algebra and a coalgebra only depends on the braided vector space (V, c_V).

For an $\mathbb{N}_0$-graded coalgebra $C = \bigoplus_{n \in \mathbb{N}_0} C(n)$ with comultiplication Δ we define

$$\Delta_{1^n} = \pi_1^{\otimes n}\Delta^{n-1} : C(n) \to C(1)^{\otimes n}$$

for all $n \in \mathbb{N}$, where π_1 is the projection onto degree 1. As endomorphisms of $V^{\otimes n}$, where V is a braided vector space, the braided symmetrizer map S_n equals the map Δ_{1^n}, where Δ is the comultiplication of the tensor algebra $T(V)$ (see

[**HS20**, Corollary 6.4.9]). We define the components of the comultiplication of an $\mathbb{N}_0$-graded coalgebra C as

$$\Delta_{i,n-i} = (\pi_i \otimes \pi_{n-i})\Delta : C(n) \to C(i) \otimes C(n-i), \tag{1.3}$$

where $n \in \mathbb{N}_0$, $0 \le i \le n$ and π_i is the projection onto degree i. We will mostly use the components $\Delta_{1,n-1}$ and $\Delta_{n-1,1}$. Proposition 2.7 and Proposition 2.10 show in which way they are useful to construct left coideal subalgebras. The following formulas are very practical to compute them in the tensor algebra $T(V)$ over a braided vector space (V, c).

$$\begin{aligned}\Delta^{T(V)}_{1,n-1} &: T(V)(n) = V^{\otimes n} \to V \otimes T(V)(n-1) = V^{\otimes n}\\ &\quad x \mapsto (1 + c_1 + c_1c_2 + \cdots + c_1c_2\cdots c_{n-1})(x),\end{aligned} \tag{1.4}$$

$$\begin{aligned}\Delta^{T(V)}_{n-1,1} &: T(V)(n) = V^{\otimes n} \to T(V)(n-1) \otimes V = V^{\otimes n}\\ &\quad x \mapsto (1 + c_{n-1} + c_{n-1}c_{n-2} + \cdots + c_{n-1}c_{n-2}\cdots{}_1)(x).\end{aligned} \tag{1.5}$$

Information about how to derive them can be found in [**HS20**, Sections 1.7–1.9] and, in particular, in [**HS20**, Theorem 1.9.1].

For the kernel of the counit ε of a coalgebra C we use the symbol $C^+ = \ker(\varepsilon|C)$. If C is a connected $\mathbb{N}_0$-graded coalgebra, then $C^+ = \bigoplus_{i\ge 1} C(i)$.

2. Introduction to left coideal subalgebras

In this section, let H be a Hopf algebra with bijective antipode S and let $V \in {}^H_H\mathcal{YD}$.

DEFINITION 2.1. A **left coideal** of $\mathcal{B}(V)$ is a subspace $C \subseteq \mathcal{B}(V)$ such that $\Delta(C) \subseteq \mathcal{B}(V) \otimes C$. A **left coideal subalgebra of $\mathcal{B}(V)$ in the category of $\mathbb{N}_0$-graded (left) H-comodules** is an $\mathbb{N}_0$-graded left H-subcomodule of $\mathcal{B}(V)$ which is a subalgebra and a left coideal.[1]

Recall from the previous section the notation $\langle\cdot\rangle$ for subalgebras of a Nichols algebra generated by subsets.

DEFINITION 2.2. Let C be a left coideal subalgebra of $\mathcal{B}(V)$ in the category of $\mathbb{N}_0$-graded H-comodules. We say that C is **generated in degree one** if $C = \langle C(1)\rangle$. We say that C can be extended in degree n if there is an element $x \in \mathcal{B}(V)(n)$ such that $x \notin C$ and $\langle C, x\rangle$ is a left coideal subalgebra in the category of $\mathbb{N}_0$-graded H-comodules. In this case we call x an **extension** of C.

To prove that a given subspace is a left coideal subalgebra, we will often use the following tools.

LEMMA 2.3. *Let C be a nonzero left coideal of $\mathcal{B}(V)$. If $C \cap V = 0$, then $C = \Bbbk 1$.*

PROOF. A proof in a much more general setting can be found in [**HS20**, Corollary 1.3.11]. □

PROPOSITION 2.4. *Let W be an $\mathbb{N}_0$-graded H-subcomodule of $\mathcal{B}(V)$. If W is a left coideal, then the subalgebra $\langle W\rangle \subseteq \mathcal{B}(V)$ is a left coideal subalgebra in the category of $\mathbb{N}_0$-graded H-comodules.*

[1] Note that each left coideal of $\mathcal{B}(V)$ contains 1 since $\mathcal{B}(V)$ is connected.

PROOF. The subalgebra generated by the $\mathbb{N}_0$-graded subspace W is obviously $\mathbb{N}_0$-graded by construction. Since W is an H-subcomodule by assumption and since multiplication in $\mathcal{B}(V)$ is an H-colinear map, the subalgebra $\langle W\rangle \subseteq \mathcal{B}(V)$ is an H-subcomodule. Indeed, for all $v_1, \dots, v_n \in W$,

$$\delta(v_1 \cdots v_n) = v_{1(-1)} \cdots v_{n(-1)} \otimes v_{1(0)} \cdots v_{n(0)} \in H \otimes \langle W\rangle$$

since $\delta(v_i) = v_{i(-1)} \otimes v_{i(0)} \in H \otimes W$ for all $i \in \{1, \dots, n\}$. Since W is an H-subcomodule,

$$c(w \otimes b) = w_{(-1)} \cdot b \otimes w_{(0)} \in \mathcal{B}(V) \otimes W$$

for all $w \in W$, $b \in \mathcal{B}(V)$ and hence $c(W \otimes \mathcal{B}(V)) \subseteq \mathcal{B}(V) \otimes W$, where c is the Yetter–Drinfeld braiding of $\mathcal{B}(V)$. Since W is a left coideal, we know that $\Delta(W) \subseteq \mathcal{B}(V) \otimes W$. It follows that

$$\begin{aligned}\Delta(WW) = \Delta(W)\Delta(W) &\subseteq (\mathcal{B}(V) \otimes W)(\mathcal{B}(V) \otimes W)\\ &\subseteq \mathcal{B}(V)c(W \otimes \mathcal{B}(V))W \subseteq \mathcal{B}(V) \otimes \langle W\rangle\end{aligned}$$

since Δ is a morphism of algebras in ${}^H_H\mathcal{YD}$. Hence $\Delta(\langle W\rangle) \subseteq \mathcal{B}(V) \otimes \langle W\rangle$. □

COROLLARY 2.5. *Let W be an H-subcomodule of V. Then the subalgebra $\langle W\rangle \subseteq \mathcal{B}(V)$ is a left coideal subalgebra in the category of $\mathbb{N}_0$-graded H-comodules.*

PROOF. Since all elements of V are primitive, we have

$$\Delta(w) = 1 \otimes w + w \otimes 1 \in \mathcal{B}(V) \otimes (W + \Bbbk 1)$$

for all $w \in W$. Hence $W + \Bbbk 1$ is a left coideal in the category of $\mathbb{N}_0$-graded H-comodules and the claim follows from Proposition 2.4. □

LEMMA 2.6. *Let $B = \bigoplus_{n\in\mathbb{N}_0} B(n)$ be an $\mathbb{N}_0$-graded coalgebra. Then*

$$(\Delta_{1^i} \otimes \mathrm{id})\Delta_{i,n-i} = (\mathrm{id}^{\otimes i-1} \otimes \Delta_{1,n-i}) \cdots (\mathrm{id} \otimes \Delta_{1,n-2})\Delta_{1,n-1}$$

as maps from $B(n)$ to $B(1)^{\otimes i} \otimes B(n-i)$ for all $n \geq 2$, $1 < i \leq n$.

PROOF. First note that

$$\pi_1^{\otimes i}\Delta^{i-1}|B(j) = \begin{cases}\Delta_{1^i} = \pi_1^{\otimes i}\Delta^{i-1} & \text{if } i = j\\ 0 & \text{if } i \neq j,\end{cases} \tag{2.1}$$

for all $i \geq 1$ and $j \geq 0$ since Δ is graded. Further,

$$(\pi_1 \otimes \pi_i)\Delta|B(j) = \begin{cases}\Delta_{1,i} & \text{if } 1+i = j\\ 0 & \text{if } 1+i \neq j,\end{cases} \tag{2.2}$$

for all $i, j \geq 0$. Using Equations (2.1) and (2.2) and the coassociativity of Δ it follows that both sides of the equation in the claim are equal to $(\pi_1^{\otimes i} \otimes \pi_{n-i})\Delta^i$. □

PROPOSITION 2.7. *Let C be an $\mathbb{N}_0$-graded subspace of $\mathcal{B}(V)$. Then C is a left coideal of $\mathcal{B}(V)$ if and only if $\Delta_{1,n-1}(C(n)) \subseteq V \otimes C(n-1)$ for all $n \in \mathbb{N}$.*

PROOF. Since $\mathcal{B}(V)$ is an $\mathbb{N}_0$-graded coalgebra, the subspace C is a left coideal if and only if

$$\Delta_{i,n-i}(C(n)) \subseteq \mathcal{B}(V)(i) \otimes C(n-i)$$

for all $n \in \mathbb{N}$ and $0 \leq i \leq n$. In particular, if C is a left coideal, it follows that $\Delta_{1,n-1}(C(n)) \subseteq V \otimes C(n-1)$ for all $n \in \mathbb{N}$. For the other implication let $n \in \mathbb{N}$. We prove $\Delta_{i,n-i}(C(n)) \subseteq \mathcal{B}(V)(i) \otimes C(n-i)$ for all $0 \leq i \leq n$. For $i = 0$ we have $\Delta_{0,n}(x) = 1 \otimes x$ for all $x \in C(n)$ since $\mathcal{B}(V)$ is a connected $\mathbb{N}_0$-graded coalgebra.

For $i = 1$ we have $\Delta_{1,n-1}(C(n)) \subseteq V \otimes C(n-1)$ by assumption. Let $1 < i \leq n$. Since $\mathcal{B}(V)$ is a strictly $\mathbb{N}_0$-graded coalgebra, the morphism Δ_{1^i} is injective (see, e.g., [**HS20**, Proposition 1.3.14]). Thus $\Delta_{i,n-i}(C(n)) \subseteq \mathcal{B}(V)(i) \otimes C(n-i)$ holds if and only if

$$(\Delta_{1^i} \otimes \mathrm{id})\Delta_{i,n-i}(C(n)) \subseteq V^{\otimes i} \otimes C(n-i).$$

Since $\Delta_{1,n-j}(C(n-j+1)) \subseteq V \otimes C(n-j)$ for all $1 \leq j \leq i$ by assumption and

$$(\Delta_{1^i} \otimes \mathrm{id})\Delta_{i,n-i} = (\mathrm{id}^{\otimes i-1} \otimes \Delta_{1,n-i}) \cdots (\mathrm{id} \otimes \Delta_{1,n-2})\Delta_{1,n-1}$$

by Lemma 2.6, the claim follows inductively. □

COROLLARY 2.8. *Let W be an H-subcomodule of V. Let $n \in \mathbb{N}$ and $x \in \mathcal{B}(V)(n)$ such that $\delta(x) \in H \otimes x$ and $\Delta_{1,n-1}(x) \in V \otimes \langle W \rangle(n-1)$. Then $\langle W + \Bbbk x \rangle$ is a left coideal subalgebra in the category of $\mathbb{N}_0$-graded H-comodules.*

PROOF. The subalgebra $\langle W \rangle$ is a coideal subalgebra in the category of $\mathbb{N}_0$-graded H-comodules by Corollary 2.5. Then $\langle W \rangle + \Bbbk x$ is a left coideal by Proposition 2.7 and $\langle W + \Bbbk x \rangle$ is a left coideal subalgebra in the category of $\mathbb{N}_0$-graded H-comodules by Proposition 2.4. □

LEMMA 2.9. *Let W be an H-subcomodule of V. Let $n \in \mathbb{N}$ and $x \in \mathcal{B}(V)(n)$ such that $\delta(x) \in H \otimes x$ and $\Delta_{n-1,1}(x) \in \mathcal{B}(V) \otimes W$. Let $L \subseteq \mathcal{B}(V)$ be the smallest left coideal in the category of $\mathbb{N}_0$-graded H-comodules containing x. If $x \notin \langle W \rangle$, then $\langle L \rangle$ is a left coideal subalgebra of $\mathcal{B}(V)$ in the category of $\mathbb{N}_0$-graded H-comodules that is not generated in degree one.*

PROOF. By Proposition 2.4, the subalgebra $\langle L \rangle$ is a left coideal subalgebra of $\mathcal{B}(V)$ in the category of $\mathbb{N}_0$-graded H-comodules. Moreover, the subalgebra $\langle L \rangle$ is not generated in degree one since $(L \cap \mathcal{B}(V))(1) \subseteq W$ and $x \notin \langle W \rangle$. □

PROPOSITION 2.10. *Let W be an H-subcomodule of V, and let $\pi : V \to V/W$ be the canonical map.*

(1) *The subspace* $\ker(\mathrm{id} \otimes \pi\pi_1)\Delta \subseteq \mathcal{B}(V)$ *is the largest $\mathbb{N}_0$-graded left coideal of $\mathcal{B}(V)$ with degree one part equal to W. It is an $\mathbb{N}_0$-graded left coideal subalgebra of $\mathcal{B}(V)$ in the category of H-comodules.*
(2) *Let C be an $\mathbb{N}_0$-graded left coideal of $\mathcal{B}(V)$ which is a subalgebra of $\mathcal{B}(V)$. If $C(1) = W$, then $\langle W \rangle \subseteq C \subseteq \ker(\mathrm{id} \otimes \pi\pi_1)\Delta$.*

PROOF.

(1) Let $\overline{W} = \ker(\mathrm{id} \otimes \pi\pi_1)\Delta \subseteq \mathcal{B}(V)$. Since $\mathcal{B}(V)$ is connected, each left coideal of $\mathcal{B}(V)$ contains 1. Moreover, each left coideal with degree one part equal to W is contained in $\overline{W}$. It is clear that $\overline{W}$ is $\mathbb{N}_0$-graded, an H-subcomodule of $\mathcal{B}(V)$, and the degree one part of $\overline{W}$ is $\ker \pi\pi_1 = W$. The coassociativity of Δ implies that $\overline{W}$ is a left coideal of $\mathcal{B}(V)$. Finally, $\overline{W}$ is a subalgebra of $\mathcal{B}(V)$: For each $x, y \in \overline{W}$ we obtain that

$$\begin{aligned}(\mathrm{id} \otimes \pi\pi_1)\Delta(xy) &= (\mathrm{id} \otimes \pi\pi_1)\big(\Delta(x)\Delta(y)\big) \\ &= (x \otimes 1)(\mathrm{id} \otimes \pi\pi_1)\Delta(y) + \big((\mathrm{id} \otimes \pi\pi_1)\Delta(x)\big)(y \otimes 1).\end{aligned}$$

The last equation uses that W is an H-subcomodule of V and that therefore $(\mathrm{id} \otimes \pi\pi_1)\big(\Delta(x)(y \otimes 1)\big) = \big((\mathrm{id} \otimes \pi\pi_1)\Delta(x)\big)(y \otimes 1)$.

(2) Follows directly from (1). □

Recall that S denotes the antipode of H. Assume that V is a finite-dimensional object in ${}^H_H\mathcal{YD}$. Then the dual vector space $V^* = \mathrm{Hom}(V, \Bbbk)$ is an object in ${}^H_H\mathcal{YD}$, where the H-module and the H-comodule structure are defined by

$$(h \cdot f)(v) = f(S(h) \cdot v), \tag{2.3}$$

$$f_{(-1)} f_{(0)}(v) = S^{-1}(v_{(-1)}) f(v_{(0)}) \tag{2.4}$$

for all $h \in H$, $f \in V^*$, $v \in V$ (see, e.g., [**HS20**, Lemm 4.2.2]). By [**HS20**, Corollary 7.2.8], there is a unique nondegenerate bilinear form $\langle \cdot, \cdot \rangle : \mathcal{B}(V^*) \otimes \mathcal{B}(V) \to \Bbbk$ extending the evaluation map $V^* \otimes V \to \Bbbk$ satisfying

$$\langle 1, 1 \rangle = 1, \tag{2.5}$$

$$\langle \mathcal{B}(V^*)(n), \mathcal{B}(V)(m) \rangle = 0 \quad \text{for all } n \neq m, \tag{2.6}$$

and for all $f, g \in \mathcal{B}(V^*)$ and $v, w \in \mathcal{B}(V)$

$$\langle fg, v \rangle = \left\langle f, v^{(2)} \right\rangle \left\langle g, v^{(1)} \right\rangle, \tag{2.7}$$

$$\langle f, vw \rangle = \left\langle f^{(2)}, v \right\rangle \left\langle f^{(1)}, w \right\rangle, \tag{2.8}$$

and for all $h \in H$, $v \in \mathcal{B}(V)$, $f \in \mathcal{B}(V^*)$

$$\langle h \cdot f, v \rangle = \langle f, S(h) \cdot v \rangle, \tag{2.9}$$

$$f_{(-1)} \langle f_{(0)}, v \rangle = S^{-1}(v_{(-1)}) \langle f, v_{(0)} \rangle. \tag{2.10}$$

Here we use Sweedler notation for the coproduct in $\mathcal{B}(V)$ ($\mathcal{B}(V^*)$, resp.) in the form $\Delta(v) = v^{(1)} \otimes v^{(2)}$ for all $v \in \mathcal{B}(V)$ ($\Delta(f) = f^{(1)} \otimes f^{(2)}$ for all $f \in \mathcal{B}(V^*)$, resp.). Note that Equations (2.7)–(2.10) are satisfied if and only if the bilinear form $\langle \cdot, \cdot \rangle$ is a Hopf pairing in ${}^H_H\mathcal{YD}$, where the H-action and the H-coaction are defined to be diagonal on $\mathcal{B}(V^*) \otimes \mathcal{B}(V)$ and trivial on $\Bbbk$ (see, e.g., [**HS20**, Lemma 4.2.]). For a subset $X \subseteq \mathcal{B}(V)$ we denote its orthogonal complement with respect to the bilinear form $\langle \cdot, \cdot \rangle$ by

$$X^\perp = \{ f \in \mathcal{B}(V^*) \mid \langle f, x \rangle = 0 \text{ for all } x \in X \}.$$

PROPOSITION 2.11. *Let V be a finite-dimensional object in ${}^H_H\mathcal{YD}$. Let C be a left coideal subalgebra of $\mathcal{B}(V)$ in the category of $\mathbb{N}_0$-graded H-comodules. Then $(C^+\mathcal{B}(V))^\perp \subseteq \mathcal{B}(V^*)$ is a left coideal subalgebra of $\mathcal{B}(V^*)$ in the category of $\mathbb{N}_0$-graded H-comodules.*

PROOF. Let $M = C^+\mathcal{B}(V)$. We first prove that $M^\perp$ is a subalgebra of $\mathcal{B}(V^*)$. Clearly, $1 \in M^\perp$ since $M \subseteq \bigoplus_{i \geq 1} \mathcal{B}(V)$. Let $n \in \mathbb{N}$, $a \in C^+(n)$ and $b \in \mathcal{B}(V)$. Since $\mathcal{B}(V)$ is a connected $\mathbb{N}_0$-graded coalgebra and C is a left coideal, we have

$$\Delta(a) \in a \otimes 1 + \mathcal{B}(V) \otimes C^+.$$

Note that C^+ is an H-subcomodule since C is an H-subcomodule and ε is H-colinear. Hence $c(C^+ \otimes \mathcal{B}(V)) \subseteq \mathcal{B}(V) \otimes C^+$ and it follows that

$$\begin{aligned}
\Delta(ab) = (ab)^{(1)} \otimes (ab)^{(2)} &\in ab^{(1)} \otimes b^{(2)} + (\mathcal{B}(V) \otimes C^+)(\mathcal{B}(V) \otimes \mathcal{B}(V)) \\
&\subset ab^{(1)} \otimes b^{(2)} + \mathcal{B}(V) \otimes M \\
&\subset M \otimes \mathcal{B}(V) + \mathcal{B}(V) \otimes M.
\end{aligned}$$

Then it follows from Equation (2.7) that

$$\langle xy, ab \rangle = \langle x, (ab)^{(2)} \rangle \langle y, (ab)^{(1)} \rangle = 0$$

for all $x, y \in M^\perp$.

Now we prove that $M^\perp$ is a left coideal. Let $x \in M^\perp$ and $m \in M$. Then $\langle x, mb\rangle = 0$ for all $b \in \mathcal{B}(V)$ and hence $\langle x^{(2)}, m\rangle\langle x^{(1)}, b\rangle = 0$ for all $b \in \mathcal{B}(V)$ by Equation (2.8). It follows that $\langle x^{(2)}, m\rangle x^{(1)} = 0$ since $\langle \cdot, \cdot\rangle$ is nondegenerate. Hence $\Delta(x) \in \mathcal{B}(V) \otimes M^\perp$.

To see, that $M^\perp$ is an H-subcomodule of $\mathcal{B}(V^*)$ note that $M = C^+M$ is an H-subcomodule since C^+ is an H-subcomodule of $\mathcal{B}(V)$ and multiplication in $\mathcal{B}(V)$ is H-colinear. Let $x \in M^\perp$ and $a \in M$. Then $\delta(a) = a_{(-1)} \otimes a_{(0)} \in H \otimes M$. Using Equation (2.10) it follows that

$$x_{(-1)}\langle x_{(0)}, a\rangle = \mathcal{S}^{-1}(a_{(-1)})\langle x, a_{(0)}\rangle = 0$$

and hence $M^\perp$ is an H-subcomodule. Finally, $M^\perp$ is $\mathbb{N}_0$-graded by Equation (2.6) and since M is $\mathbb{N}_0$-graded. □

THEOREM 2.12. *Let V be a finite-dimensional object in ${}^H_H\mathcal{YD}$ and let $V_1, V_2 \subseteq V$ be H-subcomodules such that $V = V_1 \oplus V_2$. Let $N \geq 2$ and assume that*

$$\dim \bigoplus_{i=0}^{N} (\langle V_1\rangle \otimes \langle V_2\rangle)(i) \geq \dim \bigoplus_{i=0}^{N} \mathcal{B}(V)(i).$$

Assume in addition that $\langle V_2\rangle \cong \langle {V_2}^*\rangle \subseteq \mathcal{B}(V^*)$ as $\mathbb{N}_0$-graded vector spaces, where ${V_2}^*$ is the dual H-comodule of V_2. Then $\langle V_1\rangle \subseteq \mathcal{B}(V)$ is a left coideal subalgebra in the category of $\mathbb{N}_0$-graded H-comodules that cannot be extended in a degree n for $2 \leq n \leq N$.

PROOF. The subalgebra $\langle V_1\rangle \subseteq \mathcal{B}(V)$ is a left coideal subalgebra in the category of $\mathbb{N}_0$-graded H-comodules by Corollary 2.5. Let $2 \leq n \leq N$ and assume that there is an element $x \in \mathcal{B}(V)(n)$ that extends the left coideal subalgebra $\langle V_1\rangle$ in the category of $\mathbb{N}_0$-graded H-comodules. Let $K = \langle V_1 + \Bbbk x\rangle$. Then $\dim K(l) = \dim\langle V_1\rangle(l)$ for all $l < n$, $\dim K(n) = \dim\langle V_1\rangle(n) + 1$ and $\dim K(l) \geq \dim\langle V_1\rangle(l)$ for all $l > n$. Let $M = K^+\mathcal{B}(V)$. Then M is $\mathbb{N}_0$-graded and $M(1) = V_1$ since multiplication in $\mathcal{B}(V)$ is $\mathbb{N}_0$-graded and $K^+ \subseteq \bigoplus_{i\geq 1} \mathcal{B}(V)(i)$. Hence, from $V = V_1 \oplus V_2$ and from Equation (2.6), it follows that $M^\perp(1) = {V_2}^*$. By Proposition 2.11 we know that $M^\perp \subseteq \mathcal{B}(V^*)$ is a (left coideal) subalgebra in the category of $\mathbb{N}_0$-graded H-comodules. This implies

$$\dim M^\perp(m) \geq \dim\langle {V_2}^*\rangle(m)$$

for all $m \in \mathbb{N}_0$. Moreover, there is an induced nondegenerate pairing of $\mathbb{N}_0$-graded left H-comodules $\langle \cdot, \cdot\rangle' : M^\perp \otimes \mathcal{B}(V)/M \to \Bbbk$. It follows that

$$\dim(\mathcal{B}(V)/M)(m) = \dim M^\perp(m) \geq \dim\langle {V_2}^*\rangle(m)$$

for all $m \in \mathbb{N}_0$. From [**HS20**, Corollary 6.3.10] it is known that there is an $\mathbb{N}_0$-graded left K-linear and right $\mathcal{B}(V)/M$-colinear isomorphism

$$K \otimes \mathcal{B}(V)/M \to \mathcal{B}(V).$$

It follows that

$$
\begin{aligned}
\dim \bigoplus_{i=0}^{N} \mathcal{B}(V)(i) &= \sum_{i=0}^{N}\sum_{l=0}^{i} \dim K(l) \dim(\mathcal{B}(V)/M)(i-l) \\
&\geq \sum_{i=0}^{N}\sum_{l=0}^{i} \dim K(l) \dim\langle {V_2}^* \rangle(i-l) \\
&= \sum_{i=0}^{N}\sum_{l=0}^{i} \dim K(l) \dim\langle V_2 \rangle(i-l) \\
&> \sum_{i=0}^{N}\sum_{l=0}^{i} \dim\langle V_1 \rangle(l) \dim\langle V_2 \rangle(i-l) \\
&= \dim \bigoplus_{i=0}^{N} (\langle V_1 \rangle \otimes \langle V_2 \rangle)(i) \\
&\geq \dim \bigoplus_{i=0}^{N} \mathcal{B}(V)(i),
\end{aligned} \tag{2.11}
$$

where the third equality and the last inequality follow by assumption. Hence we get a contradiction and conclude that the left coideal subalgebra $\langle V_1 \rangle$ cannot be extended in a degree n for $2 \leq n \leq N$. □

We will apply Theorem 2.12 as described in the following Corollary.

COROLLARY 2.13. *Let V be a finite-dimensional object in ${}^H_H\mathcal{YD}$. Let V_1, V_2 be H-subcomodules of V such that $V = V_1 \oplus V_2$.*

(1) *Assume that $\langle V_2 \rangle \cong \langle {V_2}^* \rangle \subseteq \mathcal{B}(V^*)$ as $\mathbb{N}_0$-graded vector spaces. Let $N \in \mathbb{N}_0$ and assume that the restricted multiplication map $\mu : (\langle V_1 \rangle \otimes \langle V_2 \rangle)(n) \to \mathcal{B}(V)(n)$ is surjective for all $n \leq N$. Then the left coideal subalgebra $\langle V_1 \rangle$ cannot be extended in a degree n for $2 \leq n \leq N$.*
(2) *Assume that $\mathcal{B}(V)$ is finite-dimensional, $\dim\langle V_2 \rangle = \dim\langle {V_2}^* \rangle$ and $\dim\langle V_1 \rangle \dim\langle V_2 \rangle = \dim \mathcal{B}(V)$. Then the left coideal subalgebra $\langle V_1 \rangle$ cannot be extended in a degree n for $2 \leq n$.*

PROOF.

(1) Follows directly from Theorem 2.12.
(2) The proof is analogous to that of Theorem 2.12 replacing the calculation (2.11) with

$$
\begin{aligned}
\dim \mathcal{B}(V) &= \dim K \dim(\mathcal{B}(V)/M) \geq \dim K \dim\langle V_2^* \rangle \\
&= \dim K \dim\langle V_2 \rangle > \dim\langle V_1 \rangle \dim\langle V_2 \rangle = \dim \mathcal{B}(V).
\end{aligned}
$$
□

3. Yetter–Drinfeld modules over groups and racks

In the first half of this section we recall known concepts and results about Yetter–Drinfeld modules which are relevant for our study. Then we provide technical tools to support our analysis of left coideal subalgebras, see also the corresponding summary at the end of the section.

From now on we are mainly dealing with Yetter–Drinfeld modules over the group algebra $H = \Bbbk G$ of an arbitrary group G. We use the notation ${}^G_G\mathcal{YD}$ for the category of Yetter–Drinfeld modules over $\Bbbk G$. Due to the category isomorphism between $\Bbbk G$-comodules and G-graded vector spaces (see, e.g., [**HS20**, Proposition 1.1.17]) in this case Yetter–Drinfeld modules in ${}^G_G\mathcal{YD}$ can equivalently be defined as G-graded vector spaces $V = \bigoplus_{g\in G} V_g$ that are also left $\Bbbk G$-modules such that

$$g \cdot V_h \subseteq V_{ghg^{-1}}$$

for all $g, h \in G$. There

$$V_g = \{v \in V \mid \delta(v) = g \otimes v\} = \{v \in V \mid \deg v = g\}$$

for all $g \in G$. The subset

$$\mathrm{supp}V = \{g \in G \mid V_g \neq 0\}$$

is called the **support** of V. For objects $V, W \in {}^G_G\mathcal{YD}$ the Yetter–Drinfeld braiding is $c_{V,W}(v \otimes w) = g \cdot w \otimes v$ for all $v \in V_g$ and $w \in W$.

For the explicit construction of all Yetter–Drinfeld modules over an arbitrary group G there is a well-known useful equivalence of categories. We briefly describe it here and refer to [**HS20**, Section 1.4] and particulary to Proposition 1.4.17 there for more details. For an element $g \in G$ we denote its conjugacy class by $\mathcal{O}_g = \{hgh^{-1} \mid h \in G\}$. Let $L \subseteq G$ be a complete set of representatives of (pairwise distinct) conjugacy classes of G. Any Yetter–Drinfeld module $M \in {}^G_G\mathcal{YD}$ has a decomposition

$$M = \bigoplus_{l\in L} \bigoplus_{s\in\mathcal{O}_l} M_s$$

into a direct sum of Yetter–Drinfeld modules $\bigoplus_{s\in\mathcal{O}_l} M_s$, $l \in L$. Let $g \in G$ and let ${}^G_G\mathcal{YD}(\mathcal{O}_g)$ be the full subcategory of Yetter–Drinfeld modules containing each $M \in {}^G_G\mathcal{YD}$ with $M = \bigoplus_{s\in\mathcal{O}_g} M_s$. We denote the centralizer of g by

$$G^g = \{h \in G \mid hg = gh\}.$$

There is an equivalence of categories between ${}^G_G\mathcal{YD}(\mathcal{O}_g)$ and the category ${}_{\Bbbk G^g}\mathcal{M}$ of left $\Bbbk G^g$-modules. The first involved functor $F : {}_{\Bbbk G^g}\mathcal{M} \to {}^G_G\mathcal{YD}(\mathcal{O}_g)$ maps a $\Bbbk G^g$-module (ρ, V) to

$$M(g, \rho) = \Bbbk G \otimes_{\Bbbk G^g} V,$$

which is a Yetter–Drinfeld module where the $\Bbbk G$-action is multiplication on the first tensor factor and with G-grading

$$\deg(h \otimes v) = hgh^{-1}$$

for all $h \in G$, $v \in V$. The decomposition of $M(g, \rho)$ is given by $M(g, \rho) = \bigoplus_{s\in\mathcal{O}_g} M(g, \rho)_s$, where $M(g, \rho)_{hgh^{-1}} = h \otimes V$ for all $h \in G$. The quasi-inverse functor ${}^G_G\mathcal{YD}(\mathcal{O}_g) \to {}_{\Bbbk G^g}\mathcal{M}$ is given by $M \mapsto M_g$.

As mentioned above, the Nichols algebra as an algebra and a coalgebra only depends on the braided vector space (V, c_V) and there may be many Yetter–Drinfeld modules over different groups that yield the same braided vector space. Braided vector spaces arising from Yetter–Drinfeld modules over groups can be described using racks and two-cocycles without referring to a special group.

A **rack** is a pair $(X, \triangleright)$, for short X, where X is a nonempty set and $\triangleright : X \times X \to X$ is a map denoted by $(x, y) \mapsto x \triangleright y$ for all $x, y \in X$ such that

(1) the map $\varphi_i : X \to X$, $x \mapsto i \triangleright x$ is bijective for all $i \in X$, and
(2) $x \triangleright (y \triangleright z) = (x \triangleright y) \triangleright (x \triangleright z)$ for all $x, y, z \in X$.

A nonempty subset $Y \subseteq X$ of a rack is a **subrack** if $(Y, \triangleright)$ is a rack. A **quandle** is a rack with $x \triangleright x = x$ for all $x \in X$. A map $f : (X, \triangleright) \to (X', \triangleright')$ is a **morphismofracks** if $f(x \triangleright y) = f(x) \triangleright' f(y)$ for all $x, y \in X$. We say that $x, y \in X$ **commute** if $x \triangleright y = y$ and $y \triangleright x = x$.

EXAMPLE 3.1. Let G be a group and let H be a subset of G which is stable under conjugation. Then H is a quandle, where $x \triangleright y = xyx^{-1}$ for all $x, y \in H$. Moreover, in such a rack H of conjugacy classes for all $x, y \in H$ we have $x \triangleright y = y$ if and only if $y \triangleright x = x$. In particular, supp$V$ is a rack for any Yetter–Drinfeld module over $\Bbbk G$.

DEFINITION 3.2. A rack $(X, \triangleright)$ is called **braided** if it is a quandle and either $x \triangleright y = y$ or $x \triangleright (y \triangleright x) = y$ for all $x, y \in X$.

We will use the important "calculation rules" for braided racks in the following two Lemmata all the time without referring to it.

LEMMA 3.3 ([**HLV12**, Lemma 2, Lemma 3]). *Let $(X, \triangleright)$ be a braided rack and let $x, y, z \in X$. Then the following hold.*

(1) *If $y \neq z$ and $x \triangleright y = z$, then $y \triangleright z = x$ and $z \triangleright x = y$.*
(2) *If $y \triangleright x = x$, then $x \triangleright y = y$.*
(3) *If $x \triangleright (y \triangleright x) = y$, then $y \triangleright (x \triangleright y) = x$.*

LEMMA 3.4 ([**HLV12**, Lemma 4]). *Let $(X, \triangleright)$ be a quandle. Then the following are equivalent.*

(1) *X is braided.*
(2) *$x \triangleright (y \triangleright x) \in \{x, y\}$ for all $x, y \in X$.*

For a rack $(X, \triangleright)$ the subgroup of the group of permutations $\mathbb{S}_X$ generated by the permutations φ_i, $i \in X$, is called **inner group** of X, and is denoted by $\mathrm{Inn}(X)$. If $\mathrm{Inn}(X)$ acts transitively on X, we say that X is **indecomposable**. For any rack $(X, \triangleright)$ we define its **enveloping group** by

$$G_X = \langle g_x \mid x \in X \rangle / (g_x g_y = g_{x \triangleright y} g_x \text{ for all } x, y \in X).$$

A rack is called **injective** if the map $X \to G_X$, $x \mapsto g_x$ is injective.

The canonical map $i : X \to G_X$, $x \mapsto g_x$ has the following universal property: If G is a group and $f : X \to G$ a morphism of racks, where G is a rack by conjugation, then there is a unique morphism of groups $h : G_X \to G$ such that $h(g_x) = f(x)$ for all $x \in X$ (see, e.g., [**AFGV11**, Section 3.1]).

There is a canonical group action of G_X on X extending the rack operation $\triangleright$ such that $g_x \triangleright y = x \triangleright y$ for all $x, y \in X$ (see, e.g., [**HLV12**, Section 1.3]).

LEMMA 3.5 ([**GnHV11**, Lemma 2.17, Lemma 2.18]). *Let X be a finite indecomposable rack. Then all permutations $\varphi_x \in \mathbb{S}_X$, where $x \in X$, have the same order n. Moreover, in the enveloping group G_X the relations ${g_x}^n = {g_y}^n$ hold and ${g_x}^n$ is a central element of G_X for all $x, y \in X$.*

For a finite indecomposable rack X we define the **finite enveloping group** of X by $\overline{G_X} = G_X / ({g_x}^n)$, where $x \in X$ and $n = \mathrm{ord}(\varphi_x)$. By Lemma 3.5, the definition is independent from the choice of $x \in X$. A proof that the group $\overline{G_X}$ is indeed finite can be found in [**GnHV11**, Lemma 2.19].

DEFINITION 3.6. Let $(X, \triangleright)$ be a rack and let

$$q : X \times X \to \Bbbk^\times, \quad (x, y) \mapsto q_{x,y}$$

be a map. Then q is called a scalar valued **two-cocycle** if

$$q_{x \triangleright y, x \triangleright z} q_{x,z} = q_{x, y \triangleright z} q_{y,z}$$

for all $x, y, z \in X$.

There is also the more general concept with (not scalar valued) two-cocycles, where the target is the automorphism group of a vector space instead of $\Bbbk^\times$ (see, e.g., [**HS20**, Definition 1.5.11]). However, we will consider Yetter–Drinfeld modules and their braidings whose G-homogeneous components are at most one-dimensional. These require only scalar valued two-cocycles.

The following Lemma is a special case of [**HS20**, Proposition 1.5.12].

LEMMA 3.7. *Let X be a nonempty set and let*

$$\triangleright : X \times X \to X, \; q : X \times X \to \Bbbk^\times$$

be maps. Let $c_q : \Bbbk X \otimes \Bbbk X \to \Bbbk X \otimes \Bbbk X$ be the linear map defined by

$$c_q(x \otimes y) = q_{x,y}(x \triangleright y) \otimes x$$

for all $x, y \in X$. Then $(\Bbbk X, c_q)$ is a braided vector space if and only if $(X, \triangleright)$ is a rack and q is a two-cocycle.

PROOF. Let $x, y, z \in X$. Then

$$(c_q)_1(c_q)_2(c_q)_1(x \otimes y \otimes z) = q_{x,z} q_{x \triangleright y, x \triangleright z}(x \triangleright y) \triangleright (x \triangleright z) \otimes q_{x,y}(x \triangleright y) \otimes x$$
$$(c_q)_2(c_q)_1(c_q)_2(x \otimes y \otimes z) = q_{y,z} q_{x, y \triangleright z} x \triangleright (y \triangleright z) \otimes q_{x,y}(x \triangleright y) \otimes x.$$

Hence c_q fulfills the Yang-Baxter equation if and only if $(X, \triangleright)$ is a rack and q is a two-cocycle. The map c_q is injective if and only if for all $x \in X$ the map $\varphi_x : X \to X$, $y \mapsto x \triangleright y$, is injective. Then it follows that c_q is bijective if and only if for all $x \in X$ the map φ_x is bijective. □

The following definition is similarly used in [**AGn03a**].

DEFINITION 3.8. Let X be a rack and $q : X \times X \to \Bbbk^\times$ a two-cocycle. Let $(\Bbbk X, c_q)$ be the associated braided vector space (as in Lemma 3.7). A **Yetter–Drinfeld realization** of $(\Bbbk X, c_q)$ over a group G is a Yetter–Drinfeld module structure over $\Bbbk G$ on $\Bbbk X$ such that the braiding c_q coincides with the Yetter–Drinfeld braiding in ${}^G_G\mathcal{YD}$ of $\Bbbk X$ and the elements of X are G-homogeneous. We call a Yetter–Drinfeld realization **faithful** if the map $X \to G$, $x \mapsto \deg(x)$ is injective.

A rack X with a two-cocycle q admits many Yetter–Drinfeld realizations over different groups. Note that the Nichols algebras of all such Yetter–Drinfeld realizations are isomorphic as algebras and coalgebras since the Yetter–Drinfeld modules are isomorphic as braided vector spaces. The following Lemma 3.9 is a variation of [**HS20**, Proposition 1.5.6]. It shows that any rack with two-cocycle can be realized as Yetter–Drinfeld module over the enveloping group of the rack. We will see in Proposition 3.12(2) that the existence of left coideal subalgebras of the associated Nichols algebra in the category of $\mathbb{N}_0$-graded $\Bbbk G$-comodules which are not generated in degree one does not depend on the choice of the group G for the realization of the braided vector space. We will mostly use the realization over $\Bbbk G_X$.

LEMMA 3.9. *Let X be a rack and $q : X \times X \to \Bbbk^\times$ a two-cocycle. Let $(\Bbbk X, c_q)$ be the associated braided vector space (as in Lemma 3.7). Then $\Bbbk X$ is a Yetter–Drinfeld module over the enveloping group G_X with G_X-action $g_x \cdot y = q_{x,y}(x \triangleright y)$ and G_X-gradation $\deg x = g_x$ for all $x, y \in X$. The braiding c_q coincides with the Yetter–Drinfeld braiding $c_{\Bbbk X}$.*

PROOF. The linear map

$$f_x : \Bbbk X \to \Bbbk X, \ y \mapsto q_{x,y}(x \triangleright y)$$

is a linear automorphism of $\Bbbk X$ for all $x \in X$. Let $x, y, z \in X$. Then

$$\begin{aligned} f_x f_y(z) &= q_{x,y\triangleright z} q_{y,z}(x \triangleright (y \triangleright z)) \\ &= q_{x\triangleright y, x\triangleright z} q_{x,z}((x \triangleright y) \triangleright (x \triangleright z)) = f_{x\triangleright y} f_x(z), \end{aligned}$$

where we used the self-distributivity of $\triangleright$ and the two-cocycle condition. Hence the group morphism $\langle g_x \mid x \in X \rangle \to \mathrm{Aut}(\Bbbk X)$, $g_x \mapsto f_x$ induces a group morphism $G_X \to \mathrm{Aut}(\Bbbk X)$. It follows that $\Bbbk X$ is a $\Bbbk G_X$-module with action $g_x \cdot y = q_{x,y}(x \triangleright y)$ for all $x, y \in X$. The Yetter–Drinfeld condition $g_x \cdot y \in (\Bbbk X)_{g_x g_y g_x^{-1}}$ for all $x, y \in X$ holds since $\deg(x \triangleright y) = g_{x\triangleright y} = g_x g_y g_x^{-1} \in G_X$. Moreover, we have

$$c_{\Bbbk X}(x \otimes y) = g_x \cdot y \otimes x = q_{x,y}(x \triangleright y) \otimes x = c_q(x \otimes y)$$

for all $x, y \in X$. □

Note that the Yetter–Drinfeld realization of a rack (with a two-cocycle) over the enveloping group is not faithful if the rack is not injective. If X is an injective rack, then $g_x \neq g_y$ for each $x, y \in X$ with $x \neq y$, hence the Yetter–Drinfeld realization over $\Bbbk G_X$ is faithful and then each G_X-homogeneous component of $\Bbbk X$ is at most one-dimensional.

LEMMA 3.10. *Let G be a group and X a subset of G which is stable under conjugation. Then X is an injective rack.*

PROOF. Let $f : X \to G$ be the inclusion map. By the universal property of the canonical map $i : X \to G_X$, there is a group homomorphism $g : G_X \to G$ such that $g \circ i = f$. Hence, the map i is injective since f is injective. □

LEMMA 3.11. *Let G be a group and V a finite-dimensional Yetter–Drinfeld module over $\Bbbk G$ such that $\dim V_g \leq 1$ for all $g \in G$. Let $X = \mathrm{supp} V$. For each $x \in X$ choose a nonzero element $v_x \in V_x$ and let $B = \{v_x \mid x \in X\}$.*

1. *For each $x \in X$ there is a map $q'_x : X \to \Bbbk^\times$ such that $x \cdot v_y = q'_x(y) v_{xyx^{-1}}$ for all $y \in X$.*
2. *X is a rack with conjugation as rack operation and $q : X \times X \to \Bbbk^\times$, $(x, y) \mapsto q'_x(y)$ is a two-cocycle.*
3. *The linear map $f : (V, c_V) \to (\Bbbk X, c_q)$, $v_x \mapsto x$ is an isomorphism of braided vector spaces. There c_V is the Yetter–Drinfeld braiding of V and $(\Bbbk X, c_q)$ is the braided vector space associated to the rack X and two-cocycle q as in Lemma 3.7.*

PROOF.

(1) Let $x \in X$. By the Yetter–Drinfeld condition we have $x \cdot v_y \in V_{xyx^{-1}}$ for all $y \in X$. Since $\dim V_{xyx^{-1}} = 1$ for all $y \in \mathrm{supp} V$ (by assumption and Example 3.1), it follows that for each $y \in X$ there is an element $q_{x,y} \in \Bbbk^\times$ such that $x \cdot v_y = q_{x,y} v_{xyx^{-1}}$. Hence the claim follows with $q'_x : X \to \Bbbk^\times$, $y \mapsto q_{x,y}$.

(2) The set $X = \mathrm{supp}V$ is a rack by Example 3.1. Note that the map $X \to B$, $x \mapsto v_x$ is an isomorphism of racks, where B is a rack with the induced rack operation $v_x \triangleright v_y = v_{xyx^{-1}}$ for all $x, y \in X$. The Yetter–Drinfeld braiding c_V of V is given by

$$c_V(v_x \otimes v_y) = x \cdot v_y \otimes v_x = q_{x,y} v_{xyx^{-1}} \otimes v_x$$

for all $x, y \in \mathrm{supp}V$. Hence the map q is a two-cocycle on $B \cong X$ by Lemma 3.7.

(3) Obviously, the linear map f is a well-defined isomorphism. Moreover,

$$\begin{aligned}(f \otimes f)c_V(v_x \otimes v_y) &= f(x \cdot v_y) \otimes f(v_x) = q'_x(y) f(v_{xyx^{-1}}) \otimes f(v_x) \\ &= q_{x,y}(x \triangleright y) \otimes x = c_q(x \otimes y)\end{aligned}$$

for all $x, y \in X$. □

PROPOSITION 3.12. *Let G be a group and V a finite-dimensional Yetter–Drinfeld module over $\Bbbk G$ such that $\dim V_g \leq 1$ for all $g \in G$. Let $X = \mathrm{supp}V$. Let V_{G_X} be the Yetter–Drinfeld realization of the underlying braided vector space over $\Bbbk G_X$.*

(1) *The Nichols algebras $\mathcal{B}(V)$ and $\mathcal{B}(V_{G_X})$ are isomorphic as $\mathbb{N}_0$-graded braided Hopf algebras.*[2] *In particular, a subspace $C \subseteq \mathcal{B}(V)$ is a left coideal subalgebra if and only if $C \subseteq \mathcal{B}(V_{G_X})$ is a left coideal subalgebra.*[3]
(2) *The Nichols algebra $\mathcal{B}(V)$ contains a left coideal subalgebra in the category of $\mathbb{N}_0$-graded $\Bbbk G$-comodules that is not generated in degree one if and only if $\mathcal{B}(V_{G_X})$ contains a left coideal subalgebra in the category of $\mathbb{N}_0$-graded $\Bbbk G_X$-comodules that is not generated in degree one.*

PROOF.

(1) The identity map $\mathrm{id} : V = (V, {}^{\Bbbk G}\delta_V) \to V_{G_X} = (V, {}^{\Bbbk G_X}\delta_V)$ is an isomorphism of braided vector spaces by construction (see Lemma 3.11(3) and Lemma 3.9). The claim follows directly from the construction of the Nichols algebra of a braided vector space. (More explicit arguments can be found in [**HS20**, Remark 7.1.4].)

(2) By Lemma 3.10, the rack X is injective. This implies that the Yetter–Drinfeld realization V_{G_X} over the enveloping group is faithful. Hence $\Bbbk G$-subcomodules of V are (as vector subspaces of V) the same as the $\Bbbk G_X$-subcomodules of V_{G_X} and the claim follows directly from Proposition 2.10 and (1).

$${}^{\Bbbk G}\delta_{\mathcal{B}(V)} = (h \otimes \mathrm{id}){}^{\Bbbk G_X}\delta_{\mathcal{B}(V)}.$$

This proves the claim. □

PROPOSITION 3.13. *Let G be a group, let $V_1, V_2 \in {}^G_G\mathcal{YD}$ be finite-dimensional and $V = V_1 \oplus V_2$. Let $c = c_V$ be the Yetter–Drinfeld braiding and assume that $c^2|V_2 \otimes V_1 \neq \mathrm{id}_{V_2 \otimes V_1}$. Then there is a left coideal subalgebra of $\mathcal{B}(V)$ in the category of $\mathbb{N}_0$-graded $\Bbbk G$-comodules that is not generated in degree one.*

PROOF. Since $c^2|V_2 \otimes V_1 \neq \mathrm{id}_{V_2 \otimes V_1}$ and V_1, V_2 are $\Bbbk G$-subcomodules, there are G-homogeneous elements $x \in V_2$ and $y \in V_1$ such that $c^2(x \otimes y) \neq x \otimes y$. Define $z = \mu(x \otimes y - c(x \otimes y)) \in \mathcal{B}(V)$. Then z is G-homogeneous and

$$\Delta_{1,1}(z) = (1 + c)(x \otimes y) - (1 + c)(c(x \otimes y)) = x \otimes y - c^2(x \otimes y) \in V_2 \otimes V_1$$

[2] Since the isomorphism is induced by the identity map $\mathrm{id} : V = (V, {}^{\Bbbk G}\delta_V) \to V_{G_X} = (V, {}^{\Bbbk G_X}\delta_V)$, we will identify $\mathcal{B}(V)$ and $\mathcal{B}(V_{G_X})$ as vector spaces without referring to the isomorphism.

[3] Here we do not require that C is a $\Bbbk G$-comodule ($\Bbbk G_X$-comodule, resp.).

by Equation (1.4). It follows that $\Delta_{1,1}(z) \neq 0$ since $c^2(x \otimes y) \neq x \otimes y$ and $z \notin \langle V_2 \rangle$ since $\Delta_{1,1}(\langle V_2 \rangle(2)) \subseteq V \otimes V_2$ by Corollary 2.5 and Proposition 2.7. Hence the subalgebra $\langle V_2 \oplus \Bbbk z \rangle \subseteq \mathcal{B}(V)$ is a left coideal in the category of $\mathbb{N}_0$-graded $\Bbbk G$-comodules that is not generated in degree one (see Corollary 2.8). □

Let G be a group, $V_1, V_2 \in {}^G_G\mathcal{YD}$ and $V = V_1 \oplus V_2$. Let $c = c_V$ be the Yetter–Drinfeld braiding. For $i \in \{1,2\}$ the inclusion $V_i \subseteq V$ induces an injective morphism $\mathcal{B}(V_i) \to \mathcal{B}(V)$ of $\mathbb{N}_0$-graded Hopf algebras in ${}^G_G\mathcal{YD}$ (see [**HS20**, Remark 1.6.19]). We identify $\mathcal{B}(V_i)$ with the image of this injective map. Let $\mu_{12} : \mathcal{B}(V_1) \otimes \mathcal{B}(V_2) \to \mathcal{B}(V)$ be the restricted multiplication map. If $c^2|V_2 \otimes V_1 = \mathrm{id}_{V_2 \otimes V_1}$, then $\mathcal{B}(V_1) \otimes \mathcal{B}(V_2)$ is a Hopf algebra in ${}^G_G\mathcal{YD}$ and μ_{12} is an isomorphism of braided Hopf algebras in ${}^G_G\mathcal{YD}$ by [**HS20**, Proposition 1.10.12].

PROPOSITION 3.14. *Let G be a group, let $V_1, V_2 \in {}^G_G\mathcal{YD}$ be finite-dimensional and $V = V_1 \oplus V_2$. Assume that $c^2|V_2 \otimes V_1 = \mathrm{id}_{V_2 \otimes V_1}$ and that $\dim V_g \leq 1$ for all $g \in G$. Let $W_1 \subseteq V_1$ and $W_2 \subseteq V_2$ be $\Bbbk G$-subcomodules and let $W = W_1 \oplus W_2$. Then the following are equivalent.*

(1) *$\langle W \rangle$ is the only $\mathbb{N}_0$-graded left coideal subalgebra K of $\mathcal{B}(V)$ with $K(1) = W$.*
(2) *$\langle W_1 \rangle$ is the only $\mathbb{N}_0$-graded left coideal subalgebra K_1 of $\mathcal{B}(V_1)$ with $K_1(1) = W_1$, and $\langle W_2 \rangle$ is the only $\mathbb{N}_0$-graded left coideal subalgebra K_2 of $\mathcal{B}(V_2)$ with $K_2(1) = W_2$.*

Note that by Proposition 2.10, the property in Proposition 3.14 (1) holds if and only if $\langle W \rangle$ is the only $\mathbb{N}_0$-graded left coideal subalgebra K of $\mathcal{B}(V)$ in the category of $\Bbbk G$-subcomodules with $K(1) = W$.

PROOF. Let $\pi : V \to V/W$ be the canonical map. By Proposition 2.10, (1) holds if and only if

$$\ker(\mathrm{id} \otimes \pi\pi_1)\Delta_{\mathcal{B}(V)} = \langle W \rangle.$$

From $c^2|V_2 \otimes V_1 = \mathrm{id}_{V_2 \otimes V_1}$ it follows that the relation $vu = \mu c(v \otimes u)$ holds in $\mathcal{B}(V)$ for all $u \in V_1$, $v \in V_2$ and that $ghg^{-1} = h$ for all $h \in \mathrm{supp} V_1$, $g \in \mathrm{supp} V_2$. From this and since $W = W_1 \oplus W_2$ and $\dim V_g \leq 1$ for all $g \in G$, we obtain

$$\langle W \rangle = \langle W_1 \rangle \langle W_2 \rangle \subseteq \mathcal{B}(V_1)\mathcal{B}(V_2).$$

The multiplication map $\mu_{12} : \mathcal{B}(V_1) \otimes \mathcal{B}(V_2) \to \mathcal{B}(V)$ is bijective by [**HS20**, Proposition 1.10.12]. This implies that

$$\langle W_1 \rangle = \langle W \rangle \cap \mathcal{B}(V_1), \qquad \langle W_2 \rangle = \langle W \rangle \cap \mathcal{B}(V_2).$$

Let $\pi_{W_1} : V_1 \to V_1/W_1$ be the canonical map. Then (1) implies that

$$\begin{aligned}\ker(\mathrm{id} \otimes \pi_{W_1}\pi_1)\Delta_{\mathcal{B}(V_1)} &= \ker(\mathrm{id} \otimes \pi\pi_1)\Delta_{\mathcal{B}(V)}|\mathcal{B}(V_1) \\ &= \mathcal{B}(V_1) \cap \langle W \rangle = \langle W_1 \rangle,\end{aligned}$$

and similarly $\ker(\mathrm{id} \otimes \pi_{W_2}\pi_2)\Delta_{\mathcal{B}(V_2)} = \langle W_2 \rangle$. Thus (1) implies (2).

Assume now (2). Note that $V = V_1 \oplus V_2$ and W are $\mathbb{Z}^2$-graded, where $\deg(V_1) = (1,0)$ and $\deg(V_2) = (0,1)$. Thus $\mathcal{B}(V)$ is $\mathbb{Z}^2$-graded and $\ker(\mathrm{id} \otimes \pi\pi_1)\Delta_{\mathcal{B}(V)}$ is $\mathbb{Z}^2$-graded. For all $k, l \geq 0$ and $x \in \mathcal{B}(V_1)(k)$, $y \in \mathcal{B}(V_2)(l)$, one has

$$(\mathrm{id} \otimes \pi\pi_1)\Delta(xy) = (\mathrm{id} \otimes \pi\pi_1)\Delta(x)(y \otimes 1) + (x \otimes 1)(\mathrm{id} \otimes \pi\pi_1)\Delta(y),$$

where $\Delta = \Delta_{\mathcal{B}(V)}$. Since $\mathcal{B}(V)$ and $\ker(\mathrm{id} \otimes \pi\pi_1)\Delta$ are $\mathbb{Z}^2$-graded and since the map $\mu_{12} : \mathcal{B}(V_1) \otimes \mathcal{B}(V_2) \to \mathcal{B}(V)$ is bijective, it follows that

$$\begin{aligned}
&\ker(\mathrm{id} \otimes \pi\pi_1)\Delta_{\mathcal{B}(V)} \\
&\quad = \bigoplus_{k,l \geq 0} \ker(\mathrm{id} \otimes \pi\pi_1)\Delta_{\mathcal{B}(V)}|\big(\mathcal{B}(V_1)(k)\,\mathcal{B}(V_2)(l)\big) \\
&\quad = \ker(\mathrm{id} \otimes \pi_{W_1}\pi_1)\Delta_{\mathcal{B}(V_1)} \cdot \ker(\mathrm{id} \otimes \pi_{W_2}\pi_1)\Delta_{\mathcal{B}(V_2)} \\
&\quad = \langle W_1 \rangle \langle W_2 \rangle,
\end{aligned}$$

where the last equation holds by (2). Since $\langle W \rangle = \langle W_1 \rangle \langle W_2 \rangle$, we conclude that (2) implies (1). □

LEMMA 3.15. *Let G be a group and V a finite-dimensional Yetter–Drinfeld module over $\Bbbk G$ such that $\dim V_g \leq 1$ for all $g \in G$. Let $X = \mathrm{supp} V$ and assume that X generates G. Then V is an irreducible Yetter–Drinfeld module if and only if X is indecomposable as rack.*

PROOF. Under the assumptions of the lemma, Yetter–Drinfeld submodules of V correspond to subracks of $\mathrm{supp} V$. This implies the claim. □

LEMMA 3.16. *Let G be a group and V a finite-dimensional Yetter–Drinfeld module over $\Bbbk G$. For all $g \in G$ define the map*

$$\alpha_g : \mathcal{B}(V) \to \mathcal{B}(V), \quad x \mapsto g \cdot x.$$

(1) *The map α_g is a well-defined isomorphism of $\mathbb{N}_0$-graded algebras and coalgebras.*
(2) *The map α_g sends G-homogeneous elements to G-homogeneous elements.*
(3) *If $K \subseteq \mathcal{B}(V)$ is a left coideal subalgebra of $\mathcal{B}(V)$ in the category of $\mathbb{N}_0$-graded $\Bbbk G$-subcomodules, then $\alpha_g(K)$ is a left coideal subalgebra of $\mathcal{B}(V)$ in the category of $\mathbb{N}_0$-graded $\Bbbk G$-subcomodules and $\alpha_g(K) \cong K$ as $\mathbb{N}_0$-graded algebras.*

PROOF.

(1) Since $\mathcal{B}(V)$ is a $\Bbbk G$-module, the map is well-defined. Since $\mathcal{B}(V)$ is an algebra and a coalgebra in the category of $\mathbb{N}_0$-graded $\Bbbk G$-Yetter–Drinfeld modules, α_g is a morphism of $\mathbb{N}_0$-graded algebras and coalgebras. Since g is invertible in $\Bbbk G$, the map is bijective (with inverse $\alpha_{g^{-1}}$).

(2) Holds obviously; and

(3) follows from (1) and (2). □

Let G be a group and V a finite-dimensional Yetter–Drinfeld module over $\Bbbk G$ such that $\dim V_g \leq 1$ for all $g \in G$. It is our aim to be able to decide whether there is a left coideal subalgebra of the Nichols algebra $\mathcal{B}(V)$ in the category of $\mathbb{N}_0$-graded $\Bbbk G$-comodules which is not generated in degree one. In this section we provided arguments which reduce the problem in the following sense: It suffices to answer the question for Yetter–Drinfeld realizations of finite indecomposable racks of conjugacy classes X with two-cocycles over the enveloping group G_X. Our result can be found in Theorem 8.8. The proof of Corollary 8.10 shows in which way the answer in the more general case then can be deduced using arguments from the current section.

4. Extensions for left coideal subalgebras in degrees 2, 3, and 4

In this section we give sufficient criteria for the existence of nontrivial extensions of left coideal subalgebras in degrees two, three and four. Our main results are Proposition 4.2, Theorem 4.3, Theorem 4.6 and Theorem 4.11, all of which describe explicit extensions. The proofs of Theorem 4.6 and Theorem 4.11 are particularly technical, but note that the assumptions in them are also very weak.

Let $(X, \triangleright)$ be a union of conjugacy classes of a group, considered as a rack. In particular,

$$x \triangleright y = y \quad \Rightarrow \quad y \triangleright x = x \tag{4.1}$$

for all $x, y \in X$. Let $q : X \times X \to \Bbbk^\times$ be a two-cocycle. Let $V = \Bbbk X$ be the corresponding Yetter–Drinfeld realization over $\Bbbk G_X$ as in Lemma 3.9 and denote the braiding by c_q. By construction, there are nonzero elements $v_x \in V_x$ such that

$$c_q(v_x \otimes v_y) = q_{x,y} v_{x \triangleright y} \otimes v_x$$

for all $x, y \in X$.

LEMMA 4.1. *Assume that X is not braided. Then there are pairwise distinct elements $a, b, c, d \in X$ such that $b \triangleright a = c$ and $c \triangleright b = d$.*

PROOF. Since X is not braided, there are distinct elements $a, b \in X$ such that $b \triangleright a \neq a$ and $b \triangleright (a \triangleright b) \neq a$. Let $c = b \triangleright a$. Then $c \neq a$. Since X is a quandle, φ_b is bijective and $a \neq b$, it follows that $c \neq b$. Let $d = c \triangleright b$. Then $d = b \triangleright (a \triangleright b)$ by self-distributivity of $\triangleright$ and the quandle property. Hence $d \neq a$ by the choice of a and b. Assume that $d = b$. Then $a \triangleright b = b$ since X is a quandle and φ_b is bijective. Since X is a rack of a union of conjugacy classes, it follows that $b \triangleright a = a$ by Equation (4.1), a contradiction to the choice of a and b. Hence $d \neq b$. Assume that $d = c$. Then $(b \triangleright a) \triangleright b = b \triangleright a$ and hence $b = b \triangleright a$ by the bijectivity of $\varphi_{b \triangleright a}$ and the quandle property. It follows that $b = a$ by the bijectivity of φ_b and the quandle property, a contradiction to the choice of a and b. Hence $d \neq c$. □

PROPOSITION 4.2. *Assume that X is not braided. Then there is a left coideal subalgebra of $\mathcal{B}(V)$ in the category of $\mathbb{N}_0$-graded $\Bbbk G_X$-comodules that can be extended in degree two.*

PROOF. By Lemma 4.1, there are pairwise distinct elements $a, b, c, d \in X$ such that $b \triangleright a = c$ and $c \triangleright b = d$. Let

$$x = v_b v_a - q_{b,a} v_c v_b \in \mathcal{B}(V).$$

Then $\Delta_{1,1}(x) = v_b \otimes v_a - q_{c,b} q_{b,a} v_d \otimes v_c$. Hence $x \neq 0$ and $\langle v_a, v_c, x \rangle$ is a left coideal subalgebra of $\mathcal{B}(V)$. Note that x is G_X-homogeneous of degree $g_b g_a$ since the relation $g_b g_a = g_{b \triangleright a} g_b = g_c g_b$ holds in G_X. Thus $\langle v_a, v_c, x \rangle$ is a $\Bbbk G_X$-comodule. We show that $x \notin \langle v_a, v_c \rangle$. The degree two component of the subalgebra generated by v_a and v_c is

$$\langle v_a, v_c \rangle(2) = \Bbbk \left\{ {v_a}^2, v_a v_c, v_c v_a, {v_c}^2 \right\}$$

and

$${v_a}^2 \in \mathcal{B}(V)_{{g_a}^2}, \ v_a v_c \in \mathcal{B}(V)_{g_a g_c}, \ v_c v_a \in \mathcal{B}(V)_{g_c g_a}, \ {v_c}^2 \in \mathcal{B}(V)_{{g_c}^2}.$$

Since $g_b g_a = g_c g_b = g_d g_c$, it follows that $g_b g_a \notin \{{g_a}^2, g_a g_c, g_c g_a, {g_c}^2\}$ and hence $x \notin \langle v_a, v_c \rangle$. Thus $\langle v_a, v_c, x \rangle$ is a left coideal subalgebra in the category of $\mathbb{N}_0$-graded $\Bbbk G_X$-comodules of $\mathcal{B}(V)$ that is not generated in degree one. □

Proposition 4.2 ensures the existence of a not primitively generated left coideal subalgebra in the category of $\mathbb{N}_0$-graded $\Bbbk G_X$-comodules in the Nichols algebra $\mathcal{B}(V)$ if X is not braided independent from the choice of the two-cocycle q.

THEOREM 4.3. *Assume that the rack X is braided.*

(1) *Let $a, b \in X$ with $a \triangleright b = b \neq a$ and let*

$$x = \mu(\mathrm{id} - c_q)(v_a \otimes v_b) \in \mathcal{B}(V).$$

If $x \neq 0$, then $\langle v_b, x\rangle$ is a left coideal subalgebra of $\mathcal{B}(V)$ in the category of $\mathbb{N}_0$-graded $\Bbbk G_X$-comodules that is not generated in degree one.

(2) *Let $a, b \in X$ with $a \triangleright b \neq b$ and let*

$$x = \mu(\mathrm{id} - c_q + {c_q}^2)(v_a \otimes v_b) \in \mathcal{B}(V).$$

If $x \neq 0$, then $\langle v_b, x\rangle$ is a left coideal subalgebra of $\mathcal{B}(V)$ in the category of $\mathbb{N}_0$-graded $\Bbbk G_X$-comodules that is not generated in degree one.

PROOF.

(1) Note first that $b \triangleright a = a$ because of Equation (4.1) and since $a \triangleright b = b$. Using Equation (1.4) it follows that

$$\begin{aligned}\Delta_{1,1}(x) &= (\mathrm{id} + c_q)(\mathrm{id} - c_q)(v_a \otimes v_b)\\ &= (\mathrm{id} - {c_q}^2)(v_a \otimes v_b)\\ &= (1 - q_{a,b}q_{b,a})v_a \otimes v_b.\end{aligned}$$

Note that x is G_X-homogeneous of degree $g_a g_b$ and hence $\Bbbk x$ is a $\Bbbk G_X$-subcomodule. Thus $\langle v_b, x\rangle$ is a left coideal subalgebra of $\mathcal{B}(V)$ in the category of $\mathbb{N}_0$-graded $\Bbbk G_X$-comodules by Corollary 2.8. By assumption, $x \neq 0$ and obviously $x \notin \langle v_b\rangle$. Hence $\langle v_b, x\rangle$ is not generated in degree one.

(2) Since X is braided and $a \triangleright b \neq b$, it follows that $(a \triangleright b) \triangleright a = b$ and $b \triangleright (a \triangleright b) = a$ by Lemma 3.3. Using this we get

$$\begin{aligned}{c_q}^3(v_a \otimes v_b) &= {c_q}^2(q_{a,b}v_{a\triangleright b} \otimes v_a) = c_q(q_{a,b}q_{a\triangleright b,a}v_b \otimes v_{a\triangleright b})\\ &= q_{a,b}q_{a\triangleright b,a}q_{b,a\triangleright b}v_{b\triangleright(a\triangleright b)} \otimes v_b = q_{a,b}q_{a\triangleright b,a}q_{b,a\triangleright b}v_a \otimes v_b.\end{aligned}$$

By Equation (1.4), it follows

$$\begin{aligned}\Delta_{1,1}(x) &= (\mathrm{id} + c_q)(\mathrm{id} - c_q + {c_q}^2)(v_a \otimes v_b)\\ &= (\mathrm{id} + {c_q}^3)(v_a \otimes v_b)\\ &= (1 + q_{a,b}q_{a\triangleright b,a}q_{b,a\triangleright b})v_a \otimes v_b.\end{aligned}$$

Note that x is G_X-homogeneous of degree $g_a g_b$ and hence $\Bbbk x$ is a $\Bbbk G_X$-subcomodule. Thus $\langle v_b, x\rangle$ is a left coideal subalgebra of $\mathcal{B}(V)$ in the category of $\mathbb{N}_0$-graded $\Bbbk G_X$-comodules by Corollary 2.8. By assumption, $x \neq 0$ and obviously $x \notin \langle v_b\rangle$. Thus $\langle v_b, x\rangle$ is not generated in degree one. □

Theorem 4.3 and Proposition 4.2 imply that if all left coideal subalgebras of $\mathcal{B}(V)$ in the category of $\mathbb{N}_0$-graded $\Bbbk G_X$-comodules are generated in degree one, then in $\mathcal{B}(V)$ the following relations hold in degree two:

$$\mu(\mathrm{id} - c_q)(v_a \otimes v_b) = 0 \quad \text{for all } a, b \in X \text{ with } a \triangleright b = b \neq a, \tag{4.2}$$

$$\mu(\mathrm{id} - c_q + {c_q}^2)(v_a \otimes v_b) = 0 \quad \text{for all } a, b \in X \text{ with } a \triangleright b \neq b. \tag{4.3}$$

We say that the Nichols algebra $\mathcal{B}(V)$ is **slim in degree two** if the relations (4.2) and (4.3) hold.

LEMMA 4.4. *Assume that the rack X is braided.*

(1) *Let $a, b \in X$ with $a \triangleright b = b \neq a$. The relation*

$$\mu(\mathrm{id} - c_q)(v_a \otimes v_b) = 0$$

holds in $\mathcal{B}(V)$ if and only if $q_{a,b}q_{b,a} = 1$ if and only if ${c_q}^2(v_a \otimes v_b) = v_a \otimes v_b$.

(2) *Let $a, b, c \in X$ with $a \triangleright b = c \neq b$. The relation*

$$\mu(\mathrm{id} - c_q + {c_q}^2)(v_a \otimes v_b) = 0$$

holds in $\mathcal{B}(V)$ if and only if $q_{a,b}q_{c,a}q_{b,c} = -1$ if and only if ${c_q}^3(v_a \otimes v_b) = -v_a \otimes v_b$.

PROOF. The claims follow from the calculations in the proof of Theorem 4.3. □

For the following considerations in degree three we need some technical calculations, which we collect in the next Lemma.

LEMMA 4.5. *Assume that the rack X is braided and that there are pairwise distinct elements $a, b, c \in X$ with $a \triangleright b \neq b$. Let $d = c \triangleright (b \triangleright a)$ and $e = d \triangleright (c \triangleright b)$. Then the following hold.*

(1) *$e = c \triangleright a$.*

(2) *Assume that c commutes with none of a, b, $b \triangleright a$. Let $S = \{a, b, d, e\}$. Then*
   (a) *$c \notin S$,*
   (b) *$b \triangleright a \notin S$,*
   (c) *$b \triangleright y \neq c$ for all $y \in S$ and*
   (d) *$b \triangleright (a \triangleright y) \neq c$ for all $y \in S$.*

PROOF.

(1) Since $a \triangleright b \neq b$ and since X is braided, we have

$$e = d \triangleright (c \triangleright b) = (c \triangleright (b \triangleright a)) \triangleright (c \triangleright b) = c \triangleright ((b \triangleright a) \triangleright b)) = c \triangleright a.$$

(2)(a) By assumption, $c \notin \{a, b\}$ and hence $c \neq e$ by (1). Moreover, $c \neq d$ since $c \neq b \triangleright a$.

(2)(b) By assumption, $b \triangleright a \notin \{a, b\}$. Moreover, $b \triangleright a \neq d$ since c does not commute with $b \triangleright a$. Finally, $b \triangleright a \neq e$ since otherwise $b \triangleright a = c \triangleright a$ by (1). Since a commutes with none of b, c and since X is braided, this would imply $b = c$.

(2)(c) Obviously, $b \triangleright a \neq c$ and $b \triangleright b = b \neq c$. Assume that $b \triangleright d = c$ or $b \triangleright e = c$. By (1) and since

$$b \triangleright d = b \triangleright (c \triangleright (b \triangleright a)) = c \triangleright (b \triangleright (c \triangleright a)),$$

it follows that $b \triangleright (c \triangleright a) = c$. Hence $(c \triangleright a) \triangleright (b \triangleright (c \triangleright a)) = a$. Since X is braided, it follows that $b = a$ or $c \triangleright a = a$ by Lemma 3.4. Since $b \neq a$ and $c \triangleright a \neq a$ by assumption, it follows that $b \triangleright d \neq c$ and $b \triangleright e \neq c$.

(2)(d) We have $b \triangleright (a \triangleright a) = b \triangleright a \neq c$ and $b \triangleright (a \triangleright b) = a \neq c$. If $c = b \triangleright (a \triangleright d) = b \triangleright ((a \triangleright c) \triangleright b)$, it would follow $c = b$ or $c = a \triangleright c$ by Lemma 3.4. If $c = b \triangleright (a \triangleright e)$, it would follow $c = b \triangleright (a \triangleright (c \triangleright a)) = b \triangleright c$. □

Recall that we have defined linear maps

$$c_i : V^{\otimes n} \to V^{\otimes n}$$
$$v_1 \otimes \ldots \otimes v_n \mapsto v_1 \otimes \ldots \otimes v_{i-1} \otimes c_q(v_i \otimes v_{i+1}) \otimes v_{i+2} \otimes \ldots \otimes v_n$$

for all $n \in \mathbb{N}$ and $1 \leq i \leq n-1$. Note that the maps c_i depend on the two-cocycle although the symbols do not indicate it. Moreover, the maps are G_X-graded (that is, $\Bbbk G_X$-comodule maps) since c_q is. The maps c_i fulfill the braid relations

$$c_i c_{i+1} c_i = c_{i+1} c_i c_{i+1} \qquad \text{for all } 1 \leq i \leq n-2, \tag{4.4}$$

$$c_i c_j = c_j c_i \qquad \text{for all } 1 \leq i, j \leq n-1, |i-j| > 1, \tag{4.5}$$

where the first relation holds since c_q fulfills the Yang–Baxter equation and the second by definition of c_i.

THEOREM 4.6. *Assume that the rack X is braided, that there are pairwise distinct elements $a, b, c \in X$ with $a \triangleright b \neq b$ and that $\mathcal{B}(V)$ is slim in degree two. Let $d = c \triangleright (b \triangleright a)$, $e = d \triangleright (c \triangleright b)$ and $S = \{v_a, v_b, v_d, v_e\}$. Define*

$$Z(a, b, c) = \mu(\mathrm{id} - c_1 c_2 + (c_1 c_2)^2)(v_c \otimes v_b \otimes v_a) \in \mathcal{B}(V),$$

and let C be the subalgebra of $\mathcal{B}(V)$ generated by $S \cup \{Z\}$. Then the following hold.

1. *The subalgebra C is a left coideal of $\mathcal{B}(V)$ in the category of $\mathbb{N}_0$-graded $\Bbbk G_X$-comodules.*
2. *Assume that c commutes with none of a, b, $b \triangleright a$. If $a \triangleright b \neq c$ or $a \triangleright c \neq b \triangleright a$, then $Z \notin \langle S \rangle$ and hence C is not generated in degree one.*

PROOF. (1) One calculates that Z is G_X-homogeneous of degree $g_c g_b g_a$. Hence, the subalgebra C is a $\Bbbk G_X$-subcomodule. Thus by Corollary 2.8 it suffices to show that $\Delta_{1,2}(Z) \in V \otimes \langle S \rangle$. Using Equation (1.4) we get

$$\begin{aligned}\Delta_{1,2}(Z) =& (\mathrm{id} \otimes \mu)(\mathrm{id} + c_1 + c_1 c_2)(\mathrm{id} - c_1 c_2 + (c_1 c_2)^2)(v_c \otimes v_b \otimes v_a) \\ =& (\mathrm{id} \otimes \mu)(\mathrm{id} - c_1^2 c_2 + (c_1 c_2)^3 + c_1 + c_1 (c_1 c_2)^2)(v_c \otimes v_b \otimes v_a).\end{aligned}$$

By the braid equation (4.4),

$$c_1 (c_1 c_2)^2 = (c_1 c_2)^2 c_2 = c_2 c_1 c_2^3.$$

Moreover, one has $c_2^3(v_c \otimes v_b \otimes v_a) = -v_c \otimes v_b \otimes v_a$ since $b \triangleright a \neq a$, and $\mathcal{B}(V)$ is slim in degree two (see Lemma 4.4 (2)). Thus

$$(c_1 + c_1 (c_1 c_2)^2)(v_c \otimes v_b \otimes v_a) = (\mathrm{id} - c_2) c_1 (v_c \otimes v_b \otimes v_a)$$

and it follows that

$$\begin{aligned}\Delta_{1,2}(Z) =& (\mathrm{id} \otimes \mu)(\mathrm{id} - c_1^2 c_2 + (c_1 c_2)^3)(v_c \otimes v_b \otimes v_a) \\ &+ (\mathrm{id} \otimes \mu)(\mathrm{id} - c_2) c_1 (v_c \otimes v_b \otimes v_a) \\ =& v_c \otimes v_b v_a - k_1 v_{d \triangleright c} \otimes v_d v_b + k_2 v_{e \triangleright (d \triangleright c)} \otimes v_e v_d \\ &+ (\mathrm{id} \otimes \mu)(\mathrm{id} - c_2) c_1 (v_c \otimes v_b \otimes v_a),\end{aligned}$$

where $k_1, k_2 \in \Bbbk$ depending on the two-cocycle. To reason $(\mathrm{id} \otimes \mu)(\mathrm{id} - c_2) c_1 (v_c \otimes v_b \otimes v_a) \in V \otimes \langle S \rangle$ we distinguish two cases and in both cases we use that $\mathcal{B}(V)$ is slim in degree two.

Case (a). Assume that $c \triangleright a = a$. Then

$$(\mathrm{id} \otimes \mu)(\mathrm{id} - c_2) c_1 (v_c \otimes v_b \otimes v_a) = (\mathrm{id} \otimes \mu)(\mathrm{id} - c_2)(q_{c,b} v_{c \triangleright b} \otimes v_c \otimes v_a) = 0$$

since Equation (4.2) holds.

Case (b). Assume that $c \triangleright a \neq a$. Then

$$\begin{aligned}(\mathrm{id}\otimes\mu)(\mathrm{id}-c_2)c_1(v_c\otimes v_b\otimes v_a) &= (\mathrm{id}\otimes\mu)(\mathrm{id}-c_2)(q_{c,b}v_{c\triangleright b}\otimes v_c\otimes v_a)\\ &= (\mathrm{id}\otimes\mu)(-{c_2}^2)(q_{c,b}v_{c\triangleright b}\otimes v_c\otimes v_a)\\ &= k_3 v_{c\triangleright b}\otimes v_a v_e \in V\otimes\langle S\rangle\end{aligned}$$

for a $k_3 \in \mathbb{k}$, where we used Equation (4.3) (applied to $c, a \in X$) in the second equality and Lemma 4.5 (1) in the last step.

In both cases it follows that $\Delta_{1,2}(Z) \in V \otimes \langle S\rangle$, which proves the claim.

(2) Let F be the linear functional on $V^{\otimes 3}$ with $F(v_c \otimes v_b \otimes v_a) = 1$ and $F(v_i \otimes v_j \otimes v_k) = 0$ for all $(i,j,k) \in X^3 \setminus \{(c,b,a)\}$. We prove that $F(\Delta_{1^3}(Z)) = 1$ but $F(\Delta_{1^3}(Y)) = 0$ for all $Y \in \langle S\rangle(3)$. It is well-known, and follows by similar arguments as in Lemma 2.6 that $\Delta_{1^3} = (\mathrm{id}\otimes\Delta_{1,1})\Delta_{1,2}$. With the result of our calculation of $\Delta_{1,2}(Z)$ in part (1) of the proof (in the case $c \triangleright a \neq a$) we get

$$\begin{aligned}\Delta_{1^3}(Z) =&(\mathrm{id}\otimes\Delta_{1,1})\Delta_{1,2}(Z)\\ =&(\mathrm{id}\otimes\Delta_{1,1})(v_c\otimes v_b v_a - k_1 v_{d\triangleright c}\otimes v_d v_b + k_2 v_{e\triangleright(d\triangleright c)}\otimes v_e v_d\\ &+ k_3 v_{c\triangleright b}\otimes v_a v_e),\end{aligned}$$

where $k_i \in \mathbb{k}$. Since $b \neq a$, we obtain $F((\mathrm{id}\otimes\Delta_{1,1})(v_c\otimes v_b v_a)) = 1$. Since X is braided, the assumption $c \triangleright (b \triangleright a) \neq b \triangleright a$ implies $d \triangleright c \neq c$ and hence $F((\mathrm{id}\otimes\Delta_{1,1})(v_{d\triangleright c}\otimes v_d v_b) = 0$. Now consider the summand

$$\begin{aligned}&F((\mathrm{id}\otimes\Delta_{1,1})(v_{e\triangleright(d\triangleright c)}\otimes v_e v_d))\\ &\qquad = F(v_{e\triangleright(d\triangleright c)}\otimes v_e\otimes v_d) + F(v_{e\triangleright(d\triangleright c)}\otimes q_{e,d}v_{e\triangleright d}\otimes v_e).\end{aligned}$$

Since c does not commute with a, it follows that $e \neq a$ by Lemma 4.5(1) and hence $F(v_{e\triangleright(d\triangleright c)}\otimes q_{e,d}v_{e\triangleright d}\otimes v_e) = 0$. By assumption, we either have $a\triangleright b \neq c$ or $a\triangleright c \neq b\triangleright a$. If $a \triangleright b \neq c$, then $b = (a\triangleright b)\triangleright a \neq c \triangleright a = e$ and thus $F(v_{e\triangleright(d\triangleright c)}\otimes v_e\otimes v_d) = 0$. If $a \triangleright c \neq b \triangleright a$, then $d \neq a$ since otherwise $c \triangleright (b \triangleright a) = a$ and $c \triangleright (a \triangleright c) = a$ and hence $b \triangleright a = a \triangleright c$. It follows that $F(v_{e\triangleright(d\triangleright c)}\otimes v_e\otimes v_d) = 0$. Finally, we get $F((\mathrm{id}\otimes\Delta_{1,1})(v_{c\triangleright b}\otimes v_a v_e)) = 0$ since $c \triangleright b \neq c$. We conclude $F(\Delta_{1^3}(Z)) = 1$.

To prove $F(\Delta_{1^3}(\langle S\rangle(3))) = \{0\}$ we show $F(\Delta_{1^3}(v_x v_y v_z)) = 0$ for all $x, y, z \in \{a, b, d, e\}$. Using Equation (1.4) we get

$$\begin{aligned}\Delta_{1^3}(v_x v_y v_z) =&(\mathrm{id}\otimes\Delta_{1,1})\Delta_{1,2}(v_x v_y v_z)\\ =&(\mathrm{id}\otimes\Delta_{1,1}\mu)(\mathrm{id}+c_1+c_1c_2)(v_x\otimes v_y\otimes v_z)\\ =&v_x\otimes v_y\otimes v_z + l_1 v_x\otimes v_{y\triangleright z}\otimes v_y\\ &+ l_2 v_{x\triangleright y}\otimes v_x\otimes v_z + l_3 v_{x\triangleright y}\otimes v_{x\triangleright z}\otimes v_x\\ &+ l_4 v_{x\triangleright(y\triangleright z)}\otimes v_x\otimes v_y + l_5 v_{x\triangleright(y\triangleright z)}\otimes v_{x\triangleright y}\otimes v_x,\end{aligned}$$

where $l_i \in \mathbb{k}$ depending on the two-cocycle. Since $c \notin \{a, b, d, e\}$ by Lemma 4.5 (2)(a), v_x can not be a multiple of v_c and it follows that

$$F(v_x\otimes v_y\otimes v_z + l_1 v_x\otimes v_{y\triangleright z}\otimes v_y) = 0.$$

The equation

$$F(v_{x\triangleright y}\otimes v_x\otimes v_z) = 0$$

holds since otherwise ($x = b$ and) $b \triangleright y = c$, but $b \triangleright y \neq c$ for all $y \in \{a, b, d, e\}$ by Lemma 4.5 (2)(c). The equation

$$F(v_{x\triangleright y}\otimes v_{x\triangleright z}\otimes v_x) = 0$$

holds since otherwise ($x = a$ and) $a \triangleright z = b$ and hence $z = b \triangleright a$, but $b \triangleright a \notin \{a, b, d, e\}$ by Lemma 4.5 (2)(b). The equation

$$F(v_{x\triangleright(y\triangleright z)} \otimes v_x \otimes v_y) = 0$$

holds since otherwise ($x = b$, $y = a$ and) $b \triangleright (a \triangleright z) = c$, but $b \triangleright (a \triangleright z) \neq c$ for all $z \in \{a, b, d, e\}$ by Lemma 4.5 (2)(d). Finally,

$$F(v_{x\triangleright(y\triangleright z)} \otimes v_{x\triangleright y} \otimes v_x) = 0$$

since otherwise $a \triangleright y = b$ and hence $y = b \triangleright a$, but $b \triangleright a \notin \{a, b, d, e\}$ by Lemma 4.5 (2)(b). We conclude $F(\Delta_{1^3}(v_x v_y v_z)) = 0$. □

LEMMA 4.7. *Assume that the rack X is braided, that there are pairwise distinct elements $a, b, c \in X$ with $a \triangleright b \neq b$ and that $\mathcal{B}(V)$ is slim in degree two. Let $d = c \triangleright (b \triangleright a)$, $e = d \triangleright (c \triangleright b)$ and $S = \{v_a, v_b, v_d, v_e\}$. Define*

$$Z = \mu(\mathrm{id} - c_1c_2 + (c_1c_2)^2)(v_c \otimes v_b \otimes v_a) \in \mathcal{B}(V),$$

and let C be the subalgebra of $\mathcal{B}(V)$ generated by $S \cup \{Z\}$. Then the following hold.

1. *If $c \triangleright b = b$ and $c \triangleright a \neq a$, then $Z \in \langle S \rangle$.*
2. *If $c \triangleright a = a$ and $c \triangleright b \neq b$, then $Z \in \langle S \rangle$.*
3. *If $c \triangleright (b \triangleright a) = b \triangleright a$ and $c \triangleright a \neq a$, then $Z \in \langle S \rangle$.*

PROOF.

(1) Our strategy is to prove that in $\mathcal{B}(V)$ the relation

$$\mu(\mathrm{id} - c_1c_2 + (c_1c_2)^2 - (c_1c_2)^3)(v_b \otimes v_a \otimes v_e) = 0 \tag{4.6}$$

holds. Note that $c_1c_2(v_b \otimes v_a \otimes v_e) \in \Bbbk v_c \otimes v_b \otimes v_a$ since

$$b \triangleright (a \triangleright e) = b \triangleright (a \triangleright (c \triangleright a)) = b \triangleright c = c, \tag{4.7}$$

where we used $e = c \triangleright a$ from Lemma 4.5 (1). Thus the claim of the Lemma follows from relation (4.6) since together with Equation (4.7), it implies

$$v_b v_a v_e - q\mu(\mathrm{id} - c_1c_2 + (c_1c_2)^2)(v_c \otimes v_b \otimes v_a) = 0 \in \mathcal{B}(V)$$

and hence $v_b v_a v_e = qZ$ in $\mathcal{B}(V)$ for a $q \in \Bbbk^\times$ and $v_b v_a v_e \in \langle S \rangle$. For the proof of Equation (4.6) we use Equation (1.4) and compute

$$\begin{aligned}
&\Delta_{1,2}\mu(\mathrm{id} - c_1c_2 + (c_1c_2)^2 - (c_1c_2)^3)(v_b \otimes v_a \otimes v_e)\\
=&(\mathrm{id} \otimes \mu)(\mathrm{id} + c_1 - c_1^2c_2 + c_1(c_1c_2)^2 - c_1(c_1c_2)^3 - (c_1c_2)^4)(v_b \otimes v_a \otimes v_e)\\
=&(\mathrm{id} \otimes \mu)(\mathrm{id} - c_1^2c_2 - c_1(c_1c_2)^3)(v_b \otimes v_a \otimes v_e)\\
&+ (\mathrm{id} \otimes \mu)(c_1 + c_1(c_1c_2)^2 - (c_1c_2)^4)(v_b \otimes v_a \otimes v_e).
\end{aligned}$$

By Lemma 4.5(1) and since X is braided, we have $a \triangleright e = c$ and hence b commutes with $a \triangleright e$ by assumption. Since $\mathcal{B}(V)$ is slim in degree two, we obtain $c_1^2c_2(v_b \otimes v_a \otimes v_e) = c_2(v_b \otimes v_a \otimes v_e)$ by Lemma 4.4(1). Using the braid relation (4.4) repeatedly we get $c_1(c_1c_2)^3 = c_2c_1^2c_2c_1^3$. Note that $a \triangleright e = c$ hence $e \neq c$ by assumption and $(b \triangleright a) \triangleright e = e$ since $a \triangleright ((b \triangleright a) \triangleright e) = b \triangleright c = c$. Thus

$$c_2c_1^2c_2c_1^3(v_b \otimes v_a \otimes v_e) = -c_2c_1^2c_2(v_b \otimes v_a \otimes v_e) = -c_2^2(v_b \otimes v_a \otimes v_e),$$

where we used that $\mathcal{B}(V)$ is slim in degree two and Lemma 4.4(2) for the first and (1) for the second equality. It follows that

$$\begin{aligned}(\mathrm{id}\otimes\mu)(\mathrm{id}-c_1^2c_2-c_1(c_1c_2)^3)(v_b\otimes v_a\otimes v_e)\\=(\mathrm{id}\otimes\mu)(\mathrm{id}-c_2+c_2^2)(v_b\otimes v_a\otimes v_e)=0,\end{aligned}$$

where we used again that $\mathcal{B}(V)$ is slim in degree two and $a\triangleright e\neq e$. Using the braid Equation (4.4) we obtain $c_1(c_1c_2)^2=c_2c_1c_2^3$ and $c_2c_1c_2^3(v_b\otimes v_a\otimes v_e)=-c_2c_1(v_b\otimes v_a\otimes v_e)$ by Lemma 4.4(2) since $a\triangleright e\neq e$. The braid relation (4.4) leads to $(c_1c_2)^4=c_1^2c_2^2c_1c_2^3$ and we have

$$c_1^2c_2^2c_1c_2^3(v_b\otimes v_a\otimes v_e)=-c_1^2c_2^2c_1(v_b\otimes v_a\otimes v_e)=-c_2^2c_1(v_b\otimes v_a\otimes v_e),$$

where the first equation holds again since $a\triangleright e\neq e$ and by Lemma 4.4(2) and the second equation holds since $(b\triangleright a)\triangleright e=e$ and by Lemma 4.4(1) since $\mathcal{B}(V)$ is slim in degree two. Note that

$$b\triangleright e=b\triangleright(c\triangleright a)=c\triangleright(b\triangleright a)=d$$

and $e=c\triangleright a\neq c\triangleright(b\triangleright a)=d$ since $a\neq b\triangleright a$. Hence $b\triangleright e\neq e$ and it follows that

$$\begin{aligned}(\mathrm{id}\otimes\mu)(c_1+c_1(c_1c_2)^2-(c_1c_2)^4)(v_b\otimes v_a\otimes v_e)\\=(\mathrm{id}\otimes\mu)(\mathrm{id}-c_2+c_2^2)c_1(v_b\otimes v_a\otimes v_e)=0,\end{aligned}$$

by Lemma 4.4(2). We conclude

$$\Delta_{1,2}\mu(\mathrm{id}-c_1c_2+(c_1c_2)^2-(c_1c_2)^3)(v_b\otimes v_a\otimes v_e)=0,$$

which implies $\mu(\mathrm{id}-c_1c_2+(c_1c_2)^2-(c_1c_2)^3)(v_b\otimes v_a\otimes v_e)=0$ since $\Delta_{1,2}$ is injective.

(2) Let $Z'=Z-\mu(c_1c_2)^3(v_c\otimes v_b\otimes v_a)$. Our strategy is to prove that in $\mathcal{B}(V)$ the relation $Z'=0$ holds. This can be shown by proving $\Delta_{1,2}(Z')=0$ with similar arguments as in (1). We only describe how the claim of the Lemma follows from the relation $Z'=0$. By the braid relation (4.4), the equation $(c_1c_2)^3=(c_2^2c_1)^2$ holds. Since $c\triangleright a=a$ and since $\mathcal{B}(V)$ is slim in degree two, we have

$$(c_2^2c_1)^2(v_c\otimes v_b\otimes v_a)=c_2^2c_1^2(v_c\otimes v_b\otimes v_a)$$

by Lemma 4.4 (1). Moreover,

$$c_2^2c_1^2(v_c\otimes v_b\otimes v_a)\in \Bbbk c_2^2(v_b\otimes v_{c\triangleright b}\otimes v_a)$$

since $(c\triangleright b)\triangleright c=b$ (since X is braided) and

$$c_2^2(v_b\otimes v_{c\triangleright b}\otimes v_a)\in \Bbbk v_b\otimes v_{((c\triangleright b)\triangleright a)\triangleright(c\triangleright b)}\otimes v_{(c\triangleright b)\triangleright a}.$$

Since a commutes with c but not with b, it does not commute with $c\otimes b$ and $((c\triangleright b)\triangleright a)\triangleright(c\triangleright b))=a$ since X is braided. Hence

$$\mu(c_1c_2)^3(v_c\otimes v_b\otimes v_a)\in \Bbbk v_bv_av_d\in\langle S\rangle.$$

Thus $Z'=0$ implies $Z=\mu(c_1c_2)^3(v_c\otimes v_b\otimes v_a)\in\langle S\rangle$.

(3) Let $Z'=Z-\mu(c_1c_2)^3(v_c\otimes v_b\otimes v_a)$. Our strategy is to prove that in $\mathcal{B}(V)$ the relation $Z'=0$ holds. This can be shown by proving $\Delta_{1,2}(Z')=0$ with similar arguments as in (1). We only describe how the claim of the Lemma follows from the relation $Z'=0$. Note that $(c_1c_2)^3=(c_1^2c_2)^2$ by the braid relation (4.4). Since $c\triangleright(b\triangleright a)=b\triangleright a$ and since $\mathcal{B}(V)$ is slim in degree two, we have

$$(c_1^2c_2)^2(v_c\otimes v_b\otimes v_a)=c_1^2c_2^2(v_c\otimes v_b\otimes v_a)$$

by Lemma 4.4 (1). Moreover,

$$c_1^2c_2^2(v_c\otimes v_b\otimes v_a)\in \Bbbk(v_{(c\rhd a)\rhd c}\otimes v_{c\rhd a}\otimes v_{b\rhd a})=\Bbbk(v_a\otimes v_e\otimes v_d)$$

and hence $\mu(c_1c_2)^3(v_c\otimes v_b\otimes v_a)\in \Bbbk v_av_ev_d\subset \langle S\rangle$. Thus $Z'=0$ implies $Z\in\langle S\rangle$.

□

In the next Lemma and the following Theorem we use abbreviating notations. For a pair of elements $x,y\in X$ we write $x\square y$ if $x\rhd y=y$ but $x\neq y$. For elements $x_1,x_2,\dots,x_n,y\in X$ we abbreviate

$$x_1x_2\cdots x_n\rhd y=x_1\rhd(x_2\rhd(\dots\rhd(x_n\rhd y))).$$

For $n\in\mathbb{N}$ the **Hurwitz action** of the braid group $\mathbb{B}_n$ on X^n is defined by

$$\sigma_i(x_1,\dots,x_n)=(x_1,\dots,x_{i-1},x_i\rhd x_{i+1},x_i,x_{i+2},\dots,x_n),$$

for all $x_1,\dots,x_n\in X$, $i\in\{1,\dots,n-1\}$. We denote the orbit of a tuple $(x_1,\dots,x_n)$ with respect to the Hurwitz action by $\mathcal{O}(x_1,\dots,x_n)$.

LEMMA 4.8. *Assume that X is braided and let $a,b,c,d\in X$ be pairwise not commuting elements except $a\square c$. Assume that $b\square(c\rhd d)$ and $a\square(b\rhd d)$. Then the Hurwitz orbit $\mathcal{O}(a,b,c,d)$ consists of the following four types of tuples*

(x) (x_1,x_2,x_3,x_4) *such that* $x_1\square x_3$, $x_2\square(x_3\rhd x_4)$ *and* $x_1\square(x_2\rhd x_4)$.
(y) (y_1,y_2,y_3,y_4) *such that* $y_2\square y_3$, $y_1\square y_4$ *and* $y_1\square(y_2\rhd(y_3\rhd y_4))$.
(z) (z_1,z_2,z_3,z_4) *such that* $z_2\square z_4$, $z_1\square(z_3\rhd z_4)$ *and* $z_1\square(z_2\rhd z_3)$.
(u) (u_1,u_2,u_3,u_4) *such that* $u_1\square u_2$, $u_3\square u_4$ *and* $u_1\square(u_2\rhd(u_3\rhd u_4))$.

The unmentioned components of the tuples do not commute.

PROOF. We only roughly sketch the proof here because it is lengthy but straightforward using the definition of a braided rack. The tuple $(a,b,c,d)\in\mathcal{O}(a,b,c,d)$ is of type (x) by assumption. As an example we show that the action of σ_1 on a tuple of type (x) leads to a tuple of type (y). If we apply σ_1 on a tuple of the form (x), we get a tuple $y=(y_1,y_2,y_3,y_4)=(x_1\rhd x_2,x_1,x_3,x_4)$. From $x_1\square x_3$ it follows directly that $y_2\square y_3$. Since $x_1\square(x_2\rhd x_4)$, we have $x_1\rhd(x_2\rhd x_4)=x_2\rhd x_4$. Applying $x_4\rhd$ on that equation we get $(x_4\rhd x_1)\rhd x_2=x_2$ since X is braided and $x_4\rhd x_2\neq x_2$. Hence

$$\begin{aligned}y_4\rhd y_1&=x_4\rhd(x_1\rhd x_2)=(x_1\rhd(x_4\rhd x_1))\rhd(x_1\rhd x_2)\\&=x_1\rhd((x_4\rhd x_1)\rhd x_2)=x_1\rhd x_2=y_1.\end{aligned}$$

The second equality holds since $x_1\square x_4$ and X is braided. Moreover, $y_1\neq y_4$ since otherwise $x_1\rhd x_2=x_4$ and hence $x_1=x_2\rhd x_4$. It follows that $y_1\square y_4$. The relation $x_2\square(x_3\rhd x_4)$ implies $(x_1\rhd(x_2\rhd x_1))\square(x_3\rhd x_4)$ and $(y_1\rhd y_2)\square(y_3\rhd y_4)$, what is equivalent to $y_1\square(y_2y_3\rhd y_4)$. It is easy to check that other components of y do not commute. In this way we proved in the other 11 cases that applying σ_1,σ_2 or σ_3 on a tuple of the form $(x),(y),(z)$ or (u) always again yields a tuple of the form $(x),(y),(z)$ or (u). Figure 1 shows the action of the generators of the braid group on the four types of tuples. □

LEMMA 4.9. *Assume that X is braided and let $a,b,c,d\in X$ be pairwise not commuting elements except $a\square c$. Assume that $b\square(c\rhd d)$ and $a\square(b\rhd d)$. Let $e=abc\rhd d$, $f=eab\rhd c$, $g=fea\rhd b$ and $S=\{b,c,d,e,f,g\}$. Then two distinct elements in S do not commute, at best $d\rhd e=e$, $c\rhd f=f$ and $b\rhd g=g$.*

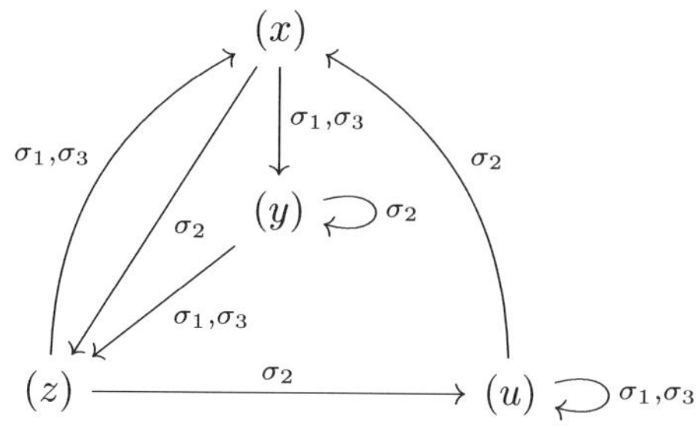

FIGURE 1. Hurwitz action on quadruples

PROOF. In X the relations

$$\begin{aligned} e &= abc \triangleright d = ac \triangleright d, \\ f &= eab \triangleright c = a \triangleright ((c \triangleright d) \triangleright (b \triangleright c)) = a \triangleright (b \triangleright d) = b \triangleright d, \\ g &= fea \triangleright b = (b \triangleright d) \triangleright ((a \triangleright (c \triangleright d)) \triangleright (a \triangleright b)) = (b \triangleright d) \triangleright (a \triangleright b) = a \triangleright d \end{aligned}$$

hold. Since $b \triangleright a \neq a$, $bc \triangleright d = c \triangleright d$ and $ac \triangleright d \neq c \triangleright d$, it follows that $b \triangleright e = b \triangleright (ac \triangleright d) \neq ac \triangleright d = e$. Since $b \triangleright d \neq d$, it follows that $b \triangleright f = bb \triangleright d \neq b \triangleright d = f$.

From $(c \triangleright d) \triangleright c = d \neq c$, $c \triangleright a = a$ and $ac \triangleright d \neq c \triangleright d$, it follows that $c \triangleright e = cac \triangleright d \neq ac \triangleright d = e$. Since $c \triangleright d \neq d$, $c \triangleright a = a$ and $a \triangleright d \neq d$, it follows that $c \triangleright g = ca \triangleright d \neq a \triangleright d = g$.

The relation $d \triangleright f \neq f$ follows from $d \triangleright b \neq b$, the relation $d \triangleright g \neq g$ from $d \triangleright a \neq a$.

The equation $g \triangleright e = e$ would imply $c \triangleright d = d$ and hence $d = c$, a contradiction. The equation $f \triangleright g = g$ would imply the contradiction $b \triangleright a = a$.

Finally, $f \triangleright e \neq e$ since otherwise $a \triangleright ((b \triangleright d) \triangleright (c \triangleright d)) = ac \triangleright d$ and hence $(b \triangleright d) \triangleright (c \triangleright d) = c \triangleright d$ which would imply $b \triangleright c = c$. □

LEMMA 4.10. *Assume that X is braided and let $a, b, c, d \in X$ be pairwise not commuting elements except $a \square c$. Assume that $b \square (c \triangleright d)$ and $a \square (b \triangleright d)$. Let $e = abc \triangleright d$, $f = eab \triangleright c$, $g = fea \triangleright b$ and $S = \{b, c, d, e, f, g\}$. Then $S^4 \cap \mathcal{O}(a, b, c, d) = \emptyset$.*

PROOF. First we prove that $abcd \triangleright x \in S$ and $abcdabcd \triangleright x = x$ for all $x \in S$. Indeed, using $e = ac \triangleright d$, $f = b \triangleright d$ and $g = a \triangleright d$ (from the proof of Lemma 4.9) it follows that

$$\begin{aligned} abcd \triangleright b &= ab \triangleright ((c \triangleright d) \triangleright (c \triangleright b))) = a \triangleright ((c \triangleright d) \triangleright c) = a \triangleright d = g, \\ abcd \triangleright c &= ab \triangleright d = b \triangleright d = f, \\ abcd \triangleright d &= abc \triangleright d = ac \triangleright d = e, \\ abcd \triangleright e &= abcd \triangleright (ac \triangleright d) = abc \triangleright ((d \triangleright c) \triangleright a), \\ &= ab \triangleright (d \triangleright a) = (b \triangleright d) \triangleright b = d, \\ abcd \triangleright f &= abcd \triangleright (b \triangleright d) = abc \triangleright b = a \triangleright c = c \text{ and} \\ abcd \triangleright g &= abcd \triangleright (a \triangleright d) = abc \triangleright a = ab \triangleright a = b. \end{aligned}$$

Since $\triangleright$ is left self-distributive,

$$\sigma_1\sigma_2\sigma_3(t_1, t_2, t_3, t_4) = (abcd \triangleright t_4, t_1, t_2, t_3)$$

for all $t = (t_1, t_2, t_3, t_4) \in \mathcal{O}(a, b, c, d)$. It follows that $S^4 \cap \mathcal{O}(a, b, c, d)$ is invariant under the action of $\sigma_1\sigma_2\sigma_3$. Now assume that there is a tuple $t \in S^4 \cap \mathcal{O}(a, b, c, d)$.

By Lemma 4.8, the tuple t is of the form $(x),(y),(z)$ or (u) and hence at least two entries of t commute. Thus t has an entry $h \in \{b,c,d\}$ and an entry $abcd \triangleright h$ by our calculations at the beginning of the proof and Lemma 4.9. But then multiple application of $\sigma_1\sigma_2\sigma_3$ on t yields a tuple $t' \in S^4 \cap \mathcal{O}(a,b,c,d)$ with two components equal to h or two components equal to $abcd \triangleright h$. This is a contradiction since tuples of the form $(x),(y),(z),(u)$ do not have two equal entries and there are no other tuples in $\mathcal{O}(a,b,c,d)$ by Lemma 4.8. We conclude $S^4 \cap \mathcal{O}(a,b,c,d) = \emptyset$. □

THEOREM 4.11. *Assume that X is braided. Let $a,b,c,d \in X$ pairwise not commuting elements except $a \square c$ and such that $b \square (c \triangleright d)$ and $a \square (b \triangleright d)$. Let $e = abc \triangleright d$, $f = eab \triangleright c$ and $g = fea \triangleright b$. Assume that $\mathcal{B}(V)$ is slim in degree two. Define the element*

$$Z(a,b,c,d) = \mu(\mathrm{id} - c_1c_2c_3 + (c_1c_2c_3)^2 - (c_1c_2c_3)^3)(v_a \otimes v_b \otimes v_c \otimes v_d)$$

of $\mathcal{B}(V)$. Let $S = \{v_b, v_c, v_d, v_e, v_f, v_g\}$ and let C be the subalgebra of $\mathcal{B}(V)$ generated by $S \cup \{Z\}$. Then the following hold.

1. *The subalgebra C is a left coideal subalgebra of $\mathcal{B}(V)$ in the category of $\mathbb{N}_0$-graded $\Bbbk G_X$-comodules.*
2. *If $e \neq d$, $f \neq c$ or $g \neq b$, then C is not generated in degree one.*

PROOF. (1) Since Z is G_X-homogeneous of degree $g_ag_bg_cg_d$, the subalgebra C is a $\Bbbk G_X$-subcomodule. By Corollary 2.8, it suffices to prove $\Delta_{1,3}(Z) \in V \otimes \langle S \rangle$. For this we use Equation (1.4) and get

$$\Delta_{1,3}(Z) = s_1 + s_2 + s_3,$$

where

$$\begin{aligned} s_1 =&(\mathrm{id} \otimes \mu)(\mathrm{id} - c_1(c_1c_2c_3) + c_1c_2(c_1c_2c_3)^2 \\ &- (c_1c_2c_3)^4)(v_a \otimes v_b \otimes v_c \otimes v_d), \\ s_2 =&(\mathrm{id} \otimes \mu)(c_1 - c_1c_2c_1c_2c_3 - c_1(c_1c_2c_3)^3)(v_a \otimes v_b \otimes v_c \otimes v_d), \\ s_3 =&(\mathrm{id} \otimes \mu)(c_1c_2 + c_1(c_1c_2c_3)^2 - c_1c_2(c_1c_2c_3)^3)(v_a \otimes v_b \otimes v_c \otimes v_d). \end{aligned}$$

The first summand s_1 lies in $V \otimes \langle S \rangle$ since

$$\begin{aligned} s_1 =&v_a \otimes v_bv_cv_d + k_1v_{e\triangleright a} \otimes v_ev_bv_c + k_2v_{f\triangleright(e\triangleright a)} \otimes v_fv_ev_b \\ &+ k_3v_{g\triangleright(f\triangleright(e\triangleright a))} \otimes v_gv_fv_e, \end{aligned}$$

where $k_i \in \Bbbk$ depend on the two-cocycle.

Next we consider the second summand s_2 of $\Delta_{1,3}(Z)$. Using the braid Equation (4.4) we get $c_1c_2c_1c_2c_3 = c_2c_1c_2^2c_3$ and

$$c_2c_1c_2^2c_3(v_a \otimes v_b \otimes v_c \otimes v_d) = c_2c_1c_3(v_a \otimes v_b \otimes v_c \otimes v_d),$$

by Lemma 4.4 (1), where we use that $b \square (c \triangleright d)$ and that $\mathcal{B}(V)$ is slim in degree two. Using the braid Equation (4.5) we conclude

$$c_1c_2c_1c_2c_3(v_a \otimes v_b \otimes v_c \otimes v_d) = c_2c_3c_1(v_a \otimes v_b \otimes v_c \otimes v_d).$$

From the braid equations we obtain $c_1(c_1c_2c_3)^3 = c_2c_3c_2c_1c_2^2c_3c_2^3$ and from Lemma 4.4 it follows that

$$c_2c_3c_2c_1c_2^2c_3c_2^3(v_a \otimes v_b \otimes v_c \otimes v_d) = -c_2c_3c_2c_1c_3(v_a \otimes v_b \otimes v_c \otimes v_d)$$

since $b \triangleright c \neq c$, $b \square (c \triangleright d)$ and $\mathcal{B}(V)$ is slim in degree two. Using again the braid equation we conclude

$$c_1(c_1c_2c_3)^3(v_a \otimes v_b \otimes v_c \otimes v_d) = (c_2c_3)^2c_1(v_a \otimes v_b \otimes v_c \otimes v_d).$$

Hence for the second summand s_2 of $\Delta_{1,3}(Z)$ above we get

$$\begin{aligned} s_2 =&(\mathrm{id} \otimes \mu)(\mathrm{id} - c_2c_3 + (c_2c_3)^2)c_1(v_a \otimes v_b \otimes v_c \otimes v_d) \\ =&(\mathrm{id} \otimes \mu)(k_4 v_{a\triangleright b} \otimes (\mathrm{id} - c_1c_2 + (c_1c_2)^2))(v_a \otimes v_c \otimes v_d) \end{aligned}$$

for some $k_4 \in \Bbbk$. Since $d \triangleright c \neq c$, $a \triangleright c = c$ and $a \triangleright d \neq d$, we have

$$\mu(\mathrm{id} - c_1c_2 + (c_1c_2)^2)(v_a \otimes v_c \otimes v_d) \in \langle v_d, v_c, v_{c\triangleright g}, v_g \rangle \subset \langle S \rangle$$

by Lemma 4.7 (3). There we used $c \triangleright g = e$ what follows from $g = a \triangleright d$ and $e = ac \triangleright d$ (see the proof of Lemma 4.9). Hence $s_2 \in V \otimes \langle S \rangle$.

Next we consider the third summand s_3 of $\Delta_{1,3}(Z)$ above. Using the braid equations we get $c_1(c_1c_2c_3)^2 = c_2c_1c_3c_2c_3^3$ and

$$c_2c_1c_3c_2c_3^3(v_a \otimes v_b \otimes v_c \otimes v_d) = -c_2c_1c_3c_2(v_a \otimes v_b \otimes v_c \otimes v_d)$$

since $c \triangleright d \neq d$ and since $\mathcal{B}(V)$ is slim in degree two. Thus finally

$$c_1(c_1c_2c_3)^2(v_a \otimes v_b \otimes v_c \otimes v_d) = -c_2c_3c_1c_2(v_a \otimes v_b \otimes v_c \otimes v_d).$$

From the braid equations we obtain

$$c_1c_2(c_1c_2c_3)^3 = c_1c_2c_1^2c_2c_3c_2c_1c_2^2c_3 = (c_2c_3)^2c_1c_2c_3^2c_2^2c_3$$

and

$$\begin{aligned} (c_2c_3)^2c_1c_2c_3^2c_2^2c_3(v_a \otimes v_b \otimes v_c \otimes v_d) &= (c_2c_3)^2c_1c_2c_3^3(v_a \otimes v_b \otimes v_c \otimes v_d) \\ &= -(c_2c_3)^2c_1c_2(v_a \otimes v_b \otimes v_c \otimes v_d) \end{aligned}$$

since $b \square (c \triangleright d)$ and $c \triangleright d \neq d$. Thus we obtain

$$\begin{aligned} s_2 =&(\mathrm{id} \otimes \mu)(\mathrm{id} - c_2c_3 + (c_2c_3)^2)c_1c_2(v_a \otimes v_b \otimes v_c \otimes v_d) \\ =&(\mathrm{id} \otimes \mu)(k_5 v_{a\triangleright(b\triangleright c)} \otimes (\mathrm{id} - c_1c_2 + (c_1c_2)^2))(v_a \otimes v_b \otimes v_d). \end{aligned}$$

Since $ab \triangleright d = b \triangleright d$, $a \triangleright d \neq d$, we have

$$\mu(\mathrm{id} - c_1c_2 + (c_1c_2)^2)(v_a \otimes v_b \otimes v_d) \in \langle v_d, v_b, v_f, v_g \rangle \subset \langle S \rangle$$

by Lemma 4.7 (3). It follows that $s_3 \in V \otimes \langle S \rangle$.

(2) First we prove that $Z \neq 0$. Let F be the linear functional on $V^{\otimes 4}$ with $F(v_a \otimes v_b \otimes v_c \otimes v_d) = 1$ and $F(v_i \otimes v_j \otimes v_k \otimes v_l) = 0$ for all $(i,j,k,l) \in X^4 \setminus \{(a,b,c,d)\}$. We prove that $F(\Delta_{1^4}(Z)) = 1$. It is well-known, and follows by similar arguments as in Lemma 2.6 that

$$\Delta_{1^4} = (\Delta_{1,1} \otimes \mathrm{id}^{\otimes 2})(\Delta_{2,1} \otimes \mathrm{id})\Delta_{3,1}.$$

For the calculation of the components $\Delta_{n,1}$, $1 \leq n \leq 3$, we will use Equation (1.5). By definition, the element Z is a linear combination of the elements $v_av_bv_cv_d$, $v_ev_av_bv_c$, $v_fv_ev_av_b$, $v_gv_fv_ev_a$ of $\mathcal{B}(V)$. Since a, b, c, d are pairwise distinct, the only summand of

$$\Delta_{3,1}(v_av_bv_cv_d) = (\mu \otimes \mathrm{id})(\mathrm{id} + c_3 + c_3c_2 + c_3c_2c_1)(v_a \otimes v_b \otimes v_c \otimes v_d)$$

such that the fourth tensor factor lies in $\Bbbk^\times v_d$ is $v_av_bv_c \otimes v_d$. With similar arguments for the third, second and first tensor factor of the summands of $(\Delta_{1,1} \otimes \mathrm{id}^{\otimes 2})(\Delta_{2,1} \otimes \mathrm{id})(v_av_bv_c \otimes v_d)$ it follows that $F(\Delta_{1^4}(v_av_bv_cv_d)) = 1$.

Now we prove $F(\Delta_{1^4}(v_e v_a v_b v_c)) = 0$. Since a, b and c are not equal to d, only the summand

$$(\mu \otimes \mathrm{id})(c_3c_2c_1)(v_e \otimes v_a \otimes v_b \otimes v_c)$$

of $\Delta_{3,1}(v_e v_a v_b v_c)$ can yield a nonzero element of $\Bbbk v_a \otimes v_b \otimes v_c \otimes v_d$ and it is only possible if $e = d$ and if $(\Delta_{1,1} \otimes \mathrm{id})\Delta_{2,1}(v_{d\triangleright a} v_{d\triangleright b} v_{d\triangleright c})$ supplies a nonzero multiple of $v_a \otimes v_b \otimes v_c$ as summand. But from the assumptions we have $d \triangleright a \neq c$, $d \triangleright b \neq c$ and $d \triangleright c \neq c$ and thus the last requirement can not be fulfilled.

Now we prove $F(\Delta_{1^4}(v_f v_e v_a v_b)) = 0$. Since a, b are not equal to d and $f = b \triangleright d \neq d$, only the summand

$$(\mu \otimes \mathrm{id})(c_3c_2)(v_f \otimes v_e \otimes v_a \otimes v_b)$$

of $\Delta_{3,1}(v_f v_e v_a v_b)$ can yield a nonzero multiple of $v_a \otimes v_b \otimes v_c \otimes v_d$ and only if $e = d$ and if $(\Delta_{1,1} \otimes \mathrm{id})\Delta_{2,1}(v_f v_{d\triangleright a} v_{d\triangleright b})$ supplies a nonzero multiple of $v_a \otimes v_b \otimes v_c$ as summand. Since $d \triangleright a \neq c$ and $d \triangleright b \neq c$, the last condition can only be fulfilled if $f = c$. In this case we would have $c = f = b \triangleright d$ and it would follow that $d = c \triangleright b$. From $d = e = ac \triangleright d$ we then would get $c \triangleright b = d = ca \triangleright d$, which would imply $b = a \triangleright d = g$. Taken all together it follows $F(\Delta_{1^4}(v_f v_e v_a v_b)) = 0$ since $d \neq e$, $c \neq f$ or $b \neq g$ by assumption.

Using $g \neq c$, similar arguments show that $F(\Delta_{1^4}(v_g v_f v_e v_a)) = 0$. We conclude that $F(\Delta_{1^4}(Z)) = 1$ and thus $Z \neq 0$. By Lemma 4.10, we know that $\{b, c, d, e, f, g\}^4 \cap \mathcal{O}(a, b, c, d) = \emptyset$. Hence

$$0 \neq \Delta_{1^4}(Z) \notin \Delta_{1^4}(\langle S \rangle(4)),$$

and therefore $Z \notin \langle S \rangle(4)$. □

5. Racks $\mathbb{T}_n$ of transpositions of $\mathbb{S}_n$

We start with a general description of the action and coaction of the dual Yetter–Drinfeld module.

LEMMA 5.1. *Let G be a group and $V \in {}^{G}_{G}\mathcal{YD}$ a finite-dimensional Yetter–Drinfeld module. Assume that $\dim V_g \leq 1$ for all $g \in G$ and let $X = \mathrm{supp} V$. For each $x \in X$ let $v_x \in V_x$ be a nonzero element and let $q_{x,y} \in \Bbbk^\times$ such that $x \cdot v_y = q_{x,y} v_{x\triangleright y}$, for all $x, y \in X$, where $x \triangleright y = xyx^{-1}$ (see Lemma 3.11 (1)). Let ${v_x}^*$, $x \in X$, be the dual basis in V^* of the basis v_x, $x \in X$, of V. Then the following explicit formulas for the $\Bbbk G$-coaction and $\Bbbk G$-action on the left dual Yetter–Drinfeld module V^* hold.*

1. *$\delta_{V^*}({v_x}^*) = x^{-1} \otimes {v_x}^*$ for all $x \in X$.*
2. *$x \cdot {v_y}^* = {q_{x,y}}^{-1} {v_{x\triangleright y}}^*$ for all $x, y \in X$.*
3. *$x^{-1} \cdot {v_y}^* = q_{x,a} {v_a}^*$ for all $x, y \in X$, where $a = x^{-1}yx \in X$.*
4. *If X is braided (as rack), then*

$$x^{-1} \cdot {v_y}^* = \begin{cases} q_{x,y} {v_y}^* & \text{if } x \triangleright y = y \\ q_{x,y\triangleright x} {v_{y\triangleright x}}^* & \text{if } x \triangleright y \neq y \end{cases}$$

for all $x, y \in X$.

PROOF. (1) Let $x \in X$. As in Equation (2.4) defined the $\Bbbk G$-coaction on V^* is given by

$$\begin{aligned} {v_x}^*{}_{(-1)}{v_x}^*{}_{(0)}(v_y) &= S^{-1}(v_{y(-1)}){v_x}^*(v_{y(0)}) \\ &= S^{-1}(y){v_x}^*(v_y) \\ &= \begin{cases} x^{-1} & \text{if } x = y \\ 0 & \text{if } x \neq y \end{cases} \end{aligned}$$

for all $y \in X$. Hence ${v_x}^*{}_{(-1)} \otimes {v_x}^*{}_{(0)} = x^{-1} \otimes {v_x}^*$.

(2) Let $x, y \in X$. As in Equation (2.3) defined the $\Bbbk G$-action on V^* is given by

$$(x \cdot {v_y}^*)(v_k) = {v_y}^*(S(x) \cdot v_k) = {v_y}^*(x^{-1} \cdot v_k)$$

for all $k \in X$. From $x \cdot (x^{-1} \cdot v_k) = v_k$ it follows that $x^{-1} \cdot v_k = {q_{x,h}}^{-1} v_h$, where $x \triangleright h = k$. Hence

$$\begin{aligned} (x \cdot {v_y}^*)(v_k) &= {v_y}^*({q_{x,h}}^{-1} v_h) \\ &= \begin{cases} {q_{x,y}}^{-1} & \text{if } x \triangleright y = k \\ 0 & \text{else} \end{cases} \end{aligned}$$

and $x \cdot {v_y}^* = {q_{x,y}}^{-1} {v_{x \triangleright y}}^*$.

(3) From $x \cdot (x^{-1} \cdot {v_y}^*) = {v_y}^*$ and since the G-homogeneous parts of V^* are one-dimensional (by assumption and part (1)) and by part (2), it follows that $x^{-1} \cdot {v_y}^* = q_{x,a} {v_a}^*$, where $x \triangleright a = y$.

(4) The claim follows from (3) and Lemma 3.3. □

DEFINITION 5.2. Let $n \in \mathbb{N}$. We denote the rack of transpositions of $\mathbb{S}_n$ with conjugation as rack operation by $\mathbb{T}_n$ and the generators of the enveloping group $G_{\mathbb{T}_n}$ by $g_{(i\ j)}$ for all $1 \leq i < j \leq n$.

REMARK 5.3. Note that $\mathbb{T}_n$ is a braided rack for all $n \in \mathbb{N}$. Indeed, if $(i\ j), (k\ l) \in \mathbb{T}_n$ such that $(i\ j) \triangleright (k\ l) \neq (k\ l)$, then $(i\ j) \triangleright (k\ l) = (k\ l) \triangleright (i\ j)$ and hence $(i\ j) \triangleright ((k\ l) \triangleright (i\ j)) = (k\ l)$. Moreover, $\mathbb{T}_n$ is indecomposable and generated by $(1\ 2), (1\ 3), \ldots, (1\ n)$.

LEMMA 5.4.

(1) *Let C be the subgroup of $G_{\mathbb{T}_n}$ generated by*
$$\{g_{(1\,2)}, g_{(i\,j)} \mid 2 < i < j \leq n\}.$$
Then $G_{\mathbb{T}_n}^{g_{(1\,2)}} = C$, where $G_{\mathbb{T}_n}^{g_{(1\,2)}}$ is the centralizer of $g_{(1\,2)} \in G_{\mathbb{T}_n}$.

(2) *Let ρ be a one-dimensional representation of the centralizer $G_{\mathbb{T}_n}^{g_{(1\,2)}}$. Then there are an element $t \in \Bbbk^\times$ and a square root of unity λ such that ρ is isomorphic to the representation*
$$\begin{aligned} \rho_{t,\lambda} : G_{\mathbb{T}_n}^{g_{(1\,2)}} &\to \operatorname{Aut}(\Bbbk v) \\ g_{(1\,2)} &\mapsto (v \mapsto tv) \\ g_{(i\,j)} &\mapsto (v \mapsto \lambda t v), \quad \textit{for all } 2 < i < j \leq n. \end{aligned}$$

PROOF. (1) By the universal property of the enveloping group, there is a group homomorphism $h : G_{\mathbb{T}_n} \to \mathbb{S}_n$ with $h(g_{(i\ j)}) = (i\ j)$ for all $1 \leq i < j \leq n$. The map h is surjective since $\mathbb{S}_n$ is generated by the set of transpositions. Moreover, the restriction of h on the conjugacy class of $g_{(1\ 2)}$ in $G_{\mathbb{T}_n}$ is injective. Hence

$h^{-1}(\mathbb{S}_n{}^{h(g_{(1\ 2)})}) = G_{\mathbb{T}_n}^{g_{(1\,2)}}$ (see, e.g., [**GnHV11**, Remark 2.21]). Since $\mathbb{S}_n{}^{(1\ 2)} = \{(1\,2), (i\,j) \mid 2 < i < j \leq n\}$ and $\ker h \subseteq C$, the claim follows.

(2) This follows since the elements $g_{(i\,j)}$, $2 < i < j \leq n$, are conjugate in $G_{\mathbb{T}_n}^{g_{(1\,2)}}$ and since $g_{(i\,j)}{}^2 = g_{(k\,l)}{}^2$ in $G_{\mathbb{T}_n}$ for all $1 \leq i < j \leq n$, $1 \leq k < l \leq n$ by Lemma 3.5. □

LEMMA 5.5. *Let V be a Yetter–Drinfeld module over $\Bbbk G_{\mathbb{T}_n}$ such that* $\operatorname{supp} V = \mathbb{T}_n$ *and* $\dim V_{g_x} = 1$ *for all $x \in \mathbb{T}_n$. Let $v_x \in V_{g_x}$ be nonzero elements and $q_{x,y} \in \Bbbk^\times$ such that $g_x \cdot v_y = q_{x,y} v_{x \triangleright y}$ for all $x, y \in \mathbb{T}_n$. Assume that $q_{x,y} = q_{x,y\triangleright x}$ for all $x, y \in \mathbb{T}_n$ with $x \triangleright y \neq y$. Then the linear map $f : V \to V^*$ with $f(v_x) = v_x{}^*$ for all $x \in \mathbb{T}_n$ is an isomorphism of braided vector spaces with respect to the Yetter–Drinfeld braidings c_V and c_{V^*}, respectively.*

PROOF. For all $x, y \in \mathbb{T}_n$ we have

$$(f \otimes f) c_V (v_x \otimes v_y) = q_{x,y} v_{x\triangleright y}{}^* \otimes v_x{}^*.$$

Using Lemma 5.1 (1) and Lemma 5.1 (4) we get

$$\begin{aligned} c_{V^*}(f \otimes f)(v_x \otimes v_y) &= g_x^{-1} \cdot v_y{}^* \otimes v_x{}^* \\ &= \begin{cases} q_{x,y} v_y{}^* \otimes v_x{}^* & \text{if } x \triangleright y = y \\ q_{x,y\triangleright x} v_{y\triangleright x}{}^* \otimes v_x{}^* & \text{if } x \triangleright y \neq y \end{cases} \\ &= q_{x,y} v_{x\triangleright y}{}^* \otimes v_x{}^*, \end{aligned}$$

where the last equation follows since for all $x, y \in \mathbb{T}_n$ either $x \triangleright y = y$ or $x \triangleright y = y \triangleright x$ hold and for all $x, y \in \mathbb{T}_n$ with $x \triangleright y \neq y$ we have $q_{x,y} = q_{x,y\triangleright x}$ by assumption. Hence the linear map f is an isomorphism of braided vector spaces. □

5.1. The rack $\mathbb{T}_3$. As described at the beginning of Section 3 a Yetter–Drinfeld module over $\Bbbk G_{\mathbb{T}_3}$ such that $\operatorname{supp} V = \mathbb{T}_3$ and $\dim V_{g_x} = 1$ for all $x \in \mathbb{T}_3$ is determined by a one-dimensional representation ρ of the centralizer $G_{\mathbb{T}_3}^{g_{(1\ 2)}}$. Let $t \in \Bbbk^\times$ and let ρ_t be the one-dimensional representation of the centralizer $G_{\mathbb{T}_3}^{g_{(1\ 2)}} = \langle g_{(1\ 2)} \rangle$ as in Lemma 5.4. Let $V = M(g_{(1\ 2)}, \rho_t)$ and

$$\begin{aligned} &v_a \in V_{g_{(1\ 2)}} \setminus \{0\}, \\ &v_b = t^{-1} g_{(2\ 3)} \cdot v_a \in V_{g_{(1\ 3)}} \qquad (5.1) \\ &v_c = t^{-1} g_{(1\ 3)} \cdot v_a \in V_{g_{(2\ 3)}}. \end{aligned}$$

Then the $\Bbbk G_{\mathbb{T}_3}$-module structure on V is given in Table 1.

TABLE 1. $\Bbbk G_{\mathbb{T}_3}$-module structure on V

$\cdot$	v_a	v_b	v_c
$g_{(1\ 2)}$	tv_a	tv_c	tv_b
$g_{(1\ 3)}$	tv_c	tv_b	tv_a
$g_{(2\ 3)}$	tv_b	tv_a	tv_c

LEMMA 5.6. *Assume that* $t = -1$ *or* $t \in \Bbbk$ *such that* $t^2 - t + 1 = 0$. *Let* $V = M(g_{(1\ 2)}, \rho_t)$ *and* v_a, v_b *and* v_c *as in Equation* (5.1). *Let* V_1, V_2 *be* $\Bbbk G_{\mathbb{T}_3}$-*subcomodules of* V *such that* $V = V_1 \oplus V_2$. *Then* $\mu : \langle V_1 \rangle \otimes \langle V_2 \rangle \to \mathcal{B}(V)$ *is surjective, where* μ *is the (restricted) multiplication in* $\mathcal{B}(V)$.

PROOF. Since $t^3 = -1$, in $\mathcal{B}(V)$ the degree two relations

$$v_a v_b - t v_c v_a + t^2 v_b v_c = 0$$

$$v_a v_c - t v_b v_a + t^2 v_c v_b = 0$$

hold by Lemma 4.4 (2). Let $S \subseteq \{v_a, v_b, v_c\}$ such that $V_1 = \Bbbk S$. Such an S exists since V_1 is a $\Bbbk G_{\mathbb{T}_3}$-subcomodule and v_a, v_b, v_c are homogeneous of different $G_{\mathbb{T}_3}$-degrees. Then $V_2 = \Bbbk S^{\complement}$, where $S^{\complement} = \{v_a, v_b, v_c\} \setminus S$. Using the degree two relations it is possible to write any monomial in $\mathcal{B}(V)$ as a linear combination of monomials which are ordered with respect to S and $S^{\complement}$, in the sense that each monomial is a product st, where $s \in \langle \Bbbk S \rangle$ and $t \in \langle \Bbbk S^{\complement} \rangle$. Thus $\mu : \langle \Bbbk S \rangle \otimes \langle \Bbbk S^{\complement} \rangle \to \mathcal{B}(V)$ is surjective. □

THEOREM 5.7. *Let* $t \in \Bbbk^\times$, $V = M(g_{(1\ 2)}, \rho_t)$ *and* v_a, v_b *and* v_c *as in Equation* (5.1).

(1) *If* $t = -1$ *or* $t^2 - t + 1 = 0$, *then all left coideal subalgebras of* $\mathcal{B}(V)$ *in the category of* $\mathbb{N}_0$-*graded* $\Bbbk G_{\mathbb{T}_3}$-*comodules are generated in degree one.*
(2) *If* $t \neq -1$ *and* $t^2 - t + 1 \neq 0$, *then* $\langle v_b, v_a v_b - t v_c v_a + t^2 v_b v_c \rangle$ *is a left coideal subalgebra of* $\mathcal{B}(V)$ *in the category of* $\mathbb{N}_0$-*graded* $\Bbbk G_{\mathbb{T}_3}$-*comodules, that is not generated in degree one.*

PROOF. (1) Let C be a (nontrivial, i.e., $C \neq \Bbbk 1$) left coideal subalgebra of $\mathcal{B}(V)$ in the category of $\mathbb{N}_0$-graded $\Bbbk G_{\mathbb{T}_3}$-comodules. Then $V \cap C \neq 0$ by Lemma 2.3 and $V_1 = V \cap C$ is a $\Bbbk G_{\mathbb{T}_3}$-subcomodule. Since $\Bbbk G_{\mathbb{T}_3}$ is cosemisimple, there exists a $\Bbbk G_{\mathbb{T}_3}$-subcomodule V_2 such that $V = V_1 \oplus V_2$. The restricted multiplication map $\mu : \langle V_1 \rangle \otimes \langle V_2 \rangle \to \mathcal{B}(V)$ is surjective by Lemma 5.6. The linear map $f : V \to V^*$ with $f(v_i) = v_i{}^*$ for all $i \in \mathbb{T}_3$ is an isomorphism of braided vector spaces by Lemma 5.5 and it induces an isomorphism $\mathcal{B}(V) \to \mathcal{B}(V^*)$ of $\mathbb{N}_0$-graded algebras. In particular, $\langle V_2 \rangle \cong \langle V_2{}^* \rangle$ as $\mathbb{N}_0$-graded algebras and the claim follows from Corollary 2.13 (1).

(2) The claim follows from Lemma 4.4 (2) and Theorem 4.3 (2) since $(1\ 2) \triangleright (1\ 3) \neq (1\ 3)$ in $\mathbb{T}_3$ and $t^3 \neq -1$ by assumption. □

REMARK 5.8. The Nichols algebra $\mathcal{B}(M(g_{(1\ 2)}, \rho_{-1}))$ associated to the rack $\mathbb{T}_3$ with constant two-cocycle -1 considered in Theorem 5.7 (1) is well-known and has dimension 12 (see, e.g., [**GnHV11**, Proposition 5.4]). It is isomorphic to the Fomin-Kirillov algebra $\mathcal{E}_3$. Its subalgebras are studied in [**BLM16**] and our result of Lemma 5.6 about the surjectivity of the multiplication map restricted on the subalgebras generated by complementary $\Bbbk G_{\mathbb{T}_3}$-subcomodules of $M(g_{(1\ 2)}, \rho_{-1})$ is proved in [**BLM16**, Corollary 4.11]. In the case where t is a primitive 6-th root of unity and the field $\Bbbk$ has characteristic 0, the Nichols algebra $\mathcal{B}(V)$ is infinite dimensional by [**HMV23**].

5.2. The racks $\mathbb{T}_4$, $\mathbb{T}_5$ and $\mathbb{T}_6$. Let $n \geq 4$, $t \in \Bbbk$ and let λ be a square root of unity. Let $\rho_{t,\lambda}$ be a one-dimensional representation of the centralizer $G_{\mathbb{T}_n}{}^{g_{(1\ 2)}}$ as in Lemma 5.4. Let $V(n, t, \lambda) = M(g_{(1\,2)}, \rho_{t,\lambda})$ and for all $1 \leq i < j \leq n$ let

$v_{(i\ j)} \in V(n,t,\lambda)_{g_{(i\ j)}}$ be nonzero elements. For all $1 \le i < j \le n$ and $1 \le k < l \le n$ define $q_{(i\ j),(k\ l)} \in \Bbbk^\times$ by

$$g_{(i\ j)} \cdot v_{(k\ l)} = q_{(i\ j),(k\ l)} v_{(i\ j) \triangleright (k\ l)}$$

(see Lemma 3.11).

LEMMA 5.9. *The following hold.*

(1) $q_{(1\ 2),(1\ 3)} q_{(2\ 3),(1\ 2)} q_{(1\ 3),(2\ 3)} = t^3$
(2) $q_{(1\ 2),(3\ 4)} q_{(3\ 4),(1\ 2)} = t^2$

PROOF. (1) There are elements $\lambda_{(1\ 3)}, \lambda_{(2\ 3)} \in \Bbbk^\times$ such that

$$v_{(1\ 3)} = \lambda_{(1\ 3)} g_{(2\ 3)} \cdot v_{(1\ 2)} \text{ and } v_{(2\ 3)} = \lambda_{(2\ 3)} g_{(1\ 3)} \cdot v_{(1\ 2)}$$

since $g_{(2\ 3)} g_{(1\ 2)} g_{(2\ 3)}^{-1} = g_{(1\ 3)}$ and $g_{(1\ 3)} g_{(1\ 2)} g_{(1\ 3)}^{-1} = g_{(2\ 3)}$ in $G_{\mathbb{T}_n}$. By definition of $M(g_{(1\,2)}, \rho_{t,\lambda})$, we have

$$\begin{aligned} g_{(1\ 2)} \cdot v_{(1\ 3)} &= \lambda_{(1\ 3)} g_{(1\ 2)} g_{(2\ 3)} \cdot v_{(1\ 2)} = t\lambda_{(1\ 3)} g_{(1\ 3)} \cdot v_{(1\ 2)} \\ &= t\lambda_{(1\ 3)} \lambda_{(2\ 3)}^{-1} v_{(2\ 3)} \\ g_{(2\ 3)} \cdot v_{(1\ 2)} &= \lambda_{(1\ 3)}^{-1} v_{(1\ 3)} \\ g_{(1\ 3)} \cdot v_{(2\ 3)} &= \lambda_{(2\ 3)} g_{(1\ 3)}^2 \cdot v_{(1\ 2)} = \lambda_{(2\ 3)} g_{(1\ 2)}^2 \cdot v_{(1\ 2)} = t^2 \lambda_{(2\ 3)} v_{(1\ 2)}. \end{aligned}$$

Hence

$$\begin{aligned} &q_{(1\ 2),(1\ 3)} = t\lambda_{(1\ 3)} \lambda_{(2\ 3)}^{-1}, \quad q_{(2\ 3),(1\ 2)} = \lambda_{(1\ 3)}^{-1}, \\ &q_{(1\ 3),(2\ 3)} = t^2 \lambda_{(2\ 3)} \end{aligned}$$

and $q_{(1\ 2),(1\ 3)} q_{(2\ 3),(1\ 2)} q_{(1\ 3),(2\ 3)} = t^3$.

(2) There is an element $\lambda_{(3\ 4)} \in \Bbbk^\times$ such that

$$v_{(3\ 4)} = \lambda_{(3\ 4)} g_{(1\ 3)} g_{(2\ 4)} \cdot v_{(1\ 2)}$$

since $g_{(1\ 3)} g_{(2\ 4)} g_{(1\ 2)} (g_{(1\ 3)} g_{(2\ 4)})^{-1} = g_{(3\ 4)}$ in $G_{\mathbb{T}_n}$. By definition of $M(g_{(1\,2)}, \rho_{t,\lambda})$, we have

$$\begin{aligned} g_{(1\ 2)} \cdot v_{(3\ 4)} &= \lambda_{(3\ 4)} g_{(1\ 2)} g_{(1\ 3)} g_{(2\ 4)} \cdot v_{(1\ 2)} \\ &= \lambda_{(3\ 4)} g_{(1\ 3)} g_{(2\ 4)} g_{(1\ 2)}^2 g_{(3\ 4)}^{-1} \cdot v_{(1\ 2)} \\ &= t\lambda^{-1} \lambda_{(3\ 4)} g_{(1\ 3)} g_{(2\ 4)} \cdot v_{(1\ 2)} = t\lambda^{-1} v_{(3\ 4)} \quad \text{and} \\ g_{(3\ 4)} \cdot v_{(1\ 2)} &= t\lambda v_{(1\ 2)}. \end{aligned}$$

Hence

$$q_{(1\ 2),(3\ 4)} = t\lambda^{-1}, \quad q_{(3\ 4),(1\ 2)} = t\lambda$$

and $q_{(1\ 2),(3\ 4)} q_{(3\ 4),(1\ 2)} = t^2$. □

THEOREM 5.10. *Assume that $t \neq -1$. Then there is a left coideal subalgebra of $\mathcal{B}(V(n,t,\lambda))$ in the category of $\mathbb{N}_0$-graded $\Bbbk G_{\mathbb{T}_n}$-comodules that is not generated in degree one.*

PROOF. If $t \neq -1$, then either $t^2 \neq 1$ or $t^3 \neq -1$. From Lemma 5.9 it follows that either

$$q_{(1\ 2),(1\ 3)} q_{(2\ 3),(1\ 2)} q_{(1\ 3),(2\ 3)} = t^3 \neq -1 \text{ or } q_{(1\ 2),(3\ 4)} q_{(3\ 4),(1\ 2)} = t^2 \neq 1.$$

Hence either

$$x = \mu(\mathrm{id} - c + c^2)(v_{(1\ 2)} \otimes v_{(1\ 3)}) \neq 0$$

or

$$y = \mu(\mathrm{id} - c)(v_{(1\ 2)} \otimes v_{(3\ 4)}) \neq 0$$

in $\mathcal{B}(V(n,t,\lambda))$ by Lemma 4.4, where c denotes the Yetter–Drinfeld braiding of $V(n,t,\lambda)$. It follows that either $\langle v_{(1\ 3)}, x\rangle$ or $\langle v_{(3\ 4)}, y\rangle$ is a left coideal subalgebra of $\mathcal{B}(V(n,t,\lambda))$ in the category of $\mathbb{N}_0$-graded $\Bbbk G_{\mathbb{T}_n}$-comodules that is not generated in degree one by Theorem 4.3. □

REMARK 5.11. For each $n > 3$ up to isomorphism there are the two Yetter–Drinfeld modules

$$V(n,-1,1) = M(g_{(1\,2)}, \rho_{-1,1}) \text{ and } V(n,-1,-1) = M(g_{(1\,2)}, \rho_{-1,-1})$$

over $\Bbbk G_{\mathbb{T}_n}$ associated to the rack $\mathbb{T}_n$ and a one-dimensional representation $\rho_{t,\lambda}$ as in Lemma 5.4 with $t = -1$ since $\lambda \in \{1,-1\}$. Both (braided vector spaces) can also be realized over $\mathbb{S}_n$ since $t^2 = 1$. They are the only irreducible Yetter–Drinfeld modules over $\Bbbk\mathbb{S}_n$ such that the corresponding Nichols algebra might be finite-dimensional (for $n = 4$ see [**AZ07**, Theorem 2.8], for $n \geq 5$ see [**AFGnV11**, Theorem 1.1]). In [**Ven12**] the two Yetter–Drinfeld modules $V(n,-1,1)$ and $V(n,-1,-1)$ are described in terms of two-cocycles on the rack $\mathbb{T}_n$. The Yetter–Drinfeld module $V(n,-1,1)$ is a realization of the rack $\mathbb{T}_n$ together with a constant two-cocycle -1. The other two-cocycle is

$$\chi(\sigma,\tau) = \begin{cases} 1 & \text{if } \sigma(i) < \sigma(j) \\ -1 & \text{if } \sigma(i) > \sigma(j) \end{cases} \tag{5.2}$$

for all transpositions σ and $\tau = (i\ j)$ with $i < j$, which corresponds to $V(n,-1,-1)$. In [**Ven12**, Theorem 3.8] it is shown that the two-cocycles -1 and χ are equivalent by twist. By [**AFGV11**, Theorem 3.8], it follows that the identity on the underlying vector spaces (and comodules) $V(n,-1,1) \to V(n,-1,-1)$ is a morphism of braided vector spaces. In particular, it follows that $\mathcal{B}(V(n,-1,1))$ and $\mathcal{B}(V(n,-1,-1))$ are isomorphic as $\mathbb{N}_0$-graded algebras and coalgebras. Moreover, the Nichols algebra $\mathcal{B}(V(n,-1,1))$ contains a left coideal subalgebra in the category of $\mathbb{N}_0$-graded $\Bbbk G_{\mathbb{T}_n}$-comodules which is not generated in degree one if and only if $\mathcal{B}(V(n,-1,1))$ contains a left coideal subalgebra in the category of $\mathbb{N}_0$-graded $\Bbbk G_{\mathbb{T}_n}$-comodules which is not generated in degree one. For $n \in \{4,5\}$ it is known that both Nichols algebras are finite dimensional with

$$\dim \mathcal{B}(V(4,-1,1)) = \dim \mathcal{B}(V(4,-1,-1)) = 576$$
$$\dim \mathcal{B}(V(5,-1,1)) = \dim \mathcal{B}(V(5,-1,-1)) = 8294400$$

(see, e.g., [**Ven12**, Example 2.3, Example 2.4]). Moreover, they are isomorphic to the Fomin-Kirillov algebra $\mathcal{E}_4$ ($\mathcal{E}_5$, resp.) by [**MS00**].

THEOREM 5.12. *Assume that $n \in \{4,5\}$, $t = -1$ and $\lambda \in \{1,-1\}$. Then all left coideal subalgebras of $\mathcal{B}(V(n,t,\lambda))$ in the category of $\mathbb{N}_0$-graded $\Bbbk G_{\mathbb{T}_n}$-comodules are generated in degree one.*

PROOF. By Remark 5.11, it suffices to prove the claim for $V(n,-1,1)$. Moreover, we can choose the nonzero elements $v_{(i\ j)} \in V(n,-1,1)_{g_{(i\ j)}}$ such that

$$g_{(i\ j)} \cdot v_{(k\ l)} = -v_{(i\ j)\triangleright(k\ l)}$$

for all $1 \le i < j \le n$ and $1 \le k < l \le n$, i.e., the corresponding two-cocycle is constant -1 (Remark 5.11). Let V_1, V_2 be $\Bbbk G_{\mathbb{T}_n}$-subcomodules of $V(n,-1,1)$ such that $V(n,-1,1) = V_1 \oplus V_2$. Then the restricted multiplication map

$$\mu : \langle V_1 \rangle \otimes \langle V_2 \rangle \to \mathcal{B}(V(n,-1,1))$$

is bijective by [**BLM16**, Corollary 4.11], where we used that $\mathcal{B}(V(n,-1,1))$ is isomorphic to the Fomin-Kirillov algebra $\mathcal{E}_n$ for $n \in \{4,5\}$ (see Remark 5.11). The linear map $f : V(n,-1,1) \to V(n,-1,1)^*$ with $f(v_i) = v_i{}^*$ for all $i \in \mathbb{T}_n$ is an isomorphism of braided vector spaces by Lemma 5.5 and it induces an isomorphism

$$\mathcal{B}(V(n,-1,1)) \to \mathcal{B}(V(n,-1,1)^*)$$

of $\mathbb{N}_0$-graded algebras. Hence $\langle V_2 \rangle \cong \langle V_2{}^* \rangle$ as $\mathbb{N}_0$-graded algebras and the claim follows from Corollary 2.13 (1) since $\mathcal{B}(V(n,-1,1))$ is finite dimensional (see Remark 5.11). □

THEOREM 5.13. *Assume that $n = 6$, $t \in \Bbbk^\times$ and $\lambda \in \Bbbk$ such that $\lambda^2 = 1$. Then there is a left coideal subalgebra of $\mathcal{B}(V(n,t,\lambda))$ in the category of $\mathbb{N}_0$-graded $\Bbbk G_{\mathbb{T}_6}$-comodules that is not generated in degree one.*

PROOF. For $t \neq -1$ the claim is proved in Theorem 5.10. Let $t = -1$. By Remark 5.11, it suffices to prove the claim for $V(6,-1,1)$. Moreover, we can choose the basis $v_{(i\ j)} \in V(6,-1,1)_{g_{(i\ j)}}$ such that $q_{(i\ j),(k\ l)} = -1$ for all $1 \le i < j \le 6$ and $1 \le k < l \le 6$, i.e., the corresponding two-cocycle is constant -1 (Remark 5.11). Let

$$C_1 = \{v_{(1\ 2)}, v_{(2\ 3)}, v_{(2\ 4)}, v_{(2\ 5)}, v_{(3\ 5)}, v_{(4\ 5)}, v_{(5\ 6)}\}$$
$$C_2 = \{v_{(1\ 3)}, v_{(1\ 4)}, v_{(1\ 5)}, v_{(1\ 6)}, v_{(2\ 6)}, v_{(3\ 4)}, v_{(3\ 6)}, v_{(4\ 6)}\}.$$

The left coideal subalgebra

$$C = \langle C_1 \rangle \subset \mathcal{B}(V(6,-1,1))$$

can be extended by an element of degree 7. We found this using GAP and the package GBNP for computing Gröbner bases ([**GAP22**], [**CKGT22**]). There we used the realization of the braided vector space $V(6,-1,1)$ over the finite enveloping group $\overline{G_{\mathbb{T}_6}} = G_{\mathbb{T}_6}/\langle {g_{(1\ 2)}}^2 \rangle$, which is isomorphic to the symmetric group $\mathbb{S}_6$ (using Proposition 3.12 (2)).[4] The extension of C in degree 7 has $\overline{G_{\mathbb{T}_6}}$-degree

$$g_{(1\ 6)} g_{(2\ 5)} g_{(3\ 4)}.$$

We got the idea to consider the subalgebra C and to look for an extension in degree 7 from the following observations. In [**BLM16**, Proposition 4.11] the Fomin-Kirillov algebra $\mathcal{E}_6$ is considered, which has the same relations as $\mathcal{B}(V(6,-1,1))$ up to degree 7. It is shown that

$$\dim \bigoplus_{i=0}^{7} (\langle C_1 \rangle \otimes \langle C_2 \rangle)(i) \neq \dim \bigoplus_{i=0}^{7} \mathcal{B}(V(6,-1,1))(i).$$

With GAP we calculated

$$\dim \bigoplus_{i=0}^{6} (\langle C_1 \rangle \otimes \langle C_2 \rangle)(i) = 85774 = \dim \bigoplus_{i=0}^{6} \mathcal{B}(V(6,-1,1))(i)$$

[4]The braided vector space $V(6,-1,1)$ can be realized as Yetter–Drinfeld module over the quotient $\overline{G_{\mathbb{T}_6}}$ since ${g_{(1\ 2)}}^2$ acts trivially in the representation $\rho_{-1,1}$.

but

$$\dim(\langle C_1\rangle \otimes \langle C_2\rangle)(7) = 228854 < \dim \mathcal{B}(V(6,-1,1))(7) = 228855.$$

From this we knew that there can not be an extension in degree smaller than 7 by Theorem 2.12 using Lemma 5.5. □

6. The rack $\mathcal{T}$ associated with vertices of the tetrahedron

Let $\mathcal{T} = \{a,b,c,d\}$ be the rack associated to the conjugacy class of (2 3 4) in the alternating group $\mathbb{A}_4$, where $a = (2\ 3\ 4)$, $b = (1\ 4\ 3)$, $c = (1\ 2\ 4)$ and $d = (1\ 3\ 2)$ (see, e.g., [**HLV12**, 7.2]). The rack structure is given explicitly in Table 2.

TABLE 2. The rack $\mathcal{T}$

$\triangleright$	a	b	c	d
a	a	c	d	b
b	d	b	a	c
c	b	d	c	a
d	c	a	b	d

In [**AGn03b**, Examples 1.3] the rack $\mathcal{T}$ is described as the rack of a tetrahedron with vertices $\{a,b,c,d\} \subset \mathbb{R}^3$ with center 0: The rack operation of an element $x \in \mathcal{T}$ is the linear map which fixes x and rotates the orthogonal plane by an angle of $2\pi/3$ with a fixed orientation (right-hand rule pointing the thumb to x).

REMARK 6.1. The rack $\mathcal{T}$ is braided (see [**HLV12**, Example 5]), indecomposable and generated by each subset with two elements.

LEMMA 6.2.

(1) [**HLV12**, Lemma 33] *The centralizer $G_{\mathcal{T}}^{g_a}$ of $g_a \in G_{\mathcal{T}}$ is the subgroup generated by $\{g_a, g_b g_d\}$. Moreover, the relation $g_a^4 = (g_b g_d)^2$ holds in $G_{\mathcal{T}}$.*
(2) *Let ρ be a one-dimensional representation of the centralizer $G_{\mathcal{T}}^{g_a}$. Then there are an element $t \in \Bbbk^\times$ and an element $\lambda \in \Bbbk^\times$ with $\lambda^2 = t^4$ such that ρ is isomorphic to the representation*

$$\begin{aligned}\rho_{t,\lambda} : G_{\mathcal{T}}^{g_a} &\to \mathrm{Aut}(\Bbbk v)\\ g_a &\mapsto (v \mapsto tv)\\ g_b g_d &\mapsto (v \mapsto \lambda v).\end{aligned}$$

Let $t \in \Bbbk^\times$, $\lambda \in \Bbbk^\times$ with $\lambda^2 = t^4$ and let $\rho_{t,\lambda}$ be a one-dimensional representation of $G_{\mathcal{T}}^{g_a}$ as in Lemma 6.2. Let $V = M(g_a, \rho_{t,\lambda})$ and

$$\begin{aligned}&v_a \in V_{g_a} \setminus \{0\},\\ &v_b = g_c \cdot v_a \in V_{g_b}\\ &v_c = g_d \cdot v_a \in V_{g_c}\\ &v_d = g_b \cdot v_a \in V_{g_d}.\end{aligned} \tag{6.1}$$

Then the $\Bbbk G_{\mathcal{T}}$-module structure on V is given in Table 3.

TABLE 3. $\Bbbk G_{\mathcal{T}}$-module structure on V

$\cdot$	v_a	v_b	v_c	v_d
g_a	tv_a	tv_c	tv_d	tv_b
g_b	v_d	tv_b	λv_a	$\frac{\lambda}{t} v_c$
g_c	v_b	$\frac{\lambda}{t} v_d$	tv_c	λv_a
g_d	v_c	λv_a	$\frac{\lambda}{t} v_b$	tv_d

LEMMA 6.3. *Let $x, y, z \in \mathcal{T}$ be pairwise distinct elements. Then there are elements $g_1, g_2, g_3 \in G_{\mathcal{T}}$ such that*

$$\{g_1 \triangleright a\} = \{x\}, \ \{g_2 \triangleright a, g_2 \triangleright b\} = \{x, y\} \text{ and } \{g_3 \triangleright a, g_3 \triangleright b, g_3 \triangleright c\} = \{x, y, z\}.$$

There $g\triangleright$ denotes the canonical action of $G_{\mathcal{T}}$ on $\mathcal{T}$ by conjugation for all $g \in G_{\mathcal{T}}$.

PROOF. One can easily convince oneself of the correctness of the claim by using the geometrical idea of the rack as vertices of a tetrahedron with rotations as action. By multiplying the elements of $G_{\mathcal{T}}$ along the paths (and their inverse elements along the opposite direction) in the subsequent diagrams it follows how the elements g_1, g_2, g_3 can be chosen suitable.

$$\begin{matrix}\{a\}, \\ \{b,c,d\}\end{matrix} \xrightarrow{g_c \triangleright} \begin{matrix}\{b\}, \\ \{a,c,d\}\end{matrix} \xrightarrow{g_a \triangleright} \begin{matrix}\{c\}, \\ \{a,b,d\}\end{matrix} \xrightarrow{g_a \triangleright} \begin{matrix}\{d\}, \\ \{a,b,c\}\end{matrix}$$

$$\begin{array}{ccccc} \{a,b\} & \xrightarrow{g_a \triangleright} & \{a,c\} & \xrightarrow{g_a \triangleright} & \{a,d\} \\ & & & & \big\downarrow {\scriptstyle g_d \triangleright} \\ \{b,c\} & \xleftarrow[g_b \triangleright]{} & \{b,d\} & \xleftarrow[g_d \triangleright]{} & \{c,d\} \end{array}$$

As indicated the claim for the sets with three elements follows directly from the claim for the sets with one element since $g\triangleright$ defines a bijection on $\mathcal{T}$ for all $g \in G_{\mathcal{T}}$. □

LEMMA 6.4. *Let $t, \lambda \in \Bbbk^\times$ with $\lambda^2 = t^4$ be the parameters of a one-dimensional representation of $G_{\mathcal{T}}^{g_a}$ as in Lemma 6.2. Let $V = M(g_a, \rho_{t,\lambda})$ be the corresponding $\Bbbk G_{\mathcal{T}}$-Yetter–Drinfeld module. Let W be a $\Bbbk G_{\mathcal{T}}$-subcomodule of V.*

(1) *If $\dim W = 1$, then there is an element $g \in G_{\mathcal{T}}$ such that $W = g \cdot V_{g_a}$.*
(2) *If $\dim W = 2$, then there is an element $g \in G_{\mathcal{T}}$ such that $W = g \cdot (V_{g_a} \oplus V_{g_b})$.*
(3) *If $\dim W = 3$, then there is an element $g \in G_{\mathcal{T}}$ such that $W = g \cdot (V_{g_a} \oplus V_{g_b} \oplus V_{g_c})$.*

PROOF. The claim follows from Lemma 6.3 since $\dim V_{g_x} = 1$ for all $x \in \mathcal{T}$. □

PROPOSITION 6.5. *Let $t \in \Bbbk$ such that $t^2 + t + 1 = 0$ and $\lambda = -t^{-1}$ be the parameters of a one-dimensional representation of $G_{\mathcal{T}}^{g_a}$ as in Lemma 6.2. Let $V = M(g_a, \rho_{t,\lambda})$ be the corresponding $\Bbbk G_{\mathcal{T}}$-Yetter–Drinfeld module. Let v_a, v_b, v_c and v_d as in Equation* (6.1).

(1) *The Nichols algebra* $\mathcal{B}(V)$ *has dimension* 5184.

(2)

$$\dim\langle V_{g_a}\rangle = 3, \quad \dim\langle V_{g_a}+V_{g_b}\rangle = 72,$$
$$\dim\langle V_{g_a}+V_{g_b}+V_{g_c}\rangle = 1728.$$

(3) *Let* W *be a* $\Bbbk G_{\mathcal{T}}$*-subcomodule of* V*. If* $\dim W = 1$*, then* $\dim\langle W\rangle = 3$*, if* $\dim W = 2$*, then* $\dim\langle W\rangle = 72$*, if* $\dim W = 3$*, then* $\dim\langle W\rangle = 1728$*.*

(4) *The linear map*

$$f : V \to V^*,$$
$$v_a \mapsto -t{v_a}^*,\ v_b \mapsto {v_c}^*,\ v_c \mapsto {v_b}^*,\ v_d \mapsto {v_d}^*$$

is an isomorphism of braided vector spaces with respect to the Yetter–Drinfeld braidings c_V *and* c_{V^*}*.*

(5) *Let* W *be a* $\Bbbk G_{\mathcal{T}}$*-subcomodule of* V^**. If* $\dim W = 1$*, then* $\dim\langle W\rangle = 3$*, if* $\dim W = 2$*, then* $\dim\langle W\rangle = 72$*, if* $\dim W = 3$*, then* $\dim\langle W\rangle = 1728$*.*

PROOF.

(1) See [**HLV12**, Prop. 36].

(2) We proved this using the GAP package GBNP for computing Gröbner bases ([**GAP22**], [**CKGT22**]). There we used the realization of the braided vector space V over the finite enveloping group $\overline{G_{\mathcal{T}}} = G_{\mathcal{T}}/\langle {g_a}^3\rangle$ (and Proposition 3.12 (2)).[5]

(3) If $\dim W = 2$, then there is an element $g \in G_{\mathcal{T}}$ such that $W = g\cdot(V_{g_a}+V_{g_b})$ by Lemma 6.4. The map

$$\alpha_g : \mathcal{B}(V) \to \mathcal{B}(V), \quad x \mapsto g\cdot x$$

is an isomorphism of $\mathbb{N}_0$-graded algebras by Lemma 3.16. Hence $\langle W\rangle \cong \langle V_{g_a}+V_{g_b}\rangle$ as $\mathbb{N}_0$-graded algebras and the claim follows from part (2). If $\dim W \in \{1,3\}$, we can use analogous arguments.

(4) Using Lemma 5.1 it follows that $\delta_{V^*}({v_x}^*) = {g_x}^{-1}\otimes {v_x}^*$ for all $x \in \mathcal{T}$ and the $\Bbbk G_{\mathcal{T}}$-action on V^* is given in Table 4. Using this it is a simple calculation to

TABLE 4. $\Bbbk G_{\mathcal{T}}$-module structure on V^*

$\cdot$	${v_a}^*$	${v_b}^*$	${v_c}^*$	${v_d}^*$
${g_a}^{-1}$	$t{v_a}^*$	$t{v_d}^*$	$t{v_b}^*$	$t{v_c}^*$
${g_b}^{-1}$	$-t^2{v_c}^*$	$t{v_b}^*$	$-t{v_d}^*$	${v_a}^*$
${g_c}^{-1}$	$-t^2{v_d}^*$	${v_a}^*$	$t{v_c}^*$	$-t{v_b}^*$
${g_d}^{-1}$	$-t^2{v_b}^*$	$-t{v_c}^*$	${v_a}^*$	$t{v_d}^*$

check that

$$(f\otimes f)c_V(v_x\otimes v_y) = c_{V^*}(f\otimes f)(v_x\otimes v_y)$$

for all $x, y \in \mathcal{T}$.

[5] The braided vector space V can be realized as Yetter–Drinfeld module over the quotient $\overline{G_{\mathcal{T}}}$ since ${g_a}^3$ acts trivially in the representation $\rho_{t,\lambda}$ since $t^3 = 1$ by assumption.

(5) The isomorphism of braided vector spaces $f : V \to V^*$ in (4) induces an isomorphism $\mathcal{B}(V) \to \mathcal{B}(V^*)$ of $\mathbb{N}_0$-graded algebras. If $W \subseteq V^*$ is a $\Bbbk G_{\mathcal{T}}$-subcomodule, then $f^{-1}(W) \subseteq V$ is a $\Bbbk G_{\mathcal{T}}$-subcomodule since f maps $G_{\mathcal{T}}$-homogeneous elements to $G_{\mathcal{T}}$-homogeneous elements. Hence the claim follows from (2). □

THEOREM 6.6. *Let $t \in \Bbbk^\times$ and $\lambda \in \Bbbk^\times$ with $\lambda^2 = t^4$ be the parameters of a one-dimensional representation of $G_{\mathcal{T}}^{g_a}$ as in Lemma 6.2. Let $V = M(g_a, \rho_{t,\lambda})$ be the corresponding Yetter–Drinfeld module over $\Bbbk G_{\mathcal{T}}$ and denote its braiding by c_V. Let v_a, v_b, v_c and v_d as in Equation (6.1). Then the following hold.*

1. *If $t^6 \neq 1$, then $\langle v_b, v_a v_b - t v_c v_a + t v_b v_c\rangle$ is a left coideal subalgebra of $\mathcal{B}(V)$ in the category of $\mathbb{N}_0$-graded $\Bbbk G_{\mathcal{T}}$-comodules, that is not generated in degree one.*
2. *If $t = -1$ and $\lambda = 1$ or $t^2 - t + 1 = 0$ and $\lambda = -t^{-1}$, and $z = v_c v_b v_a - \lambda v_a v_c v_b + \lambda^2 v_b v_a v_c$, then $\langle v_a, v_b, z\rangle$ is a left coideal subalgebra of $\mathcal{B}(V)$ in the category of $\mathbb{N}_0$-graded $\Bbbk G_{\mathcal{T}}$-comodules that is not generated in degree one.*
3. *If $t = 1$ and $\lambda = -1$, then the left coideal subalgebra $\langle v_a, v_b\rangle$ of $\mathcal{B}(V)$ has an extension z in degree 11, where $z \in \langle v_a, v_b, v_c\rangle(11) \subset \mathcal{B}(V)(11)$.*
4. *If $t^2 + t + 1 = 0$ and $\lambda = -t^{-1}$, then all left coideal subalgebras of $\mathcal{B}(V)$ in the category of $\mathbb{N}_0$-graded $\Bbbk G_{\mathcal{T}}$-comodules are generated in degree one.*

PROOF. (1) We have

$$\begin{aligned}\Delta_{1,1}\left(\mu\left(1 - c_V + {c_V}^2\right)(v_a \otimes v_b)\right) &= \Delta_{1,1}\left(v_a v_b - t v_c v_a + t v_b v_c\right)\\ &= (1 + t\lambda)\, v_a \otimes v_b\end{aligned}$$

and $1 + t\lambda \neq 0$ since $t^6 \neq 1$ and $\lambda^2 = t^4$ by assumption. Hence the claim follows from Theorem 4.3 (2).

(2) One calculates that the element z is $G_{\mathcal{T}}$-homogeneous of degree $g_c g_a$. In $\mathcal{B}(V)$ the relation

$$v_c v_a - v_b v_c + \lambda v_a v_b = 0$$

holds. Indeed, we have

$$\Delta_{1,1}(v_c v_a - v_b v_c + \lambda v_a v_b) = (1 + \lambda t) v_c \otimes v_a$$

and $1 + \lambda t = 0$ for the considered parameters λ and t. Using this relation and Equation (1.4) we get

$$\begin{aligned}\Delta_{1,2}(z) &= v_c \otimes (1 + \lambda^3) v_b v_a + v_d \otimes (\lambda t^{-1} v_c v_a - \lambda t v_a v_b + \lambda^2 v_b v_c)\\ &= v_c \otimes (1 + \lambda^3) v_b v_a \in V \otimes \langle v_a, v_b\rangle(2).\end{aligned}$$

Hence by Corollary 2.8, it follows that $\langle v_a, v_b, z\rangle$ is a left coideal subalgebra. Moreover, it follows that $z \neq 0$ since $(1 + \lambda^3) v_b v_a \neq 0$. Note that $z \in \langle v_a, v_b\rangle$ if and only if $\Delta_{1^3}(z) \in \Delta_{1^3}(\langle v_a, v_b\rangle(3))$. Thus, since $\Delta_{1^3}(z) \neq 0$, it suffices to prove that the Hurwitz orbit of the triple $(c, b, a) \in \mathcal{T}^3$ does not contain a triple with only a and b as entries. One checks that the Hurwitz orbit of $(c, b, a) \in \mathcal{T}^3$ is the union of the $\sigma_1\sigma_2$-orbits

$$\begin{gathered}\{(a,c,b),(b,a,c),(c,b,a)\}, \{(a,b,d),(d,a,b),(b,d,a)\},\\ \{(a,d,c),(c,a,d),(d,c,a)\}, \{(b,c,d),(d,b,c),(c,d,b)\}.\end{gathered}$$

(3) We found this extension using the GAP package GBNP for computing Gröbner bases ([**GAP22**], [**CKGT22**]). There we used the realization of the braided vector space V over the finite enveloping group $\overline{G_\mathcal{T}} = G_\mathcal{T}/\langle {g_a}^3\rangle$ (and Proposition 3.12 (2)).[6]

(4) Let C, $C \neq \Bbbk 1$, be a left coideal subalgebra of $\mathcal{B}(V)$ in the category of $\mathbb{N}_0$-graded $\Bbbk G_\mathcal{T}$-comodules. Then $V \cap C \neq 0$ by Lemma 2.3 and $V_1 = V \cap C$ is a $\Bbbk G_\mathcal{T}$-subcomodule by assumption. Since $\Bbbk G_\mathcal{T}$ is cosemisimple, there exists a $\Bbbk G_\mathcal{T}$-subcomodule V_2 such that $V = V_1 \oplus V_2$. Then $\dim\langle V_1\rangle \dim\langle V_2\rangle = \dim \mathcal{B}(V)$ by Proposition 6.5 (1) and (3) and $\dim\langle V_2\rangle = \dim\langle {V_2}^*\rangle$ by Proposition 6.5 (3) and (5). Hence the claim follows from Corollary 2.13 (2). □

REMARK 6.7. The Nichols algebra in Theorem 6.6 (2) with parameters $t = -1$ and $\lambda = 1$ appeared in [**AGn03b**, Theorem 6.15]. It has dimension 72.

7. The rack $\mathcal{B}$ associated with the faces of a cube

Let $\mathcal{B} = \{a, b, c, d, e, f\}$ be the rack associated to the conjugacy class of 4-cycles of the symmetric group $\mathbb{S}_4$, where $a = (2\ 3\ 4\ 5)$, $b = (1\ 5\ 6\ 3)$, $c = (1\ 2\ 6\ 4)$, $d = (1\ 3\ 6\ 5)$, $e = (1\ 4\ 6\ 2)$ and $f = (2\ 5\ 4\ 3)$ (see, e.g., [**GnHV11**, Example 2.2]). The rack structure is explicitly given in Table 5.

TABLE 5. The rack $\mathcal{B}$

$\triangleright$	a	b	c	d	e	f
a	a	c	d	e	b	f
b	e	b	a	d	f	c
c	b	f	c	a	e	d
d	c	b	f	d	a	e
e	d	a	c	f	e	b
f	a	e	b	c	d	f

The rack $\mathcal{B}$ has also a geometric interpretation, where $\{a, b, c, d, e, f\}$ represents the set of faces of a cube with center 0 (see, e.g., [**AGn03b**, Example 1.3, Example 3.5]). The action of an element $x \in \mathcal{B}$ is given by $\pi/2$-rotation of the cube around the center of face x with a fixed orientation (right-hand rule pointing the thumb to x, see Figure 2).

REMARK 7.1. The rack $\mathcal{B}$ is braided (see [**HLV12**, Example 5]), indecomposable and generated by each subset with two not-commuting elements.

LEMMA 7.2.

(1) [**GnHV11**, Lemma 5.10] *The centralizer $G_\mathcal{B}^{g_a}$ of $g_a \in G_\mathcal{B}$ is the abelian subgroup generated by $\{g_a, g_f\}$. Moreover, the relation ${g_a}^4 = {g_f}^4$ holds in $G_\mathcal{B}$.*

(2) *There is a group isomorphism*
$$G_\mathcal{B}^{g_a} \to \mathbb{Z}^2/4(e_1 - e_2),\ a \mapsto e_1, f \mapsto e_2,$$
where e_1, e_2 are generators of $\mathbb{Z}^2$.

[6]The braided vector space V can be realized as Yetter–Drinfeld module over the quotient $\overline{G_\mathcal{T}}$ since ${g_a}^3$ acts trivially in the representation $\rho_{t,\lambda}$ since $t^3 = 1$ by assumption.

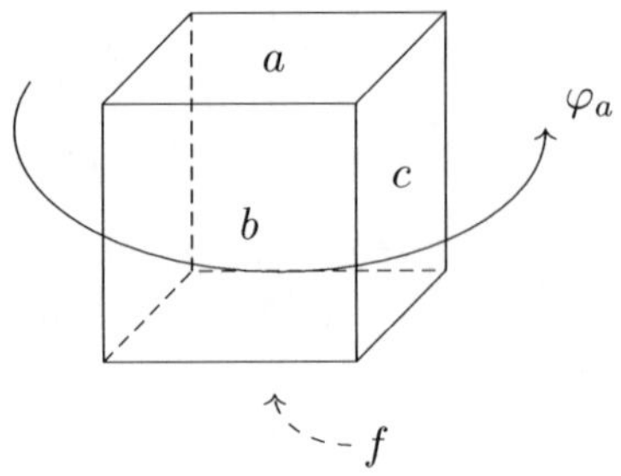

FIGURE 2. The rack $\mathcal{B}$ as rack of the faces of a cube

(3) *Let ρ be a one-dimensional representation of the centralizer $G_{\mathcal{B}}^{g_a}$. Then there are an element $t \in \Bbbk^\times$ and an element $\lambda \in \Bbbk^\times$ with $\lambda^4 = 1$ such that ρ is isomorphic to the representation*

$$\begin{aligned} \rho_{t,\lambda} : \ G_{\mathcal{B}}^{g_a} &\to \operatorname{Aut}(\Bbbk v) \\ g_a &\mapsto (v \mapsto tv) \\ g_f &\mapsto (v \mapsto \lambda tv). \end{aligned}$$

PROOF. (2) Since $G_{\mathcal{B}}^{g_a}$ is abelian and ${g_a}^4 = {g_f}^4$ by (1), it suffices to prove that ${g_a}^k \neq {g_f}^l$ for all $1 < k, l < 4$. Since $g_a \neq g_f$, it suffices to prove that ${g_a}^2 \neq {g_f}^2$. To see this we consider the subset $M = \{A, B, C, D, E, F\}$ of the general linear group $\mathrm{GL}(2, \mathbb{F}_9)$, where

$$A = \begin{pmatrix} 1 & 0 \\ 0 & i \end{pmatrix}, \qquad B = -(1+i)\begin{pmatrix} 1 & -i \\ -i & 1 \end{pmatrix},$$
$$C = -(1+i)\begin{pmatrix} 1 & -1 \\ 1 & 1 \end{pmatrix}, \qquad D = -(1+i)\begin{pmatrix} 1 & i \\ i & 1 \end{pmatrix},$$
$$E = -(1+i)\begin{pmatrix} 1 & 1 \\ -1 & 1 \end{pmatrix}, \qquad F = \begin{pmatrix} i & 0 \\ 0 & 1 \end{pmatrix},$$

where $i^2 = -1$.[7] By direct calculations one obtains that they form a rack, where the rack operation is conjugation, and that the map

$$\begin{aligned} \mathcal{B} \to M, \quad a &\mapsto A, \quad b \mapsto B, \quad c \mapsto C, \\ d &\mapsto D, \quad e \mapsto E, \quad f \mapsto F \end{aligned}$$

is a rack isomorphism. Hence from $A^2 \neq F^2$ in $\mathrm{GL}(2, \mathbb{F}_9)$, it follows that ${g_a}^2 \neq {g_f}^2$ in $G_{\mathcal{B}}^{g_a}$.

(3) Follows from (2). □

Let $t \in \Bbbk^\times$, $\lambda \in \Bbbk^\times$ with $\lambda^4 = 1$ and let $\rho_{t,\lambda}$ be a one-dimensional representation of $G_{\mathcal{B}}^{g_a}$ as in Lemma 7.2 (3). Let $V = M(g_a, \rho_{t,\lambda})$ and

$$\begin{aligned} & v_a \in V_{g_a} \setminus \{0\}, && v_b = g_c \cdot v_a \in V_{g_b}, && v_c = g_d \cdot v_a \in V_{g_c} \\ & v_d = g_e \cdot v_a \in V_{g_d}, && v_e = g_b \cdot v_a \in V_{g_e}, && v_f = {g_b}^2 \cdot v_a \in V_{g_f}. \end{aligned} \tag{7.1}$$

The $\Bbbk G_{\mathcal{B}}$-module structure on V is given in Table 6.

[7]The set $M \subset \mathrm{GL}(2, \mathbb{F}_9)$ is a conjugacy class and a generating set of the unitary group $U(2,3)$. The Hermitian form on the vector space ${\mathbb{F}_9}^2$ is defined with respect to the order two automorphism $\mathbb{F}_9 \to \mathbb{F}_9$, $x \mapsto x^3$.

TABLE 6. $\Bbbk G_{\mathcal{B}}$-module structure on V

$\cdot$	v_a	v_b	v_c	v_d	v_e	v_f
g_a	tv_a	tv_c	tv_d	tv_e	tv_b	$t\lambda v_f$
g_b	v_e	tv_b	$t^2\lambda v_a$	$t\lambda v_d$	v_f	$t^2\lambda^3 v_c$
g_c	v_b	λv_f	tv_c	$t^2\lambda v_a$	$t\lambda v_e$	$t^2\lambda^2 v_d$
g_d	v_c	$t\lambda v_b$	$\lambda^2 v_f$	tv_d	$t^2\lambda v_a$	$t^2\lambda v_e$
g_e	v_d	$t^2\lambda v_a$	$t\lambda v_c$	$\lambda^3 v_f$	tv_e	$t^2 v_b$
g_f	$t\lambda v_a$	$t\lambda v_e$	$t\lambda v_b$	$t\lambda v_c$	$t\lambda v_d$	tv_f

LEMMA 7.3. *Let $x_1, x_2, \ldots, x_5 \in \mathcal{B}$ be pairwise distinct elements.*

(1) *There is an element $g \in G_{\mathcal{B}}$ such that $\{g \triangleright a\} = \{x_1\}$.*
(2) *If $x_1 \triangleright x_2 = x_2$, then there is an element $g \in G_{\mathcal{B}}$ such that $\{g \triangleright a, g \triangleright f\} = \{x_1, x_2\}$.*
(3) *If $x_1 \triangleright x_2 \neq x_2$, then there is an element $g \in G_{\mathcal{B}}$ such that $\{g \triangleright a, g \triangleright b\} = \{x_1, x_2\}$.*
(4) *$x_1 \triangleright x_2 = x_2$ then there is an element $g \in G_{\mathcal{B}}$ such that $\{g \triangleright a, g \triangleright f, g \triangleright b\} = \{x_1, x_2, x_3\}$.*
(5) *If x_1, x_2 and x_3 do pairwise not commute, then there is an element $g \in G_{\mathcal{B}}$ such that $\{g \triangleright a, g \triangleright b, g \triangleright c\} = \{x_1, x_2, x_3\}$.*
(6) *$x_1 \triangleright x_2 = x_2$ and $x_3 \triangleright x_4 \neq x_4$ then there is an element $g \in G_{\mathcal{B}}$ such that $\{g \triangleright a, g \triangleright f, g \triangleright b, g \triangleright c\} = \{x_1, x_2, x_3, x_4\}$.*
(7) *If $x_1 \triangleright x_2 = x_2$ and $x_3 \triangleright x_4 = x_4$, then there is an element $g \in G_{\mathcal{B}}$ such that $\{g \triangleright a, g \triangleright f, g \triangleright b, g \triangleright d\} = \{x_1, x_2, x_3, x_4\}$.*
(8) *There is an element $g \in G_{\mathcal{B}}$ such that $\{g \triangleright a, g \triangleright b, g \triangleright c, g \triangleright d, g \triangleright e\} = \{x_1, x_2, x_3, x_4, x_5\}$.*

PROOF. One can easily convince oneself of the correctness of the claim by using the geometrical idea of the rack as faces of a cube with rotations as action (see above). Note that commuting elements of $\mathcal{B}$ belong to opposite faces. Hence in order not to be too detailed we give the explicit proof only for (4). Note first that by (2) there is an element $h \in G_{\mathcal{B}}$ such that $\{x_1, x_2\} = \{h \triangleright a, h \triangleright f\}$. Since $x_3 \notin \{x_1, x_2\}$, it follows that $h^{-1} \triangleright x_3 \in \{b, c, d, e\}$. Since a and f are fixed by g_a and since g_a acts transitively on $\{b, c, d, e\}$, there exists $k \geq 0$ such that $g = g_a^k h$ and $\{g \triangleright a, g \triangleright f, g \triangleright b\} = \{x_1, x_2, x_3\}$. □

LEMMA 7.4. *Let $t \in \Bbbk^\times$, $\lambda \in \Bbbk^\times$ with $\lambda^4 = 1$ and $\rho_{t,\lambda}$ a one-dimensional representation of $G_{\mathcal{B}}^{g_a}$ as in Lemma 7.2 (3). Let $V = M(g_a, \rho_{t,\lambda})$. Let W be a $\Bbbk G_{\mathcal{B}}$-subcomodule of V. Let $S = \{x \in \mathcal{B} \mid W \cap V_{g_x} \neq 0\}$. Note that $|S| = \dim W$ since $\dim V_{g_x} = 1$ for all $x \in \mathcal{B}$ and W is a $\Bbbk G_{\mathcal{B}}$-subcomodule.*

(1) *If $\dim W = 1$, then there is an element $g \in G_{\mathcal{B}}$ such that $W = g \cdot V_{g_a}$.*
(2) *Assume that $\dim W = 2$. If the elements of S commute, then there is an element $g \in G_{\mathcal{B}}$ such that $W = g \cdot (V_{g_a} \oplus V_{g_f})$. If the elements of S do not commute, then there is an element $g \in G_{\mathcal{B}}$ such that $W = g \cdot (V_{g_a} \oplus V_{g_b})$.*
(3) *Assume that $\dim W = 3$. If two elements of S commute, then there is an element $g \in G_{\mathcal{B}}$ such that $W = g \cdot (V_{g_a} \oplus V_{g_b} \oplus V_{g_f})$. If the elements of*

S are pairwise not commuting, then there is an element $g \in G_{\mathcal{B}}$ such that $W = g \cdot (V_{g_a} \oplus V_{g_b} \oplus V_{g_c})$.

(4) *Assume that $\dim W = 4$. If precisely two elements of S commute, then there is an element $g \in G_{\mathcal{B}}$ such that $W = g \cdot (V_{g_a} \oplus V_{g_b} \oplus V_{g_c} \oplus V_{g_f})$. If S consists of two pairs of commuting elements, then there is an element $g \in G_{\mathcal{B}}$ such that $W = g \cdot (V_{g_a} \oplus V_{g_b} \oplus V_{g_d} \oplus V_{g_f})$.*

(5) *If $\dim W = 5$, then there is an element $g \in G_{\mathcal{B}}$ such that $W = g \cdot \oplus_{x\in\{a,b,c,d,e\}} V_{g_x}$.*

PROOF. The claim follows from Lemma 7.3 since $\dim V_x = 1$ for all $x \in \mathcal{B}$. □

PROPOSITION 7.5. *Let $t = -1$ and $\lambda = 1$ be the parameters of a one-dimensional representation of $G_{\mathcal{B}}^{g_a}$ as in Lemma 7.2. Let $V = M(g_a, \rho_{t,\lambda})$ be the corresponding $\Bbbk G_{\mathcal{B}}$-Yetter–Drinfeld module. Let v_a, v_b, v_c, v_d, v_e and v_f as in Equation (7.1). Let W be a $\Bbbk G_{\mathcal{B}}$-subcomodule of V. Let $S = \{x \in \mathcal{B} \mid W \cap V_{g_x} \neq 0\}$.*

(1) *The Nichols algebra $\mathcal{B}(V)$ has dimension 576.*

(2)

$$\begin{aligned}
&\dim\langle V_{g_a}\rangle = 2,\\
&\dim\langle V_{g_a} + V_{g_b}\rangle = 6, \quad \dim\langle V_{g_a} + V_{g_f}\rangle = 4,\\
&\dim\langle V_{g_a} + V_{g_b} + V_{g_c}\rangle = 24, \quad \dim\langle V_{g_a} + V_{g_b} + V_{g_f}\rangle = 24,\\
&\dim\langle V_{g_a} + V_{g_b} + V_{g_c} + V_{g_d}\rangle = 96,\\
&\dim\langle V_{g_a} + V_{g_b} + V_{g_d} + V_{g_f}\rangle = 144,\\
&\dim\langle V_{g_a} + V_{g_b} + V_{g_c} + V_{g_d} + V_{g_e}\rangle = 288.
\end{aligned}$$

(3) *If $\dim W = 1$, then $\dim\langle W\rangle = 2$.*

(4) *Assume that $\dim W = 2$. If the elements of S commute, then $\dim\langle W\rangle = 4$. If the elements of S do not commute, then $\dim\langle W\rangle = 6$.*

(5) *Assume that $\dim W = 3$. If two elements of S commute, then $\dim\langle W\rangle = 24$. If the elements of S do pairwise not commute, then $\dim\langle W\rangle = 24$.*

(6) *Assume that $\dim W = 4$. If precisely two elements of S commute, then $\dim\langle W\rangle = 96$. If S consists of two pairs of commuting elements, then $\dim\langle W\rangle = 144$.*

(7) *If $\dim W = 5$, then $\dim\langle W\rangle = 288$.*

(8) *The linear map*

$$\begin{aligned}
F : V \to V^*, \quad & v_c \mapsto {v_e}^*, v_e \mapsto {v_c}^*\\
& v_x \mapsto {v_x}^* \text{ for } x \in \{a, b, d, f\}
\end{aligned}$$

is an isomorphism of the braided vector spaces V and V^* with respect to the Yetter–Drinfeld braidings c_V and c_{V^*}.

(9) *Let $W^* \subseteq V^*$ be the dual $\Bbbk G_{\mathcal{B}}$-subcomodule of W and $\langle W^*\rangle \subset \mathcal{B}(V^*)$ the subalgebra generated by W^*. Then $\dim\langle W\rangle = \dim\langle W^*\rangle$.*

PROOF. (1) See, e.g., [**GnHV11**, Proposition 5.11] or [**AGn03b**, Theorem 6.12].

(2) We proved this using the GAP package GBNP for computing Gröbner bases ([**GAP22**], [**CKGT22**]). There we used the realization of the braided vector space V over the finite enveloping group $\overline{G_{\mathcal{B}}} = G_{\mathcal{B}}/\langle {g_a}^4\rangle$ (and Proposition 3.12 (2)).[8]

[8]The braided vector space V can be realized as Yetter–Drinfeld module over the quotient $\overline{G_{\mathcal{B}}}$ since ${g_a}^4$ acts trivially in the representation $\rho_{t,\lambda}$ since $t^4 = 1$ by assumption.

(3) If $\dim W = 1$, then there is an element $g \in G_{\mathcal{B}}$ such that $W = g \cdot V_{g_a}$ by Lemma 7.4. The map

$$\alpha_g : \mathcal{B}(V) \to \mathcal{B}(V), \quad x \mapsto g \cdot x$$

is an isomorphism of $\mathbb{N}_0$-graded algebras by Lemma 3.16. Hence $\langle W\rangle \cong \langle V_{g_a}\rangle$ as $\mathbb{N}_0$-graded algebras and the claim follows from part (2).

(4)–(7) The arguments are analogous to those in part (3).

(8) Using Lemma 5.1 it follows that $\delta_{V^*}({v_x}^*) = {g_x}^{-1} \otimes {v_x}^*$ for all $x \in \mathcal{B}$ and the $\Bbbk G_{\mathcal{B}}$-action on V^* is given in Table 7. Using this it is a simple calculation to

TABLE 7. $\Bbbk G_{\mathcal{B}}$-module structure on V^*

$\cdot$	${v_a}^*$	${v_b}^*$	${v_c}^*$	${v_d}^*$	${v_e}^*$	${v_f}^*$
${g_a}^{-1}$	$-{v_a}^*$	$-{v_e}^*$	$-{v_b}^*$	$-{v_c}^*$	$-{v_d}^*$	$-{v_f}^*$
${g_b}^{-1}$	${v_c}^*$	$-{v_b}^*$	${v_f}^*$	$-{v_d}^*$	${v_a}^*$	${v_e}^*$
${g_c}^{-1}$	${v_d}^*$	${v_a}^*$	$-{v_c}^*$	${v_f}^*$	$-{v_e}^*$	${v_b}^*$
${g_d}^{-1}$	${v_e}^*$	$-{v_b}^*$	${v_a}^*$	$-{v_d}^*$	${v_f}^*$	${v_c}^*$
${g_e}^{-1}$	${v_b}^*$	${v_f}^*$	$-{v_c}^*$	${v_a}^*$	$-{v_e}^*$	${v_d}^*$
${g_f}^{-1}$	${v_a}^*$	$-{v_c}^*$	$-{v_d}^*$	$-{v_e}^*$	$-{v_b}^*$	$-{v_f}^*$

check that

$$(F \otimes F)c_V(v_x \otimes v_y) = c_{V^*}(F \otimes F)(v_x \otimes v_y)$$

for all $x, y \in \mathcal{B}$ and hence the map $F : V \to V^*$ is a braided isomorphism.

(9) Define the linear map

$$\widetilde{F} : V \to V, \quad v_c \mapsto v_e, v_e \mapsto v_c$$
$$v_x \mapsto v_x \text{ for all } x \in \{a, b, d, f\}.$$

Then $\dim\langle \widetilde{F}(W)\rangle = \dim\langle W\rangle$ since by (3)-(7) the dimensions only depend on the dimension of W and the number of pairs of commuting elements of S, but F is a bijection and the number of pairs of commuting elements of S equals the number of pairs of commuting elements of $\{x \in \mathcal{B} \mid \widetilde{F}(W) \cap V_{g_x} \neq 0\}$. Moreover, $W^* = F(\widetilde{F}(W))$ by definition. The isomorphism of braided vector spaces $F : V \to V^*$ in (8) induces an isomorphism $\mathcal{B}(V) \to \mathcal{B}(V^*)$ of $\mathbb{N}_0$-graded algebras. It follows that

$$\dim\langle W^*\rangle = \dim\langle F(\widetilde{F}(W))\rangle = \dim\langle \widetilde{F}(W)\rangle = \dim\langle W\rangle. \qquad \square$$

THEOREM 7.6. *Let $t \in \Bbbk^\times$, $\lambda \in \Bbbk^\times$ with $\lambda^4 = 1$ be the parameters of a one-dimensional representation of $G_{\mathcal{B}}^{g_a}$ as in Lemma 7.2. Let $V = M(g_a, \rho_{t,\lambda})$ be the corresponding $\Bbbk G_{\mathcal{B}}$-Yetter–Drinfeld module. Let v_a, v_b, v_c, v_d, v_e and v_f as in Equation (7.1). Then the following hold.*

1. *If $t \neq -\lambda$, then $\langle v_b, v_a v_b - t v_c v_a + t v_b v_c\rangle$ or $\langle v_f, v_a v_f - t\lambda v_f v_a\rangle$ is a left coideal subalgebra in $\mathcal{B}(V)$ in the category of $\mathbb{N}_0$-graded $\Bbbk G_{\mathcal{B}}$-comodules, that is not generated in degree one.*
2. *If $t = 1$ and $\lambda = -1$, then there is an extension of the left coideal subalgebra $\langle v_a, v_b, v_c\rangle$ in degree 11. The extended left coideal subalgebra is contained in $\langle v_a, v_b, v_c, v_d\rangle$.*
3. *If $\lambda^2 = -1$ and $t = -\lambda$, then there is an extension of the left coideal subalgebra $\langle v_a, v_b, v_c\rangle$ in degree 7 in the subalgebra $\langle v_a, v_b, v_c, v_d\rangle$.*

(4) *If $t = -1$ and $\lambda = 1$, then all left coideal subalgebras of $\mathcal{B}(V)$ in the category of $\mathbb{N}_0$-graded $\Bbbk G_{\mathcal{B}}$-comodules are generated in degree one.*

PROOF. (1) We have

$$\begin{aligned}\Delta_{1,1}\left(\mu\left(1-c_V+{c_V}^2\right)(v_a\otimes v_b)\right)&=\Delta_{1,1}\left(v_av_b-tv_cv_a+tv_bv_c\right)\\&=\left(1+t^3\lambda\right)v_a\otimes v_b\end{aligned}$$

and

$$\begin{aligned}\Delta_{1,1}\left(\mu\left(1-c_V\right)(v_a\otimes v_f)\right)&=\Delta_{1,1}\left(v_av_f-t\lambda v_fv_a\right)\\&=\left(1-t^2\lambda^2\right)v_a\otimes v_f.\end{aligned}$$

The claim follows from Theorem 4.3 since $1+t^3\lambda\neq 0$ or $1-t^2\lambda^2\neq 0$ since $t\neq-\lambda$ by assumption.

(2), (3) We found these extensions using the GAP package GBNP for computing Gröbner bases ([**GAP22**], [**CKGT22**]). There we used the realization of the braided vector space V over the finite enveloping group $\overline{G_{\mathcal{B}}} = G_{\mathcal{B}}/\langle {g_a}^4\rangle$ (and Proposition 3.12(2)).[9]

(4) Let C, $C\neq \Bbbk 1$, be a left coideal subalgebra of $\mathcal{B}(V)$ in the category of $\mathbb{N}_0$-graded $\Bbbk G_{\mathcal{B}}$-comodules. Then $V\cap C\neq 0$ by Lemma 2.3 and $V_1 = V\cap C$ is a $\Bbbk G_{\mathcal{B}}$-subcomodule by assumption. Since $\Bbbk G_{\mathcal{B}}$ is cosemisimple, there exists a $\Bbbk G_{\mathcal{B}}$-subcomodule V_2 such that $V = V_1\oplus V_2$. By Proposition 7.5 (9), we have $\dim\langle V_2\rangle = \dim\langle {V_2}^*\rangle$. For $i\in\{1,2\}$ let $S_i = \{x\in\mathcal{B}\mid V_i\cap V_{g_x}\neq 0\}$. We now consider different cases depending on the dimension of V_1.

(i) $\dim V_1 = 1$. Then $\dim V_2 = 5$ and it follows that

$$\dim\langle V_1\rangle\dim\langle V_2\rangle = 2\cdot 288 = 576 = \dim\mathcal{B}(V)$$

by Proposition 7.5(3), (7).

(iia) $\dim V_1 = 2$ and the elements of S_1 commute. Then $\dim V_2 = 4$ and $S_2 = \mathcal{B}\backslash S_1$ consists of two pairs of commuting elements. It follows that $\dim\langle V_1\rangle\dim\langle V_2\rangle = 4\cdot 144 = 576 = \dim\mathcal{B}(V)$ by Proposition 7.5(4), (6).

(iib) $\dim V_1 = 2$ and the elements of S_1 do not commute. Then $\dim V_2 = 4$ and exactly two elements of $S_2 = \mathcal{B}\setminus S_1$ commute. It follows that $\dim\langle V_1\rangle\dim\langle V_2\rangle = 6\cdot 96 = 576 = \dim\mathcal{B}(V)$ by Proposition 7.5(4), (6).

(iii) $\dim V_1 = 3$. Then $\dim V_2 = 3$. It follows that $\dim\langle V_1\rangle\dim\langle V_2\rangle = 24\cdot 24 = 576 = \dim\mathcal{B}(V)$ by Proposition 7.5(5).

If $\dim V_1 > 3$, then it follows that $\dim\langle V_1\rangle\dim\langle V_2\rangle = 576 = \dim\mathcal{B}(V)$ by changing the roles of V_1 and V_2. Hence the claim follows from Corollary 2.13(2). □

8. Racks with special subracks

Let $(X,\triangleright)$ be a finite indecomposable rack of a conjugacy class of a group and $q: X\times X\to\Bbbk^\times$ a two-cocycle. Let $V = \Bbbk X$ be the corresponding Yetter–Drinfeld realization over $\Bbbk G_X$ and $v_x\in V_x\setminus\{0\}$, $x\in X$, such that $g_x\cdot v_y = q_{x,y}v_{x\triangleright y}$ for all $x, y\in X$. Denote the braiding by c_q and let $\mathcal{B}(V)$ be the corresponding Nichols algebra.

Our tactic in this section is to assume that all left coideal subalgebras of $\mathcal{B}(V)$ in the category of $\mathbb{N}_0$-graded $\Bbbk G_X$-comodules are generated in degree one and to

---

[9]The braided vector space V can be realized as Yetter–Drinfeld module over the quotient $\overline{G_{\mathcal{B}}}$ since ${g_a}^4$ acts trivially in the representation $\rho_{t,\lambda}$ since $t^4 = 1$ by assumption.

argue then that X is either $\mathbb{T}_n$, $\mathcal{T}$ or $\mathcal{B}$, where $n \in \mathbb{N}$. In our main Theorem 8.8 we combine this result with our results about the racks $\mathbb{T}_n$, $\mathcal{T}$ and $\mathcal{B}$ in Sections 5–7.

LEMMA 8.1. *Assume that all left coideal subalgebras of $\mathcal{B}(V)$ in the category of $\mathbb{N}_0$-graded $\Bbbk G_X$-comodules are generated in degree one. Let $x, y \in X$ with $x \triangleright y \neq y$ and let Y be the subrack of X generated by x and y. Then Y is isomorphic to $\mathbb{T}_3$, $\mathcal{T}$ or $\mathcal{B}$.*

PROOF. Since all left coideal subalgebras of $\mathcal{B}(V)$ are generated in degree one, it follows that X is braided by Proposition 4.2 and that $\mathcal{B}(V)$ is slim in degree two by Theorem 4.3. We distinguish three cases.

(1) Assume that $x \triangleright (x \triangleright y) = y$. Since X is braided, this implies that $x \triangleright y = y \triangleright x$ and $Y = \{x, y, x \triangleright y\}$. Hence there is a rack isomorphism $Y \to \mathbb{T}_3$ such that
$$x \mapsto (1\ 2),\ y \mapsto (1\ 3).$$
(2) Assume that $x \triangleright (x \triangleright y) \neq y$ and $x \triangleright (x \triangleright (x \triangleright y) = y$. Since X is braided, this implies that $x \triangleright (x \triangleright y) = y \triangleright x$ and $Y = \{x, y, x \triangleright y, y \triangleright x\}$. Hence there is a rack isomorphism $Y \to \mathcal{T}$ such that
$$x \mapsto a = (2\ 3\ 4),\ y \mapsto b = (1\ 4\ 3)$$
(see Table 2).
(3) Assume that $x \triangleright (x \triangleright y) \neq y$ and $x \triangleright (x \triangleright (x \triangleright y)) \neq y$. Since X is braided, this implies that $x \triangleright (x \triangleright y) \neq y \triangleright x$. Let $z = x \triangleright (x \triangleright y)$. Then $z \neq y \triangleright x$ and since all left coideal subalgebras of $\mathcal{B}(V)$ in the category of $\mathbb{N}_0$-graded $\Bbbk G_X$-comodules are generated in degree one, it follows from Theorem 4.6 that z commutes with at least one of the elements x, y, $x \triangleright y$. Since $(x \triangleright y) \triangleright z = x \neq z$ and $z \triangleright x = x \triangleright y \neq x$, this implies that $y \triangleright z = z$. Since X is braided, it follows that $x \triangleright z = y \triangleright x$ and $x^4 \triangleright y = y$. Moreover, $x \triangleright (y \triangleright (y \triangleright x)) = y \triangleright (y \triangleright x)$ and $y^3 \triangleright x = x \triangleright y$ since X is braided. Hence Y has 6 elements and degree 4 and there is a rack isomorphism $Y \to \mathcal{B}$ such that
$$x \mapsto a = (2\ 3\ 4\ 5),\ y \mapsto b = (1\ 5\ 6\ 3)$$
(see Table 5). □

PROPOSITION 8.2. *Assume that X is not generated by less than 3 elements and that X contains the rack $\mathcal{T}$ as subrack. Then there is a primitively generated left coideal subalgebra of $\mathcal{B}(V)$ in the category of $\mathbb{N}_0$-graded $\Bbbk G_X$-subcomodules which can be extended in a degree ≤ 3.*

PROOF. Assume that $\mathcal{B}(V)$ does not contain a left coideal subalgebra in the category of $\mathbb{N}_0$-graded $\Bbbk G_X$-subcomodules that has an extension in degree two. Then X is braided by Proposition 4.2 and $\mathcal{B}(V)$ is slim in degree two by Theorem 4.3. Since the subrack $\mathcal{T}$ is generated by each two of its elements but X is not generated by less than 3 elements by assumption, there is an element $z \in X$ with $z \notin \mathcal{T}$. Since X is indecomposable, we can choose z such that it does not commute with at least one element of $\mathcal{T}$. Then z commutes with at most one element of $\mathcal{T}$. Indeed, if z would commute with more than one element of $\mathcal{T}$, it would commute with any element of $\mathcal{T}$ since $\mathcal{T}$ is generated by each two of its elements. Hence there are elements $x, y \in \mathcal{T}$ such that $x \triangleright y \neq z$ and z commutes with none of x, y and $y \triangleright x$. By Theorem 4.6, it follows that the left coideal subalgebra

$\langle v_x, v_y, v_{z\triangleright(y\triangleright x)}, v_{z\triangleright x}\rangle \subset \mathcal{B}(V)$ in the category of $\mathbb{N}_0$-graded $\Bbbk G_X$-comodules can be extended in degree three by the element $Z(x,y,z)$. □

PROPOSITION 8.3. *Assume that X is not generated by less than 3 elements and that X contains the rack $\mathcal{B}$ as subrack. Then there is a primitively generated left coideal subalgebra of $\mathcal{B}(V)$ in the category of $\mathbb{N}_0$-graded $\Bbbk G_X$-subcomodules that can be extended in a degree ≤ 4.*

PROOF. Assume that $\mathcal{B}(V)$ does not contain a left coideal subalgebra in the category of $\mathbb{N}_0$-graded $\Bbbk G_X$-subcomodules that has an extension in degree two. Then X is braided by Proposition 4.2 and $\mathcal{B}(V)$ is slim in degree two by Theorem 4.3. Recall that $\mathcal{B}$ is generated by every two not commuting elements of $\mathcal{B}$. Since X is not generated by less than 3 elements and since X is indecomposable, there is an element $z \in X \setminus \mathcal{B}$ which does not commute with at least one element of $\mathcal{B}$. We distinguish two cases.

(1) Assume that $z\triangleright x \neq x$ for all $x \in \mathcal{B}$. Choose elements $a, b \in \mathcal{B}$ with $a\triangleright b \neq b$. Then $a \triangleright b \neq z$ since $z \notin \mathcal{B}$. Moreover, the element z commutes with none of a, b, $b \triangleright a$ by assumption in the considered case. By Theorem 4.6, it follows that the left coideal subalgebra $\langle v_a, v_b, v_{z\triangleright(b\triangleright a)}, v_{z\triangleright a}\rangle \subset \mathcal{B}(V)$ in the category of $\mathbb{N}_0$-graded $\Bbbk G_X$-subcomodules can be extended in degree 3 by the element $Z(a,b,z)$.

(2) Assume that there is an element $a \in \mathcal{B}$ such that $z \triangleright a = a$. Let $b \in \mathcal{B}$ such that $a\triangleright b \neq b$. Then $z\triangleright b \neq b$ since otherwise z would commute with both elements of the generating set $\{a, b\}$ of $\mathcal{B}$ and hence with all elements of $\mathcal{B}$. Let $c = a \triangleright b$. Then $a\triangleright c \neq c$, $a\triangleright c \neq b$ and $b\triangleright (a\triangleright c) = a\triangleright c$ since these relations hold for each two not commuting elements a, b of the cube rack $\mathcal{B}$. Moreover, $z\triangleright c \neq c$ since $z\triangleright a = a$, $z\triangleright b \neq b$ and $a\triangleright b \neq b$. Since $b\triangleright c = a$, it follows that $z\triangleright (b\triangleright c) = b\triangleright c$ by assumption and $z \neq b \triangleright c$ since $b \triangleright c \in \mathcal{B}$ but $z \notin \mathcal{B}$. Then by Theorem 4.11, there is a left coideal subalgebra of $\mathcal{B}(V)$ in the category of $\mathbb{N}_0$-graded $\Bbbk G_X$-subcomodules that can be extended by the element $Z(z,b,a,c)$ of degree 4 since $zba \triangleright c \neq c$. Indeed, if $zba \triangleright c = c$, then $z \triangleright (a \triangleright c) = c$ and hence $a \triangleright (z \triangleright c) = c$. It follows $z \triangleright c = b$ and hence $z = c \triangleright b \in \mathcal{B}$ but $z \notin \mathcal{B}$. □

LEMMA 8.4. *Assume that all left coideal subalgebras of $\mathcal{B}(V)$ in the category of $\mathbb{N}_0$-graded $\Bbbk G_X$-comodules are generated in degree one. Assume that the subrack of X generated by x and y is isomorphic to $\mathbb{T}_3$ for all $x, y \in X$ with $x \triangleright y \neq y$. Then for all pairwise not commuting elements $x, y, z \in X$ the equality $x \triangleright (y \triangleright z) = y \triangleright z$ holds.*

PROOF. Since all left coideal subalgebras of $\mathcal{B}(V)$ in the category of $\mathbb{N}_0$-graded $\Bbbk G_X$-comodules are generated in degree one, X is braided by Proposition 4.2 and $\mathcal{B}(V)$ is slim in degree two by Theorem 4.3. Assume that there are pairwise not commuting elements $x, y, z \in X$ such that $x \triangleright (y \triangleright z) \neq y \triangleright z$. Then $x \triangleright z \neq z$, $x \triangleright y \neq y$ and $x \triangleright (y \triangleright z) \neq y \triangleright z$. Moreover, $y \triangleright z = z \triangleright y$ since the subrack generated by z and y is the rack $\mathbb{T}_3$ by assumption. Hence $x \neq z \triangleright y$ and by Theorem 4.6 there is a left coideal subalgebra of $\mathcal{B}(V)$ in the category of $\mathbb{N}_0$-graded $\Bbbk G_X$-comodules that is not generated in degree one. This is a contradiction to the assumption. □

LEMMA 8.5. *Assume that all left coideal subalgebras of $\mathcal{B}(V)$ in the category of $\Bbbk G_X$-comodules are generated in degree one. Assume that the subrack of X generated by x and y is isomorphic to $\mathbb{T}_3$ for all $x, y \in X$ with $x \triangleright y \neq y$. Let*

$x, a, b, c \in X$ be pairwise distinct elements such that $a \triangleright b = c$ and $x \triangleright a \neq a$. Then x commutes with precisely one element of $\{b, c\}$.

PROOF. Since all left coideal subalgebras of $\mathcal{B}(V)$ in the category of $\mathbb{N}_0$-graded $\Bbbk G_X$-comodules are generated in degree one, X is braided by Proposition 4.2 and hence $a = b \triangleright c$. Assume that $x \triangleright b = b$ and $x \triangleright c = c$. Then $x \triangleright a = x \triangleright (b \triangleright c) = b \triangleright c = a$, but $x \triangleright a \neq a$ by assumption. Hence x commutes with at most one element of $\{b, c\}$. Assume that $x \triangleright b \neq b$ and $x \triangleright c \neq c$. Then $x \triangleright (b \triangleright c) = b \triangleright c$ by Lemma 8.4. But this is a contradiction to the assumption $x \triangleright a \neq a$. □

REMARK 8.6. Assume that all left coideal subalgebras of $\mathcal{B}(V)$ in the category of $\mathbb{N}_0$-graded $\Bbbk G_X$-comodules are generated in degree one. Let Y be a subrack of X, $q' : Y \times Y \to \Bbbk$ the restriction of q on $Y \times Y$ and $V' = \Bbbk Y$ the corresponding Yetter–Drinfeld realization over $\Bbbk G_Y$. Then all left coideal subalgebras of $\mathcal{B}(V')$ in the category of $\mathbb{N}_0$-graded $\Bbbk G_Y$-comodules are generated in degree one.

PROPOSITION 8.7. *Assume that X is generated by $n \geq 2$ elements and that any $n-1$ elements of X generate a proper subrack. Assume that all left coideal subalgebras of $\mathcal{B}(V)$ in the category of $\mathbb{N}_0$-graded $\Bbbk G_X$-comodules are generated in degree one. Assume that the subrack of X generated by x and y is isomorphic to $\mathbb{T}_3$ for all $x, y \in X$ with $x \triangleright y \neq y$. Then X is the rack of transpositions $\mathbb{T}_{n+1}$.*

PROOF. Since all left coideal subalgebras of $\mathcal{B}(V)$ in the category of $\mathbb{N}_0$-graded $\Bbbk G_X$-comodules are generated in degree one, X is braided by Proposition 4.2. We prove the claim by induction.

For $n = 2$ the claim holds by assumption. To clarify the idea of the proof, we additionally show the proof for the case $n = 3$. Choose $x, y \in X$ such that $x \triangleright y \neq y$ and let X_2 be the subrack of X generated by x and y. Then $X_2 \cong \mathbb{T}_3$ by assumption. Hence we can identify the elements of X_2 with transpositions $x = (1\ 2)$, $y = (1\ 3)$ and $z = (2\ 3)$. Since X is generated by 3 elements and since X is indecomposable, there exists an element $a \in X \setminus X_2$ that does not commute with at least one element of X_2. Hence a does not commute with precisely two elements of X_2 by Lemma 8.5. Without loss of generality we may assume that a does not commute with $(1\ 2)$ and $(1\ 3)$. It follows that there is a rack isomorphism $X \to \mathbb{T}_4$ such that

$$(1\ 2) \mapsto (1\ 2),\ \ (1\ 3) \mapsto (1\ 3),\ \ a \mapsto (1\ 4).$$

Let $n \geq 4$ and let X_{n-1} be a subrack of X generated by $n-1$ elements. Then $X_{n-1} \cong \mathbb{T}_n$ by induction hypothesis and Remark 8.6. We identify X_{n-1} with the rack $\mathbb{T}_n$ and consider the elements of X_{n-1} as transpositions, i.e.,

$$X_{n-1} = \mathbb{T}_n = \{(i\ j) \mid 1 \leq i < j \leq n\}.$$

Let $a \in X \setminus \mathbb{T}_n$ such that X is generated by $\mathbb{T}_n \cup \{a\}$. Since X is indecomposable, we can assume that a does not commute with at least one element of $\mathbb{T}_n$. Let

$$T = \{y \in \mathbb{T}_n \mid a \triangleright y \neq y\} \text{ and}$$
$$T^{\mathsf{C}} = \mathbb{T}_n \setminus T = \{y \in \mathbb{T}_n \mid a \triangleright y = y\}.$$

Then T^{C} is a subrack since for $x, y \in T^{\mathsf{C}}$ we have $a \triangleright x = x$ and $a \triangleright y = y$ and hence $a \triangleright (x \triangleright y) = x \triangleright y$. This means that we obtain a union of complete subgraphs if we regard the elements of $\{1, \ldots, n\}$ as vertices and connect vertices i, j if and only if the element a commutes with the transposition $(i\ j)$. We will visualise these graphs in figures. For example, in the above case $n = 3$ one has two complete subgraphs and $T^{\complement} = \{(2\ 3)\}$ as in Figure 3. Note that the element a itself is not shown in the illustrations. The figures are intended to show which transpositions of $\mathbb{T}_n$ commute with a and which do not.

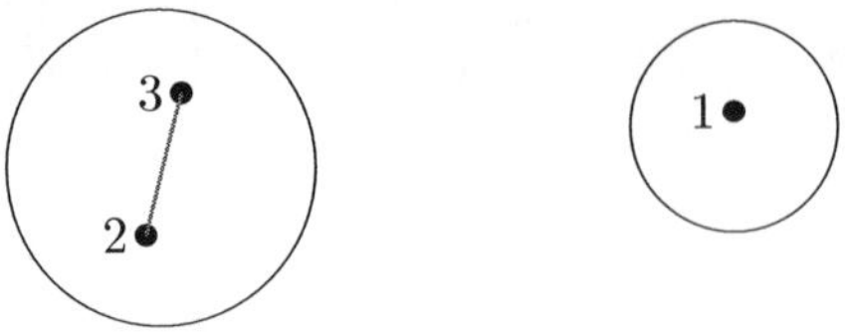

FIGURE 3. $n = 3$, a commutes with $(2\ 3)$

Formally, we describe the decomposition as follows. There are a number $p \in \mathbb{N}$ and a disjoint decomposition

$$\dot\cup_{\alpha=1}^{p} M_\alpha = \{1, \dots, n\}$$

such that $T^{\complement} = \cup_{\alpha=1}^{p}\{(i\ j) \mid i, j \in M_\alpha, i < j\}$. Since the element a does not commute with at least one element of $\mathbb{T}_n$, it follows $p > 1$. We are going to prove that $p = 2$ and that one of M_1, M_2 has cardinality 1.

Claim 1: There is at most one $\alpha \in \{1, \dots, p\}$ such that $|M_\alpha| \geq 2$.

Proof of claim 1: Assume that there are $\alpha, \beta \in \{1, \dots, p\}$ and pairwise distinct $i, j, k, l \in \{1, \dots, n\}$ such that $i, j \in M_\alpha$ and $k, l \in M_\beta$. Then $(i\ j), (k\ l) \in T^{\complement}$ and $\{(i\ k), (i\ l), (j\ k), (j\ l)\} \subset T$. In Figure 4 we indicate this constellation: a commutes with $(i\ j)$ and $(k\ l)$ but with no element from $(i\ k)$, $(i\ l)$, $(j\ k)$ and $(j\ l)$. There elements of T are represented by a dotted line. We only plotted relevant points.

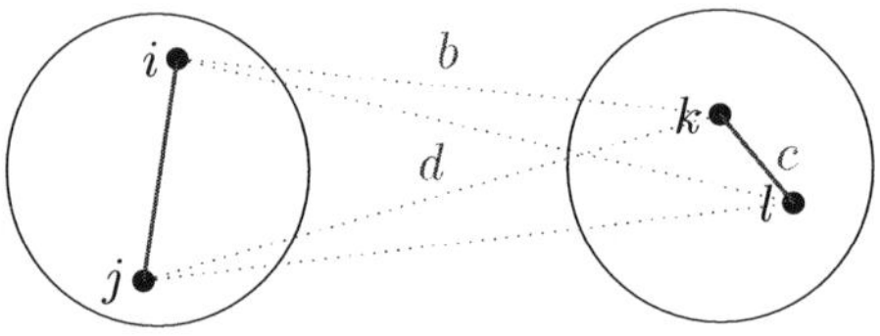

FIGURE 4. In this situation there would be an extendable left coideal subalgebra.

Let $b = (i\ k)$, $c = (k\ l)$ and $d = (j\ k)$. Then a, b, c, d are pairwise not commuting elements except $a \triangleright c = c$ but $a \neq c$ since $a \notin \mathbb{T}_n$. The elements fulfill the following relations:

$$\begin{aligned}
&c \triangleright d = (j\ l) \neq b,\\
&b \triangleright (c \triangleright d) = (i\ k) \triangleright (j\ l) = (j\ l) = c \triangleright d,\\
&a \triangleright (b \triangleright d) = a \triangleright (i\ j) = (i\ j) = b \triangleright d \quad \text{and}\\
&b \triangleright d \neq a \text{ since } a \notin \mathbb{T}_n, (i\ j) \in \mathbb{T}_n.
\end{aligned}$$

Moreover, $a \triangleright (b \triangleright (c \triangleright d)) \neq d$ since otherwise we would have

$$a \triangleright (b \triangleright (c \triangleright d)) = a \triangleright (j\ l) = (j\ k).$$

Then $(j\ l)\triangleright(j\ k) = a$ since X is braided. But $a \notin \mathbb{T}_n$. Hence by Theorem 4.11, there is a left coideal subalgebra of $\mathcal{B}(V)$ in the category of $\mathbb{N}_0$-graded $\Bbbk G_X$-comodules that is not generated in degree one. This is a contradiction to the assumption.

Claim 2: $p = 2$.

Proof of claim 2: Assume that $p \geq 3$. Let $i \in M_1$, $j \in M_2$ and $k \in M_3$. Then $(i\ j)$, $(j\ k)$, $(i\ k) \in T$ and $(i\ j) \triangleright (j\ k) = (i\ k)$. This a contradiction to Lemma 8.5.

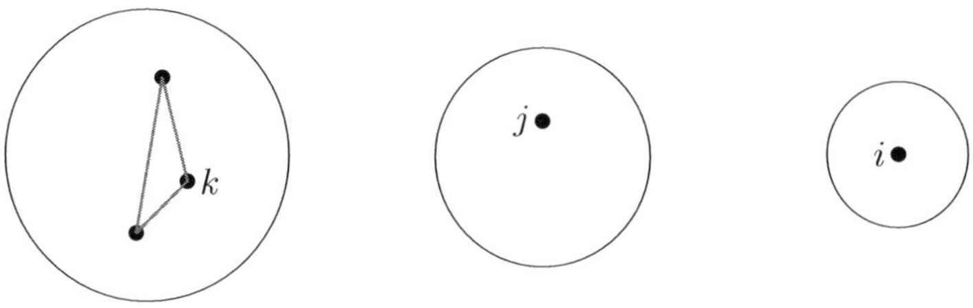

FIGURE 5. Impossible constellation

From claims 1 and 2 it follows that $p = 2$ and one of M_1, M_2 has cardinality 1.

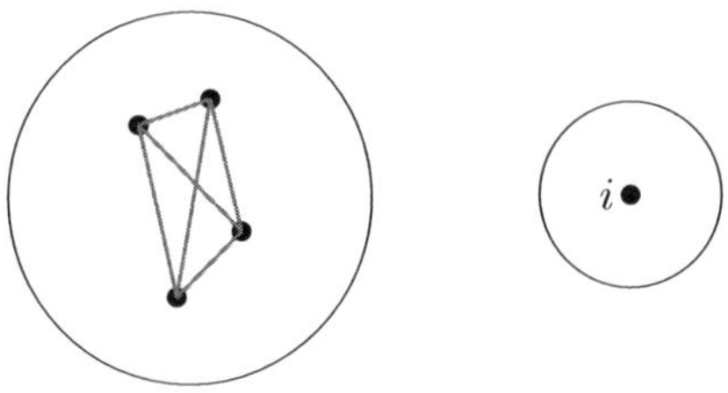

FIGURE 6. Possible constellation

Hence there is an element $i \in \{1, \dots, n\}$ such that for all $k, l \in \{1, \dots, n\}$, $k < l$, the element a commutes with $(k\ l)$ if and only if $k = i$ or $l = i$. Without loss of generality we may assume that $i = 1$. It follows that there is a rack isomorphism $X \to \mathbb{T}_{n+1}$ such that

$$a \mapsto (1\ n+1),\ (1\ j) \mapsto (1\ j) \text{ for all } j \in \{2, \dots, n\}. \qquad \square$$

Now all preparations are made for the proof of the following main Theorem.

THEOREM 8.8. *Let $(X, \triangleright)$ be a finite indecomposable rack of a conjugacy class of a group, $q : X \times X \to \Bbbk^\times$ a two-cocycle and $(\Bbbk X, c_q)$ the corresponding braided vector space. Let $V = \Bbbk X$ be its Yetter–Drinfeld realization over $\Bbbk G_X$. Then all left coideal subalgebras of $\mathcal{B}(V)$ in the category of $\mathbb{N}_0$-graded $\Bbbk G_X$-comodules are generated in degree one if and only if V appears in the following list (up to isomorphism).*

1. *$V = \Bbbk\mathbb{T}_3 = M(g_{(1\ 2)}, \rho_t)$ with $t = -1$ or $t^2 - t + 1 = 0$, where ρ_t is the linear character on the centralizer $G_{\mathbb{T}_3}^{g_{(1\ 2)}} = \langle g_{(1\ 2)} \rangle$ with $\rho_t(g_{(1\ 2)}) = t$. The corresponding $G_{\mathbb{T}_3}$-action on V is explicitly given in Table 1.*
2. *$V = \Bbbk\mathbb{T}_4 = M(g_{(1\ 2)}, \rho_{t,\lambda})$ with $t = -1$ and $\lambda \in \{1, -1\}$, where $\rho_{t,\lambda}$ is the linear character on the centralizer $G_{\mathbb{T}_4}^{g_{(1\ 2)}} = \langle g_{(1\ 2)}, g_{(3\ 4)} \rangle$ with*

$\rho_{t,\lambda}(g_{(1\ 2)}) = t$ and $\rho_{t,\lambda}(g_{(3\ 4)}) = \lambda t$. The possible two-cocycles are either the constant two-cocycle -1 or the two-cocycle $q = \chi$ given in Equation (5.2), respectively.

(3) *$V = \Bbbk\mathbb{T}_5 = M(g_{(1\ 2)}, \rho_{t,\lambda})$ with $t = -1$ and $\lambda \in \{1, -1\}$, where $\rho_{t,\lambda}$ is the linear character on the centralizer $G_{\mathbb{T}_5}^{g_{(1\ 2)}} = \langle g_{(1\ 2)}, g_{(3\ 4)}, g_{(3\ 5)}, g_{(4\ 5)}\rangle$ with $\rho_{t,\lambda}(g_{(1\ 2)}) = t$ and $\rho_{t,\lambda}(g_{(i\ j)}) = \lambda t$ for all $2 < i < j \leq 5$. The possible two-cocycles are either the constant two-cocycle -1 or the two-cocycle $q = \chi$ given in Equation (5.2) respectively.*

(4) *$V = \Bbbk\mathcal{T} = M(g_a, \rho_{t,\lambda})$ with $t \in \Bbbk$ such that $t^2 + t + 1 = 0$ and $\lambda = -t^{-1}$, where $\rho_{t,\lambda}$ is the linear character on the centralizer $G_{\mathcal{T}}^{g_a} = \langle g_a, g_b g_d\rangle$ with $\rho_{t,\lambda}(g_a) = t$ and $\rho_{t,\lambda}(g_b g_d) = \lambda$. The $G_{\mathcal{T}}$-action on V is explicitly given in Table 3.*

(5) *$V = \Bbbk\mathcal{B} = M(g_a, \rho_{t,\lambda})$ with $t = -1$ and $\lambda = 1$, where $\rho_{t,\lambda}$ is the linear character on the centralizer $G_{\mathcal{B}}^{g_a} = \langle g_a, g_f\rangle$ with $\rho_{t,\lambda}(g_a) = t$ and $\rho_{t,\lambda}(g_f) = \lambda t$. The $G_{\mathcal{B}}$-action on V is explicitly given in Table 6.*

(6) *$V = \Bbbk X$, where X is the rack with one element and q is an arbitrary two-cocycle.*

PROOF. First assume that all left coideal subalgebras of $\mathcal{B}(V)$ in the category of $\mathbb{N}_0$-graded $\Bbbk G_X$-comodules are generated in degree one. The rack X is braided by Proposition 4.2. Assume that X is not the rack with one element. Then there are two elements $x, y \in X$ with $x \triangleright y \notin \{x, y\}$ since X is indecomposable. Hence X includes $\mathbb{T}_3$, $\mathcal{T}$ or $\mathcal{B}$ as subrack by Lemma 8.1. It follows that X is $\mathbb{T}_3$, $\mathcal{T}$ or $\mathcal{B}$ if X is generated by two elements. If $X = \mathbb{T}_3$, it follows from Theorem 5.7 that the only possible Yetter–Drinfeld structures on $\Bbbk\mathbb{T}_3$ such that all left coideal subalgebras of $\mathcal{B}(V)$ in the category of $\mathbb{N}_0$-graded $\Bbbk G_{\mathbb{T}_3}$-subcomodules are generated in degree one are those in (1). For $X = \mathcal{T}$ it follows from Theorem 6.6 that the only choices for suitable Yetter–Drinfeld structures are those in (4). For $X = \mathcal{B}$ it follows from Theorem 7.6 that the only choice for a suitable Yetter–Drinfeld structure is that in (5). Assume that X is not generated by less than 3 elements. Proposition 8.2 and Proposition 8.3 imply that X does not contain the rack $\mathcal{T}$ or the rack $\mathcal{B}$ as subrack. If X is generated by $n \geq 3$ elements (and not by less than n elements) and if X does not contain $\mathcal{T}$ or $\mathcal{B}$ as a subrack, then for all $x, y \in X$ with $x \triangleright y \neq y$ the subrack generated by x and y is isomorphic to $\mathbb{T}_3$ by Lemma 8.1 and hence X is the rack of transpositions of $\mathbb{S}_{n+1}$ by Proposition 8.7. Then if $n \in \{4, 5\}$, the claim follows from Theorem 5.10. If $n \geq 6$, the claim follows from Theorem 5.13 and from Remark 8.6.

The other implication follows in case (1) from Theorem 5.7(1), in cases (2) and (3) from Theorem 5.12, in case (4) from Theorem 6.6(4) and in case (5) from Theorem 7.6(4). In case (6) it is obvious. □

REMARK 8.9. The corresponding Nichols algebras $\mathcal{B}(V)$ to the braided vector spaces V in (2)–(5) and to that with $t = -1$ in (1) of Theorem 8.8 were already known to be finite-dimensional (see Remark 5.8 for (1), Remark 5.11 for (2) and (3), Proposition 6.5 for (4) and Proposition 7.5 for (5)). The Nichols algebras of the braided vector spaces in Theorem 8.8(1) with $t^2 - t + 1 = 0$ are infinite-dimensional by [**HMV23**]: By Theorem 1.6 therein, if the base field has characteristic 0, then the cocycle of an absolutely irreducible Yetter–Drinfeld module of dimension three with finite-dimensional Nichols algebra is constant -1. The 72-dimensional Nichols algebra in Theorem 6.6(2) and Remark 6.7 is an example for a finite-dimensional Nichols algebra that contains a left coideal subalgebra which is not generated in degree one.

COROLLARY 8.10. *Let G be a group and V a Yetter–Drinfeld module over $\Bbbk G$ such that $\dim V_g \leq 1$ for all $g \in G$. Then all left coideal subalgebras of $\mathcal{B}(V)$ in the category of $\mathbb{N}_0$-graded $\Bbbk G$-comodules are generated in degree one if and only if there are braided subspaces*[10] *$V_1, \dots, V_n \subseteq V$ such that*

(1) *V_i is a $\Bbbk G$-subcomodule for all $i \in \{1, \dots, n\}$,*
(2) *$V = \bigoplus_{i=1}^n V_i$,*
(3) *${c_V}^2|V_i \otimes V_j = \mathrm{id}_{V_i \otimes V_j}$ for all $i, j \in \{1, \dots, n\}$ with $i \neq j$, and*
(4) *V_i is isomorphic to one of the braided vector spaces in (1)-(6) of Theorem 8.8 as braided vector space for all $i \in \{1, \dots, n\}$.*

PROOF. First assume that all left coideal subalgebras of $\mathcal{B}(V)$ in the category of $\mathbb{N}_0$-graded $\Bbbk G$-comodules are generated in degree one. Let $V = \bigoplus_{k=1}^m W_k$ be the decomposition of V into its irreducible $\Bbbk G$-Yetter–Drinfeld submodules. Then ${c_V}^2|W_k \otimes W_l = \mathrm{id}_{W_k \otimes W_l}$ for all $k, l \in \{1, \dots, m\}$ with $k \neq l$ by Proposition 3.13 and for all $1 \leq k \leq m$ all left coideal subalgebras of $\mathcal{B}(W_k)$ in the category of $\mathbb{N}_0$-graded $\Bbbk G$-comodules are generated in degree one by Proposition 3.14. For all $1 \leq k \leq m$ let $X_k = \mathrm{supp} W_k \subseteq G$ be the underlying rack[11] and $\widetilde{W_k}$ the $\Bbbk G_{X_k}$-Yetter–Drinfeld realization of the underlying braided vector space $\Bbbk X_k = W_k$ (see Lemma 3.11 and Lemma 3.9 for the construction). Then for all $1 \leq k \leq m$ all left coideal subalgebras of $\mathcal{B}(\widetilde{W_k})$ in the category of $\mathbb{N}_0$-graded $\Bbbk G_{X_k}$-comodules are generated in degree one by Proposition 3.12(2). If $\widetilde{W_k}$ is irreducible as Yetter–Drinfeld module over $\Bbbk G_{X_k}$, then X_k is indecomposable by Lemma 3.15. By Theorem 8.8, it follows that $\widetilde{W_k}$ is isomorphic to one of the braided vector spaces mentioned in (1)–(6) of Theorem 8.8 as braided vector space. If $\widetilde{W_k}$ is reducible as Yetter–Drinfeld module over $\Bbbk G_{X_k}$, then we decompose it into a direct sum of its irreducible $\Bbbk G_{X_k}$-Yetter–Drinfeld submodules and repeat the steps of the proof until all summands are irreducible. Note that $\Bbbk G_{X_k}$-Yetter–Drinfeld submodules of $\widetilde{W_k}$ are also $\Bbbk G$-subcomodules of V by construction.

For the other implication assume that there are braided subspaces $V_1, \dots, V_n \subseteq V$ that fulfill the conditions (1)–(4). Let $X = \mathrm{supp} V$ and let V_{G_X} be the $\Bbbk G_X$-Yetter–Drinfeld realization of the braided vector space V. By Proposition 3.12(2) it suffices to prove that all left coideal subalgebras of $\mathcal{B}(V_{G_X})$ in the category of $\mathbb{N}_0$-graded $\Bbbk G_X$-comodules are generated in degree one. The construction of V_{G_X} is described in Lemma 3.11 and Lemma 3.9. In particular, the $\Bbbk G$-subcomodules V_i are also $\Bbbk G_X$-subcomodules and the braidings c_V and $c_{V_{G_X}}$ coincide. The assumptions $\dim V_g \leq 1$ for all $g \in G$ and ${c_V}^2|V_i \otimes V_j = \mathrm{id}_{V_i \otimes V_j}$ for all $i, j \in \{1, \dots, n\}$ with $i \neq j$ imply that $V_h = V_{ghg^{-1}}$ for all $g, h \in X$ with g, h belonging to supports of different direct summands V_i of V. Since X generates G_X and since $\dim V_g \leq 1$ for all $g \in G$ by assumption, it follows that the subcomodules V_i are $\Bbbk G_X$-Yetter–Drinfeld submodules of V_{G_X}. By Theorem 8.8, for all $1 \leq i \leq n$ all left coideal subalgebras of $\mathcal{B}(V_i)$ in the category of $\mathbb{N}_0$-graded $\Bbbk G_{\mathrm{supp} V_i}$-comodules are generated in degree one. By Proposition 3.12(2) it follows that all left coideal subalgebras of $\mathcal{B}(V_i)$ in

[10]A subspace $W \subseteq V$ is a braided subspace if $c_V(W \otimes W) = W \otimes W$.

[11]Note that the racks X_k may be decomposable since it is not assumed that $\mathrm{supp} W_k$ generates G.

the category of $\mathbb{N}_0$-graded $\Bbbk G_X$-comodules are generated in degree one. Then the claim follows from Proposition 3.14 since $c_V{}^2|V_i \otimes V_j = \mathrm{id}_{V_i \otimes V_j}$ for all $1 \leq i, j \leq n$ with $i \neq j$ and $\dim V_g \leq 1$ for all $g \in G$ by assumption. □

References

[AFGnV11] N. Andruskiewitsch, F. Fantino, M. Graña, and L. Vendramin, *Finite-dimensional pointed Hopf algebras with alternating groups are trivial*, Ann. Mat. Pura Appl. (4) **190** (2011), no. 2, 225–245, DOI 10.1007/s10231-010-0147-0. MR2786171

[AFGV11] N. Andruskiewitsch, F. Fantino, G. A. García, and L. Vendramin, *On Nichols algebras associated to simple racks*, Groups, algebras and applications, Contemp. Math., vol. 537, Amer. Math. Soc., Providence, RI, 2011, pp. 31–56, DOI 10.1090/conm/537/10565. MR2799090

[AGn03a] N. Andruskiewitsch and M. Graña, *Examples of liftings of Nichols algebras over racks*, AMA Algebra Montp. Announc. (2003), Paper 1, 6. Théories d'homologie, représentations et algèbres de Hopf. MR2065444

[AGn03b] N. Andruskiewitsch and M. Graña, *From racks to pointed Hopf algebras*, Adv. Math. **178** (2003), no. 2, 177–243, DOI 10.1016/S0001-8708(02)00071-3. MR1994219

[AHS10] N. Andruskiewitsch, I. Heckenberger, and H.-J. Schneider, *The Nichols algebra of a semisimple Yetter-Drinfeld module*, Amer. J. Math. **132** (2010), no. 6, 1493–1547. MR2766176

[AS98] N. Andruskiewitsch and H.-J. Schneider, *Lifting of quantum linear spaces and pointed Hopf algebras of order p^3*, J. Algebra **209** (1998), no. 2, 658–691, DOI 10.1006/jabr.1998.7643. MR1659895

[AZ07] N. Andruskiewitsch and S. Zhang, *On pointed Hopf algebras associated to some conjugacy classes in $\mathbb{S}_n$*, Proc. Amer. Math. Soc. **135** (2007), no. 9, 2723–2731, DOI 10.1090/S0002-9939-07-08880-6. MR2317945

[BLM16] J. Blasiak, R. I. Liu, and K. Mészáros, *Subalgebras of the Fomin-Kirillov algebra*, J. Algebraic Combin. **44** (2016), no. 3, 785–829, DOI 10.1007/s10801-016-0688-4. MR3552907

[CKGT22] A. Cohen, J. Knopper, and T. GAP Team, *GBNP, computing Gröbner bases of noncommutative polynomials, Version 1.0.5*, `https://gap-packages.github.io/gbnp/`, Mar 2022, Refereed GAP package.

[GAP22] The GAP Group, *GAP – Groups, Algorithms, and Programming, Version 4.12.2*, 2022.

[GnHV11] M. Graña, I. Heckenberger, and L. Vendramin, *Nichols algebras of group type with many quadratic relations*, Adv. Math. **227** (2011), no. 5, 1956–1989, DOI 10.1016/j.aim.2011.04.006. MR2803792

[Hec06] I. Heckenberger, *The Weyl groupoid of a Nichols algebra of diagonal type*, Invent. Math. **164** (2006), no. 1, 175–188, DOI 10.1007/s00222-005-0474-8. MR2207786

[Hec09] I. Heckenberger, *Classification of arithmetic root systems*, Adv. Math. **220** (2009), no. 1, 59–124, DOI 10.1016/j.aim.2008.08.005. MR2462836

[HLV12] I. Heckenberger, A. Lochmann, and L. Vendramin, *Braided racks, Hurwitz actions and Nichols algebras with many cubic relations*, Transform. Groups **17** (2012), no. 1, 157–194, DOI 10.1007/s00031-012-9176-7. MR2891215

[HMV23] I. Heckenberger, E. Meir, and L. Vendramin, *Finite-dimensional Nichols algebras of simple Yetter-Drinfeld modules (over groups) of prime dimension*, Adv. Math. **444** (2024), Paper No. 109637, 30, DOI 10.1016/j.aim.2024.109637. MR4729697

[HS10] I. Heckenberger and H.-J. Schneider, *Root systems and Weyl groupoids for Nichols algebras*, Proc. Lond. Math. Soc. (3) **101** (2010), no. 3, 623–654, DOI 10.1112/plms/pdq001. MR2734956

[HS13] I. Heckenberger and H.-J. Schneider, *Right coideal subalgebras of Nichols algebras and the Duflo order on the Weyl groupoid*, Israel J. Math. **197** (2013), no. 1, 139–187, DOI 10.1007/s11856-012-0180-3. MR3096611

[HS20] I. Heckenberger and H.-J. Schneider, *Hopf algebras and root systems*, Mathematical Surveys and Monographs, vol. 247, American Mathematical Society, Providence, RI, [2020] ©2020. MR4164719

[HS22] I. Heckenberger and K. Schäfer, *A decomposition theorem for pointed braided Hopf algebras*, Comm. Algebra **51** (2023), no. 9, 3904–3915, DOI 10.1080/00927872.2023.2190407. MR4605249

[Kha05] V. K. Kharchenko, *Braided version of Shirshov-Witt theorem*, J. Algebra **294** (2005), no. 1, 196–225, DOI 10.1016/j.jalgebra.2005.03.034. MR2179722

[Kha08] V. K. Kharchenko, *PBW-bases of coideal subalgebras and a freeness theorem*, Trans. Amer. Math. Soc. **360** (2008), no. 10, 5121–5143, DOI 10.1090/S0002-9947-08-04483-8. MR2415067

[MS00] A. Milinski and H.-J. Schneider, *Pointed indecomposable Hopf algebras over Coxeter groups*, New trends in Hopf algebra theory (La Falda, 1999), Contemp. Math., vol. 267, Amer. Math. Soc., Providence, RI, 2000, pp. 215–236, DOI 10.1090/conm/267/04272. MR1800714

[Nic78] W. D. Nichols, *Bialgebras of type one*, Comm. Algebra **6** (1978), no. 15, 1521–1552, DOI 10.1080/00927877808822306. MR506406

[ST16] H.-J. Schneider and B. Torrecillas, *A braided version of some results of Skryabin*, Comm. Algebra **44** (2016), no. 1, 205–217, DOI 10.1080/00927872.2014.974253. MR3413681

[Ven12] L. Vendramin, *Nichols algebras associated to the transpositions of the symmetric group are twist-equivalent*, Proc. Amer. Math. Soc. **140** (2012), no. 11, 3715–3723, DOI 10.1090/S0002-9939-2012-11215-8. MR2944712

Fachbereich Mathematik und Informatik, Philipps-Universität Marburg, Hans-Meerwein-Str. 6, 35032 Marburg, Germany
Email address: `heckenberger@mathematik.uni-marburg.de`

Fachbereich Mathematik und Informatik, Philipps-Universität Marburg, Hans-Meerwein-Str. 6, 35032 Marburg, Germany
Email address: `schaef7a@mathematik.uni-marburg.de`

Contemporary Mathematics
Volume **814**, 2025
https://doi.org/10.1090/conm/814/16304

On Hopf algebras of dimension p^n in characteristic p

Siu-Hung Ng and Xingting Wang

This paper is dedicated to the memory of Earl J. Taft.

ABSTRACT. Let $\Bbbk$ be an algebraically closed field of characteristic $p > 0$. We study the general structures of p^n-dimensional Hopf algebras over $\Bbbk$ with p^{n-1} group-like elements or a primitive element generating a p^{n-1}-dimensional Hopf subalgebra. As applications, we have proved that Hopf algebras of dimension p^2 over $\Bbbk$ are pointed or basic for $p \leq 5$, and provided a list of characterizations of the Radford algebra $R(p)$. In particular, $R(p)$ is the unique nontrivial extension of $\Bbbk[C_p]^*$ by $\Bbbk[C_p]$, where C_p is the cyclic group of order p. In addition, we have proved a vanishing theorem for some 2nd Sweedler cohomology group and investigated the extensions of p-dimensional Hopf algebras. All these extensions have been identified and shown to be pointed.

1. Introduction

Hopf algebras are generalizations of groups and Lie algebras. In the last few decades, they have played an essential role in invariants of knots, links, and 3-manifolds via their representations. In particular, the representation categories of finite-dimensional Hopf algebras provide fundamental examples of finite tensor categories (see, e.g., [**3, 9**]).

Let $\Bbbk$ be an algebraically closed field and p a prime number. It has long been a program classifying finite-dimensional Hopf algebras over $\Bbbk$ of characteristic zero. Kaplansky first conjectured that the only p-dimensional Hopf algebra in characteristic zero is the group algebra, later proved by Zhu [**43**]. For p^2-dimensional Hopf algebras, Masuoka proved that the semisimple ones are group algebras [**19**]. At that time, the only known nonsemisimple Hopf algebras of dimension p^2 were the Taft algebras, which were introduced by Taft in [**38**] and are described as follows.

DEFINITION 1.1 ([**38**]). Let $\Bbbk$ be a field, and $\omega \in \Bbbk$ a primitive nth root of unity for some positive integer n. The *Taft algebra* $T_n(\omega)$ is the free algebra $\Bbbk\langle x, g\rangle$ subject to the relations

$$g^n = 1, \quad x^n = 0, \quad xg = \omega gx.$$

2020 *Mathematics Subject Classification.* Primary 16T05.

Key words and phrases. Hopf algebra, positive characteristic, classification, cleft extension.

The first author was partially supported by NSF grant DMS1664418. The second author was partially supported by Simons collaboration grant #688403 and Air Force Office of Scientific Research grant FA9550-22-1-0272. This research began in the summer of 2022 during the second author's visit to the first author at Louisiana State University.

The Hopf algebra structure on the Taft algebra $T_p(\omega)$ is given by

$$\Delta(g) = g \otimes g, \qquad \epsilon(g) = 1, \qquad S(g) = g^{-1},$$
$$\Delta(x) = x \otimes 1 + g \otimes x, \qquad \epsilon(x) = 0, \qquad S(x) = -g^{-1}x.$$

Montgomery, in several international conferences, asked the following (cf. [**28**, Section 5]).

QUESTION 1.2. Suppose the base field $\Bbbk$ is algebraically closed of characteristic zero, and p is a prime number. Are the Taft algebras $T_p(\omega)$ the only nonsemisimple Hopf algebras in dimension p^2?

The first author of this paper gave an affirmative answer to Montgomery's question in [**24**].

Meanwhile, the classification of finite-dimensional Hopf algebras over an algebraically closed field $\Bbbk$ of positive characteristic p remains wide open. For any semisimple and cosemisimple Hopf algebra in positive characteristic, Etingof and Gelaki showed that it could be lifted to a semisimple Hopf algebra in characteristic zero [**6**]. Hence it yields the same classification result. Recently, the classification of the p-dimensional Hopf algebras over $\Bbbk$ in characteristic p has been accomplished by the authors of this paper in [**26**]. Etingof later generalizes this result to finite tensor categories in [**5**].

Turning to the p^2-dimensional Hopf algebras, the pointed ones are classified in [**39**]. Based on this classification, there is only one *pointed* noncommutative and noncocommutative p^2-dimensional Hopf algebra. This Hopf algebra was discovered by Radford in [**31**] and is described as follows.

DEFINITION 1.3 ([**31**]). Let $\Bbbk$ be an algebraically closed field of characteristic $p > 0$. The *Radford algebra* $R(p)$ is the free algebra $\Bbbk\langle x, g\rangle$ subject to the relations

$$g^p = 1, \quad x^p = x, \quad [g, x] = g(g-1).$$

The Hopf algebra structure on $R(p)$ is given by

$$\Delta(g) = g \otimes g, \qquad \epsilon(g) = 1, \qquad S(g) = g^{-1},$$
$$\Delta(x) = x \otimes 1 + g \otimes x, \qquad \epsilon(x) = 0, \qquad S(x) = -g^{-1}x.$$

It is clear that the Radford algebra could be over any field of characteristic p. However, we generally assume $\Bbbk$ is algebraically closed and the notation $R(p)$ is defined under this assumption.

The first author of this paper, on many occasions, asked the following question.

QUESTION A. Let $\Bbbk$ be an algebraically closed field of characteristic $p > 0$. Is the Radford algebra $R(p)$ the unique noncommutative and noncocommutative p^2-dimensional Hopf algebra over $\Bbbk$ up to isomorphism?

So far, there is no example of p^2-dimensional Hopf algebra H over $\Bbbk$ such that both H and H^* are not pointed. An affirmative answer to the following question will lead to a complete classification of p^2-dimensional Hopf algebras over $\Bbbk$, and hence an affirmative answer to Question A.

QUESTION B. Let $\Bbbk$ be an algebraically closed field of characteristic $p > 0$. Must every p^2-dimensional Hopf algebra over $\Bbbk$ be pointed?

In this paper, most of our results are stated over an algebraically closed base field $\Bbbk$ of positive characteristic p. In the first part of our paper, we study the general structures of the p^n-dimensional Hopf algebras over $\Bbbk$ with $n \geq 2$. We first show that any noncosemisimple p^n-dimensional Hopf algebra over $\Bbbk$ with p^{n-1} group-like elements must be pointed. Along this line of thoughts, one would like to understand those p^n-dimensional Hopf algebras H which admit a p^{n-1}-dimensional Hopf subalgebra $\Bbbk[x]$ generated by a primitive element x. In this case, assuming x is semisimple and H is nonsemisimple, we show that H has exactly p^{n-1} indecomposable projectives whose dimensions are equal to p. Moreover, H is involutive if it is unimodular. When H is p^2-dimensional, we are able to show if it contains a primitive element, then its coradical H_0 cannot have dimension greater than p. As an application, we answer Questions A and B affirmatively for $p \leq 5$, which completes the classification of all p^2-dimensional Hopf algebras over $\Bbbk$ for $p \leq 5$.

For the second part of our paper, we introduce the notion of twist equivalence for cleft extensions to classify the p^2-dimensional Hopf algebras, which are extensions of p-dimensional ones. We show the second Sweedler cohomology $H^2_{sw}(\Bbbk[G]^*, A)$ is trivial for any elementary abelian p-group G which acts as automorphisms on a finite-dimensional commutative algebra A over $\Bbbk$. Consequently, we describe the corresponding cleft extensions in terms of twist equivalence. Independent of the classification result of pointed Hopf algebras of dimension p^2 [**39**], we provide a list of characterizations of the Radford algebra $R(p)$. In particular, $R(p)$ is the unique nontrivial Hopf algebra extension of $\Bbbk[C_p]^*$ by $\Bbbk[C_p]$, where C_p denotes the (multiplicative) cyclic group of order p. Finally, we use the classification result [**39**] to identify all the p^2-dimensional Hopf algebras with a normal Hopf subalgebra of dimension p.

2. Preliminaries and background

Let $\Bbbk$ be a ground field of arbitrary characteristic. We use H to denote a finite-dimensional Hopf algebra over $\Bbbk$ with counit $\epsilon : H \to \Bbbk$, comultiplication $\Delta : H \to H \otimes H$ and antipode $S : H \to H$. The dual vector space H^* of H is also a Hopf algebra with the structure induced from that of H. The *augmentation ideal* of H is defined to be $H^+ := \ker \epsilon$. We use Sweedler's notation to write $\Delta(h) = \sum_{(h)} h_1 \otimes h_2$ for all $h \in H$. When the context is clear, we may suppress $\sum_{(h)}$. The readers are referred to [**27**] and [**37**] for more details on Hopf algebras and some fundamental results.

By an H-module, we always mean a left H-module of finite dimension over $\Bbbk$. For any two H-modules V and W, their tensor product $V \otimes W$ over the base field $\Bbbk$ admits a natural action of H via Δ. The *left dual* V^* of an H-module V is the $\Bbbk$-linear space $V^* = \mathrm{Hom}_{\Bbbk}(V, \Bbbk)$ equipped with the H-module structure given by

$$(hf)(v) = f(S(h)v), \quad \text{for all } h \in H,\ f \in V^*,\ v \in V.$$

Given an algebra automorphism σ of H, one can twist the H-action of any H-module V to obtain another H-module ${}_\sigma V$. More precisely, ${}_\sigma V = V$ as vector spaces with the σ-twisted action given by

$$h \cdot_\sigma v = \sigma(h)v, \quad \text{for all } h \in H,\ v \in V.$$

It is immediate to see that ${}_\sigma(\cdot)$ defines a $\Bbbk$-linear equivalence on the category H-mod of left H-modules. Moreover, the natural isomorphism between V and V^{**}

of finite-dimensional vector spaces defines an isomorphism ${}_{S^2}V \cong V^{**}$ of $\Bbbk$-linear monoidal functors.

We denote by $\mathrm{Irr}(H)$ a complete set of nonisomorphic simple H-modules. For any $V \in \mathrm{Irr}(H)$, the *projective cover* of V is denoted by $P(V)$. Since H is a Frobenius algebra [**15**], we have (see [**4**, 61.13])

$$\dim H = \sum_{V \in \mathrm{Irr}(H)} \dim V \cdot \dim P(V). \tag{2.1}$$

For any H-module M and $V \in \mathrm{Irr}(H)$, we define $[M : V]$ to be the multiplicity of V appearing as a composition factor of M. Let $V, W \in \mathrm{Irr}(H)$. Since $V^* \otimes P(W)$ is projective, it is a direct sum of indecomposable projective H-modules. More precisely,

$$\begin{aligned} V^* \otimes P(W) &= \bigoplus_{U \in \mathrm{Irr}(H)} \dim\left(\mathrm{Hom}_H(V^* \otimes P(W), U)\right) \cdot P(U) \\ &= \bigoplus_{U \in \mathrm{Irr}(H)} [V \otimes U : W] \cdot P(U). \end{aligned} \tag{2.2}$$

Recall that a *group-like element* in a Hopf algebra H is a nonzero element g such that $\Delta(g) = g \otimes g$. The group-like elements in H form a subgroup of the group of units in H, denoted by $G(H)$. Note that the set of 1-dimensional characters of H coincides with the set of group-like elements of the dual Hopf algebra H^*, which means $G(H^*) = \mathrm{Hom}_{\mathrm{Alg}}(H, \Bbbk)$. For any $\beta \in G(H^*)$, we denote by $\Bbbk_\beta$ the corresponding H-module with the underlying $\Bbbk$-linear space equal to $\Bbbk$. In particular, the trivial H-module determined by the counit map $\epsilon : H \to \Bbbk$ will simply be denoted by $\Bbbk$. Moreover, we define an injective group homomorphism $R : G(H^*) \to \mathrm{Aut}_{\mathrm{Alg}}(H)$ such that

$$R(\beta)(h) := h \leftharpoonup \beta = \sum_{(h)} \beta(h_1)\, h_2, \text{ for all } \beta \in G(H^*) \text{ and } h \in H. \tag{2.3}$$

For any H-module V, one can check directly that

$$\Bbbk_\beta \otimes V \cong {}_{R(\beta)}V \quad \text{as } H\text{-modules.} \tag{2.4}$$

An element $x \in H$ is called *skew primitive* if $\Delta(x) = x \otimes g + h \otimes x$ for some group-like elements g and h in H. In this case, x is also called a (g, h)*-primitive element.* An $(1, 1)$-primitive element is simply called a *primitive element.* For any group-like elements g, h of H, we use $P_{g,h}(H)$ to denote the subspace of all (g, h)-primitive elements in H. In particular, the subspace $P(H) = P_{1,1}(H)$ of primitive elements of H is closed under the commutator bracket, and hence a Lie subalgebra of H. Moreover, if the base field $\Bbbk$ has characteristic p, then $P(H)$ is a p*-restricted Lie algebra*, where the restricted map $[p] : P(H) \to P(H)$ is given by the pth power map in H (see [**14**]).

The sum of simple subcoalgebras of a coalgebra C is called the *coradical* of C and denoted by C_0. A Hopf algebra is called *pointed* if $H_0 = \Bbbk[G(H)]$, and called *connected* if $H_0 = \Bbbk 1_H$. A finite-dimensional Hopf algebra H is often called *basic* if H^* is pointed. In particular, H is *local*, which means H^+ is the Jacobson radical of H, if and only if H^* is connected. The endomorphism ring $\mathrm{End}_H(V)$ of any indecomposable H-module V is a local ring (see [**1, 29**]). The following trivial lemma is a reminder of these terminologies.

LEMMA 2.1 ([**40**, Lemma 2.2]). *The following statements concerning a finite-dimensional Hopf algebra H with counit ϵ are equivalent.*

(i) *H is local;*
(ii) $\ker \epsilon$ *is a nilpotent ideal;*
(iii) $\ker \epsilon$ *is the Jacobson radical of H;*
(iv) *H^* is connected.*

Moreover, H is basic if, and only if, H^* is pointed.

Finally, we recall some basic notations on extensions of finite-dimensional Hopf algebras. In the sense of [**18**, Definition 1.3], a short exact sequence of finite-dimensional Hopf algebras K, A and H, denoted by

$$1 \to K \xrightarrow{\iota} A \xrightarrow{\pi} H \to 1, \tag{2.5}$$

consists of Hopf algebra homomorphisms ι and π such that ι is injective, π is surjective, $\iota(K)$ is a normal Hopf subalgebra of A and $\ker \pi = \iota(K^+)A$ (see [**18**, Lemma-Definition 1.1]). In this case, we say that A is an *extension* of H by K.

LEMMA 2.2. *For a short exact sequence of finite-dimensional Hopf algebras described in* (2.5)*, the following hold true.*

(i) *Suppose K is local. Then* $\mathrm{Irr}(H) = \mathrm{Irr}(A)$*, and so H is basic if and only if A is basic.*
(ii) *Suppose H is connected. Then,* $\mathrm{Irr}(K^*) = \mathrm{Irr}(A^*)$*, and so K is pointed if and only if A is pointed.*

In particular, we have

(iii) *A is local if and only if both K and H are local.*
(iv) *A is connected if and only if both K and H are connected.*

PROOF. Without loss of generality, we may assume ι is just the inclusion map and so K is a normal Hopf subalgebra of A.

(i) Since K is local, K^+ is nilpotent. The normality of K in A implies that $I := \ker \pi = K^+A = AK^+$ is also a nilpotent ideal of A. Therefore, simple A-modules are simple modules over $A/I \cong H$. Hence the result follows by definition.

(iii) If H and K are local, then A is local by the proof of (i). Now suppose A is local. Then A^+ is nilpotent, and so is K^+. Since $H^+ = \pi(A^+)$, H^+ is also nilpotent. So K and H are local by Lemma 2.1.

(ii) and (iv) follow directly from (i) and (iii) by duality. □

3. Hopf algebras of dimension p^n with p^{n-1} group-like elements

In this section, we discuss the pointedness of finite-dimensional Hopf algebras with large group-like elements. Though some of our results may apply to arbitrary characteristics, we mainly focus on the positive characteristic case.

For any module V over a finite-dimensional Hopf algebra H, we define

$$\mathrm{stab}(V) = \{\beta \in G(H^*) \mid \Bbbk_\beta \otimes V \cong V \text{ as } H\text{-modules}\}.$$

Obviously, $\mathrm{stab}(V)$ is a subgroup of $G(H^*)$. The following result was established over a base field $\Bbbk$ of characteristic zero in [**7**, Proposition 2.5] and [**25**, Lemma 1.4]. We extend its proof to positive characteristics by applying group cohomology for subgroups of $\mathrm{stab}(V)$ for any simple H-module V. We first observe the following cohomology vanishing lemma.

LEMMA 3.1. *Let $\Bbbk$ be an algebraically closed field of characteristic $p > 0$, and G a p-group. For any finite-dimensional commutative semisimple algebra A over $\Bbbk$ on which G acts as algebra automorphisms, $H^2(G, A^\times)$ is trivial.*

PROOF. For any $\sigma \in Z^2(G, A^\times)$, $\sigma^{|G|}$ is a coboundary by [**42**, Theorem 6.5.8], and so $\sigma^{|G|} = \partial f$ for some function $f : G \to A^\times$. Since A is commutative semisimple, $A = \oplus_{i=1}^n \Bbbk e_i$ where $\{e_1, \dots, e_n\}$ is the complete set of primitive orthogonal idempotents for A. Note that for any $a \in A$, the polynomial equation $X^{|G|} - a = 0$ has a unique solution in A since $\Bbbk$ has characteristic p and G is a p-group. Therefore, there exists $\overline{f} : G \to A^\times$ such that $\overline{f}^{|G|} = f$. Thus, $(\sigma/\partial\overline{f})^{|G|} = 1$ or $\sigma = \partial\overline{f}$. Hence, $H^2(G, A^\times)$ is trivial. □

LEMMA 3.2. *Let H be a finite-dimensional Hopf algebra over an algebraically closed field $\Bbbk$, and V a simple H-module. Suppose there is a subgroup $G \le \operatorname{stab}(V)$ such that $H^2(G, \Bbbk^\times)$ is trivial. Then V admits a Hopf module structure in ${}_{\Bbbk[G]}\mathcal{M}^{H^*}$, and hence is a free $\Bbbk[G]$-module. In particular, if* (i) $\operatorname{char}(\Bbbk) = 0$ *and G is a cyclic group, or* (ii) $\operatorname{char}(\Bbbk) = p$ *and G is a p-group, then $|G| \mid \dim(V)$.*

PROOF. In view of (2.4), we have H-module isomorphisms

$$_{R(\beta)}V \cong \Bbbk_\beta \otimes V \cong V \quad \text{for } \beta \in G\,.$$

Let $F_\beta : {}_{R(\beta)}V \to V$ be such an isomorphism of H-modules for each $\beta \in G$, which are not canonical. Since $R(\epsilon) = \mathrm{id}_H$, we choose $F_\epsilon = \mathrm{id}_V$. Note that $F_\beta \in \mathrm{GL}(V)$ and ${}_{R(\beta)}({}_{R(\beta')}V) = {}_{R(\beta\beta')}V$ (or $\Bbbk_\beta \otimes \Bbbk_{\beta'} \cong \Bbbk_{\beta\beta'}$) for any $\beta, \beta' \in G$. We find $F_\beta \circ F_{\beta'}$ and $F_{\beta\beta'} : {}_{R(\beta\beta')}V \to V$ are H-module isomorphisms. By Schur's lemma,

$$F_\beta \circ F_{\beta'} = \gamma(\beta, \beta') F_{\beta\beta'}$$

for some $\gamma(\beta, \beta') \in \Bbbk^\times$. The associativity of composition

$$(F_\beta \circ F_{\beta'}) \circ F_{\beta''} = F_\beta \circ (F_{\beta'} \circ F_{\beta''}),$$

implies $\gamma(\beta', \beta'')\gamma(\beta, \beta'\beta'') = \gamma(\beta\beta', \beta'')\gamma(\beta, \beta')$ for any $\beta, \beta', \beta'' \in G$. Since $F_\epsilon = \mathrm{id}_V$, $\gamma(\beta, \epsilon) = \gamma(\epsilon, \beta) = 1$. Therefore, γ defines a normalized 2-cocycle in $Z^2(G, \Bbbk^\times)$. Since $H^2(G, \Bbbk^\times)$ is trivial, γ is a coboundary, namely, there exists a map $f : G \to \Bbbk^\times$ such that $\gamma(\beta, \beta') = f(\beta)f(\beta')/f(\beta\beta')$ with $f(\epsilon) = 1$ for all $\beta, \beta' \in G$. Now we define the $\Bbbk$-linear map $\eta : \Bbbk[G] \to \mathrm{End}_\Bbbk(V)$ by

$$\eta(\beta) = F_\beta / f(\beta) \quad \text{for any } \beta \in G.$$

By the definitions of γ and f, it is straightforward to check that η is indeed an $\Bbbk$-algebra homomorphism. Therefore, V is a left $\Bbbk[G]$-module via η, i.e., $\beta v = \eta(\beta)(v)$ for any $v \in V$ and $\beta \in G$.

Next, the left H-action on V defines a right H^*-coaction ρ on V, namely $hv = \sum_{(v)} v_1(h)v_0$ for $h \in H$ and $v \in V$, where $\rho(v) = \sum_{(v)} v_0 \otimes v_1 \in V \otimes H^*$. This makes V simultaneously a left $\Bbbk[G]$-module and a right H^*-comodule. We claim that V is a Hopf module in ${}_{\Bbbk[G]}\mathcal{M}^{H^*}$, and hence by the Nichols–Zoeller theorem (cf. [**23**]), V is a free $\Bbbk[G]$-module. It suffices to show that

$$\sum_{(\beta v)} (\beta v)_0 \otimes (\beta v)_1 = \sum_{(v)} \beta v_0 \otimes \beta v_1 \quad \text{or} \quad \sum_{(\beta v)} (\beta v)_1(h)(\beta v)_0 = \sum_{(v)} (\beta v_1)(h) \beta v_0$$

for all $\beta \in G$, $v \in V$ and $h \in H$. Note that $\eta(\beta)$ is an H-module isomorphism from ${}_{R(\beta)}V$ to V. So our claim follows from this calculation:

$$\begin{aligned}\sum_{(\beta v)}(\beta v)_1(h)(\beta v)_0 &= h(\beta v) = \beta\left(R(\beta)(h)v\right) = \beta\left(\sum_{(v)} v_1(R(\beta)(h))v_0\right) \\ &= \beta\left(\sum_{(v)} v_1\left(\sum_{(h)}\beta(h_1)h_2\right)v_0\right) = \sum_{(v)}\left(\sum_{(h)}\beta(h_1)v_1(h_2)\right)\beta v_0 \\ &= \sum_{(v)}(\beta v_1)(h)(\beta v_0).\end{aligned}$$

If G is a cyclic group and $\mathrm{char}(\Bbbk) = 0$, $H^2(G, \Bbbk^\times)$ is well known to be trivial (cf. [**42**, Theorem 6.2.2]). If $\mathrm{char}(\Bbbk) = p$ and G is a p-group, then $H^2(G, \Bbbk^\times)$ is trivial by Lemma 3.1. Now, the final assertion follows directly from the first statement. □

Our next lemma is derived from the proof of [**26**, Lemma 1.3].

LEMMA 3.3. *Let $\Bbbk$ be an algebraically closed field of characteristic $p > 0$, H a nonsemisimple Hopf algebra of dimension p^n over $\Bbbk$ for some integer $n \geq 1$, and P an indecomposable projective H-module. Suppose $P \cong \Bbbk_\beta \otimes P^{**}$ for some $\beta \in G(H^*)$. Then,*

(i) $\dim P$ *is a positive even integer if $p = 2$;*
(ii) $\dim P \geq p$ *if* $\dim P$ *and p are odd.*

PROOF. Note that $\Bbbk_\beta \otimes P^{**} \cong {}_\sigma P$, where $\sigma = R(\beta) \circ S^2 = S^2 \circ R(\beta)$. So there exists an H-module isomorphism $\phi : P \to {}_\sigma P$. By the celebrated formula of Radford [**30**], $S^{4p^n} = \mathrm{id}_H$. Hence $\sigma^{2p^n} = R(\beta)^{2p^n} \circ S^{4p^n} = R(\beta^{2p^n}) \circ S^{4p^n} = \mathrm{id}_H$. Thus, ϕ^{2p^n} is an H-module automorphism of P. Since P is indecomposable, $\mathrm{End}_H(P)$ is a local $\Bbbk$-algebra and so $\phi^{2p^n} - c\,\mathrm{id}_P$ is nilpotent for some $c \in \Bbbk^\times$. By rescaling, we can assume that both ϕ^{2p^n} and ϕ^2 are unipotent. Thus, the eigenvalues of ϕ are ± 1 if $\mathrm{char}(\Bbbk) = p > 2$, and are 1 if $p = 2$.

We claim $\mathrm{tr}(\phi) = 0$. If β is trivial, it follows from [**26**, Lemma 1.1]. If β is nontrivial, it follows from [**25**, Lemma 1.2]. Thus, if $p = 2$, $\mathrm{tr}(\phi) = 0$ implies $\dim P$ is a positive multiple of 2.

Now, we assume $p > 2$ and $\dim P$ is odd. Let $d_\pm$ be the multiplicities of the eigenvalues ± 1 for ϕ, respectively. Since

$$0 = \mathrm{tr}(\phi) = d_+ - d_-,$$

$|d_+ - d_-| = mp$ for some nonnegative integer m. If $m = 0$, then $\dim(P) = 2d_+$, which is even. Since $\dim(P)$ is odd, $m \geq 1$ and hence $\dim(P) = d_+ + d_- \geq p$. □

THEOREM 3.4. *Let $\Bbbk$ be an algebraically closed field of characteristic $p > 0$ and H a nonsemisimple Hopf algebra of dimension p^n over $\Bbbk$ for some integer $n \geq 1$. If $|G(H^*)| = p^{n-1}$, then H^* is pointed.*

PROOF. We first consider $p = 2$. Suppose H^* is not pointed. Then the sum of all the simple subcoalgebras of H^* with dimensions > 1 is a nontrivial subcoalgebra

C of H^*. Note that C is a free $\Bbbk[G(H^*)]$-module since it is a ${}_{\Bbbk[G(H^*)]}\mathcal{M}^{H^*}$ Hopf module. Thus,

$$\dim(H_0^*) = |G(H^*)| + \dim C \geq 2|G(H^*)| = 2^n = \dim H^* .$$

Hence H^* is cosemisimple, or H is semisimple. This proves the statement for $p = 2$.

Now, we assume $p > 2$. By Lemma 2.1, it suffices to show that $\mathrm{Irr}(H) = \{\Bbbk_\beta \mid \beta \in G(H^*)\}$. It is easy to see that $P(\Bbbk_\beta \otimes V) \cong \Bbbk_\beta \otimes P(V)$ and $P(V^{**}) \cong P(V)^{**}$ for any $V \in \mathrm{Irr}(H)$. In particular, $\dim P(\Bbbk) = \dim P(\Bbbk_\beta)$ for all $\beta \in G(H^*)$.

If $\dim P(\Bbbk)$ is odd, then $\dim P(\Bbbk) \geq p$ by Lemma 3.3. So (2.1) implies that

$$p^n = \dim H \geq \sum_{\beta \in G(H^*)} \dim \Bbbk_\beta \cdot \dim P(\Bbbk_\beta) \geq p^{n-1} p .$$

Hence, $\dim P(\Bbbk) = p$ and $\mathrm{Irr}(H) = \{\Bbbk_\beta \mid \beta \in G(H^*)\}$.

To complete the proof of the theorem, it suffices to show that $\dim P(\Bbbk)$ must be odd. Let $K = \langle S^2 \rangle \subset \mathrm{Aut}_\Bbbk(H)$. Since $S^{4p^n} = \mathrm{id}_H$, we have $|K|$ divides $2p^n$. For $\beta \in G(H^*)$, $R(\beta)$ is an algebra automorphism of H via (2.3). Thus, both $G(H^*)$ and K act on $\mathrm{Irr}(H)$ by twisting the H-actions on irreducible modules through their associated algebra automorphisms of H.

Since $R(\beta) \circ S^2 = S^2 \circ R(\beta)$ for any $\beta \in G(H^*)$, the above two actions commute with each other. Thus, the set of K-orbits in $\mathrm{Irr}(H)$ admits a $G(H^*)$-action induced by its action of $\mathrm{Irr}(H)$. Let $\mathcal{O}_1, \ldots, \mathcal{O}_r$ be the K-orbits, and $V_1, \ldots, V_r$ be representative irreducible H-modules in these K-orbits, respectively. Then (2.1) implies that

$$p^n = \dim H = \sum_{i=1}^{r} |\mathcal{O}_i| \dim(V_i) \dim P(V_i). \tag{3.1}$$

Since $\dim H$ is odd, there exists some index i_0 such that $|\mathcal{O}_{i_0}|$, $\dim V_{i_0}$ and $\dim P(V_{i_0})$ are all odd. Without loss of generality, we may assume $i_0 = 1$.

If we could show that $\dim V_1 = 1$, then $V_1 \cong \Bbbk_\beta$ for some $\beta \in G(H^*)$, and hence $\dim P(\Bbbk) = \dim P(\Bbbk_\beta)$ is odd. Therefore, it suffices to show that $\dim V_1 = 1$.

Since $|\mathcal{O}_1|$ is odd and $|\mathcal{O}_1| \mid |K| \mid 2p^n$, $|\mathcal{O}_1| = p^k$ for some nonnegative integer $k \leq n$. Let

$$G = \{\beta \in G(H^*) \mid \Bbbk_\beta \otimes V_1 \in \mathcal{O}_1\}$$

and G_1 be the stabilizer of V_1. Obviously, $G_1 \subseteq G$, and $|G_1| \mid \dim(V_1)$ by Lemma 3.2. Since the K-action and the $G(H^*)$-action on $\mathrm{Irr}(H)$ commute, $\mathcal{O}_1$ is closed under the action of G and K acts transitively on the set of G-obits in $\mathcal{O}_1$. Consequently, all the G-orbits in $\mathcal{O}_1$ have the same size $\frac{|G|}{|G_1|}$ and we have

$$\ell \cdot \frac{|G|}{|G_1|} = |\mathcal{O}_1| = p^k, \tag{3.2}$$

where ℓ is the number of G-orbits in $\mathcal{O}_1$. In particular, ℓ is a power of p.

Now suppose $\dim V_1 > 1$. Let $\{\beta\mathcal{O}_1 \mid \beta \in G(H^*)\} = \{\mathcal{O}_1, \ldots, \mathcal{O}_t\}$ for some nonnegative integer $t \leq r$ by relabelling the indices of the K-orbits. Then (3.1) and

(3.2) imply that

$$\begin{aligned} p^n = \dim(H) &\geq \sum_{\beta\in G(H^*)} \dim(\Bbbk_\beta) \dim P(\Bbbk_\beta) + \sum_{1\leq i\leq t} |\mathcal{O}_i| \dim(V_i) \dim P(V_i) \\ &= |G(H^*)| \dim P(\Bbbk) + \frac{|G(H^*)|}{|G|} |\mathcal{O}_1| \dim(V_1) \dim P(V_1) \\ &= p^{n-1} \left(\dim P(\Bbbk) + \ell \dim(P(V_1)) \frac{\dim(V_1)}{|G_1|} \right). \end{aligned}$$

Therefore,

$$p = \dim P(\Bbbk) + \ell \dim(P(V_1)) \frac{\dim(V_1)}{|G_1|} > \ell \dim P(V_1) \tag{3.3}$$

and hence $\ell < p$. It follows from (3.2) that $\ell = 1$, or G acts transitively on $\mathcal{O}_1$. Since both V_1 and V_1^{**} are in the K-orbit $\mathcal{O}_1$, there exists $\beta \in G$ such that $V_1^{**} \cong \Bbbk_\beta \otimes V_1$. Hence $\Bbbk_{\beta^{-1}} \otimes P(V_1)^{**} \cong P(V_1)$. It follows from Lemma 3.3 that $\dim P(V_1) \geq p$, which contradicts (3.3). Therefore, $\dim V_1 = 1$. □

The dual statement of the preceding theorem is stated as follows.

THEOREM 3.5. *Let $\Bbbk$ be an algebraically closed field of characteristic $p > 0$ and H a non-cosemisimple Hopf algebra of dimension p^n over $\Bbbk$ for some integer $n \geq 1$. If $|G(H)| = p^{n-1}$, then H is pointed.*

4. Finite-dimensional Hopf algebras with primitive elements

In this section, we turn our attention to the existence of primitive elements in finite-dimensional Hopf algebras over an algebraically closed field $\Bbbk$. It is well-known that finite-dimensional Hopf algebras in characteristic zero do not possess nonzero primitive elements [**34**, Proposition 1(b)]. One may assume, a priori, all results about primitive elements are asserted over positive characteristics.

We first observe the following lemma regarding extensions of simple H-modules, and we simply denote H^+ by $\mathfrak{m}$.

LEMMA 4.1. *Let H be a finite-dimensional Hopf algebra over an algebraically closed field $\Bbbk$, and L any simple H-module. Denote by $\mathfrak{n} = Ann(L)$ the annihilator ideal of L in H. Then we have*

$$\dim L \cdot \dim \operatorname{Ext}^1_H(L, \Bbbk) \; = \; \dim(\mathfrak{m} \cap \mathfrak{n}/\mathfrak{m}\mathfrak{n}).$$

PROOF. Applying $\operatorname{Ext}^1_H(-, \Bbbk)$ to the short exact sequence $0 \to \mathfrak{n} \to H \to H/\mathfrak{n} \to 0$, we obtain the exact sequence

$$0 \to \operatorname{Hom}_H(H/\mathfrak{n}, \Bbbk) \to \operatorname{Hom}_H(H, \Bbbk) \to \operatorname{Hom}_H(\mathfrak{n}, \Bbbk) \to \operatorname{Ext}^1_H(H/\mathfrak{n}, \Bbbk) \to 0.$$

Note that $H/\mathfrak{n} = H/Ann(L) \cong \bigoplus_{\dim L} L$ as H-modules. Also we have

$$\operatorname{Hom}_H(V, \Bbbk) \; \cong \; \operatorname{Hom}_\Bbbk(V/\mathfrak{m}V, \Bbbk) \; \cong \; (V/\mathfrak{m}V)^*$$

as vector spaces for any H-module V. Thus,

$$\begin{aligned}
\dim L\cdot\dim \operatorname{Ext}^1_H(L,\Bbbk) &= \dim \operatorname{Ext}^1_H(H/\mathfrak{n},\Bbbk)\\
&= \dim \operatorname{Hom}_H(H/\mathfrak{n},\Bbbk)+\dim \operatorname{Hom}_H(\mathfrak{n},\Bbbk)-\dim \operatorname{Hom}_H(H,\Bbbk)\\
&= \dim(H/(\mathfrak{m}+\mathfrak{n}))+\dim(\mathfrak{n}/\mathfrak{m}\mathfrak{n})-\dim(H/\mathfrak{m})\\
&= \dim(\mathfrak{n}/\mathfrak{m}\mathfrak{n})-\dim((\mathfrak{m}+\mathfrak{n})/\mathfrak{m})\\
&= \dim(\mathfrak{n}/\mathfrak{m}\mathfrak{n})-\dim(\mathfrak{n}/\mathfrak{m}\cap\mathfrak{n})\\
&= \dim(\mathfrak{m}\cap\mathfrak{n}/\mathfrak{m}\mathfrak{n}).
\end{aligned}$$

□

Note that if $L=\Bbbk$, the condition $\Bbbk$ being algebraically closed in the preceding lemma can be dropped, which applies to the following two generally known lemmas (e.g., see [**11**]). We give their proofs for the sake of completeness.

LEMMA 4.2. *Let H be a finite-dimensional Hopf algebra over any field $\Bbbk$. Then*

$$\dim \operatorname{Ext}^1_H(\Bbbk,\Bbbk) = \dim(\mathfrak{m}/\mathfrak{m}^2) = \dim P(H^*).$$

PROOF. The first equality in the statement follows from Lemma 4.1. Next, consider the subcoalgebra $C=\pi^*(H/\mathfrak{m}^2)^*$ of H^*, where $\pi: H\to H/\mathfrak{m}^2$ is the natural algebra surjection. For $f\in C$, $\Delta_{H^*}(f)(\mathfrak{m}\otimes\mathfrak{m})=f(\mathfrak{m}^2)=0$. This implies that $\Delta_{H^*}(f)\in\epsilon\otimes H^*+H^*\otimes\epsilon$.

Let $I=1^\perp\cap C$. Then $\dim I=\dim H-\dim\mathfrak{m}^2-1=\dim(\mathfrak{m}/\mathfrak{m}^2)$. We will complete the proof by showing $I=P(H^*)$. For $f\in I$, $\Delta_{H^*}(f)=\epsilon\otimes u+v\otimes\epsilon+c\epsilon\otimes\epsilon$ for some $c\in\Bbbk$ and $u,v\in I$. The counit property of Δ_{H^*} implies $f=v+c\epsilon=u+c\epsilon$. Since $f,u,v\in 1^\perp$, $c=0$ and $f=u=v$. Thus, $f\in P(H^*)$ and hence $I\subseteq P(H^*)$. Conversely, if $f\in P(H^*)\setminus\{0\}$, then $f(1)=0$ and $\Delta_{H^*}(f)=\epsilon\otimes f+f\otimes\epsilon$. For any $x,y\in\mathfrak{m}$, $f(xy)=\Delta_{H^*}(f)(x\otimes y)=0$. Thus, $f\in I$ and hence $P(H^*)\subseteq I$. □

We recall the associated graded algebra

$$\operatorname{gr}_{\mathfrak{m}} H := \bigoplus_{i=0}^{\infty}\mathfrak{m}^i/\mathfrak{m}^{i+1} \tag{4.1}$$

with respect to the $\mathfrak{m}$-adic filtration in H.

LEMMA 4.3. *Let H be a finite-dimensional Hopf algebra over $\Bbbk$ of characteristic $p>0$. Then $\bigcap_{i\ge 0}\mathfrak{m}^i$ is a Hopf ideal of H with the quotient Hopf algebra $H/\bigcap_{i\ge0}\mathfrak{m}^i$ of dimension p^n for some $n\ge\dim P(H^*)$. In particular, if $\operatorname{gr}_{\mathfrak{m}}H$ is commutative, then so is $H/\bigcap_{i\ge0}\mathfrak{m}^i$.*

PROOF. The assertion that $\bigcap_{i\ge0}\mathfrak{m}^i$ is a Hopf ideal of H and that $\operatorname{gr}_{\mathfrak{m}}H$ is a graded Hopf algebra follows directly from [**11**, Lemma 3.3]. The graded algebra $\operatorname{gr}_{\mathfrak{m}}H$ is connected, cocommutative, and generated by degree one elements, which are primitive. Thus, $\operatorname{gr}_{\mathfrak{m}}H$ is isomorphic to the restricted universal enveloping algebra $u(\mathfrak{g})$ (cf. [**37**, 13.2.3],[**21**, Theorem 1]), where $\mathfrak{g}=P(\operatorname{gr}_{\mathfrak{m}}H)\supseteq\mathfrak{m}/\mathfrak{m}^2$. Hence we have

$$\dim H/\bigcap_{i\ge0}\mathfrak{m}^i=\dim\operatorname{gr}_{\mathfrak{m}}H=p^{\dim\mathfrak{g}}\ge p^{\dim(\mathfrak{m}/\mathfrak{m}^2)}=p^{\dim P(H^*)}$$

by Lemma 4.2. The last conclusion is [**11**, Lemma 3.5]. □

PROPOSITION 4.4. *Let H be a finite-dimensional Hopf algebra over an algebraically closed field $\Bbbk$ of characteristic $p > 0$, which contains a nonzero primitive element x. Suppose $C = \sum_i C_i$ is a sum of simple subcoalgebras of H.*

(i) *If $xC \subset C$, then any simple right C-comodule M admits a left $\Bbbk[x]$-action such that M is a Hopf module in ${}_{\Bbbk[x]}\mathcal{M}^H$. In particular, $\dim(\Bbbk[x]) \mid \dim(M)$ and hence $\dim(\Bbbk[x])^2 \mid \dim(C_i)$ for all i.*

(ii) *If C is simple and $p \nmid \dim(C)$, then $\Bbbk[x]C$ is a free left $\Bbbk[x]$-module of rank $\dim(C)$.*

PROOF.

(i) Since $xC \subset C$, we can let x act on the dual algebra C^* via usual dual module action, namely $(xg)(c) = g(S(x)c) = -g(xc)$ for any $c \in C, g \in C^*$. Furthermore, one can check that x acts as a $\Bbbk$-derivation on C^* for x is primitive. Since C^* is separable, $HH^1(C^*, C^*)$ is trivial. Therefore, x acts as an inner derivation on C^*, which means there exists $f_x \in C^*$ such that $xg = [f_x, g] = \operatorname{ad}(f_x)(g)$ for all $g \in C^*$. Note that $f_x \in C^*$ is uniquely determined by x up to the center of C^*. Since $C^* = \bigoplus_i C_i^*$ as algebra, $f_x = \sum_i f_i$ where $f_i \in C_i^*$, where f_i is uniquely determined up to a scalar multiple of $1_{C_i^*}$. By definition we obtain that $xc = \sum_{(c)} f_x(c_2)c_1 - f_x(c_1)c_2 = \sum_{(c)} f_i(c_2)c_1 - f_i(c_1)c_2$ for $c \in C_i$. In particular, $xC_i \subset C_i$ for each i.

Suppose M is a simple right comodule of C. Then M is also a simple right comodule of C_i for some i, and hence a C_i^*-module given by

$$g \cdot m = \sum_{(m)} m_0\, g(m_1)$$

for all $m \in M$ and $g \in C_i^*$. We claim that one can choose a suitable $f_i \in C_i^*$ to define an $\Bbbk[x]$-module action on M given by $x \cdot m := f_i \cdot m$ for $m \in M$. Since $\Bbbk[x]$ is primitively generated, so $\Bbbk[x] \cong u(\mathfrak{g})$ with $\mathfrak{g} = \operatorname{span}_{\Bbbk}(x, x^p, x^{p^2}, \ldots)$. Thus x satisfies the following relation in $\Bbbk[x]$

$$x^{p^n} + a_{n-1}x^{p^{n-1}} + \cdots + a_1 x^p + a_0 x = 0 \tag{4.2}$$

for some coefficients $a_0, \ldots, a_{n-1} \in \Bbbk$. Then

$$\begin{aligned} 0 &= \operatorname{ad}(f_x)^{p^n} + a_{n-1}\operatorname{ad}(f_x)^{p^{n-1}} + \cdots + a_1(\operatorname{ad} f_x)^p + a_0 \operatorname{ad}(f_x) \\ &= \operatorname{ad}\left(f_x^{p^n} + a_{n-1}f_x^{p^{n-1}} + \cdots + a_1 f_x^p + a_0 f_x\right), \end{aligned}$$

which is equivalent to

$$0 = \operatorname{ad}\left(f_j^{p^n} + a_{n-1}f_j^{p^{n-1}} + \cdots + a_1 f_j^p + a_0 f_j\right) \quad \text{for all } j.$$

This implies that $f_j^{p^n} + a_{n-1}f_j^{p^{n-1}} + \cdots + a_1 f_j^p + a_0 f_j = \lambda_j\, 1_{C^*}$ for some scalar $\lambda_j \in \Bbbk$. For each j, replacing f_j by $f_j + \eta_j\, 1_{C_j^*}$ with $\eta_j \in \Bbbk$ satisfying $\eta_j^{p^n} + a_{n-1}\eta_j^{p^{n-1}} + \cdots + a_1\eta_j^p + a_0\eta_j = -\lambda_j$, we have f_x satisfy the same relation as x in (4.2). This makes the $\Bbbk[x]$-action:

$$x \cdot m = \sum_m m_0 f_i(m_1) = \sum_m m_0 f_x(m_1), \qquad \text{for } m \in M,$$

well-defined on M.

With this $\Bbbk[x]$-module action on M, one can check directly that M is a Hopf module in ${}_{\Bbbk[x]}\mathcal{M}^H$. This amounts to show that

$$\sum_{(u\cdot m)} (u\cdot m)_0 \otimes (u\cdot m)_1 \;=\; \sum_{(u),(m)} u_1\cdot m_0 \otimes u_2 m_1$$

for any $u \in \Bbbk[x]$ and $m \in M$. Since $\Bbbk[x]$ is generated by x, it suffices to check the above equation for $u = x$. Indeed, we have

$$\begin{aligned}\sum_{(x),(m)} x_1\cdot m_0 \otimes x_2 m_1 &= \sum_{(m)} x\cdot m_0 \otimes m_1 + m_0 \otimes x m_1\\ &= \sum_{(m)} m_0 \otimes f_x(m_1)m_2 + m_0 \otimes (f_x(m_2)m_1 - f_x(m_1)m_2)\\ &= \sum_{(m)} m_0 \otimes m_1 f_x(m_2)\\ &= \sum_{(x\cdot m)} (x\cdot m)_1 \otimes (x\cdot m)_2.\end{aligned}$$

Therefore, M is a free $\Bbbk[x]$-module by the Nichols-Zoeller Theorem. So $\dim(\Bbbk[x]) \mid \dim(M)$, and hence $\dim(\Bbbk[x])^2 \mid \dim(M)^2 = \dim(C_i)$.

(ii) It follows from (i) that $xC \not\subseteq C$. Then, the comultiplication Δ defines a natural nontrivial C-C-bicomodule structure on $(xC+C)/C$. Since every simple C-C-bicomodule has dimension $\dim(C)$ and $\dim(xC+C)/C \le \dim(C)$, we find $\dim(xC+C)/C = \dim(C)$. Hence $xC + C = xC \oplus C$. By repeating the argument, one can inductively show that

$$\Bbbk[x]\,C \;=\; \bigoplus_{i=0}^{n} x^i\,C \quad\text{and}\quad \dim(x^i\,C) \;=\; \dim(C)$$

where $x^iC \not\subseteq C \oplus xC \oplus \cdots \oplus x^{i-1}C$ for all $i \le n$. Thus, $\dim(C) \mid \dim(\Bbbk[x]C)$. It is clear that $\Bbbk[x]C$ is a Hopf module in ${}_{\Bbbk[x]}\mathcal{M}^H$, and so $\Bbbk[x]C$ is a free $\Bbbk[x]$-module. In particular $\dim(\Bbbk[x]) \mid \dim(\Bbbk[x]C)$. Therefore, $\dim(\Bbbk[x])\dim(C) \mid \dim(\Bbbk[x]C) \le \dim(\Bbbk[x])\dim(C)$. Hence, $\dim(\Bbbk[x])\dim(C) = \dim(\Bbbk[x]C)$. Therefore, the $\Bbbk[x]C$ is a free $\Bbbk[x]$-module of rank $\dim(C)$. □

COROLLARY 4.5. *If H is a p^2-dimensional Hopf algebra over an algebraically closed field $\Bbbk$ of characteristic $p > 0$, which admits a nonzero primitive element x, then* $\dim(H_0) \le p$.

PROOF. Let $K = \Bbbk[x]$. If $K = H$, then $H_0 = \Bbbk$, and the statement holds vacuously. Suppose $K \ne H$. Then $\dim(K) = p$ by Nichols-Zoeller Theorem since K is a Hopf subalgebra of H. Let C be any simple subcoalgebra of H. It is clear that $p \nmid \dim C$ for $\dim H = p^2$. By Proposition 4.4 (ii), $\dim(KC) = p\dim(C)$. Note that

$$C \subseteq C + xC \subseteq C + xC + x^2C \subseteq \cdots \subseteq KC$$

is a coalgebra filtration on KC. So [**27**, Lemma 5.3.4] implies that KC is an irreducible subcoalgebra of H with $(KC)_0 = C$ (see [**27**, Definition 5.6.1]). Thus,

$$\sum_C KC \;=\; \bigoplus_C KC,$$

where the sum is over the set of all simple subcoalgebra C of H. Hence

$$p^2 = \dim(H) \geq \dim\left(\bigoplus_C KC\right) = p\left(\sum_C \dim C\right) = p \dim H_0,$$

which yields that $p \geq \dim(H_0)$. $\square$

PROPOSITION 4.6. *Let H be a nonsemisimple p^n-dimensional Hopf algebra over an algebraically closed field $\Bbbk$ of characteristic $p > 2$. If H admits a semisimple primitive element x such that $\dim \Bbbk[x] = p^{n-1}$, then $S^{2p} = \mathrm{id}$ and He is a p-dimensional indecomposable projective H-module for any primitive idempotent $e \in \Bbbk[x]$. In addition, if H is unimodular, then $S^2 = \mathrm{id}$.*

PROOF. Write $K = \Bbbk[x]$, a connected Hopf subalgebra of H generated by the primitive element x. By [**21**, Theorem 1], $K \cong u(P(K))$. Because x is semisimple, by Hochschild's theorem [**13**], $K \cong \Bbbk[G]^*$ as Hopf algebras, where $G \cong C_p^{n-1}$. In particular, K admits a complete set of orthogonal idempotents $e_1, \ldots, e_{p^{n-1}}$ and $S^2(e_i) = e_i$ for all i. Since H is a free K-module of rank p, $\dim(He_i) = p$ for all i. Since H is not semisimple, $\sum_{(\Lambda)} S(\Lambda_2)S^2(\Lambda_1) = \epsilon(\Lambda)1 = 0$, where Λ is a nontrivial left integral of H. By [**32**, Lemma 3(b)], $\mathrm{tr}(r(e_i) \circ S^2) = 0$ for all i, where $r(e_i)$ denotes the operator on H given by the right multiplication of e_i. Hence $\mathrm{tr}(r(e_i) \circ S^2) = \mathrm{tr}(S^2|_{He_i}) = 0$. By Radford's S^4-formula, we know $S^{4p^n} = \mathrm{id}$ and $(S^4 - \mathrm{id})^{p^n} = 0$. Then we can write $S^2|_{He_i} = \pm\,\mathrm{id}_{He_i} + N_i$ for some nilpotent operator N_i of He_i. However, $e_i \in He_i$ and $S^2(e_i) = e_i$. Therefore, $S^2|_{He_i} = \mathrm{id}_{He_i} + N_i$ for all i. Since $\dim He_i = p$, $N_i^p = 0$ and so $S^{2p}|_{He_i} = \mathrm{id}_{He_i}$. Therefore, $S^{2p} = \mathrm{id}$. Note that, for each i, $(He_i)^{**} \cong_{S^2} (He_i) \cong HS^2(e_i) = He_i$.

Now we show the He_i is indecomposable. Assume to the contrary. Then for any indecomposable summand P of He_i, $1 < \dim P < p$ since H is nonsemisimple. Since $(He_i)^{**} \cong He_i$, $D^j(P)$ is a summand of He_i for $j = 0, \ldots, p-1$, where $D(V) = V^{**}$. This implies $P^{**} \cong P$ for all indecomposable summands of He_i. Since $\dim He_i$ is odd, there must be an indecomposable summand of He_i, which has an odd dimension, but this contradicts Lemma 3.3(ii).

For the last assertion, it suffices to show that $S^4 = \mathrm{id}$. If H contains a nontrivial group-like element g, then $S^2 = \mathrm{id}$ since it holds for g, x which generate H. On the other hand, if $G(H) = 1$, then H^* is unimodular. It follows immediately from Radford's S^4-formula such that $S^4 = \mathrm{id}$ since both H and H^* are unimodular. $\square$

5. Affirmative answer to Questions A and B for $p \leq 5$

As an application of our previous results, in this section, we give an affirmative answer to our main Questions A and B for primes $p \leq 5$.

The ensuing result is a natural extension of [**26**, Lemma 2.3] to dimension p^2. Note that the group algebra $\Bbbk[G]$ for any p-group G of order p^n is a local p^n-dimensional Hopf algebra over $\Bbbk$ of characteristic p.

LEMMA 5.1. *Let $\Bbbk$ be an algebraically closed field of characteristic $p > 2$. For any non-local Hopf algebra of dimension p^2, we have $|\mathrm{Irr}(H)| > 2$. Moreover, if $p = 5$, then $|\mathrm{Irr}(H)| > 3$.*

PROOF. Set $\mathrm{Irr}(H) = \{\Bbbk = V_0, \ldots, V_n\}$ with $|\mathrm{Irr}(H)| = 1 + n \geq 2$. For simplicity, we write $(\dim V_i, \dim P(V_i)) = (d_i, D_i)$, $N_{ij}^k = [V_i \otimes V_j : V_k]$ for all

$0 \le i,j,k \le n$. Note that $[V_i \otimes V_j : V_k] = \dim \operatorname{Hom}_H(P(V_k), V_i \otimes V_j)$ and that (2.1) and (2.2) can be written as

$$p^2 = \sum_{j=0}^{n} d_j D_j \,, \tag{5.1}$$

$$d_i D_k = \sum_{j=0}^{n} N_{ij}^k D_j \,. \tag{5.2}$$

Suppose $n=1$. Since $p>2$, we have $d_1>1$ otherwise $G(H^*)=2 \nmid p^2$. Thus (5.1) and (5.2) implies that

$$p^2 = D_0 + d_1 D_1 \text{ and } d_1 D_1 = D_0 + N_{11}^1 D_1.$$

We get $D_1 \mid D_0 \mid p^2$. Hence $D_1 = p$ and $D_0 = kp$ for some $k \ge 1$. Since $d_1 = p-k < D_1$, we must have $D_1 \ge 2d_1$, which implies that $k > p/2$. So $N_{11}^1 = (d_1 D_1 - D_0)/D_1 = (p-2k) < 0$, a contradiction!

Next suppose $n=2$ and $p=5$. We first treat the case $V_1^* \cong V_2$ (hence $V_2^* \cong V_1$). This implies that $d_1 = d_2 > 1$ and $D_1 = D_2$. Moreover, (5.1) and (5.2) implies that

$$p^2 = D_0 + 2d_1 D_1,\ d_1 D_1 = d_2 D_1 = D_0 + (N_{11}^1 + N_{12}^1) D_1.$$

Similar calculation yields that $D_1 = p$, $D_0 = kp$ for $k \ge 1$ and $d_1 = (p-k)/2 \ge 2$. Since $p=5$, we must have $k=1$ and $d_1 = 2$, $D_0 = D_1 = 5$. However,

$$4 = \dim(V_2 \otimes V_1) = N_{21}^0 + 2\cdot(N_{21}^1 + N_{21}^2) = N_{21}^0 + 4 \cdot N_{21}^1 \ge 1 + 4 \cdot N_{21}^1$$

since $V_2 \otimes V_1$ is self-dual and $N_{21}^0 \ge 1$. This forces $N_{21}^1 = 0$ and $N_{21}^0 = 4$ but this leads to the absurd inequality

$$2p = d_2 D_0 \ge N_{21}^0 D_1 = 4p\,.$$

It remains to treat the case such that $V_i^* \cong V_i$ for $i = 1,2$. Note that we still have $d_1, d_2 > 1$. For all $0 \le i,j,k \le 2$, we have

$$N_{ij}^k = [V_i \otimes V_j : V_k] = [(V_i \otimes V_j)^* : V_k^*] = [V_j^* \otimes V_i^* : V_k^*] = [V_j \otimes V_i : V_k] = N_{ji}^k.$$

Subtracting (5.2) for $i=k=2$ from (5.1) and letting $\{i,k\} = \{1,2\}$ in (5.2), we get

$$p^2 = (2d_2 - N_{22}^2) D_2 + (d_1 - N_{21}^2) D_1, \tag{5.3}$$

$$N_{11}^2 D_1 = (d_1 - N_{12}^2) D_2, \tag{5.4}$$

$$N_{22}^1 D_2 = (d_2 - N_{21}^1) D_1. \tag{5.5}$$

If $d_1 = N_{21}^2$, then (5.3) implies that $D_2 = p$ and $d_2 = (p + N_{22}^2)/2 \ge \frac{p+1}{2} > D_2/2$. Therefore, $d_2 = D_2$ or V_2 is projective, but this is absurd as $d_2 D_2 = p^2$. By the same argument, $d_2 = N_{21}^1$ is also impossible. Therefore, $d_1 > N_{21}^2$ and $d_2 > N_{21}^1$. By (5.3), $\gcd(D_1, D_2) \mid p^2$. If $\gcd(D_1, D_2) = 1$, (5.4) and (5.5) imply that $D_1 | d_1 - N_{12}^2$ and $D_2 | d_2 - N_{21}^1$. Since $D_i \ge d_i > 0$, $N_{12}^2 = N_{21}^1 = N_{12}^1 = 0$. Therefore, all the composition factor of $V_1 \otimes V_2$ are isomorphic to V_0, and hence

$$0 \ne \operatorname{Hom}_H(V_1 \otimes V_2 : \Bbbk) \cong \operatorname{Hom}_H(V_1, V_2)$$

which is absurd. Thus, we must have $\gcd(D_1, D_2) = p$. By (5.1) again, we conclude that $D_0 = D_1 = D_2 = 5$ and $d_1 = d_2 = 2$. This implies $[P(V_i) : V_0] = 1$ for $i = 1, 2$. By duality, $[P(V_0) : V_i] = 1$ for $i = 1, 2$, which leads

$$5 = \dim P(V_0) = [P(V_0) : V_0] + 2([P(V_0) : V_1] + [P(V_0) : V_2]) \geq 2 + 2 \cdot 2 = 6,$$

a contradiction again! □

In [**22**, **35**], the quantum coordinate ring $\mathcal{O}_q(SL_2)$ was used to show the nonexistence of S-invariant 4-dimensional simple subcoalgebras in Hopf algebras of low dimension in characteristic 0. We extend their results to p^2-dimensional Hopf algebras in characteristic $p > 2$.

LEMMA 5.2. *Let $\Bbbk$ be an algebraically closed field of characteristic $p > 2$, and H a nonsemisimple Hopf algebra of dimension p^2 over $\Bbbk$ such that H and H^* are not pointed. Then H and H^* are unimodular, and hence $S^4 = \mathrm{id}$. If, in addition, $S^2 \neq \mathrm{id}$, then H has no S-invariant 4-dimensional simple subcoalgebras.*

PROOF. Since H is not semisimple and H^* is not pointed, $G(H^*)$ is trivial by Theorem 3.4, and hence H is unimodular. We claim that H^* is also unimodular. It is true if H^* is semisimple. On the other hand, if H^* is not semisimple, then $G(H)$ has to be trivial; otherwise, H would be pointed by Theorem 3.4. In particular, H^* is unimodular. Thus by Radford's S^4-formula, $S^4 = \mathrm{id}$.

Now suppose $S^2 \neq \mathrm{id}$ and assume that H has a 4-dimensional simple subcoalgebra C such that $S(C) = C$. Let K be the subalgebra of H generated by C, which is a noncocommutative Hopf subalgebra of H. Since the p-dimensional Hopf algebras over $\Bbbk$ are cocommutative (cf. [**26**] or Theorem 7.1), $\dim K \neq p$. Hence $K = H$ by the Nichols-Zoeller theorem. As a consequence, $\mathrm{ord}(S^2) = \mathrm{ord}(S^2|_C) = 2$.

Note that the argument of [**35**, Theorem 1.4(a)] remains valid when $\mathrm{char}(\Bbbk) = p \nmid \mathrm{ord}(S^2) = 2$. Hence we can choose a basis $\{e_{ij}\}_{1 \leq i,j \leq 2}$ of C such that $\Delta(e_{ij}) = \sum_{1 \leq k \leq 2} e_{ik} \otimes e_{kj}$, $\epsilon(e_{ij}) = \delta_{ij}$ and $S|_C$ is given by

$$\begin{pmatrix} S(e_{11}) & S(e_{12}) \\ S(e_{21}) & S(e_{22}) \end{pmatrix} = \begin{pmatrix} e_{22} & -\mathbf{i}e_{12} \\ \mathbf{i}e_{21} & e_{11} \end{pmatrix},$$

where $\mathbf{i} \in \Bbbk$ satisfies $\mathbf{i}^2 = -1$. By the antipode axiom, the following relations hold in H:

$$e_{12}e_{11} = \mathbf{i}e_{11}e_{12}, \quad e_{22}e_{12} = \mathbf{i}e_{12}e_{22}, \quad e_{21}e_{11} = \mathbf{i}e_{11}e_{21}, \tag{5.6}$$

$$e_{22}e_{21} = \mathbf{i}e_{21}e_{22}, \quad e_{12}e_{21} = e_{21}e_{12}, \quad e_{11}e_{22} + e_{12}e_{21} = 1, \tag{5.7}$$

$$e_{11}e_{22} - e_{22}e_{11} = -2\mathbf{i}e_{12}e_{21}. \tag{5.8}$$

As a consequence, H is a quotient of the quantum coordinate ring $\mathcal{O}_q(SL_2)$ over $\Bbbk$ with $q = \mathbf{i}$. Consider the subalgebra D of H generated by $e_{11}^2, e_{12}^2, e_{21}^2, e_{22}^2$. It is straightforward to check that D is a commutative Hopf subalgebra with $\Delta(e_{ij}^2) = \sum_{1 \leq k \leq 2} e_{ik}^2 \otimes e_{kj}^2$, $\epsilon(e_{ij}^2) = \delta_{ij}$ and

$$\begin{pmatrix} S(e_{11}^2) & S(e_{12}^2) \\ S(e_{21}^2) & S(e_{22}^2) \end{pmatrix} = \begin{pmatrix} e_{22}^2 & -e_{12}^2 \\ -e_{21}^2 & e_{11}^2 \end{pmatrix}.$$

Thus D is a quotient of the commutative Hopf algebra $\mathcal{O}(SL(2))$ over $\Bbbk$. It is clear that $D \neq H$ since H is not involutive. So $\dim D = p$ or 1. We have two cases to consider according to the classification of p-dimensional Hopf algebras over $\Bbbk$ (cf. [**26**] or Theorem 7.1): (1) $D \cong \Bbbk[x]/(x^p - x)$ with x being primitive. Since x is semisimple primitive, we know $S^2 = \mathrm{id}$ by Proposition 4.6, a contradiction. (2) $D \cong \Bbbk[x]/(x^p)$ with x being primitive, $D \cong \Bbbk[C_p]$ or $D = \Bbbk$. Then $(D^+)^p = 0$ and so $y^p = 0$ for $y \in \{e_{11}^2 - 1, e_{12}^2, e_{21}^2, e_{22}^2 - 1\}$. Note that $Hy = yH$ is nilpotent for $y \in \{e_{21}, e_{12}\}$. Thus, $Jac(H) \supseteq (e_{12}, e_{21})$. It follows from the relations (5.6) and (5.7) that $H/Jac(H)$ is commutative. Thus, every simple H-module is 1-dimensional; hence, H^* is pointed, which is a contradiction. □

The following result is simply the dual version of the preceding lemma.

LEMMA 5.3. *Let $\Bbbk$ be an algebraically closed field of characteristic $p > 2$, and H a non-cosemisimple Hopf algebra of dimension p^2 such that H and H^* are not pointed. Then H and H^* are unimodular and hence $S^4 = \mathrm{id}$. If, in addition, $S^2 \neq \mathrm{id}$, then H has no self-dual 2-dimensional simple modules.*

Now, we prove the main result of this section.

THEOREM 5.4. *Let $\Bbbk$ be an algebraically closed field of positive characteristic $p \leq 5$, and H a p^2-dimensional Hopf algebra over $\Bbbk$. Then H or H^* is pointed.*

PROOF. The statement is obvious when $p = 2$. Now we consider $p = 3$. By [**6**, Cor. 3.2(i)], H cannot be both semisimple and cosemisimple since $\dim H = 0$ in $\Bbbk$. Without loss of generality, we assume H to be nonsemisimple by duality. Suppose that H^* is not pointed. In particular, H is not local. Hence $|\operatorname{Irr}(H)| \geq 3$ by Lemma 5.1, and $G(H^*)$ is trivial by Theorem 3.4. Since $\dim H = 9$, a simple dimension argument yields that $\operatorname{Irr}(H) = \{V_0, V_1, V_2\}$ where $\dim V_0 = 1$ and $\dim V_1 = \dim V_2 = 2$ with $P(V_i) = V_i$ for all i, which contradicts to the fact that H is nonsemisimple. Therefore, H^* is pointed.

It remains to deal with the case for $p = 5$. We may still assume H to be nonsemisimple. Suppose H and H^* are not pointed. Then $G(H^*)$ is trivial by Theorem 3.4 and $S^4 = \mathrm{id}$ by Lemma 5.2. We consider the cases (i) $\operatorname{ord}(S^2) = 2$ and (ii) $\operatorname{ord}(S^2) = 1$, separately.

(i) Suppose $\operatorname{ord}(S^2) = 2$. Write $\operatorname{Irr}(H^*) = \{\Bbbk = V_0, V_1, \ldots, V_n\}$. Since H is not pointed, $n \geq 3$ by Lemma 5.1 and $|G(H)| = 1$ or 5. The group $G(H)$ can be identified with the 1-dimensional H^*-modules $\Bbbk_g$ for $g \in G(H)$. We first claim that H^* is also nonsemisimple. If not, then

$$25 = 1 + \sum_{d_i \geq 2} d_i^2 \quad \text{or} \quad 25 = 5 + \sum_{d_i \geq 2} d_i^2$$

respectively for $|G(H)| = 1$ or 5. By enumeration, $d_i \in \{2, 4\}$ and there must be i such that $d_i = 2$. Let $V \in \operatorname{Irr}(H^*)$ such that $\dim(V) = 2$. Since H^* is semisimple, $[V^* \otimes V : \Bbbk] = 1$, and so $V^* \otimes V = \Bbbk \oplus W$ for some H-module W of dimension 3. Therefore, W is not simple and $[W : \Bbbk_g] \geq 1$ for some nontrivial $g \in G(H)$. Hence $[V : V \otimes \Bbbk_g] = 1$ or, equivalently, $V \otimes \Bbbk_g \cong V$. Since g generates $G(H)$, $G(H) \leq \operatorname{stab}(V)$. By Lemma 3.2, $5 \mid \dim(V)$, a contradiction! Therefore, H^* is also nonsemisimple. Since H is not pointed, $G(H)$ is also trivial by Theorem 3.4.

We next claim H must have a nonzero primitive element. Suppose not. By Lemma 4.2, $\operatorname{Ext}^1_{H^*}(\Bbbk, \Bbbk) = 0$. Then $[P(V_0) : V_i] > 0$ for some $i > 0$, say $i = 1$. If $\dim V_1 = 2$, we may assume $V_2 = V_1^*$ since H^* has no 2-dimensional self-dual simple module by Lemma 5.3. Since $P(V_0)$ is also self-dual, $[P(V_0) : V_2] > 0$ and so $\dim P(V_0) \geq 6$. In this case, $[P(V_1) : V_0], [P(V_2) : V_0] \geq 1$. Therefore,

$\dim P(V_1), \dim P(V_2) \geq 5$, and hence

$$25 \geq \dim P(V_0) + \sum_{i=1}^{2} \dim(V_i) \dim P(V_i) \geq 6 + 2 \cdot 5 \cdot 2 = 26,$$

a contradiction! Therefore, $\dim V_1 \geq 3$. This implies that $\dim P(V_0) \geq 5$ and $\dim P(V_1) \geq 7$. Thus,

$$25 \geq \dim P(V_0) + \dim(V_1) \dim P(V_1) = 5 + 3 \cdot 7 = 26, \text{ a contradiction again!}$$

Therefore, H must have a nonzero primitive element.

Let $x \in H$ be a nonzero primitive element. Then $\Bbbk[x]$ is commutative Hopf subalgebra of H. Since H^* is not pointed, $\Bbbk[x]$ is of dimension p. It follows from Corollary 4.5 that $\dim H_0 \leq 5$. Since $\dim(V_i) \geq 2$ for $i \geq 1$ and $n \geq 3$,

$$5 \geq \dim(H_0) = 1 + \sum_{i=1}^{n} \dim(V_i)^2 \geq 1 + 3 \cdot 2^2 = 13,$$

a contradiction! Therefore, H or H^* must be pointed if $\operatorname{ord}(S^2) = 2$.

(ii) Suppose $S^2 = \mathrm{id}$. Write $\operatorname{Irr}(H) = \{\Bbbk = V_0, V_1, \dots, V_n\}$. Since H^* is not pointed, $\dim V_i \geq 2$ for $i > 0$ and $n \geq 3$ by Lemma 5.1. Since H is not semisimple, $\sum_{(\Lambda)} S(\Lambda_2)S^2(\Lambda_1) = \epsilon(\Lambda)1 = 0$, where Λ is a nontrivial left integral of H. By [**32**, Lemma 3(b)], $\operatorname{tr}(r(a) \circ S^2) = \operatorname{tr}(r(a)) = 0$ for all $a \in H$, where $r(a)$ denotes the operator on H given by the right multiplication of a. As a consequence, for any indecomposable projective module He for some primitive idempotent $e \in H$, we have $\dim He = \operatorname{tr}(r(e)) = 0$ in $\Bbbk$, and hence $p \mid \dim He$. In particular, $\dim P(V_i) \geq p$. Now, we have

$$p^2 = \sum_{i=0}^{n} \dim V_i \cdot \dim P(V_i) \geq (1 + 2n)p$$

and so $n \leq \frac{p-1}{2} = 2$, a contradiction. Therefore, H^* must be pointed if $S^2 = \mathrm{id}$. □

According to [**39**] or Theorem 7.1, the p^2-dimensional pointed Hopf algebras are closed under duality, which is stated in Corollary 7.7. The following theorem is an immediate consequence of Theorem 5.4 and Corollary 7.7, which provides affirmative answers to both Questions A and B for $p \leq 5$.

THEOREM 5.5. *Let $\Bbbk$ be an algebraically closed field of positive characteristic $p \leq 5$, and H a p^2-dimensional Hopf algebra over $\Bbbk$. Then H is pointed.*

6. Twist equivalence of cleft extensions

In this section, we introduce the notion of twist equivalence for cleft extensions and crossed products. As an application, we classify certain cleft extensions, up to twist equivalence, over group algebras of elementary abelian p-groups and their duals over an algebraically closed field $\Bbbk$ of characteristic p. The readers are referred to [**27**, Ch. 7] for definitions and results on cleft extensions.

Let B be a right H-comodule algebra and A a subalgebra of B. The extension $A \subseteq B$ is called an *H-cleft extension* if $B^{coH} = A$ and there exists a right H-comodule map $\gamma : H \to B$ which is convolution invertible, and γ is called a *cleft map*. By [**33**, Theorem 2.2], if $\pi : A \to H$ is a surjection of Hopf algebras, then $A^{co\pi} \subset A$ is an H-cleft extension. In particular, for the short exact sequence (2.5), A is an H-cleft extension of $\iota(K)$.

According to [**27**, Theorem 7.2.2], H-cleft extensions can be characterized by crossed products, which we now recall. Let H be a Hopf algebra that *measures* an algebra A, namely there is a $\Bbbk$-linear map $H \otimes A \to A$, given by $h \otimes a \mapsto h \cdot a$ such that $h \cdot 1 = \epsilon(h)1$ and $h \cdot (ab) = \sum_{(h)}(h_1 \cdot a)(h_2 \cdot b)$, for all $h \in H$ and $a, b \in A$. In this paper, we simply call the triple $(H, A, \cdot)$ a *measurement.* It is worth noting that A may not be an H-module algebra as the H-action on A may not be associative.

Given a measurement $(H, A, \cdot)$ and a convolution invertible map $\sigma \in \mathrm{Hom}_{\Bbbk}(H \otimes H, A)$, one can construct the *crossed product* $A\#_\sigma H$ of A with H, which is the vector space $A \otimes H$ equipped with multiplication given by

$$(a\#h)(b\#k) = \sum_{(h),(k)} a(h_1 \cdot b)\sigma(h_2, k_1)\#h_3k_2 \tag{6.1}$$

for all $h, k \in H$ and $a, b \in A$. In this case, the associativity of (6.1) with identity element $1_A\#1_H$ is equivalent to the fact that A is a *twisted* H-module and σ is a 2-cocycle (see [**27**, Lemma 7.1.2] and the definitions therein) of the measurement $(H, A, \cdot)$. Note that the crossed product $A\#_\sigma H$ canonically admits a left A-module and a right H-comodule structure. In particular, $A\#_\sigma H$ is an H-cleft extension of A with the cleft map $\gamma : H \to A\#_\sigma H$, $h \mapsto 1_A\#h$, under the identification $A \equiv A\#1_H$ (cf. [**27**, Proposition 7.2.7]).

Recall that one can twist a left A-module M by an algebra automorphism φ of A to ${}_\varphi M$. By duality, if ϕ is a coalgebra automorphism of a coalgebra C, and V is a right C-comodule with the C-coaction $\rho : V \to V \otimes C$, V^ϕ denotes the right C-comodule with the C-coaction given by $(\mathrm{id}_V \otimes \phi) \circ \rho$.

DEFINITION 6.1. Let H be a Hopf algebra and A an algebra.

(i) We say two H-cleft extensions B, B' of A are *twist equivalent*, if there exist an algebra automorphism φ of A, a Hopf algebra automorphism ϕ of H and an algebra isomorphism $f : B \to {}_\varphi(B')^\phi$ that is a left A-module and a right H-comodule map. In this case, f is called a *(ϕ, φ)-twist isomorphism* or simply a *twist isomorphism* of the H-cleft extensions.

(ii) Two crossed products $A\#_\sigma H$ and $A\#'_{\sigma'}H$ are said to be *twist equivalent* if they are twist equivalent as H-cleft extensions of A.

(iii) In particular, if $\varphi = \mathrm{id}_A$ and $\phi = \mathrm{id}_H$, we will say an *equivalent* (resp. *isomorphism*) instead of twist equivalent (resp. *twist isomorphism*) of H-cleft extensions or crossed products as in (i) and (ii).

One can easily see that twist equivalence is an equivalence relation on the set of all H-cleft extensions of A.

The following result is a natural extension of [**27**, Theorem 7.3.4] to the context of twist equivalence. We include its proof for the sake of completeness.

PROPOSITION 6.2. *Let $(H, A, \cdot)$ and $(H, A, \bullet)$ be measurements, σ, σ' be 2-cocycles of the measurements respectively, and $A\#_\sigma H$, $A\#'_{\sigma'}H$ their corresponding crossed products. Suppose*

$$\Phi : A\#_\sigma H \to {}_\varphi(A\#'_{\sigma'}H)^\phi$$

is a (ϕ, φ)-twist isomorphism for some algebra automorphism φ of A and some Hopf automorphism ϕ of H. Then there exists a convolution invertible map $u \in \mathrm{Hom}_{\Bbbk}(H, A)$ satisfying $u(1_H) = 1_A$ with the convolution inverse $\overline{u}$ such that for all $a \in A$, $h, k \in H$,

(i) $\Phi(a\#h) = \sum_{(h)} \varphi(au(h_1))\#'\phi^{-1}(h_2)$,

(ii) $h \bullet a = \varphi\left(\sum_{(h)} \overline{u}(\phi(h_1))\ (\phi(h_2)\cdot\varphi^{-1}(a))\ u(\phi(h_3))\right)$,

(iii) $\sigma'(h,k) = m\left(\overline{u}(\phi(h_1))\#\phi(h_2))(\overline{u}(\phi(k_1))\#\phi(k_2))\right)$ *where the* $\Bbbk$*-linear map* $m : A\#_\sigma H \to A$ *is given by* $m(a\#h) = \varphi(au(h))$.

Conversely given a convolution invertible map $u \in \mathrm{Hom}_\Bbbk(H,A)$ *with* $u(1_H) = 1_A$ *such that* (ii) *and* (iii) *hold, then the map* Φ *in* (i) *is a* (φ,ϕ)*-twist isomorphism.*

PROOF. We define $u \in \mathrm{Hom}_\Bbbk(H,A)$ by $u(h) = (\varphi^{-1}\otimes\epsilon)\Phi(1_A\#h)$ for $h \in H$. Then $\varphi(u(h)) = (\mathrm{id}\otimes\epsilon)\Phi(1_A\#h)$ and $\Phi(a\#h) = \varphi(a)\Phi(1_A\#h)$ as Φ is a left A-module map. Since Φ is a right H-comodule map, we have

$$(\mathrm{id}\otimes\Delta)\circ\Phi = (\Phi\otimes\phi^{-1})\circ(\mathrm{id}\otimes\Delta)\,.$$

By applying $(\mathrm{id}\otimes m_H)\circ(\mathrm{id}\otimes\epsilon 1_H\otimes\mathrm{id})$ to both sides of this equation, we get

$$\begin{aligned}\Phi(1\#h) &= (\mathrm{id}\otimes m_H)\circ(\mathrm{id}\otimes\epsilon 1_H\otimes\mathrm{id})\left(\sum\Phi(1_A\#h_1)\otimes\phi^{-1}(h_2)\right)\\ &= \sum(\mathrm{id}\otimes m_H)(\varphi(u(h_1))\#'1_H\otimes\phi^{-1}(h_2))\\ &= \sum\varphi(u(h_1))\#'\phi^{-1}(h_2),\end{aligned}$$

which proves the formula in (i). Note that $\Phi(1_A\#1_H) = 1_A\#1_H$ implies that $u(1_H) = 1_A$.

It is immediate to see that $\Phi^{-1} : A\#'_{\sigma'}H \to {}_{\varphi^{-1}}(A\#_\sigma H)^{\phi^{-1}}$ is a (φ^{-1},ϕ^{-1})-twist isomorphism. By (i), there exists $v \in \mathrm{Hom}_\Bbbk(H,A)$ given by $v(h) = (\varphi\otimes\epsilon)\Phi^{-1}(1\#'h)$ such that $\Phi^{-1}(a\#'h) = \sum\varphi^{-1}(av(h_1))\#\phi(h_2)$. We claim that $\overline{u} = \varphi^{-1}\circ v\circ\phi$ is the convolution inverse of u. This is because

$$\begin{aligned}1_A\#h = \Phi^{-1}\Phi(1_A\#h) &= \Phi^{-1}\left(\sum\varphi(u(h_1))\#'\phi^{-1}(h_2)\right)\\ &= \sum u(h_1)\varphi^{-1}(v(\phi^{-1}(h_2)))\#h_3.\end{aligned}$$

By applying $\mathrm{id}_A\otimes\epsilon$ to this equation, we see that $u*(\varphi^{-1}\circ v\circ\phi^{-1}) = \epsilon 1_A$ in $\mathrm{Hom}_\Bbbk(H,A)$. Similarly, we see $v*(\varphi\circ u\circ\phi) = \epsilon 1_A$, which is equivalent to $(\varphi^{-1}\circ v\circ\phi^{-1})*u = \epsilon 1_A$. Thus, $\overline{u} = \varphi^{-1}\circ v\circ\phi^{-1}$ is the convolution inverse of u.

Now we consider the equation

$$\Phi^{-1}((1_A\#'h)(1_A\#'k)) = \Phi^{-1}(1_A\#'h)\Phi^{-1}(1_A\#'k).$$

We apply (6.1) to obtain

$$\begin{aligned}&\sum\varphi^{-1}\left(\sigma'(h_1,k_1)v(h_2k_2)\right)\#\phi(h_3k_3)\\ =&\sum\varphi^{-1}(v(h_1))\left(\phi(h_2)\cdot\varphi^{-1}(v(k_1))\right)\sigma(\phi(h_3),\phi(k_2))\#\phi(h_4k_3).\end{aligned}$$

This proves (iii) after applying m to the equation.

Applying $\mathrm{id}\otimes\epsilon$ to $\Phi^{-1}((1_A\#'h)(b\#'1_H)) = \Phi^{-1}(1_A\#'h)\Phi^{-1}(b\#1_H)$ or

$$\sum\varphi^{-1}\left((h_1\bullet b)v(h_2)\right)\#\phi(h_3) = \sum\varphi^{-1}(v(h_1))\left(\phi(h_2)\cdot\varphi^{-1}(b)\right)\#\phi(h_3),$$

we find

$$\sum\varphi^{-1}(h_1\bullet b)\varphi^{-1}(v(h_2)) = \varphi^{-1}(v(h_1))\left(\phi(h_2)\cdot\varphi^{-1}(b)\right).$$

Note that $\overline{u}\circ\phi = \varphi^{-1}\circ v$ and $\varphi^{-1}\circ v$ is the convolution inverse of $u\circ\phi$. Inverting $\varphi^{-1}\circ v$ gives (ii).

Finally, the converse follows by direct computation. $\square$

Throughout the remainder of this section, we will assume that H is a cocommutative Hopf algebra and A is a commutative algebra. In this setting, a crossed product $A\#_\sigma H$ is defined as an H-module algebra A equipped with a 2-cocycle $\sigma : H \otimes H \to A$, which can be placed within the context of Sweedler's cohomology $H^\bullet_{sw}(H, A)$, as described in [**36**]. In particular, the measurement $(H, A, \cdot)$ simply means an H-module algebra.

For any integer $n \geq 0$, denote by $\operatorname{Reg}^n(H, A) \subseteq \operatorname{Hom}_{\Bbbk}(H^{\otimes n}, A)$ the set of all convolution invertible linear maps, which forms an abelian subgroup under the convolution product. The coface operator $d^i : \operatorname{Reg}^n(H, A) \to \operatorname{Reg}^n(H, A)$ is defined as follows:

$$d^i(f)(h_0 \otimes \cdots \otimes h_n) = \begin{cases} h_0 f(h_1 \otimes \cdots \otimes h_n) & \text{for } i = 0, \\ f(h_0 \otimes \cdots \otimes h_{i-1}h_i \otimes \cdots \otimes h_n) & \text{for } 1 \leq i \leq n, \\ f(h_0 \otimes \cdots \otimes h_{n-1})\epsilon(h_n) & \text{for } i = n+1 \end{cases}$$

for any $f \in \operatorname{Reg}^n(H, A)$ and $h_0, \ldots, h_n \in H$. This gives rise to a cochain complex $(\operatorname{Reg}^\bullet(H, A), \partial)$ whose differential ∂ is defined by

$$\partial(f) = d^0(f) * d^1(f^{-1}) * \cdots * d^{n+1}(f^{\pm 1})$$

for any $f \in \operatorname{Reg}^n(H, A)$ where $*$ denotes the convolution product and f^{-1} means the convolution inverse of f. The *Sweedler cohomology* $H^\bullet_{sw}(H, A)$ *of* H *with coefficients in* A is defined to be the cohomology of the above cochain complex. In particular,

$$H^n_{sw}(H, A) := \ker \partial^n / \operatorname{Im} \partial^{n-1},$$

for any $n \geq 0$ under the convention ∂^{-1} is trivial.

As noted in [**36**], one can use the normalized subcomplex $(\operatorname{Reg}^\bullet_+(H, A), \partial)$ of $(\operatorname{Reg}^\bullet(H, A), \partial)$ to compute $H^\bullet_{sw}(H, A)$, where $\operatorname{Reg}^n_+(H, A)$ consists of those convolution invertible maps $f \in \operatorname{Hom}_{\Bbbk}(H^{\otimes n}, A)$ satisfying $f(x_1 \otimes \cdots \otimes x_n) = \epsilon(x_1 \cdots x_n)1_A$ whenever $1_H \in \{x_1, \ldots, x_n\}$. In particular, $\sigma \in \operatorname{Reg}^2_+(H, A)$ is a 2-cocycle if and only if $d^0(\sigma) * d^1(\sigma^{-1}) * d^2(\sigma) * d^3(\sigma^{-1}) = \epsilon \otimes \epsilon \otimes \epsilon 1_A$. Since $\operatorname{Reg}^2_+(H, A)$ is an abelian group, $\sigma \in \operatorname{Reg}^2_+(H, A)$ is a 2-cocycle if and only if

$$\sum_{(h),(k),(m)} h_1 \cdot \sigma(k_1, m_1)\, \sigma(h_2, k_2 m_2) = \sum_{(h),(k)} \sigma(h_1, k_1)\, \sigma(h_2 k_2, m) \tag{6.2}$$

for any $h, k, m \in H$.

The first assertion of our following result is well-known since Sweedler's second cohomology classifies all cleft extensions up to equivalence (see [**36**, Theorem 8.6]).

PROPOSITION 6.3. *Let* H *be a cocommutative Hopf algebra and* A *a commutative* H*-mdoule algebra.*

(i) *If* $H^2_{sw}(H, A)$ *is trivial, then the crossed product* $A\#_\sigma H$ *for any Sweedler 2-cocycle* σ *is equivalent to the smash product* $A\#H$.

(ii) *Suppose* $H^2_{sw}(H, A)$ *is trivial for any* H*-module algebra structure on* A. *Then every right* H*-cleft extension of* A *is equivalent to a smash product* $A\#H$, *where the* H*-module action on* A *is determined by a cleft map.*

(iii) *Suppose* $H^2_{sw}(H, A)$ *is trivial for any* H*-module algebra structure on* A. *For any two* H*-actions* $\cdot, \bullet : H \otimes A \to A$, *the corresponding smash products* $A\#H$ *and* $A\#'H$ *are twist equivalent if and only if there is some algebra automorphism* φ *of* A *and some Hopf automorphism* ϕ *of* H *such that*

$$h \bullet a \;=\; \varphi\left(\phi(h) \cdot \varphi^{-1}(a)\right) \tag{6.3}$$

for all $h \in H$ and $a \in A$.

PROOF.

(i) Since $H^2_{sw}(H,A)$ is trivial, the Sweedler 2-cocycle $\sigma : H \otimes H \to A$ is a 2-coboundary, namely there is some $u \in \mathrm{Reg}^1_+(H,A)$ such that $u(1_H) = 1_A$ and

$$\sigma(h,k) = \sum (h_1 \cdot u(k_1))\, \overline{u}(h_2k_2)\, u(h_3),$$

where $\overline{u}$ is the convolution inverse of u. Then by applying Proposition 6.2 in case of $\varphi, \phi = \mathrm{id}$, one can check that $f : A\#_\sigma H \to A\#H$ defined by $f(a\#h) = \sum_{(h)} au(h_1)\#h_2$ with inverse $f^{-1}(a\#h) = \sum_{(h)} a\overline{u}(h_1)\#h_2$ for any $a \in A$ and $h \in H$ is an isomorphism of H-cleft extensions.

(ii) By [**27**, Theorem 7.2.2], every H-cleft extension of A is equivalent to $A\#_\sigma H$ as H-cleft extensions of A for some Sweedler's 2-cocycle $\sigma \in \mathrm{Reg}^2_+(H,A)$ of the H-module algebra A with its H-action determined by a cleft map. Since $H^2_{sw}(H,A)$ is trivial, the statement follows from (i).

(iii) Suppose $\Phi : A\#H \to {}_\varphi(A\#'H)^\phi$ is a twist isomorphism of smash products for some algebra automorphism φ of A and some Hopf algebra automorphism ϕ of H. By Theorem 6.2, there is some $u \in \mathrm{Reg}^1_+(H,A)$ with $u(1_H) = 1_A$ and convolution inverse $\overline{u}$ such that

$$h \bullet a = \varphi\left(\sum_{(h)} \overline{u}(\phi(h_1))\, \left(\phi(h_2)\cdot\varphi^{-1}(a)\right)\, u(\phi(h_3))\right) = \varphi\left(\phi(h)\cdot\varphi^{-1}(a)\right),$$

where the last equality follows from the cocommutativity of H and the commutativity of A. Moreover, by letting $u = \epsilon 1_A$ and $\sigma = \sigma' = \epsilon 1_A \otimes \epsilon 1_A$, one can check (iii) in Theorem 6.2 holds. This implies our result. □

It was observed in [**12**, Theorem 5.10] and [**8**, Proposition 5.7] that $H^2_{sw}(\Bbbk[C_p^n]^*, \Bbbk)$ is trivial for any positive integer n. Our next result extends this vanishing property to nontrivial coefficients.

THEOREM 6.4. *Let $\Bbbk$ be an algebraically closed field of characteristic $p > 0$, and $K = \Bbbk[C_p^n]^*$ for some positive integer n. For any finite-dimensional commutative K-module algebra A over $\Bbbk$, $H^2_{sw}(K,A)$ is trivial.*

PROOF. Note that $\Bbbk[C_p^n]^*$ is a connected semisimple Hopf algebra. By the Hochschild theorem [**13**], we have $\Bbbk[C_p^n]^* \cong \Bbbk[x]/(x^{p^n} - x)$ as Hopf algebras, where x is primitive. Now, we simply assume $K = \Bbbk[x]/(x^{p^n} - x)$ with x primitive. Suppose $\sigma \in \mathrm{Reg}^2_+(K,A)$. Then $\sigma(1\otimes 1) = 1$ and $\sigma(x^i \otimes 1) = \sigma(1 \otimes x^i) = 0$ for $i \geq 1$. We claim that there exists $f \in \mathrm{Reg}^1_+(K,A)$ such that

$$(\sigma * \partial f^{-1})(x^i \otimes x) = 0, \tag{6.4}$$

for $0 \leq i \leq p^n - 1$. We proceed with the proof of the above claim in several steps.

(i) We first find $f \in \mathrm{Reg}^1_+(K,A)$, where (6.4) holds for $0 \leq i \leq p^n - 2$. Note that (6.4) holds trivially for $i = 0$ and one can rewrite (6.4) as $(\sigma * d^1 f)(x^i \otimes x) = (d^0 f * d^2 f)(x^i \otimes x)$ or

$$f(x^{i+1}) = -\sum_{j=1}^{i} \binom{i}{j} \sigma(x^j \otimes x) f(x^{i-j}) + \sum_{j=0}^{i} \binom{i}{j} (x^j \cdot f(x)) f(x^{i-j}) \tag{6.5}$$

for $1 \le i \le p^n - 2$. For any $a \in A$, we can define $f(1) = 1$, $f(x) = a$, and set $f(x^{i+1})$ inductively by (6.5) for $1 \le i \le p^n - 2$. Then (6.4) holds for $0 \le i \le p^n - 2$.

(ii) By (i), we may assume $\sigma(x^i \otimes x) = 0$ for all $0 \le i \le p^n - 2$. Then we set $h = x, k = x^{p^n-1}, m = x$ in (6.2) to get

$$\sum_{i,l=0}^{1}\sum_{j=0}^{p^n-1}\binom{p^n-1}{j}(x^i \cdot \sigma(x^j, x^l))\sigma(x^{1-i}, x^{p^n-j-l})$$
$$= \sum_{i=0}^{1}\sum_{j=0}^{p^n-1}\binom{p^n-1}{j}\sigma(x^i, x^j)\sigma(x^{p^n-i-j}, x)\,,$$

which is equivalent to

$$x \cdot \sigma(x^{p^n-1}, x) + \sigma(x, x) = \sigma(x, x)\,.$$

Therefore, $x \cdot \sigma(x^{p^n-1}, x) = 0$, namely, $\sigma(x^{p^n-1}, x) \in A^K$.

(iii) Now for any $f \in \mathrm{Reg}^1_+(K, A^K)$ that satisfies (6.4) for $0 \le i \le p^n - 2$, (6.5) yields $f(x^{i+1}) = f(x)^{i+1}$ for $0 \le i \le p^n - 2$ and when $i = p^n - 1$ it is equal to

$$f(x) = f(x^{p^n}) = -\sigma(x^{p^n-1} \otimes x) + f(x)f(x^{p^n-1}) = -\sigma(x^{p^n-1} \otimes x) + f(x)^{p^n}.$$

(iv) We show that the polynomial equation $F(X) = 0$, where $F(X) = X^{p^n} - X - \sigma(x^{p^n-1}, x) \in A^K[X]$ has a solution in A^K. Since A^K is finite-dimensional commutative over an algebraically closed field $\Bbbk$, the equation $F(X)$ has solution in $A^K/J(A^K) = \Bbbk \times \cdots \times \Bbbk$, where $J(A^K)$ denotes the Jacobson radical of A^K. Note A^K is complete with respect to its $J(A^K)$-adic filtration. Since $F'(X) = -1 \neq 0$ in A^K, we conclude that $F(X) = 0$ has a solution in A^K by Hensel's lifting lemma. We choose $\alpha \in A^K$ as a solution that $F(\alpha) = 0$.

(v) We construct $f \in \mathrm{Reg}^1_+(K, A^K)$ by setting $f(1) = 1$, $f(x) = \alpha$ and $f(x^{i+1}) = f(x)^{i+1}$ for $1 \le i \le p^n - 2$. By the previous discussion, it is straightforward to check that $(\sigma * \partial f^{-1})(x^i, x) = 0$ for all $0 \le i \le p^n - 1$. This proves our claim.

Now up to a coboundary, we can assume that $\sigma(x^t, x) = 0$ for all nonnegative integers t. By setting $h = x, k = x^t, m = x$ in (6.2) again, we find

$$\sum_{i,l=0}^{1}\sum_{j=0}^{t}\binom{t}{j}(x^i \cdot \sigma(x^j, x^l))\sigma(x^{1-i}, x^{t-j-l+1})$$
$$= \sum_{i=0}^{1}\sum_{j=0}^{t}\binom{t}{j}\sigma(x^i, x^j)\sigma(x^{t-i-j+1}, x)\,,$$

which implies that $\sigma(x, x^{t+1}) = 0$. Therefore, $\sigma(x, m) = 0$ for all $m \in K$. Inductively, suppose t is a positive integer such that $\sigma(x^j, m) = 0$ for all $m \in K$ whenever $1 \le j \le t$. By setting $h = x, k = x^t$ in (6.2), we find

$$x \cdot \sigma(x^t, m) = \sigma(x^{t+1}, m)$$

for all $m \in K$. Therefore, $\sigma(x^{t+1}, m) = 0$ for all $m \in K$. Thus, σ is trivial. This completes our proof. □

REMARK 6.5. The assumption that $\Bbbk$ is algebraically closed is essential for Theorem 6.4. For instance, if $\Bbbk$ is of characteristic $p = 2$ (not necessarily algebraically closed), then [**12**, Theorem 1.1] states that $H^2_{sw}(\Bbbk[C_p^n]^*, \Bbbk) = (\Bbbk/\{x+x^2 \mid x \in \Bbbk\})^{\oplus n}$.

An immediate application of Theorem 6.4 is the classification of certain cleft extensions up to twist equivalence over $\Bbbk[C_p^n]$ and their duals.

PROPOSITION 6.6. *Let $\Bbbk$ be an algebraically closed field of characteristic $p > 0$, and $H = \Bbbk[C_p^n]^*$ for some positive integer n. Then any H-cleft extension of a finite-dimensional commutative local algebra A over $\Bbbk$ is twist equivalent to some smash product $A\#H$. Moreover, there are exactly 3 twist inequivalent smash products $A\#H$, $A\#'H$, and $A\#''H$, and their H-actions on A are respectively given by*

$$x \cdot t \;=\; 0,\; t,\; t+1,$$

where x is a primitive generator of H satisfying $x^{p^n} = x$, and t is a nilpotent generator of A. The last action $x \cdot t = t+1$ only occurs when $p \mid \dim A$.

PROOF. By Theorem 6.4, $H^2_{sw}(H, A)$ is trivial for any H-action on A. So it suffices to classify all possible H-actions on A, which yields twist inequivalent smash products $A\#H$ by Proposition 6.3. Again, by Hochschild's theorem, we may assume $H = \Bbbk[x]$ with some primitive generator x satisfying $x^{p^n} = x$. Thus any H-action on A is determined by the action of x on A as a $\Bbbk$-derivation satisfying $x^{p^n} = x$. Denote by $\mathbb{F}_q$ the finite field of $q = p^n$ elements. Since the operator x is semisimple, there is a decomposition $A = \bigoplus_{i \in \mathbb{F}_q} A_i$ where A_i is the eigenspace of x for the eigenvalue $i \in \mathbb{F}_q$. It is immediate to see $A_iA_j \subseteq A_{i+j}$ for $i, j \in \mathbb{F}_q$.

Since A is a finite-dimensional local commutative algebra over $\Bbbk$, $A = \Bbbk[t]/(t^m)$ with $m = \dim A$ and t a nilpotent generator of A. Then $t = \sum_{i\in\mathbb{F}_q} t_i$ where $t_i \in A_i$, and there exists $s \in \mathbb{F}_q$ such that $t_s = \sum_{i=0}^{m-1} a_i t^i$ with $a_i \in \Bbbk$ and $a_1 \neq 0$. Thus, $t_s = a_0 + tu$ for some invertible element $u \in A$, and hence $t_s^i \in t^i u^i + F_{i-1}$ where F_ℓ is the $\Bbbk$-linear span of $\{t^0, \ldots, t^\ell\}$. This implies $\{t_s^0, \ldots, t_s^{m-1}\}$ is a $\Bbbk$-basis for A and $1, t_s$ generate A as $\Bbbk$-algebra.

If $s = 0$, then $A = A_0$ and so $x \cdot t = 0$. Now, we assume $s \neq 0$. Recall that $t_s = a_0 + tu$ for some invertible element $u \in A$ and $x \cdot t_s = st_s$. We first consider the case for $a_0 \neq 0$ and set $t' = tu/a_0$. Then $x \cdot t' = s(t'+1)$. Consider the algebra automorphism φ of A defined by $\varphi(t) = t'$ and the Hopf algebra automorphism ϕ of H defined by $\phi(x) = x/s$. Since $\varphi^{-1}(\phi(x)\cdot\varphi(t)) = t+1$, the smash product in this is twist equivalent to $A\#''H$, with the H-action given by $x \cdot t = t+1$ by (6.3). Since $x \cdot (t^m) = mt^{m-1}(x\cdot t)$, it is easy to see that the action $x \cdot t = t+1$ is not well-defined if $p \nmid m$.

Now we deal with the case for $a_0 = 0$. By considering the algebra automorphism φ of A defined by $\varphi(t) = tu$ and the Hopf algebra automorphism ϕ of H defined by $\phi(x) = x/s$, we can assume that $x \cdot t = t$ up to twist equivalence of smash products by (6.3) again.

The smash product $A\#H$ given by the trivial H-action is commutative, and $A\#'H \cong \Bbbk\langle x, t\rangle/(x^{p^n} - x, t^m, [x,t] = t)$ and $A\#''H \cong \Bbbk\langle x,t\rangle/(x^{p^n} - x, t^m, [x,t] = t+1)$ as algebras. But the former has p^n 1-dimensional characters while the latter has none. Therefore, they are twist inequivalent. $\square$

Regarding the second H-action described in the preceding proposition, when $H = \Bbbk[C_p]^*$ and $A = H^* = \Bbbk[C_p]$, the given smash product is the Heisenberg double of the group algebra $\Bbbk[C_p]$, as shown in the following example.

EXAMPLE 6.7 ([**27**, Ex. 4.1.10]). Let H be any finite-dimensional Hopf algebra. There is a left H-action $\rightharpoonup$ on the dual Hopf algebra H^* given by

$$h \rightharpoonup f = \sum_{(h)} f_2(h) f_1$$

for all $h \in H, f \in H^*$. Under this H-action, H^* is a left H-module algebra. The corresponding smash product $H^*\#H$ is the *Heisenberg double* of H^*, denoted by $\mathcal{H}(H^*)$. Note that $\mathcal{H}(H^*)$ is isomorphic to $\mathrm{End}_{\Bbbk}(H^*)$ as algebras (cf. [**27**, 9.4.3]) and it is an H-cleft extension of H^*.

COROLLARY 6.8. *Let $\Bbbk$ be an algebraically closed field $\Bbbk$ of characteristic $p > 0$ and $H = \Bbbk[C_p]^*$. Then any H-cleft extension of the group algebra $\Bbbk[C_p]$ is twist equivalent to one of the following smash products:*

(1) *The tensor product algebra: $\Bbbk[C_p] \otimes \Bbbk[C_p]^*$.*
(2) *The Heisenberg double: $\mathcal{H}(\Bbbk[C_p])$.*
(3) *The smash product: $\Bbbk[C_p]\#\Bbbk[C_p]^*$, where the $\Bbbk[C_p]^*$-action on $\Bbbk[C_p]$ is given by $x \cdot g = g - 1$ with $C_p = \langle g \rangle$ and $x \in \Bbbk[C_p]^*$ a primitive generator.*

PROOF. Note that $H \cong \Bbbk[x]/(x^p - x)$ as Hopf algebras, where x is primitive. Our result follows from Theorem 6.6 since $\Bbbk[C_p] \cong \Bbbk[t]/(t^p)$ as algebras with $t = g - 1$. Case (1) corresponds to the trivial action $x \cdot t = 0$. For Case (2), the H-action on $\Bbbk[C_p]$ is given by $x \cdot t = t + 1$ or $x \cdot g = g$, which coincides with the natural action $\rightharpoonup$ of $\Bbbk[C_p]^*$ on $\Bbbk[C_p]$. By Example 6.7, the corresponding smash product $\Bbbk[C_p]\#\Bbbk[C_p]^*$ is the Heisenberg double $\mathcal{H}(\Bbbk[C_p])$. Case (3) corresponds to the action $x \cdot t = t$ or $x \cdot g = g - 1$. □

PROPOSITION 6.9. *Let $\Bbbk$ be an algebraically closed field of characteristic $p > 0$, and $H = \Bbbk[G]$ for some elementary abelian p-group G. Then any H-cleft extension of the semisimple algebra $A = \Bbbk[x]/(x^p - x)$ is twist equivalent to either $A \otimes H$ or $\mathcal{H}(\Bbbk[C]^*) \otimes \Bbbk[G']$ for some subgroups C, G' of G such that $C \cong C_p$ and $C \oplus G' = G$, where $\Bbbk[G]$ is naturally identified with $\Bbbk[C] \otimes \Bbbk[G']$ and A is identified with $\Bbbk[C]^*$. In particular, if $G = C_p$, then any H-cleft extension of $A = H^*$ is twist equivalent to one of the smash products below:*

(1) *The tensor product algebra: $\Bbbk[C_p]^* \otimes \Bbbk[C_p]$.*
(2) *The Heisenberg double: $\mathcal{H}(\Bbbk[C_p]^*)$.*

PROOF. We write $A = \bigoplus_{i \in \mathbb{Z}_p} \Bbbk\, e_i$ where $\{e_0, \ldots, e_{p-1}\}$ is a complete set of primitive orthogonal idempotents of A and $x = \sum_{i \in \mathbb{Z}_p} i e_i$. By [**36**, Theorem 3.1], we have

$$H^2_{sw}(\Bbbk[G], A) = H^2(G, A^\times).$$

It follows from Lemma 3.1 that $H^2_{sw}(\Bbbk[G], A)$ is trivial for any G-action on A.

Now, it suffices to classify all possible $\Bbbk[G]$-actions on A up to automorphisms of A and H by Proposition 6.3. It is clear that $\mathrm{Aut}_{\Bbbk}(A)$ is in one-to-one correspondence with the permutation group S_p of $\{e_0, \ldots, e_{p-1}\}$. One can identify $\varphi \in \mathrm{Aut}_{\Bbbk}(A)$ with the permutation in S_p, again denoted by φ, via $\varphi(e_i) = e_{\varphi(i)}$. Hence any $\Bbbk[G]$-action on A is determined by a group homomorphism $f : G \to S_p$. If f is trivial, then the $\Bbbk[G]$-action on A is trivial, so the smash product is $A \otimes H$. If f is nontrivial, then $\mathrm{Im}(f)$ is a nontrivial elementary p-subgroup of S_p. Since $p^2 \nmid p!$, every nontrivial p-subgroup of S_p is a Sylow p-subgroup isomorphic to C_p. In particular, $\mathrm{Im}(f) \cong C_p$. Hence $G' = \ker(f)$ is an index p subgroup of G, and there exists a cyclic subgroup $C \cong C_p$ of G such that $G = C \oplus G'$ and $f|_C$ is injective. Since $\mathrm{Im}(f)$ is a Sylow p-subgroup of S_p, there is some automorphism φ of A (or S_p) such that $\varphi \circ f(g) \circ \varphi^{-1}$ sends every e_i to e_{i-1}, where g is a generator of C. By (6.3), we can assume that $g \cdot e_i = e_{i-1}$ for all $i \in \mathbb{Z}_p$. Hence

$$g \cdot x = \sum_{i \in \mathbb{Z}_p} i(g \cdot e_i) = \sum_{i \in \mathbb{Z}_p} (i-1)e_{i-1} + \sum_{i \in \mathbb{Z}_p} e_i = x + 1.$$

Note the above action of g on A coincides with the natural action $\rightharpoonup$ of $\Bbbk[C_p]$ on $A = \Bbbk[C]^*$. Therefore, it gives the corresponding smash product

$$A\#H \cong A\#(\Bbbk[C] \otimes \Bbbk[G']) \cong (A\#\Bbbk[C]) \otimes \Bbbk[G'] \cong \mathcal{H}(\Bbbk[C]^*) \otimes \Bbbk[G'].$$

Finally, it is clear that these smash products are not twist equivalent. $\square$

7. Short exact sequences of p-dimensional Hopf algebras

In the last section, we will provide a list of the Radford algebra $R(p)$ characterizations and identify all p^2-dimensional Hopf algebra extensions of p-dimensional ones. In particular, these extensions are all pointed.

We first recall the classification of p-dimensional Hopf algebras over an algebraically closed field $\Bbbk$ of characteristic p. There are only two nonisomorphic p-restricted Lie algebras of dimension 1 generated by x, and they are given by (i) $\mathfrak{a}_1$: $x^p = x$; (ii) $\mathfrak{a}_2$: $x^p = 0$. Note that the restricted universal enveloping algebra $u(\mathfrak{a}_1) = \Bbbk[x]/(x^p - x)$ is isomorphic to the dual group algebra $\Bbbk[C_p]^*$.

THEOREM 7.1 ([**26**, Theorem 2.1]). *Let $\Bbbk$ be an algebraically closed field of characteristic $p > 0$ and H a p-dimensional Hopf algebra over $\Bbbk$. Then $H \cong \Bbbk[C_p]$, $u(\mathfrak{a}_1)$ or $u(\mathfrak{a}_2)$.*

Independent of the classification of p^2-dimensional pointed Hopf algebras, we provide some characterizations of the Radford algebra $R(p)$ as the unique p^2-dimensional Hopf algebra over $\Bbbk$ which satisfies any of the equivalent conditions in the following theorem.

THEOREM 7.2. *Let $\Bbbk$ be an algebraically closed field of characteristic $p > 0$, and H a p^2-dimensional Hopf algebra over $\Bbbk$. The following statements are equivalent:*

(i) *H is isomorphic to the Radford algebra $R(p)$;*
(ii) *H is noncommutative, pointed and $G(H)$ is not trivial;*
(iii) *H is noncommutative, and it admits a normal Hopf subalgebra isomorphic to $\Bbbk[C_p]$;*
(iv) *H is a nontrivial extension, i.e., not isomorphic to the tensor Hopf algebra $\Bbbk[C_p] \otimes \Bbbk[C_p]^*$, that fits into the exact sequence*

$$1 \to \Bbbk[C_p] \xrightarrow{\iota} H \xrightarrow{\pi} \Bbbk[C_p]^* \to 1. \tag{7.1}$$

In particular, $R(p) \cong R(p)^*$.

PROOF. It is clear that (i) implies (ii), (iii), and (iv). It remains to show the converse implications.

(ii) $\Rightarrow$ (i) Suppose H is pointed and noncommutative with nontrivial $G(H)$. Then $G(H) \cong C_p$ since H is noncommutative and $\dim H = p^2$. Therefore, $\Bbbk[G(H)] = H_0 \subsetneq H_1$ or H admits a nontrivial $(1, a)$-primitive element x for some $a \in G(H)$.

Let g be a generator of $G(H)$. Then g, x generate a Hopf subalgebra of H of dimension $> p$. Thus, g, x generate H as an algebra by the Nichols-Zoeller theorem. Since $\dim H = p^2$, $P_{1,a}(H)$ is spanned by $x, a-1$ and is a $G(H)$-module by conjugation. Let $N = \Bbbk(a-1)$, which is a trivial $G(H)$-submodule of $P_{1,a}(H)$. Note that N could be zero if $a=1$ but $P_{1,g}(H)/N$ is a nonzero trivial $G(H)$-module. Thus $gxg^{-1} = x + \alpha(a-1)$ for some $\alpha \in \Bbbk$. If $\alpha(a-1) = 0$, then H generated by g, x would be commutative. Therefore, $\alpha(a-1) \neq 0$ which is equivalent to $\alpha \in \Bbbk^\times$ and $G(H) = \langle a \rangle$. Without loss of generality, one can assume $a = g$ and replace the $(1,g)$-primitive element x by x/α to obtain the relation $[g,x] = g(g-1)$ or $[x,g] = g - g^2$. In particular, H_0 is normal in H. Since $[x, g^j] = jg^{j-1}[x,g] = j(g^j - g^{j+1})$ for $0 \le j \le p-1$, the set of eigenvalues of $\mathrm{ad}(x)|_{H_0}$ is $\mathbb{Z}_p$ and so the characteristic polynomial of $\mathrm{ad}(x)|_{H_0}$ is $X^p - X$. Thus, $\mathrm{ad}(x^p)|_{H_0} = \mathrm{ad}(x)^p|_{H_0} = \mathrm{ad}(x)|_{H_0}$. Thus, $x^p - x$ lies in the center of H.

Let $L = H/(H_0^+H)$ and $\pi : H \to L$ be the natural surjection. We claim that $L \cong u(\mathfrak{a}_1)$. It is clear that L is generated by the nonzero primitive element $\pi(x)$, and so $L \cong u(\mathfrak{a}_i)$ for $i = 1$ or 2. If $L \cong u(\mathfrak{a}_2)$, then H is local by Lemma 2.2(iii). This implies that $x^{p^2} \in (H^+)^{p^2} = 0$, which contradicts to the fact that $\mathrm{ad}(x^{p^2}) = \mathrm{ad}(x)^{p^2} = \mathrm{ad}(x) \neq 0$ on H_0. Therefore, $L \cong u(\mathfrak{a}_1)$.

Let $\overline{x}$ be a primitive generator for L such that $\overline{x}^p = \overline{x}$. Since $\pi(x)$ is a nonzero primitive element of L, $\beta\pi(x) = \overline{x}$ for some $\beta \in \Bbbk^\times$ and $\{x^j \mid j = 0, \ldots, p-1\}$ is a $\Bbbk$-linearly independent subset in H. Define the $\Bbbk$-linear map $\gamma : L \to H$ by $\gamma(\overline{x}^i) = (\beta x)^i$ for $0 \le i \le p-1$. On can check directly that γ is a left L-comodule map with the convolution inverse $\overline{\gamma} : L \to H$ given by $\overline{\gamma}(\overline{x}^i) = (-\beta x)^i$ for $0 \le i \le p-1$. Thus H is spanned by $\{x^i g^j \mid i,j = 0, \ldots, p-1\}$ as H is a left $\Bbbk[C_p]^*$-cleft extension of $\Bbbk[C_p]$, and hence it forms a basis for H as $\dim H = p^2$.

Now, we can write $x^p - x = \sum_{0 \le j \le p-1} f_j(x) g^j$ for some unique polynomial $f_j(x)$ in x of degree $\le p-1$. We find

$$
\begin{aligned}
0 = [x, x^p - x] &= \sum_{0 \le j \le p-1} f_j(x)[x, g] = \sum_{0 \le j \le p-1} j f_j(x)(g^j - g^{j+1}) \\
&= f_1(x) g + \left(\sum_{j=1}^{p-2} ((j+1) f_{j+1}(x) - j f_j(x)) g^{j+1} \right) + f_{p-1}(x).
\end{aligned}
$$

This implies $f_j(x) = 0$ for all $j = 1, \ldots, p-1$. Thus $x^p - x = f_0(x)$. Applying π to this equality, we find

$$
(\beta^{-p} - \beta^{-1})\overline{x} = \beta^{-p}\overline{x}^p - \beta^{-1}\overline{x} = \pi(x^p - x) = \pi(f_0(x)) = f_0(\beta^{-1}\overline{x}),
$$

or equivalently, $h(\overline{x}) = 0$ where $h(X) = f_0(\beta^{-1}X) - (\beta^{-p} - \beta^{-1})X \in \Bbbk[X]$ has degree $\le p-1$. Thus, $h(X) = 0$ or $f_0(X) = (\beta^{(1-p)} - 1)X$. Now, we find

$$
0 = [x^p - x, g] = [(\beta^{(1-p)} - 1)x, g] = (\beta^{(1-p)} - 1) g(1-g)
$$

which forces $\beta^{(1-p)} - 1 = 0$. Hence $f_0(x) = 0$ or $x^p - x = 0$ in H. Therefore, H is isomorphic to the Radford algebra $R(p)$.

(iii) $\Rightarrow$ (ii) Suppose H is noncommutative and admits a normal Hopf subalgebra $K \cong \Bbbk[C_p]$. In view of Theorem 3.4, it suffices to show that H^* is nonsemisimple. We have an exact sequence of Hopf algebras:

$$
1 \to L^* \to H^* \to K^* \to 1, \quad \text{where } L = H/K^+H\,.
$$

Suppose H^* is semisimple. Then so is L^*. It follows from Theorem 7.1 that $L^* \cong u(\mathfrak{a}_1)$. Since K^* and L^* are connected, so is H^* by Lemma 2.2(ii). However,

this implies $H \cong \Bbbk[C_p \times C_p]$ or $\Bbbk[C_{p^2}]$ by [**20**, Theorem 0.1]. This contradicts the noncommutativity of H.
(iv) $\Rightarrow$ (iii) Let H be a commutative Hopf algebra which fits into the exact sequence (7.1). It suffices to show that H is a trivial extension. Note that H^* satisfies the same exact sequence. Therefore, both H and H^* are nonsemisimple, so H and H^* are pointed by Theorem 3.4. Then H^* is also commutative; otherwise, $H^* \cong R(p)$ as Hopf algebras by (ii), which would imply H is noncommutative. Since $H_0 \subsetneq H_1$, there exists a nontrivial primitive element $x \in H$ and $\dim P(H) = 1$. It is clear that $\Bbbk[x]$ maps surjectively onto $\Bbbk[C_p]^*$ and hence $\Bbbk[x] \cong \Bbbk[C_p]^*$ as Hopf algebras. Since H is generated by $G(H)$ and x as an algebra, $H \cong \Bbbk[G(H)] \otimes \Bbbk[x] \cong \Bbbk[C_p] \otimes \Bbbk[C_p]^*$ as Hopf algebras.

The last assertion follows directly from the equivalence of (i) and (iv). □

Recently, Gelaki considered in [**10**, Ex. 6.14] the Drinfeld twist

$$J = \sum_{i=0}^{p-1} \frac{x(x-1)\cdots(x-i+1) \otimes y^i}{i!} \tag{7.2}$$

of the restricted universal enveloping algebra $u(\mathfrak{g}_5)$ of the 2-dimensional nonabelian p-restricted Lie algebra $\mathfrak{g}_5$ spanned by x, y such that $[x, y] = y$, $x^p = x$ and $y^p = 0$. It is pointed out that the corresponding twisted Hopf algebra $u(\mathfrak{g}_5)^J$ is noncommutative and noncocommutative of dimension p^2 in characteristic p. Now, we show that this noncommutative and noncocommutative Hopf algebra is isomorphic to the Radford algebra $R(p)$ due to the preceding theorem. The readers are referred to [**17**] for the notions of a Drinfeld twist J of a Hopf algebra H, the corresponding twisted Hopf algebra H^J, and gauge equivalence of Hopf algebras.

COROLLARY 7.3. *Let $\Bbbk$ be an algebraically closed field of characteristic $p > 0$ and $\mathfrak{g}_5$ the 2-dimensional nonabelian p-restricted Lie algebra over $\Bbbk$ described above. For any Drinfeld twist J of $u(\mathfrak{g}_5)$ such that $u(\mathfrak{g}_5)^J$ is noncocommutative, $u(\mathfrak{g}_5)^J \cong R(p)$ as Hopf algebras. In particular, $u(\mathfrak{g}_5)$ and $R(p)$ are gauge equivalent.*

PROOF. Note that $u(\mathfrak{g}_5) = \Bbbk\langle x, y\rangle/(x^p - x, y^p, [x, y] - y)$ as Hopf algebras where x, y are primitive. Thus $(y) = u(\mathfrak{g}_5)y$ is a nilpotent ideal of $u(\mathfrak{g}_5)$ and $u(\mathfrak{g}_5)/(y) \cong \Bbbk[x]/(x^p - x)$ is semisimple and commutative. Therefore, (y) is the Jacobson radical of $u(\mathfrak{g}_5)$ and $u(\mathfrak{g}_5)^*$ is pointed with $|G(u(\mathfrak{g}_5)^*)| = p$. Let J be a Drinfeld twist of $u(\mathfrak{g}_5)$ such that its associated twisted Hopf algebra $H = u(\mathfrak{g}_5)^J$ is noncocommutative or H^* is noncommutative. Since $H \cong u(\mathfrak{g}_5)$ as algebras, $H^* = u(\mathfrak{g}_5)^*$ as coalgebras. Therefore, H^* is pointed with $|G(H^*)| = p$. Now, it follows from Theorem 7.2(ii) that $H^* \cong R(p)$, and so $H \cong R(p)$ as $R(p)$ is self-dual. The last assertion follows from the existence of such a Drinfeld twist J given by (7.2) according to [**10**, Ex. 6.14]. □

REMARK 7.4. The Radford algebra first appeared in the work of Radford as an example of pointed Hopf algebra of dimension p^2 with antipode of order $2p$ (see [**31**]). Taft pointed out that it is the smash product of the commutative and cocommutative Hopf algebras $\Bbbk[C_p]$ and $\Bbbk[C_p]^*$, and its antipode is not semisimple. Theorem 7.2 provides several characterizations of the Radford algebra. In particular, it is the unique nontrivial extension of $\Bbbk[C_p]^*$ by $\Bbbk[C_p]$.

Finally, we would like to classify all the extensions of p-dimensional Hopf algebras using the classification theorem of p^2-dimensional Hopf algebras. Recall that there are five nonisomorphic 2-dimensional p-restricted Lie algebras $\mathfrak{g}_i$ with basis elements x, y (cf. [**41**, Appendix A]). Four of them are abelian, and they are given by

$$\mathfrak{g}_1 : x^p = x, y^p = y;\ \mathfrak{g}_2 : x^p = x, y^p = 0;\ \mathfrak{g}_3 : x^p = y, y^p = 0;\ \mathfrak{g}_4 : x^p = 0, y^p = 0.$$

The p-restricted Lie algebra $\mathfrak{g}_5$ introduced in the remark preceding Corollary 7.3 is the unique 2-dimensional nonabelian p-restricted Lie algebra up to isomorphism.

THEOREM 7.5 ([**41**, Theorem 7.4][**39**, Theorem 3.3]). *Let $\Bbbk$ be an algebraically closed field of characteristic $p > 0$, and H a pointed Hopf algebra of dimension p^2. Then H is isomorphic to one of the following 14 nonisomorphic Hopf algebras:*

(1) *the group algebra $\Bbbk[G]$, where $G \cong C_{p^2}$ or $C_p \times C_p$;*
(2) *the dual group algebra $\Bbbk[C_{p^2}]^*$;*
(3) *the restricted universal enveloping algebra $u(\mathfrak{g}_i)$ with $1 \le i \le 5$;*
(4) *the dual restricted universal enveloping algebra $u(\mathfrak{g}_i)^*$ with $i = 3, 5$;*
(5) *the tensor product Hopf algebras $\Bbbk[C_p] \otimes u(\mathfrak{a}_1)$ and $\Bbbk[C_p] \otimes u(\mathfrak{a}_2)$;*
(6) *the Radford algebra $R(p)$;*
(7) *the connected and local Hopf algebra $\mathcal{G}_p(2) := \Bbbk[x]/(x^{p^2})$ with $\Delta(x) = x \otimes 1 + 1 \otimes x + \sum_{i=1}^{p-1} \frac{x^{pi}}{i!} \otimes \frac{x^{p(p-i)}}{(p-i)!}$.*

REMARK 7.6. There are relationships between these Hopf algebras listed in Theorem 7.5.

(i) We have the isomorphisms of Hopf algebras $u(\mathfrak{g}_1)^* \cong \Bbbk[C_p \times C_p]$ and $u(\mathfrak{g}_2)^* \cong \Bbbk[C_p] \otimes u(\mathfrak{a}_2)$.
(ii) The Hopf algebras $u(\mathfrak{g}_4)$, $\Bbbk[C_p] \otimes u(\mathfrak{a}_1)$, $R(p)$ and $\mathcal{G}_p(2)$ are self-dual.
(iii) The Radford algebra $R(p)$ has a $\Bbbk$-basis $\{x^i g^j\}_{0 \le i,j \le p-1}$. Let $\gamma, \chi \in R(p)^*$ defined by $\gamma(x^i g^j) = (-1)^i$, $\chi(x^i g^j) = -j(-1)^i$. One can check directly that γ is a character of $R(p)$, and χ is (ϵ, γ)-primitive which satisfy $\gamma^p = 1, \chi^p = \chi$ and $[\gamma, \chi] = \gamma(\gamma - 1)$. The assignment $g \mapsto \gamma, x \mapsto \chi$ extends to a Hopf algebra isomorphism from $R(p)$ to $R(p)^*$.
(iv) Let $\{\delta_i\}_{i=0}^{p^2-1}$ denote the dual basis of $\{x_i\}_{i=0}^{p^2-1}$ for $(\Bbbk[x]/(x^{p^2}))^*$. One can check directly that $(\delta_p)^p = \delta_1$ and $\Delta(\delta_p) = \delta_p \otimes 1 + 1 \otimes \delta_p + \sum_{i=1}^{p-1} \frac{(\delta_1)^i}{i!} \otimes \frac{(\delta_1)^{p-i}}{(p-i)!}$. Therefore, $\mathcal{G}_2(p)^* = \Bbbk[\delta_p]$ and the map $x \mapsto \delta_p$ extends to a Hopf algebra isomorphism from $\mathcal{G}_p(2)$ and $\mathcal{G}_p(2)^*$. Note that $\mathcal{G}_p(2)$ is not primitively generated, and its comultiplication of x appears in the 2-step iterated Hopf-Ore extensions (IHOEs) in positive characteristic [**2**].
(v) The Hopf algebras $R(p)$ and $u(\mathfrak{g}_5)$ are gauge equivalent by Corollary 7.3.

The following corollary is an immediate consequence of the above remark.

COROLLARY 7.7. *The pointed Hopf algebras over $\Bbbk$ of dimension p^2 are closed under duality.*

Now, we can provide an affirmative answer to Question B for those p^2-dimensional Hopf algebras which are extensions of p-dimensional ones.

THEOREM 7.8. *Let H be a p^2-dimensional Hopf algebra over an algebraically closed field $\Bbbk$ of characteristic $p > 0$. Then H admits a normal Hopf subalgebra of dimension p if and only if H is pointed. In particular, every pointed p^2-dimensional Hopf algebra is an extension of p-dimensional ones.*

PROOF. If H is pointed, it follows from the classification Theorem 7.5 that H has a normal Hopf subalgebra of dimension p. Conversely, suppose H has a normal Hopf subalgebra K of dimension p. It suffices to show that H or H^* is pointed by Corollary 7.7. Now, H fits into a short exact sequence of Hopf algebras

$$1 \to K \to H \to L \to 1,$$

where $L = H/K^+H$ is another p-dimensional Hopf algebra by the Nichols-Zoeller theorem. By Theorem 7.1, $L, K \cong \Bbbk[C_p]$, $u(\mathfrak{a}_1)$ or $u(\mathfrak{a}_2)$. If $L \cong u(\mathfrak{a}_i)$, then L is connected and so H is pointed by Lemma 2.2(ii). On the other hand, if $K \cong \Bbbk[C_p]$ or $u(\mathfrak{a}_2)$, then K is local and so H^* is pointed by Lemma 2.2(i). The remaining case is that $K \cong u(\mathfrak{a}_1)$ and $L \cong \Bbbk[C_p]$. In particular, H is an L-cleft extension of K. It follows from Proposition 6.9 that $H \cong K \otimes L$ as $\Bbbk$-algebras, which is commutative. Thus, H^* is pointed. □

As a conclusion of our section, we list all possible nontrivial exact sequences of Hopf algebras

$$1 \to K \to H \to L \to 1,$$

where K and L are p-dimensional Hopf algebras over $\Bbbk$.

TABLE 1. Nontrivial exact sequences $1 \to K \to H \to L \to 1$

K	H	L
$\Bbbk[C_p]$	$\Bbbk[C_{p^2}]$	$\Bbbk[C_p]$
$\Bbbk[C_p]$	$R(p)$	$u(\mathfrak{a}_1)$
$\Bbbk[C_p]$	$u(\mathfrak{g}_5)^*$	$u(\mathfrak{a}_2)$
$u(\mathfrak{a}_1)$	None	$\Bbbk[C_p]$
$u(\mathfrak{a}_1)$	$\Bbbk[C_{p^2}]^*$	$u(\mathfrak{a}_1)$
$u(\mathfrak{a}_1)$	None	$u(\mathfrak{a}_2)$
$u(\mathfrak{a}_2)$	None	$\Bbbk[C_p]$
$u(\mathfrak{a}_2)$	$u(\mathfrak{g}_5)$	$u(\mathfrak{a}_1)$
$u(\mathfrak{a}_2)$	$u(\mathfrak{g}_3)$, $u(\mathfrak{g}_3)^*$, $\mathcal{G}_p(2)$	$u(\mathfrak{a}_2)$

REMARK 7.9. The absence of a nontrivial extension of $\Bbbk[C_p]$ by $\Bbbk[C_p]^*$ can be proven without using Theorem 7.5. If H is an extension, then H^* can also fit into that same extension. By Proposition 6.9, both H and H^* are the tensor product algebra $\Bbbk[C_p]^* \otimes \Bbbk[C_p]$. In particular, H admits a nonzero primitive element x and a group-like element g of order p. Since H is generated by the commuting elements g, x, $H \cong \Bbbk[x] \otimes \Bbbk[C_p]$ as Hopf algebras. Since $\dim P(H) = 1$ and $\Bbbk[C_p]^*$ admits a nonzero primitive element $x \in \Bbbk[C_p]^*$, we have $\Bbbk[x] \cong \Bbbk[C_p]^*$.

References

[1] F. W. Anderson and K. R. Fuller, *Rings and categories of modules*, 2nd ed., Graduate Texts in Mathematics, vol. 13, Springer-Verlag, New York, 1992, DOI 10.1007/978-1-4612-4418-9. MR1245487

[2] K. A. Brown and J. J. Zhang, *Iterated Hopf Ore extensions in positive characteristic*, J. Noncommut. Geom. **16** (2022), no. 3, 787–837, DOI 10.4171/jncg/453. MR4506529

[3] C. Kassel, *Quantum groups*, Graduate Texts in Mathematics, vol. 155, Springer-Verlag, New York, 1995, DOI 10.1007/978-1-4612-0783-2. MR1321145

[4] C. W. Curtis and I. Reiner, *Representation theory of finite groups and associative algebras*, Pure and Applied Mathematics, Vol. XI, Interscience Publishers (a division of John Wiley & Sons, Inc.), New York-London, 1962. MR144979

[5] P. Etingof, *Frobenius-Perron dimensions of integral $\mathbb{Z}_+$-rings and applications*, Algebr. Represent. Theory **23** (2020), no. 5, 2059–2078, DOI 10.1007/s10468-019-09924-1. MR4140065

[6] P. Etingof and S. Gelaki, *On finite-dimensional semisimple and cosemisimple Hopf algebras in positive characteristic*, Internat. Math. Res. Notices **16** (1998), 851–864, DOI 10.1155/S107379289800052X. MR1643702

[7] P. Etingof and S. Gelaki, *On Hopf algebras of dimension pq*, J. Algebra **277** (2004), no. 2, 668–674, DOI 10.1016/j.jalgebra.2003.06.003. MR2067625

[8] P. Etingof and S. Gelaki, *Finite symmetric tensor categories with the Chevalley property in characteristic 2*, J. Algebra Appl. **20** (2021), no. 1, Paper No. 2140010, 18, DOI 10.1142/S0219498821400107. MR4209964

[9] P. Etingof, S. Gelaki, D. Nikshych, and V. Ostrik, *Tensor categories*, Mathematical Surveys and Monographs, vol. 205, American Mathematical Society, Providence, RI, 2015, DOI 10.1090/surv/205. MR3242743

[10] S. Gelaki, *Module categories over affine group schemes*, Quantum Topol. **6** (2015), no. 1, 1–37, DOI 10.4171/QT/59. MR3335005

[11] K. R. Goodearl and J. J. Zhang, *Noetherian Hopf algebra domains of Gelfand-Kirillov dimension two*, J. Algebra **324** (2010), no. 11, 3131–3168, DOI 10.1016/j.jalgebra.2009.11.001. MR2732991

[12] P. Guillot, *Examples of Sweedler cohomology in positive characteristic*, Comm. Algebra **43** (2015), no. 5, 2174–2200, DOI 10.1080/00927872.2014.889146. MR3316844

[13] G. Hochschild, *Representations of restricted Lie algebras of characteristic p*, Proc. Amer. Math. Soc. **5** (1954), 603–605, DOI 10.2307/2032042. MR66361

[14] N. Jacobson, *Lie algebras*, Interscience Tracts in Pure and Applied Mathematics, No. 10, Interscience Publishers (a division of John Wiley & Sons, Inc.), New York-London, 1962. MR143793

[15] R. G. Larson and M. E. Sweedler, *An associative orthogonal bilinear form for Hopf algebras*, Amer. J. Math. **91** (1969), 75–94, DOI 10.2307/2373270. MR240169

[16] M. Hilgemann and S.-H. Ng, *Hopf algebras of dimension $2p^2$*, J. Lond. Math. Soc. (2) **80** (2009), no. 2, 295–310, DOI 10.1112/jlms/jdp026. MR2545253

[17] Y. Kashina, S. Montgomery, and S.-H. Ng, *On the trace of the antipode and higher indicators*, Israel J. Math. **188** (2012), 57–89, DOI 10.1007/s11856-011-0092-7. MR2897723

[18] A. Masuoka, *Coideal subalgebras in finite Hopf algebras*, J. Algebra **163** (1994), no. 3, 819–831, DOI 10.1006/jabr.1994.1047. MR1265867

[19] A. Masuoka, *The p^n theorem for semisimple Hopf algebras*, Proc. Amer. Math. Soc. **124** (1996), no. 3, 735–737, DOI 10.1090/S0002-9939-96-03147-4. MR1301036

[20] A. Masuoka, *Semisimplicity criteria for irreducible Hopf algebras in positive characteristic*, Proc. Amer. Math. Soc. **137** (2009), no. 6, 1925–1932, DOI 10.1090/S0002-9939-09-09863-3. MR2480272

[21] J. P. May, *Some remarks on the structure of Hopf algebras*, Proc. Amer. Math. Soc. **23** (1969), 708–713, DOI 10.2307/2036615. MR246938

[22] S. Natale, *Hopf algebras of dimension 12*, Algebr. Represent. Theory **5** (2002), no. 5, 445–455, DOI 10.1023/A:1020504123567. MR1935855

[23] W. D. Nichols and M. B. Zoeller, *A Hopf algebra freeness theorem*, Amer. J. Math. **111** (1989), no. 2, 381–385, DOI 10.2307/2374514. MR987762

[24] S.-H. Ng, *Non-semisimple Hopf algebras of dimension p^2*, J. Algebra **255** (2002), no. 1, 182–197, DOI 10.1016/S0021-8693(02)00139-4. MR1935042

[25] S.-H. Ng, *Hopf algebras of dimension pq. II*, J. Algebra **319** (2008), no. 7, 2772–2788, DOI 10.1016/j.jalgebra.2007.08.003. MR2397407

[26] S.-H. Ng and X. Wang, *Hopf algebras of prime dimension in positive characteristic*, Bull. Lond. Math. Soc. **51** (2019), no. 3, 459–465, DOI 10.1112/blms.12242. MR3964500

[27] S. Montgomery, *Hopf algebras and their actions on rings*, CBMS Regional Conference Series in Mathematics, vol. 82, Published for the Conference Board of the Mathematical Sciences, Washington, DC; by the American Mathematical Society, Providence, RI, 1993, DOI 10.1090/cbms/082. MR1243637

[28] S. Montgomery, *Classifying finite-dimensional semisimple Hopf algebras*, Trends in the representation theory of finite-dimensional algebras (Seattle, WA, 1997), Contemp. Math., vol. 229, Amer. Math. Soc., Providence, RI, 1998, pp. 265–279, DOI 10.1090/conm/229/03323. MR1676225

[29] R. S. Pierce, *Associative algebras*, Studies in the History of Modern Science, vol. 9, Springer-Verlag, New York-Berlin, 1982. Graduate Texts in Mathematics, 88. MR674652

[30] D. E. Radford, *The order of the antipode of a finite dimensional Hopf algebra is finite*, Amer. J. Math. **98** (1976), no. 2, 333–355, DOI 10.2307/2373888. MR407069

[31] D. E. Radford, *Operators on Hopf algebras*, Amer. J. Math. **99** (1977), no. 1, 139–158, DOI 10.2307/2374012. MR437581

[32] D. E. Radford and H.-J. Schneider, *On the even powers of the antipode of a finite-dimensional Hopf algebra*, J. Algebra **251** (2002), no. 1, 185–212, DOI 10.1006/jabr.2001.9127. MR1900281

[33] H.-J. Schneider, *Normal basis and transitivity of crossed products for Hopf algebras*, J. Algebra **152** (1992), no. 2, 289–312, DOI 10.1016/0021-8693(92)90034-J. MR1194305

[34] D. Ştefan, *Hopf subalgebras of pointed Hopf algebras and applications*, Proc. Amer. Math. Soc. **125** (1997), no. 11, 3191–3193, DOI 10.1090/S0002-9939-97-04143-9. MR1423335

[35] D. Ştefan, *Hopf algebras of low dimension*, J. Algebra **211** (1999), no. 1, 343–361, DOI 10.1006/jabr.1998.7602. MR1656583

[36] M. E. Sweedler, *Cohomology of algebras over Hopf algebras*, Trans. Amer. Math. Soc. **133** (1968), 205–239, DOI 10.2307/1994939. MR224684

[37] M. E. Sweedler, *Hopf algebras*, Mathematics Lecture Note Series, W. A. Benjamin, Inc., New York, 1969. MR252485

[38] E. J. Taft, *The order of the antipode of finite-dimensional Hopf algebra*, Proc. Nat. Acad. Sci. U.S.A. **68** (1971), 2631–2633, DOI 10.1073/pnas.68.11.2631. MR286868

[39] L. Wang and X. Wang, *Classification of pointed Hopf algebras of dimension p^2 over any algebraically closed field*, Algebr. Represent. Theory **17** (2014), no. 4, 1267–1276, DOI 10.1007/s10468-013-9446-7. MR3228487

[40] X. Wang, *Locality criteria for cocommutative Hopf algebras*, Comm. Algebra **42** (2014), no. 12, 5180–5191, DOI 10.1080/00927872.2013.833213. MR3223634

[41] X. Wang, *Connected Hopf algebras of dimension p^2*, J. Algebra **391** (2013), 93–113, DOI 10.1016/j.jalgebra.2013.06.008. MR3081624

[42] C. A. Weibel, *An introduction to homological algebra*, Cambridge Studies in Advanced Mathematics, vol. 38, Cambridge University Press, Cambridge, 1994, DOI 10.1017/CBO9781139644136. MR1269324

[43] Y. Zhu, *Hopf algebras of prime dimension*, Internat. Math. Res. Notices **1** (1994), 53–59, DOI 10.1155/S1073792894000073. MR1255253

Department of Mathematics, Louisiana State University, Baton Rouge, Louisiana 70803

Email address: `rng@math.lsu.edu`

Department of Mathematics, Louisiana State University, Baton Rouge, Louisiana 70803

Email address: `xingtingwang@math.lsu.edu`

Contemporary Mathematics
Volume **814**, 2025
https://doi.org/10.1090/conm/814/16305

Categorical Morita equivalence of pointed fusion categories, and subgroups

Peter Schauenburg

ABSTRACT. Work of Naidu (Comm. Algebra **35** (2007), no. 11, 3544–3565) answers completely the question as to when two pointed fusion categories $\mathcal{C}(G,\omega)$ and $\mathcal{C}(G',\omega')$ are categorically Morita equivalent. The constructions that relate the groups G, G' and three-cocycles ω, ω' can be rather complicated to handle in concrete examples, however. We discuss simple necessary criteria for Morita equivalence based on the subgroups of G and G' and the restricted cocycles. While neither the groups, nor the subgroup lattices, nor the cocycles need to coincide, we find that every fusion subcategory of $\mathcal{C}(G,\omega)$ has a Morita equivalent copy related to $\mathcal{C}(G',\omega')$, albeit not as a subcategory but a subquotient. If the subcategory corresponds to a Sylow subgroup, it is in fact Morita equivalent to a subcategory in $\mathcal{C}(G',\omega')$.

1. Introduction

As part of any effort to classify fusion categories, it is of course a vital ingredient to distinguish fusion categories. The present paper deals with a special case of distinguishing categories up to categorical Morita equivalence, with techniques limited to group-theoretical categories. By definition, a group-theoretical fusion category is one that is Morita equivalent to a pointed fusion category. Thus being able to distinguish pointed fusion categories up to Morita equivalence amounts to the same thing as being able to do the same for group-theoretical categories (provided one knows a pointed category Morita equivalent to the group-theoretical category in question). In the main example for our new techniques we will treat, the problem solved in [**5**] was to distinguish a certain collection of group theoretical fusion categories up to equivalence of fusion categories. We will prove the stronger statement that the categories in question are even *Morita inequivalent.* It will turn out that this a priori harder task is easier to achieve.

One possible technique to distinguish categories is to compute numerical invariants of the categories or specific objects in them, and to observe that the invariants take distinct values. The invariants used in [**5**] are the Frobenius-Schur indicators of individual objects, as well as the group of invertible objects (an algebraic rather than numerical invariant). The calculations pertain to certain group-theoretical fusion categories $\mathcal{C}(G,\omega,H,1)$ where G is a finite group, ω a $\mathbb{C}^\times$-valued three-cocycle

2020 *Mathematics Subject Classification.* 18M20.

on G, and H a subgroup such that the restriction of ω to H is trivial (the coboundary of the last parameter 1). For these categories formulas for the indicators [**7**] as well as descriptions of the group of invertible objects [**3**] are available. Part of the difficulty of the calculations lies in the fact that the explicit formulas used rely on the knowledge of ω as an "explicit cocycle", in other words a cocycle on the bar resolution of the group G, or a cocycle in the sense of a map $G^3 \to \mathbb{C}^\times$ satisfying a cocycle condition. In order to build a group-theoretical category $\mathcal{C}(G, \omega, H, \theta)$ the restriction of ω to H^3 has to be cohomologically trivial, and the knowledge of an explicit cochain θ on H whose coboundary is that restriction is also necessary to do the actual calculation (which may begin by modifying ω so as to allow $\theta = 1$). In fact the use of "explicit" cocycles is rather typical for situations where group cohomology plays into constructions in the theory of quantum groups or tensor category theory. For example, the entire question as to when two pointed fusion categories are Morita equivalent is settled [**6**] in terms of calculations based on such cocycles. The most efficient and elegant techniques from group cohomology theory, on the other hand, often give information on the abstract cohomology groups, as well as maps between such abstract cohomology groups. Powerful theorems and arguments are available for this which rarely refer to the bar resolution, or any resolution at all. Even if one refers to the powerful computational possibilities of GAP/HAP to calculate cohomology information without using elegant theoretical methods by hand, finding an explicit cocycle and dealing with it is an additional cumbersome step.

The main theoretical result of the present paper is a statement on the relation between the lattices of fusion subcategories of two Morita equivalent pointed fusion categories. Note however that if two pointed fusion categories $\mathcal{C}(G, \omega)$ and $\mathcal{C}(G', \omega')$ are Morita equivalent, the lattices of subgroups of G and G' can be different. What can be said is a lot weaker than a direct correspondence, and we doubt that it is of great theoretical interest. Instead, we view Theorem 4.1 as a possibility to sidestep explicit cocycle calculations and to exclude a Morita equivalence by considerations that pertain instead to the subgroup structure of the groups involved, and information on the cohomology group level. We will illustrate how to apply this with respect to one subgroup in Section 3, then with respect to two subgroups at once in Section 4. Finally we prove the most catchy consequence of our results on general subgroups in Section 5: A Morita equivalence between $\mathcal{C}(G, \omega)$ and $\mathcal{C}(G', \omega')$ is inherited by every Sylow subgroup of G (and the corresponding Sylow subgroup of G'). This is combined with the example calculations of the preceding sections to treat the example from [**5**] as well as give information on a larger type of example suggested but not treated there. While the result on Sylow subgroups has a particularly clean statement (The Morita equivalence classes of the subcategories associated to Sylow subgroups are categorical Morita invariants of a pointed fusion category), the weaker and less easily stated general results are still useful: In particular they are used to treat the question of Morita equivalence between the Sylow subgroup categories themselves in our example.

2. Preliminaries

Throughout the paper we deal with pointed fusion categories defined over the complex numbers; for general terminology we refer to [**2**].

A fusion subcategory is always meant to be a full replete abelian subcategory closed under duality and tensor products.

We denote by $\mathcal{C}(G,\omega)$ the pointed fusion category associated to a finite group G and a three-cocycle $\omega \in\in H^3(G,\mathbb{C}^\times)$, the double element symbol meaning that ω is an element (representative) of a cohomology class in the indicated cohomology group. Thus $\mathcal{C}(G,\omega)$ is the category of finite dimensional complex vector spaces graded over G with associativity constraint given by ω. By $\mathcal{C}(G,\omega,H,\theta)$ we denote the group-theoretical fusion category determined by G and ω as before together with a subgroup $H \leq G$ and a two-cochain $\theta\colon H^2 \to \mathbb{C}^\times$ such that $d\theta = \omega|_{H^3}$. This is the dual category with respect to the left $\mathcal{C}(G,\omega)$-module category of right modules over a twisted group algebra $\mathbb{C}_\theta[H]$ in the tensor category $\mathcal{C}(G,\omega)$. For our purposes the following description is more handy: $\mathcal{C}(G,\omega,H,\theta)$ is the category ${}_{\mathbb{C}_\theta[H]}\mathcal{C}(G,\omega)_{\mathbb{C}_\theta[H]}$ of bimodules inside $\mathcal{C}(G,\omega)$ over the twisted group ring. Without changing the category (up to fusion category equivalence) one can always replace ω by a cohomologous cocycle and θ by the trivial cochain 1.

As a very special case of the general theory of categorical Morita equivalence a fusion category $\mathcal{D}$ is Morita equivalent to $\mathcal{C}(G,\omega)$ iff there exist a subgroup $H \leq G$ and θ as above such that $\mathcal{D} \cong \mathcal{C}(G,\omega,H,\theta)$. Equivalently there is ω' cohomologous to ω and $H \leq G$ such that $\mathcal{D} \cong \mathcal{C}(G,\omega',H,1)$.

3. Morita equivalence and subgroups

In this section we consider the simplest instance of the general question treated in the present paper: If two pointed fusion categories are Morita equivalent, what can be said about the fusion subcategories associated to subgroups? It is too much to ask that said subcategories should also turn out to be Morita equivalent. In fact there is not even a one-to-one correspondence between subgroups in the two groups underlying the categories in consideration. The following result is a corollary or rather partial statement of a more involved version to be seen in Section 4.

THEOREM 3.1. *Let $\mathcal{C},\mathcal{C}'$ be categorically Morita equivalent pointed fusion categories. If $\mathcal{D} \subset \mathcal{C}$ is a fusion subcategory, then there is a fusion subcategory $\mathcal{D}' \subset \mathcal{C}'$ and an essentially surjective tensor functor $\mathcal{D}' \to \tilde{\mathcal{D}}$ with $\tilde{\mathcal{D}}$ a pointed fusion category Morita equivalent to $\mathcal{D}$.*

Equivalently, consider finite groups G and G' with three-cocycles $\omega \in\in H^3(G,\mathbb{C}^\times)$ and $\omega' \in\in H^3(G,\mathbb{C}^\times)$ such that $\mathcal{C}(G,\omega)$ and $\mathcal{C}(G',\omega')$ are categorically Morita equivalent. Then for any subgroup $H \subset G$ there exists a subgroup $H' \subset G'$, a group epimorphism $H' \to \tilde{H}$ and $\tilde{\omega} \in\in H^3(\tilde{H},\mathbb{C}^\times)$ such that $\operatorname{res}^{G'}_{H'}[\omega'] = \inf^{H'}_{\tilde{H}}[\tilde{\omega}]$, and $\mathcal{C}(\tilde{H},\tilde{\omega})$ and $\mathcal{C}(H,\operatorname{res}^G_H \omega)$ are categorically Morita equivalent.

The theorem gives a necessary condition for Morita equivalence, albeit perhaps not a particularly strong one, since not the subgroup H' but only an epimorphic image is Morita equivalent to H. Consider for example the semidirect product $G = (C_2 \times C_2) \rtimes C_2 \cong D_8$ (where the action is by swapping factors). It is known that $\mathcal{C}(G,1)$ and $\mathcal{C}(G,\omega)$ are not Morita equivalent if ω is inflated from the nontrivial cocycle on the acting copy of C_2. To see this one can consult [**4**] where nonabelian doubles of groups of order 8 are classified using in part tabulated numerical invariants. One can also distinguish the categories "by hand" using second indicators in the Drinfeld center. It is tempting to use Theorem 3.1 on the acting subgroup $C_2 \subset G$ considered in $\mathcal{C}(G,\omega)$. It is not Morita equivalent to any pointed fusion category with trivial cocycle, but unfortunately C_2 is an epimorphic image of C_4, which of course occurs as a subgroup of G, and the inflated cocycle from C_2 to C_4 is cohomologically trivial.

Nonetheless Theorem 3.1 can be made to work on (a generalization of) this example. We do not claim that this is a particularly clever or efficient way to treat this example, but work it out to demonstrate that our result *can* be used to distinguish Morita inequivalent pointed fusion categories despite its apparent weakness.

LEMMA 3.2. *Let p be a prime and $q = p^r$. Consider $V = \mathbb{F}_q^2$ with the natural action of group*

$$K = \begin{pmatrix} 1 & \mathbb{F}_q \\ 0 & 1 \end{pmatrix}.$$

Then K also acts componentwise on V^n and we consider $G = V^n \rtimes K$ with the projection $\pi_K \colon G \to K$. We denote

$$E = \begin{pmatrix} \mathbb{F}_q \\ 0 \end{pmatrix} \subset V$$

the subspace of invariant elements under the action of K.

Consider a subgroup $H' \subset G$ and an epimorphism $f \colon H' \to \mathbb{F}_q^m$ with $m \geq n+1$.

1. *If $H' \not\subset \operatorname{Ker}(\pi) = V^n$, then $m = n+1$ and $E^n \subset H'$. If moreover $p = 2$, then $E^n \times K \subset H'$ and $H' = \operatorname{Ker}(\pi|_{H'}) \rtimes K$.*
2. *For every group epimorphism $\nu \colon \mathbb{F}_q^m \to \mathbb{F}_q$ there is a group epimorphism $\theta \colon \mathbb{F}_q \to \mathbb{F}_p$ such that $\theta\nu f$ splits.*

PROOF. In what follows, dimension is the dimension of $\mathbb{F}_p$-vector spaces. We consider V^n and K as subgroups of G.

For the first part, assume that $H' \not\subset \operatorname{Ker}(\pi)$; put $W := H' \cap V^n$. We note that for any $a = (v, \alpha) \in H' \setminus \operatorname{Ker}(\pi)$, conjugation by a acts like α on V^n, hence W is invariant under α. With $\tilde{\alpha} := \alpha - 1$, we have $E^n = \operatorname{Ker}(\tilde{\alpha}) = \operatorname{Im}(\tilde{\alpha})$, and since the target of f is abelian, $f(axa^{-1}) = f(x)$, and so $f\alpha = f$ on W, or $\operatorname{Im}(\tilde{\alpha}|_W) \subset \operatorname{Ker}(f|_W)$. Let $B \subset H'$ be a set mapped bijectively by π onto a $\mathbb{F}_p$-basis of $\pi(H')$. Then $\langle W \cup B \rangle = H'$, and so $\langle f(W) \cup f(B) \rangle = \mathbb{F}_q^m$. With $b := \dim f(B) \leq r$ we conclude that $\dim f(W) \geq rm - b$. Now for any α as above,

$$\begin{aligned} rm - r \leq rm - b \leq \dim f(W) & \\ &= \dim W - \dim \operatorname{Ker}(f|_W) \\ &\leq \dim W - \dim \operatorname{Im}(\tilde{\alpha}|_W) \\ &= \dim \operatorname{Ker}(\tilde{\alpha}|_W) \\ &= \dim(W \cap E^n) \leq rn \leq rm - r. \end{aligned}$$

Thus we have equality overall, in particular $b = r$, $n = m - 1$, $E^n \subset W$ and $\operatorname{Ker}(f|_W) = \operatorname{Im}(\tilde{\alpha}|_W)$.

Now assume $p = 2$, and let $\alpha \in K$. There is $v \in W$ such that $a = (v, \alpha) \in H'$. Since $(\tilde{\alpha}v, e) = a^2 \in \operatorname{Ker}(f)$, we have $\tilde{\alpha}v \in \operatorname{Ker}(f|W) = \operatorname{Im}(\tilde{\alpha}|_W)$, say $\tilde{\alpha}(v) = \tilde{\alpha}(w)$ with $w \in W$. Then $v + w \in \operatorname{Ker}\tilde{\alpha} = \operatorname{Im}\tilde{\alpha} = E^n \subset W$, and so $v \in W$, which implies $(e, \alpha) = (v, e)a \in H'$. This proves $K \subset H'$ and so the remaining two claims in (1).

For part (2) we need to show that there is an element of order p not in the kernel of νf. If $p \neq 2$, one can check that $\exp(G) = p$, so there is nothing to show, and the same is true if $p = 2$ but $H' \subset \operatorname{Ker}(\pi)$. So if we assume all elements of order p are in $\operatorname{Ker}(\nu f)$ then $p = 2$ and $H' \not\subset \operatorname{Ker}(\pi)$. We have $H' = W \rtimes K$ by part

(1), and since $\nu f(W) = 0$, we have $\nu f(H') = \nu f(K)$; since K has exponent p, we have found elements of H' not in the kernel of νf after all. □

EXAMPLE 3.3. Let p be a prime and $q = p^r$. Consider a three-cohomology class $[\omega_0]$ with coefficients in $\mathbb{C}^\times$ on the group

$$\begin{pmatrix} 1 & \mathbb{F}_q \\ 0 & 1 \end{pmatrix} \cong C_p^r$$

with the property that its restriction to every cyclic subgroup is nontrivial.

Endow

$$G := (\mathbb{F}_q^2)^n \rtimes \begin{pmatrix} 1 & \mathbb{F}_q \\ 0 & 1 \end{pmatrix}$$

with the pulled back cocycle ω on G. Then $\mathcal{C}(G,\omega)$ is not categorically Morita equivalent to $\mathcal{C}(G,1)$.

PROOF. Consider $H = E^n \times K \subset G$. The restriction $\omega|_H$ is the same as the cocycle pulled back to H from ω_0 on K. It is easy to check that any category Morita equivalent to $\mathcal{C}(H,\omega)$ is equivalent to $\mathcal{C}(H,\omega)$. Assume that $\mathcal{C}(G,\omega)$ and $\mathcal{C}(G,1)$ are Morita equivalent. Then by Theorem 3.1 there is a subgroup $H' \subset G$ and an epimorphism $H' \to H$ such that the pullback of ω to H' is trivial. But by part (2) of the preceding Lemma, this is impossible. □

REMARK 3.4. The example presented above for any p is only useful at all as a reasonable applications of our results if $p = 2$. In fact if $p \neq 2$, we have already mentioned that the group $G = V^n \rtimes K$ has exponent p. If we try to construct a group-theoretical fusion category $\mathcal{C}(G, A, \omega, \mu)$, we need an abelian normal subgroup A such that the restriction of ω to A is cohomologically trivial. If A were not contained in $\mathrm{Ker}(\pi)$, then A would have a subgroup of order p mapped injectively by π into K, contradicting the condition on ω. But if A is a subgroup of V^n, then it should not be problematic to determine the structure of $\mathcal{C}(G, A, \omega, 1)$ directly (although we do not attempt to do this here) without using Theorem 3.1. If $p = 2$ on the other hand, there are abelian normal subgroups A not in $\mathrm{Ker}(\pi)$ but where ω pulls back to the trivial cocycle on A. The corresponding pointed categories were determined (for $q = p$) in the paper **[5]** that motivated our examples, and that will be discussed in Section 5. The calculations in **[5]** however needed more detailed handling of explicit cocycles $G \times G \times G \to \mathbb{C}^\times$.

4. Several subgroups

To illustrate the mechanism of the tool presented in this section, we will discuss Example 3.3 again. We were able to show that $\mathcal{C}(G,\omega)$ and $\mathcal{C}(G,1)$ are not Morita equivalent by considering a subcategory of $\mathcal{C}(G,\omega)$. This subcategory could not be (up to Morita equivalence) the epimorphic image of a subcategory of $\mathcal{C}(G,1)$, because we were able to show the cocycle cannot "disappear" from pulling back due to the specific nature of subgroup which forces a certain property of the epimorphism. We could try to do the same considering the same subgroup as a subcategory in $\mathcal{C}(G,1)$. But the subgroup is just isomorphic to $\mathbb{F}_q^{n+1}$, now with a trivial cocycle. A subgroup isomorphic to this, also endowed with the trivial cocycle, is easily found in $\mathcal{C}(G,\omega)$, within V^n. So to treat the example by considering a subgroup, it was essential to consider the subgroup as a subcategory in the pointed category with nontrivial associator and showing it cannot be a subquotient of the category with trivial associator. Considering the same subgroup in the category with trivial associator would have been useless.

We'll prove again below that $\mathcal{C}(G,\omega)$ and $\mathcal{C}(G,1)$ are not Morita equivalent, by considering not one, but simultaneously two subgroups of G considered as subcategories in $\mathcal{C}(G,1)$.

The general statement behind that technique concerns the entire lattice of fusion subcategories of a pointed category (i.e. the lattice of subgroups of the underlying group).

THEOREM 4.1. *Let $\mathcal{C} = \mathcal{C}(G,\omega)$ and $\mathcal{C}' = \mathcal{C}(G',\omega')$ be categorically Morita equivalent fusion categories. Then there is a join-preserving monotonic map $\mathcal{P}$ from the lattice of subgroups of G to that of G', and for each subgroup $H \subset G$ there exist a pointed fusion category $\mathcal{T}(H) = \mathcal{C}(\tilde{H},\tilde{\omega})$ categorically Morita equivalent to $\mathcal{C}(H, \operatorname{res}^G_H(\omega))$ and an epimorphism $\nu\colon \mathcal{P}(H) \to \mathcal{T}(H)$ such that $\operatorname{res}^{G'}_{\mathcal{P}(H)}(\omega') = \nu^*\tilde{\omega}$.*

For the time being at least, we will only use a particularly simple consequence.

COROLLARY 4.2. *Let G, G', ω, ω' be as in Theorem 3.1. Let H_1, H_2 be subgroups of G which together generate G. Then there are subgroups $H_1', H_2' \leq G'$ generating G', groups $\tilde{H}_i$, epimorphisms $f_i\colon H_i' \to \tilde{H}_i$ and cocycles $\tilde{\omega}_i \in\in H^3(\tilde{H}_i, \mathbb{C}^\times)$ such that $\operatorname{res}^{G'}_{H_i'}[\omega'] = \inf^{H_i'}_{\tilde{H}_i}[\tilde{\omega}_i]$, and $\mathcal{C}(\tilde{H}_i,\tilde{\omega}_i)$ and $\mathcal{C}(H_i, \operatorname{res}^G_{H_i}\omega)$ are categorically Morita equivalent.*

EXAMPLE 4.3. Let G be as in Lemma 3.2. If $\mathcal{C}(G,\omega)$ is categorically Morita equivalent to $\mathcal{C}(G,1)$, then $\omega|_K$ is trivial.

PROOF. Assume that $\mathcal{C}(G,\omega)$ and $\mathcal{C}(G,1)$ are Morita equivalent. We will apply the Corollary with $\mathcal{C}(G,1)$ taking the role of the unprimed category. We consider the two subgroups $H_1 = E^n \times K$ and $H_2 = V^n$ in G. Since both H_i are products of cyclic groups C_p, the only pointed fusion category Morita equivalent to $\mathcal{C}(H_i,1)$ is $\mathcal{C}(H_i,1)$ itself, so that in applying the Corollary we can take $\tilde{H}_i = H_i$ and $\tilde{\omega}_i = 1$. We get two subgroups H_i' and epimorphisms $f_i\colon H_i' \to H_i$. Since H_1' and H_2' together generate G, at least one of the two subgroups is not contained in $V^n = \operatorname{Ker}(\pi)$. That subgroup, say H_i', contains K. It follows that the restriction of ω to K is the restriction of the pullback of the trivial cocycle on H_i to H_i'. □

5. Sylow subgroups

We make an easy preliminary remark that shows that the theorem following it makes sense as stated.

REMARK 5.1. Let $\mathcal{C} = \mathcal{C}(G,\omega)$ be a pointed fusion category. If S, S' are conjugate subgroups of G, then $\mathcal{C}(S, \operatorname{res}^G_S\omega)$ and $\mathcal{C}(S', \operatorname{res}^G_{S'}\omega)$ are equivalent pointed fusion categories, since the action of G on itself by conjugation induces the identity on the cohomology groups. In particular, this applies if H, H' are p-Sylow subgroups of G.

THEOREM 5.2. *The categorical Morita equivalence classes of the subcategories corresponding to the Sylow subgroups in a pointed fusion category are categorical Morita invariants.*

Stated less succinctly but more transparently: If $\mathcal{C} = \mathcal{C}(G,\omega)$ and $\mathcal{C}' = \mathcal{C}(G',\omega')$ are Morita equivalent pointed fusion categories, p is a prime, and $H \subset G$ and $H' \subset$

G' are the respective p-Sylow subgroups, then $\mathcal{C}(H, \operatorname{res}^G_H \omega)$ and $\mathcal{C}(H', \operatorname{res}^{G'}_{H'} \omega')$ are Morita equivalent.

PROOF. The group order being a Morita invariant, the groups have the same order, and their p-Sylow subgroups also have the same order, say p^n. By Theorem 3.1 there is a subgroup $H'' \subset G$, an epimorphism $H'' \to \tilde{H}$ and a cocycle $\tilde{\omega}$ on $\tilde{H}$ such that $\mathcal{C}(H'', \operatorname{res}^G_{H''} \omega) \to \mathcal{C}(\tilde{H}, \tilde{\omega})$ is a tensor functor and $\mathcal{C}(\tilde{H}, \tilde{\omega})$ is Morita equivalent to $\mathcal{C}(H', \omega')$. Since $\tilde{H}$ has order p^n, it follows that H'' contains a Sylow subgroup S of G mapping isomorphically onto $\tilde{H}$, and $\mathcal{C}(S, \operatorname{res}^G_S \omega)$ is equivalent to $\mathcal{C}(\tilde{H}, \tilde{\omega})$. □

Before discussing an application of Theorem 5.2, let us advertise Theorem 3.1 and Corollary 4.2 again: While Theorem 5.2 clearly has a stronger conclusion, it is moot for the case of a p-group, where we were nonetheless able to draw information from those results for an example above. Moreover, in the examples we treat below, Theorem 3.1 will still be used.

The examples we treat feature in work of Izumi and Tucker. In [**5**], the authors show how to explicitly construct all near-group fusion categories with a nonabelian group of invertible objects as group-theoretical categories. A key ingredient in their approach is to prove the inequivalence of the group-theoretical categories constructed from certain group-theoretical data (a group, a subgroup, a cocycle). The authors achieve this by considering Frobenius-Schur indicators and the group structure of the respective groups of invertible objects. We note that explicit calculations with cocycles (in the sense of functions of three group elements) are needed in the process, both to describe the group structure through results of Gelaki and Naidu [**3**], and to calculate certain indicators using [**7**]. For our approach we need less detailed information (on the cohomology classes rather than the explicit cocycles) and we prove more (the categories are categorically Morita inequivalent rather than only inequivalent). Also we will treat some examples of similar type with relative ease, one of them also taken from [**5**].

EXAMPLE 5.3. Consider $G_n = V^n \rtimes \mathrm{SL}(2, \mathbb{F}_2)$, where $V = \mathbb{F}_2^2$ and the action on each copy is the natural one. We will take for granted that $H^3(\mathrm{SL}(2, \mathbb{F}_2), \mathbb{C}^\times) = \langle \omega \rangle \cong C_6$, where the cocycle ω restricts to the generating cocycles in $H^3(C_2, \mathbb{C}^\times) \cong \mathbb{Z}/2\mathbb{Z}$ and $H^3(C_3, \mathbb{C}^\times) \cong \mathbb{Z}/3\mathbb{Z}$, for the Sylow subgroups in $\mathrm{SL}(2, \mathbb{F}_2) \cong S_3$, which are cyclic. As in [**5**] we inflate ω to G_n and denote the result again by ω. One task achieved in [**5**] is to show that certain six group-theoretical categories associated to G_n, the six cocycles ω^l for $l = 0, \dots, 5$, and a certain subgroup $H_n \subset G_n$ are pairwise nonequivalent; we will show instead that the six pointed fusion categories $\mathcal{C}(G_n, \omega^l)$ are even Morita inequivalent. In fact the subcategory corresponding to the 3-Sylow subgroup is $\mathcal{C}(C_3, \omega^l)$. Two of these categories are Morita equivalent if and only if they are equivalent if and only if the values of l are congruent modulo 3. A 2-Sylow subgroup of G_n is the group $V^n \rtimes K$ considered in Example 3.3 (taking $p = q = 2$), with the cocycle obtained as the inflation of the restriction of ω^l to $K \cong C_2$. We have already seen in Example 3.3 and again in Example 4.3 that two of these categories are Morita equivalent iff the values of l have the same parity. Combining the criteria coming from the 2- and 3-Sylow subgroups yields that two categories $\mathcal{C}(G_n, \omega^l)$ can only be Morita equivalent if the values of l are congruent modulo 6.

The next example is suggested as an interesting object for further investigation in [**5**]; we will discuss what information on it our methods provide the following example.

EXAMPLE 5.4. Consider $G = V^n \rtimes \mathrm{SL}(2,\mathbb{F}_4)$ with $V = \mathbb{F}_4^2$. We will not dwell on the computation of $H^3(\mathrm{SL}(2,\mathbb{F}_4),\mathbb{C}^\times) \cong \mathbb{Z}/2\mathbb{Z}\oplus\mathbb{Z}/3\mathbb{Z}\oplus\mathbb{Z}/5\mathbb{Z}$. In fact we admit that we simply asked GAP/HAP retrieve the result for us. We claim that the inflations to G of the 30 cohomology classes yield at least 18 Morita inequivalent pointed fusion categories. In fact let ω be a generating cohomology class. The Morita equivalence class of $\mathcal{C}(G,\omega^\ell)$ determines those of the p-Sylow subgroups. Since $|\,\mathrm{SL}(2,\mathbb{F}_4)| = 3\cdot 5\cdot 4$, the 3- and 5-Sylow subgroups are cyclic. Classical results on group cohomology [**1**, Thm.10.1] (which certainly would feature in an explicit calculation of the cohomology of the special linear group that we skipped) imply that the restrictions of ω generate the cohomology groups of the respective Sylow subgroups $H^3(C_p,\mathbb{C}^\times)\cong\mathbb{Z}/p\mathbb{Z}$ for $p=3,5$. The categories $\mathcal{C}(C_p,\omega^\ell)$ and $\mathcal{C}(C_p,\omega^k)$ are Morita equivalent iff they are equivalent iff k and ℓ are both multiples of p, or if both are prime to p and their quotient is a square modulo p. Thus the Morita equivalence class of $\mathcal{C}(G,\omega^\ell)$ determines ℓ modulo 3, and whether ℓ is congruent to 0, or congruent to 1 or 4, or congruent to 2 or 3 modulo 5. Treating the parity of ℓ is slightly more complicated. Either a calculation using classical methods from [**1**], or a more detailed questioning of GAP/HAP reveals that ω on $\mathrm{SL}(2,\mathbb{F}_4)$ restricts to a nontrivial cohomology class on the 2-Sylow subgroup

$$\begin{pmatrix}1 & \mathbb{F}_4\\ 0 & 1\end{pmatrix}\cong C_2^2.$$

Thus ω on G restricts to the cohomology class on the 2-Sylow subgroup $V^n\rtimes K$ studied in Lemma 3.2 which is the pullback of a nontrivial cocycle on K. By Example 4.3 (taking $p=2$ and $q=4$) this is enough to show that the Morita class of $\mathcal{C}(G,\omega^\ell)$ determines the parity of ℓ. Note that Theorem 3.1 instead of Corollary 4.2 would not have been sufficient for this reasoning. It is true (upon still further inspection which we will not give in detail) that ω on C_2^2 does have the property that its restriction to every cyclic subgroup is nontrivial; with this, Example 3.3 would also allow us to conclude.

6. Proof of the main result

PROOF OF THEOREM 3.1. Any fusion category Morita equivalent to $\mathcal{C}(G,\omega)$ is of the form $\mathcal{C}' = {}_{\mathbb{C}[A]}\mathcal{C}(G,\omega)_{\mathbb{C}[A]}$, the category of $\mathbb{C}[A]$-bimodules in the category $\mathcal{C}(G,\omega)$. Here we have assumed that $\omega|_{A^3}=1$ which is no loss of generality up to cohomology. For $\mathcal{C}'$ to be pointed it is necessary that A is normal and abelian. Now let H be a subgroup of G. Because A is normal, $AH = HA$ is a subgroup of G. We consider the pointed subcategory $\mathcal{C}'' = {}_{\mathbb{C}[A]}\mathcal{C}(AH,\omega)_{\mathbb{C}[A]}$. Recall that objects in $\mathcal{C}''$ are vector spaces graded by AH with a two-sided action of the algebra $\mathbb{C}[A]$ inside the monoidal category $\mathcal{C}(AH,\omega)$. We define a functor $\mathcal{F}\colon \mathcal{C}'' \to {}_{\mathbb{C}[A\cap H]}\mathcal{C}(H,\omega)_{\mathbb{C}[A\cap H]}$ by $\mathcal{F}(M) = M_H = \oplus_{h\in H} M_h$. There is an obvious coherent natural transformation

$$\mathcal{F}(M)\underset{\mathbb{C}[A\cap H]}{\otimes}\mathcal{F}(N)\to\mathcal{F}(M\underset{\mathbb{C}[A]}{\otimes}N) \tag{6.1}$$

induced by the inclusions $\mathcal{F}(X)\to X$. For any $M\in\mathcal{C}(G,A,\omega,1)$, and any section Q of the right cosets of A in G (i.e. $QA=G$), we have that the right $\mathbb{C}[A]$ module structure induces an isomorphism $M_Q \otimes \mathbb{C}[A] \to M$. We can choose Q so that $HA = (Q \cap H)A$. Then $(M_Q \otimes \mathbb{C}[A])_H = M_{Q\cap H} \otimes \mathbb{C}[A \cap H] = M_Q \otimes \mathbb{C}[A \cap H]$ for $M \in \mathcal{C}''$, and so $M = M_Q \otimes \mathbb{C}[A \cap H] \underset{\mathbb{C}[A\cap H]}{\otimes} \mathbb{C}[A]$.

$$\begin{aligned}\mathcal{F}(M \underset{\mathbb{C}[A]}{\otimes} N) &= \mathcal{F}(M_Q \otimes \mathbb{C}[A \cap H] \underset{\mathbb{C}[A\cap H]}{\otimes} \mathbb{C}[A] \underset{\mathbb{C}[A]}{\otimes} N) \\ &= \mathcal{F}(M_Q \otimes \mathbb{C}[A \cap H] \underset{\mathbb{C}[A\cap H]}{\otimes} N) \\ &= M_Q \otimes \mathbb{C}[A \cap H] \underset{\mathbb{C}[A\cap H]}{\otimes} N_H \\ &= \mathcal{F}(M) \underset{\mathbb{C}[A\cap H]}{\otimes} \mathcal{F}(N).\end{aligned}$$

It remains to show that $\mathcal{F}$ is essentially surjective. Now if a simple of $\mathcal{C}''$ is described by a projective character of $\mathrm{Stab}_A(qA)$ with $q \in Q$, then clearly $\mathcal{F}(M) \in \mathcal{C}(H, A \cap H, \omega, 1)$ is described by the restriction of that character to $\mathrm{Stab}_{A\cap H}\, q(A \cap H)$. The fact that $\mathcal{C}(G, H, \omega, 1)$ is pointed entails that the relevant cocycle governing the projective characters of stabilizers of right H-cosets is cohomologically trivial, and so they remain on any subgroup. Thus the projective characters we need are just ordinary characters twisted by cocycles. But any character of the subgroup $\mathrm{Stab}_{A\cap H}\, q(A \cap H)$ can be extended to a character of the abelian group $\mathrm{Stab}_A(qA)$.

Thus far we have proved Theorem 3.1; it remains to observe that the assignment $H \mapsto {}_{\mathbb{C}[A]}\mathcal{C}(AH, \omega)_{\mathbb{C}[A]}$ from subgroups of G to fusion subcategories of $\mathcal{C}(G', \omega')$ is monotonic and join preserving because the assigment $H \to AH$ between subgroups of G is. □

References

[1] H. Cartan and S. Eilenberg, *Homological algebra*, Princeton University Press, Princeton, NJ, 1956. MR77480

[2] P. Etingof, S. Gelaki, D. Nikshych, and V. Ostrik, *Tensor categories*, Mathematical Surveys and Monographs, vol. 205, American Mathematical Society, Providence, RI, 2015, DOI 10.1090/surv/205. MR3242743

[3] S. Gelaki and D. Naidu, *Some properties of group-theoretical categories*, J. Algebra **322** (2009), no. 8, 2631–2641, DOI 10.1016/j.jalgebra.2009.05.047. MR2559854

[4] C. Goff, G. Mason, and S.-H. Ng, *On the gauge equivalence of twisted quantum doubles of elementary abelian and extra-special 2-groups*, J. Algebra **312** (2007), no. 2, 849–875, DOI 10.1016/j.jalgebra.2006.10.022. MR2333187

[5] M. Izumi and H. Tucker, *Algebraic realization of noncommutative near-group fusion categories*, Algebra Number Theory **15** (2021), no. 5, 1077–1093, DOI 10.2140/ant.2021.15.1077. MR4283097

[6] D. Naidu, *Categorical Morita equivalence for group-theoretical categories*, Comm. Algebra **35** (2007), no. 11, 3544–3565, DOI 10.1080/00927870701511996. MR2362670

[7] P. Schauenburg, *A higher Frobenius-Schur indicator formula for group-theoretical fusion categories*, Comm. Math. Phys. **340** (2015), no. 2, 833–849, DOI 10.1007/s00220-015-2437-2. MR3397032

IMB UMR 5584, Université Bourgogne Europe, CNRS, F-21000 Dijon, France
Email address: peter.schauenburg@u-bourgogne.fr

Contemporary Mathematics
Volume **814**, 2025
https://doi.org/10.1090/conm/814/16306

On classification of Hopf superalgebras of low dimension

Taiki Shibata and Ryota Wakao

ABSTRACT. We examine the inverse procedure of the Radford–Majid bosonization for Hopf superalgebras and give a handy method for enumerating Hopf superalgebras whose bosonization is isomorphic to a given Hopf algebra. As an application, we classify Hopf superalgebras of dimension up to 5 and give examples of higher dimensions.

1. Introduction

The classification problem of finite-dimensional Hopf algebras has been actively studied by many researchers since it was proposed by Kaplansky in 1975; see the survey [**4**]. Due to the importance of supersymmetry in mathematics and mathematical physics, we are rather interested in classification problem of finite-dimensional *Hopf superalgebras* over an algebraically closed field $\Bbbk$ of characteristic zero. Among them, super-(co)commutative ones have been fully classified by the following result: According to Kostant [**9**, Theorem 3.3] (see also [**3**, Corollary 2.3.5]), a finite-dimensional super-cocommutative Hopf superalgebra can be decomposed into a semidirect product of the group algebra $\Bbbk\Gamma$ of a finite group Γ and the exterior superalgebra $\bigwedge(V)$ over a finite-dimensional $\Bbbk\Gamma$-module V.

In 2014, Aissaoui and Makhlouf [**1**] gave a complete list of Hopf superalgebras of dimension up to 4 using a computer software and found some nontrivial Hopf superalgebras (see Example 3.5). However, for higher dimensions, a systematic study has not yet been done, and the classification problem is widely open. In this paper, we use an approach different from [**1**], classify Hopf superalgebras of dimension up to 5, and give new examples. The key ingredient of our approach is the *bosonization* technique introduced by Radford [**18**] and Majid [**10**] (see Section 2.3). This technique involves reducing Hopf superalgebras $\mathcal{H}$ to ordinary Hopf algebras $\widehat{\mathcal{H}}$, which allows us to apply various (classification) results obtained in ordinary settings.

Let us explain our approach in detail. Let A be a Hopf algebra over $\Bbbk$. It has been known that there is a one-to-one correspondence between the isomorphism classes of the set of all split epimorphisms $A \to \Bbbk\mathbb{Z}_2$ of Hopf algebras and the isomorphism classes of Hopf algebras $\mathcal{H}$ in the category of $\mathbb{Z}_2$-Yetter–Drinfeld modules ${}^{\Bbbk\mathbb{Z}_2}_{\Bbbk\mathbb{Z}_2}\mathcal{YD}$ such that $\widehat{\mathcal{H}} \cong A$. We show that the set of all split epimorphisms $A \to \Bbbk\mathbb{Z}_2$

2020 *Mathematics Subject Classification.* Primary 16T05, 17A70.
Key words and phrases. Hopf superalgebra, bosonization, classification.

is parameterized by the set

$$\mathsf{AD}(A) = \{(g,\alpha) \in \mathbf{G}(A) \times \mathbf{G}(A^\circ) \mid \mathsf{ord}(g) = 2, \mathsf{ord}(\alpha) = 2, \alpha(g) = -1\},$$

whose element is called an *admissible datum* for A in this paper (Definition 4.8). Here, for a Hopf algebra K, we have denoted by $\mathbf{G}(K)$ (resp., K°) the group of group-like elements of K (resp., the finite dual of K). Given an admissible datum (g,α), we let $\pi_{(g,\alpha)}$ be the corresponding split epimorphism $A \to \Bbbk\mathbb{Z}_2$, see (4.1) for the precise definition. We show that the Hopf algebra $\mathcal{H}$ in ${}^{\Bbbk\mathbb{Z}_2}_{\Bbbk\mathbb{Z}_2}\mathcal{YD}$ corresponding to $\pi_{(g,\alpha)}$ is a Hopf superalgebra whose $\bar{1}$-part is nonzero if and only if

$$\alpha \rightharpoonup a \leftharpoonup \alpha = gag \quad (a \in A) \quad \text{and} \quad g \notin Z(A),$$

where $Z(A)$ is the center of A. We define $\mathsf{SD}(A)$ to be the subset of $\mathsf{AD}(A)$ consisting of elements satisfying this condition and call its element a *super-datum* for A (Definition 4.16). The group of Hopf algebra automorphisms of A naturally acts on $\mathsf{SD}(A)$. A key observation is that there is a one-to-one correspondence between the orbits of $\mathsf{SD}(A)$ and the isomorphism classes of the set of all Hopf superalgebras $\mathcal{H}$ such that $\mathcal{H}_{\bar{1}} \neq 0$ and $\widehat{\mathcal{H}} \cong A$. Moreover, the Hopf superalgebra $\mathcal{H}$ is semisimple (resp., pointed) if and only if A is semisimple (resp., pointed).

In this paper, we study Hopf superalgebras by utilizing the above bijection. One of our applications classifies finite-dimensional Hopf superalgebras of prime dimensions. By the classification results of Hopf algebras of dimension $2p$ for a prime number p by Masuoka [**11**] and Ng [**17**], we immediately get the following result.

THEOREM 1.1 (= Theorems 5.1 and 5.2). *Up to isomorphism, $\bigwedge(\Bbbk)$ is the only Hopf superalgebra of dimension 2 whose $\bar{1}$-part is nonzero. If p is an odd prime number, then a Hopf superalgebra of dimension p is purely even, that is, its $\bar{1}$-part is zero (thus such a Hopf superalgebra is isomorphic to $\Bbbk\mathbb{Z}_p$ by Zhu* [**22**]*).*

We fix a primitive fourth root $\zeta_4 \in \Bbbk$ of unity. By the above theorem and the classification results of Hopf algebras of dimension 4 obtained by Masuoka [**12**] and Ştefan [**21**], we classify Hopf superalgebras of dimension up to 5 and determine their duals as follows.

THEOREM 1.2 (= Theorems 5.5, 5.8 and 5.10). *Let $\mathcal{H}$ be a Hopf superalgebra of dimension less than or equal to 5. If $\mathcal{H}_{\bar{1}} \neq 0$, then $\mathcal{H}$ is isomorphic to one of the following Hopf superalgebras.*

Note that $\mathcal{H}_4^{(2)} \cong \Bbbk\mathbb{Z}_2 \otimes \bigwedge(z)$ and $\mathcal{A}_4(\zeta_4) \ncong \mathcal{A}_4(-\zeta_4)$.

In our forthcoming papers, we will give a complete list of Hopf superalgebras of dimension up to 10. We will not address the classification of Hopf superalgebras of dimensions higher than 5 in this paper, but, provide new and intriguing examples. One example arises from the classification result of semisimple Hopf algebras of dimension 12 by Fukuda [**6**]. Using his result, we prove:

THEOREM 1.3 (= Theorem 5.12). *A semisimple Hopf superalgebra $\mathcal{H}$ of dimension 6 with $\mathcal{H}_{\bar{1}} \neq 0$ is isomorphic to the self-dual Hopf superalgebra $\mathcal{A}_6$ (see Proposition 5.11 for the definition).*

Hopf algebras of dimension 16 has been classified by García and Vay in [**7**]. According to their result, a nonsemisimple nonpointed self-dual Hopf algebra of this dimension is isomorphic to one of two Hopf algebras given in Călinescu, Dăscălescu, Masuoka and Menini [**5**]. In response, we obtain Theorem 1.4.

TABLE 1

dim	**Hopf superalgebras $\mathcal{H}$ with $\mathcal{H}_{\bar{1}} \neq 0$**	**structures**	**notes**
2	$\bigwedge(z) = \Bbbk\langle z \mid z^2 = 0\rangle$ with $\lvert z\rvert = 1$	z: odd primitive	pointed self-dual
3	None		
4	$\mathcal{H}_4^{(1)} = \bigwedge(z_1, z_2) = \Bbbk\langle z_1, z_2 \mid z_i z_j = -z_j z_i\rangle$ with $\lvert z_1\rvert = \lvert z_2\rvert = 1$	z_1, z_2: primitive	pointed self-dual
	$\mathcal{H}_4^{(2)} = \Bbbk\langle g, z \mid g^2 = 1, z^2 = 0, gz = zg\rangle$ with $\lvert g\rvert = 0, \lvert z\rvert = 1$	g: group-like z: odd primitive	pointed self-dual
	$\mathcal{H}_4^{(3)} = \Bbbk\langle g, z \mid g^2 = 1, z^2 = 0, gz = zg\rangle$ with $\lvert g\rvert = 0, \lvert z\rvert = 1$	g: group-like z: odd g-skew primitive	pointed $(\mathcal{H}_4^{(3)})^* \cong \mathcal{H}_4^{(4)}$
	$\mathcal{H}_4^{(4)} = \Bbbk\langle g, z \mid g^2 = 1, z^2 = 0, gz = -zg\rangle$ with $\lvert g\rvert = 0, \lvert z\rvert = 1$	g: group-like z: odd primitive	pointed $(\mathcal{H}_4^{(4)})^* \cong \mathcal{H}_4^{(3)}$
	$\mathcal{A}_4(\zeta_4), \mathcal{A}_4(-\zeta_4)$	Example 3.5	semisimple self-dual
5	None		

THEOREM 1.4 (= Theorem 5.18). *A nonsemisimple Hopf superalgebra $\mathcal{H}$ of dimension 8 such that neither $\mathcal{H}$ nor $\mathcal{H}^*$ is pointed and $\mathcal{H}_{\bar{1}} \neq 0$ is isomorphic to either one of pairwise nonisomorphic eight Hopf superalgebras*

$$\mathcal{K}_8(\zeta;\epsilon,\eta) \qquad (\zeta \in \{\zeta_4, -\zeta_4\},\ \epsilon,\eta \in \{0,1\})$$

given in Proposition 5.17. *Moreover, for each $\zeta \in \{\zeta_4, -\zeta_4\}$, we have*

$$\mathcal{K}_8(\zeta;0,0)^* \cong \mathcal{K}_8(\zeta;0,0),\ \ \mathcal{K}_8(\zeta;1,1)^* \cong \mathcal{K}_8(\zeta;1,1),\ \ \mathcal{K}_8(\zeta;0,1)^* \cong \mathcal{K}_8(\zeta;1,0)$$

as Hopf superalgebras.

Organization of the paper. The paper is organized as follows. In Section 2 we review the definition and properties of Yetter–Drinfeld modules (Section 2.2) and the bosonization technique (Section 2.3) introduced by Majid [**10**] and Radford [**18**].

In Section 3 we review the definition and properties of Hopf superalgebras. We first see that the category of superspaces is naturally embedded in the category of left $\Bbbk\mathbb{Z}_2$-Yetter–Drinfeld category ${}^{\Bbbk\mathbb{Z}_2}_{\Bbbk\mathbb{Z}_2}\mathcal{YD}$ (Section 3.1). We next see some examples of Hopf superalgebras (Section 3.2). We recall from [**3**] a construction of Hopf superalgebras (Section 3.3). For a finite-dimensional Hopf superalgebra, as in the nonsuper situation, we see that its dual superspace becomes a finite-dimensional Hopf superalgebra again, called the *dual Hopf superalgebra* (Section 3.4).

In Section 4 we apply the Radford–Majid bosonization to Hopf superalgebras and study its properties (Section 4.1). We say that a Hopf superalgebra $\mathcal{H}$ is a *super-form* of a Hopf algebra A if the bosonization $\widehat{\mathcal{H}}$ of $\mathcal{H}$ is isomorphic to A as Hopf algebras (Definition 4.4). For a given Hopf algebra A, in Section 4.2 we introduce the notion of *admissible data* for A (Definition 4.8) and show that there is one-to-one correspondence between the isomorphism classes of the set of all admissible data for A and the set of all isomorphism classes of Hopf algebras $\mathcal{H}$ in ${}^{\Bbbk\mathbb{Z}_2}_{\Bbbk\mathbb{Z}_2}\mathcal{YD}$ such that the bosonization of $\mathcal{H}$ is isomorphic to A (Proposition 4.11). In Section 4.3, we give a criterion for such $\mathcal{H}$ to be a Hopf superalgebra (Propositions 4.14 and 4.15). If an admissible datum for A satisfies the conditions given in Propositions 4.14 and 4.15, we call it a *super-datum* for A (Definition 4.16).

In Section 5, as a demonstration of our method, we classify Hopf superalgebras of dimension up to 5 (Sections 5.1, 5.2, and 5.3). The proofs of Theorems 1.3 and 1.4 are given in Section 5.4 and in Section 5.5, respectively.

2. The Radford–Majid bosonization

Throughout this paper, we work over a field $\Bbbk$. In Sections 3 and 4, we suppose $\Bbbk$ to be of characteristic $\neq 2$. In Section 5, $\Bbbk$ is supposed to be an algebraically closed field of characteristic zero. The unadorned $\otimes$ is the tensor product $\otimes_\Bbbk$ over $\Bbbk$.

In this section, we review the bosonization (biproduct) technique introduced by Radford [**18**] and Majid [**10**]. In the following, we fix a Hopf algebra $H = (H; m_H, u_H, \Delta_H, \varepsilon_H, S_H)$, where m_H, u_H, Δ_H, ε_H and S_H are the multiplication, the unit, the comultiplication, the counit and the antipode of H, respectively. As usual, we denote the unit of H by $1_H := u_H(1)$ and we sometimes write 1_H just 1 (omitting the subscript H) for simplicity. We freely use the Heyneman–Sweedler notation such as $\Delta_H(h) = h_{(1)} \otimes h_{(2)}$ for $h \in H$. Let H^+ denote the kernel of the counit ε_H of H.

2.1. Coinvariant subspaces. A right H-module M is called a *right H-Hopf module* if M is simultaneously a right H-comodule satisfying $(m.h)_{(0)} \otimes (m.h)_{(1)} = m_{(0)}.h_{(1)} \otimes m_{(1)}h_{(2)}$ for all $h \in H, m \in M$. Here, $m.h$ denotes the right H-action of $h \in H$ on $m \in M$ and $M \to M \otimes H; m \mapsto m_{(0)} \otimes m_{(1)}$ denotes the right H-coaction on M. Let $\mathcal{M}_H^H$ (resp., $\mathcal{V}$) denote the category of all right H-Hopf modules (resp., all vector spaces). For $V \in \mathcal{V}$, we naturally regard $V \otimes H$ as a right H-Hopf module (by $\mathsf{id}_V \otimes m_H$ and $\mathsf{id}_V \otimes \Delta_H$), and hence we get a functor $\mathcal{V} \to \mathcal{M}_H^H; V \mapsto V \otimes H$.

Let M be a a right H-Hopf module. We denote the *H-coinvariant subspace* of M by $M^{\mathsf{co}(H)} := \{m \in M \mid m_{(0)} \otimes m_{(1)} = m \otimes 1_H\}$. One sees that the map

$$M^{\mathsf{co}(H)} \otimes H \xrightarrow{\cong} M; \quad v \otimes h \longmapsto v.h \tag{2.1}$$

is a natural isomorphism in $\mathcal{M}_H^H$ whose inverse is given by $m \mapsto E_M(m_{(0)}) \otimes m_{(1)}$, where

$$E_M : M \longrightarrow M^{\mathsf{co}(H)}; \quad m \longmapsto m_{(0)}.S_H(m_{(1)}). \tag{2.2}$$

This shows that the functor $\mathcal{M}_H^H \to \mathcal{V}; M \mapsto M^{\mathsf{co}(H)}$ is an equivalence whose quasi-inverse is $\mathcal{V} \to \mathcal{M}_H^H; V \to V \otimes H$ (the fundamental theorem of Hopf modules).

Let M be a right H-Hopf module. We set $\overline{M}^H := M/MH^+$ and denote the canonical quotient map by $\overline{(-)} : M \to \overline{M}^H; m \mapsto \overline{m}$. One sees that there is another equivalence $\mathcal{M}_H^H \approx \mathcal{V}$ given by $M \mapsto \overline{M}^H$. The associated natural isomorphism in $\mathcal{M}_H^H$ is given by

$$M \xrightarrow{\cong} \overline{M}^H \otimes H; \quad m \longmapsto \overline{m_{(0)}} \otimes m_{(1)}. \tag{2.3}$$

Indeed, the inverse is given as $\overline{m} \otimes h \mapsto E_M(m).h$.

A natural isomorphism between the two equivalences is given by

$$M^{\mathsf{co}(H)} \xrightarrow{\cong} \overline{M}^H; \quad v \longmapsto \overline{v} \quad \text{and} \quad \overline{M}^H \xrightarrow{\cong} M^{\mathsf{co}(H)}; \quad \overline{m} \longmapsto E_M(m) \tag{2.4}$$

which are inverse of each other. If we identify $M^{\mathsf{co}(H)}$ and $\overline{M}^H$ through these isomorphisms, then (2.1) and (2.3) are seen to be inverses of each other.

2.2. Yetter–Drinfeld categories. In this section we suppose that the antipode S_H of H is bijective. A left H-module V is called a *left H-Yetter–Drinfeld module* if V is simultaneously a left H-comodule satisfying the following compatible condition.

$$(h.v)_{(-1)} \otimes (h.v)_{(0)} = h_{(1)}v_{(-1)}S_H(h_{(3)}) \otimes h_{(2)}.v_{(0)} \quad (h \in H, v \in V).$$

Here, $h.v$ denotes the left H-action of $h \in H$ on $v \in V$ and $V \to H \otimes V; v \mapsto v_{(-1)} \otimes v_{(0)}$ denotes the left H-coaction on V. Let ${}_H^H\mathcal{YD}$ denote the category of all left H-Yetter–Drinfeld modules.

The category ${}_H^H\mathcal{YD}$ has a structure of a monoidal category inherited from the category of left H-modules and that of left H-comodules. Moreover, one sees that ${}_H^H\mathcal{YD}$ forms a braided monoidal category with braiding

$$c_{V,W} : V \otimes W \longrightarrow W \otimes V; \quad v \otimes w \longmapsto v_{(-1)}.w \otimes v_{(0)} \quad (V, W \in {}_H^H\mathcal{YD}). \tag{2.5}$$

Therefore, we may consider *(co)algebras in ${}_H^H\mathcal{YD}$* and *Hopf algebras in ${}_H^H\mathcal{YD}$*. Note that, in some literature, a Hopf algebra in ${}_H^H\mathcal{YD}$ is simply called a *braided Hopf algebra* (if H is obvious). For latter use, we let ${}_H^H\mathcal{YDH}$ denote the category consisting of all Hopf algebras in ${}_H^H\mathcal{YD}$.

Let $\mathscr{T}_H$ denote the category defined by the following.

- (objects) An object is a triplet $(A;\iota,\pi)$ consisting of a Hopf algebra A and Hopf algebra maps $\iota : H \to A, \pi : A \to H$ satisfying $\pi \circ \iota = \mathsf{id}_H$.
- (morphisms) A morphism $(A;\iota,\pi) \to (A';\iota',\pi')$ is a Hopf algebra map $\varphi : A \to A'$ which satisfies $\varphi \circ \iota = \iota'$ and $\pi' \circ \varphi = \pi$.

In [**5**], an object in $\mathscr{T}_H$ is called a *Hopf algebra triple* over H.

Let $(A;\iota,\pi)$ be an object in $\mathscr{T}_H$. We denote the Hopf algebra structure of A by $A = (A; m_A, u_A, \Delta_A, \varepsilon_A, S_A)$. Then $m_A \circ (\mathsf{id}_A \otimes \iota)$ and $(\mathsf{id}_A \otimes \pi) \circ \Delta_A$ make A into an object in $\mathcal{M}_H^H$.

Let $\mathcal{B}$ be the H-coinvariant subspace $A^{\mathsf{co}(H)}$ of A. It is easy to see that $\mathcal{B}$ is a subalgebra of A and $\mathcal{B}$ is a left H-comodule via $(\pi \otimes \mathsf{id}_A) \circ \Delta_A$, and hence $\mathcal{B}$ becomes a left H-comodule algebra. Since AH^+ is a coideal of A, the quotient $\overline{A}^H$ $(= A/AH^+)$ is naturally a coalgebra and $\overline{A}^H$ is a left H-module via $H \otimes \overline{A}^H \to \overline{A}^H; h \otimes \overline{a} \mapsto \overline{\iota(h)a}$. In this way, $\overline{A}^H$ becomes a left H-module coalgebra.

Through the identification (2.4), we see that the left H-comodule algebra $\mathcal{B}$ $(= \overline{A}^H)$ is simultaneously a left H-module coalgebra. One sees that the induced left H-module structure on $\mathcal{B}$ coincides with the adjoint action

$$H \otimes \mathcal{B} \longrightarrow \mathcal{B}; \quad h \otimes b \longmapsto h \triangleright b := \iota(h_{(1)}) b \iota(S_H(h_2))$$

and the induced coalgebra structure $\Delta_{\mathcal{B}}$ on $\mathcal{B}$ is

$$\Delta_{\mathcal{B}} : \mathcal{B} \longmapsto \mathcal{B} \otimes \mathcal{B}; \quad b \longmapsto E_A(b_{(1)}) \otimes b_{(2)} = b_{(1)}(\iota \circ S_H \circ \pi)(b_{(2)}) \otimes b_{(3)},$$

where E_A is given by (2.2). Here, we write $\Delta_A(b) = b_{(1)} \otimes b_{(2)}$ for $b \in \mathcal{B}$ $(\subset A)$. A direct calculation shows that $\Delta_{\mathcal{B}}$ is an algebra map in ${}^H_H\mathcal{YD}$. Moreover, we have the following.

PROPOSITION 2.1. *The left H-comodule algebra $\mathcal{B} = A^{\mathsf{co}(H)}$ becomes an object in ${}^H_H\mathcal{YDH}$ by the following structures.*

- *(left H-action) $H \otimes \mathcal{B} \to \mathcal{B}$; $h \otimes b \mapsto h \triangleright b$.*
- *(comultiplication) $\Delta_{\mathcal{B}} : \mathcal{B} \to \mathcal{B} \otimes \mathcal{B}$; $b \mapsto b_{(1)}(\iota \circ S_H \circ \pi)(b_{(2)}) \otimes b_{(3)}$.*
- *(counit) $\varepsilon_{\mathcal{B}} : \mathcal{B} \to \Bbbk$; $b \mapsto \varepsilon_A(b)$.*
- *(antipode) $S_{\mathcal{B}} : \mathcal{B} \to \mathcal{B}$; $b \mapsto (\iota \circ \pi)(b_{(1)}) S_A(b_{(2)})$.*

In the following, we simply denote the Hopf algebra $\mathcal{B}$ in ${}^H_H\mathcal{YD}$ of the above proposition by $A^{\mathsf{co}(\iota,\pi)}$ and call it the *coinvariant subalgebra* of A with respect to (ι,π).

For a morphism $\varphi : (A;\iota,\pi) \to (A';\iota',\pi')$ in $\mathscr{T}_H$, one sees that the restriction of $\varphi : A \to A'$ induces a morphism $A^{\mathsf{co}(\iota,\pi)} \to A'^{\mathsf{co}(\iota',\pi')}$ in ${}^H_H\mathcal{YDH}$. Thus, we get a functor

$$\mathcal{F} : \mathscr{T}_H \longrightarrow {}^H_H\mathcal{YDH}; \quad (A;\iota,\pi) \longmapsto A^{\mathsf{co}(\iota,\pi)}.$$

2.3. The Radford–Majid bosonization. In this section, we also suppose that the antipode S_H of H is bijective.

Let $\mathcal{B} = (\mathcal{B}; m_{\mathcal{B}}, u_{\mathcal{B}}, \Delta_{\mathcal{B}}, \varepsilon_{\mathcal{B}}, S_{\mathcal{B}})$ be a Hopf algebra in ${}^H_H\mathcal{YD}$. The following result is due to [**10, 18**].

THEOREM 2.2. *The following structure makes $\mathcal{B} \# H := \mathcal{B} \otimes H$ into a Hopf algebra. For $b, b' \in \mathcal{B}$ and $h, h' \in H$,*

- *(multiplication) $(b \# h)(b' \# h') := b(h_{(1)}.b') \# h_{(2)}h'$,*
- *(unit) $1_{\mathcal{B}\#H} := 1_{\mathcal{B}} \# 1_H$,*

- *(comultiplication)* $\Delta_{\mathcal{B}\#H}(b\#h) := (b_{(1)}\#(b_{(2)})_{(-1)}h_{(1)})\otimes((b_{(2)})_{(0)}\#h_{(2)})$,
- *(counit)* $\varepsilon_{\mathcal{B}\#H}(b\# h) := \varepsilon_{\mathcal{B}}(b)\varepsilon_H(h)$,
- *(antipode)* $S_{\mathcal{B}\#H}(b\# h) := (1_{\mathcal{B}}\# S_H(b_{(-1)}h))(S_{\mathcal{B}}(b_{(0)})\# 1_H)$.

Here, we denote $b\otimes h$ by $b\# h$ for $b\in\mathcal{B}$ and $h\in H$. Moreover, if we set

$$\pi_{\mathcal{B}\#H}(b\# h) = \varepsilon_{\mathcal{B}}(b)h \quad \text{and} \quad \iota_{\mathcal{B}\#H}(h) = 1_{\mathcal{B}}\# h \quad (b\in\mathcal{B}, h\in H),$$

then $\mathcal{B}\mapsto(\mathcal{B}\# H;\iota_{\mathcal{B}\#H},\pi_{\mathcal{B}\#H})$ gives rise to a functor $\mathcal{G}: {}^H_H\mathcal{YDH}\to\mathscr{T}_H$.

The Hopf algebra $\mathcal{B}\#H$ constructed above is called the *bosonization* (or *biproduct*) of $\mathcal{B}$ by H. Since $(\mathcal{B}\# H)^{\mathsf{co}(H)}\cong\mathcal{B}$, one easily sees that $\mathcal{F}\circ\mathcal{G}\simeq\mathsf{id}$.

Let $(A;\iota,\pi)$ be an object in $\mathscr{T}_H$. As we have seen in Section 2.2, $\mathcal{B} := A^{\mathsf{co}(\iota,\pi)}$ has a structure of left H-module coalgebra, and hence the isomorphism given in (2.1) is now an isomorphism

$$\mathcal{B}\rtimes H \xrightarrow{\cong} A; \quad b\otimes h\longmapsto b\iota(h)$$

of right H-comodule algebras, where $\mathcal{B}\rtimes H$ is the smash product of $\mathcal{B}$ and H. Similarly, the isomorphism given in (2.3) is now an isomorphism

$$A\xrightarrow{\cong}\overline{A}^H \blacktriangleright\!\!< H; \quad a\longmapsto\overline{a_{(1)}}\otimes\pi(a_{(2)})$$

of right H-module coalgebras, where $\overline{A}^H \blacktriangleright\!\!< H$ is the smash coproduct of $\overline{A}^H$ and H. As a result, we get a Hopf algebra isomorphism

$$\mathcal{B}\# H\xrightarrow{\cong} A; \quad b\otimes h\longmapsto b\iota(h) \tag{2.6}$$

through the identification $\mathcal{B}=\overline{A}^H$ given in (2.4). This proves $\mathcal{G}\circ\mathcal{F}\simeq\mathsf{id}$.

We have thus proved the following theorem, which is a categorical reformulation of Radford's famous results [**18**, Theorem 1–3]. In fact, the reformulation is due to Călinescu, Dăscălescu, Masuoka, and Menini [**5**, Proposition 1.1] who, however, did not give a proof.

THEOREM 2.3. *$\mathcal{G}: {}^H_H\mathcal{YDH}\to\mathscr{T}_H$ is an equivalence with quasi-inverse $\mathcal{F}$.*

For $(A;\iota,\pi),(A;\iota',\pi')\in\mathscr{T}_H$, we denote by $(\iota,\pi)\approx(\iota',\pi')$ if there exists a Hopf algebra automorphism $\varphi: A\to A$ such that $\iota'=\varphi\circ\iota$ and $\pi=\pi'\circ\varphi$. Then $\approx$ is an equivalence relation and we immediately get the following result.

PROPOSITION 2.4. *Let A be a fixed Hopf algebra. There exists a one-to-one correspondence between the following two sets.*

- *The isomorphism classes of Hopf algebras $\mathcal{H}$ in ${}^H_H\mathcal{YD}$ such that $\mathcal{B}\# H$ is isomorphic to A as a Hopf algebra.*
- *The equivalence classes of pairs (ι,π) such that $(A;\iota,\pi)\in\mathscr{T}_H$.*

2.4. Duals of coinvariant subalgebras. In this section, we suppose that H is finite-dimensional. Let $\mathscr{T}_H^{\mathsf{fd}}$ (resp., ${}^H_H\mathcal{YDH}^{\mathsf{fd}}$) denote the full subcategory of $\mathscr{T}_H$ (resp., ${}^H_H\mathcal{YDH}$) consisting of all finite-dimensional objects. The assignment $(A;\iota,\pi)\mapsto(A^*;\pi^*,\iota^*)$ gives a category anti-equivalence between $\mathscr{T}_H^{\mathsf{fd}}$ and $\mathscr{T}_{H^*}^{\mathsf{fd}}$.

In the following, we fix $\mathcal{B}\in{}^H_H\mathcal{YDH}^{\mathsf{fd}}$ and put $(A;\iota,\pi):=\mathcal{G}(\mathcal{B})\in\mathscr{T}_H^{\mathsf{fd}}$. Note that $\mathcal{B}=A^{\mathsf{co}(H)}=\mathcal{F}(A;\iota,\pi)$. Since A^* is naturally a right H^*-module (resp., right H^*-comodule) by $\Delta_A^*\circ(\mathsf{id}\otimes\pi^*)$ (resp., $(\mathsf{id}\otimes\iota^*)\circ m_A^*$), we may consider $A^{*\mathsf{co}(H^*)}$ and $\overline{A^*}^{H^*}$, see Section 2.1 for the notations. One easily sees that $(\overline{A}^H)^*\cong A^{*\mathsf{co}(H^*)}$ and $\mathcal{B}^*\cong\overline{A^*}^{H^*}$.

Recall that the left H-module coalgebra structure on $\overline{A}^H$ and the left H-comodule algebra structure on $\mathcal{B}$ are given as

$$H \otimes \overline{A}^H \longrightarrow \overline{A}^H;\ h \otimes \overline{a} \longmapsto \overline{\iota(h)a} \quad \text{and} \quad \mathcal{B} \longrightarrow H \otimes \mathcal{B};\ b \longmapsto \pi(b_{(1)}) \otimes b_{(2)},$$

respectively. By taking the linear dual, we see that $\overline{A^*}^{H^*}$ and $(A^*)^{\mathrm{co}(H^*)}$ become a left H^*-comodule algebra and a left H^*-module coalgebra, respectively.

By the above structures and the identification (2.4), we see that $A^{*\mathrm{co}(H^*)}$ ($= \overline{A^*}^{H^*}$) coincides with $\mathcal{F}(A^*; \pi^*, \iota^*) = A^{*\mathrm{co}(\pi^*,\iota^*)}$ in ${}^{H^*}_{H^*}\mathcal{YDH}$. Therefore, under the identification (2.4), the duals of the left H-Yetter-Drinfeld structure maps $H \otimes \mathcal{B} \to \mathcal{B}$ and $\mathcal{B} \to H \otimes \mathcal{B}$ of $\mathcal{B}$ make $\mathcal{B}^* \cong (\overline{A}^H)^* \cong A^{*\mathrm{co}(H^*)}$ into a Hopf algebra in ${}^{H^*}_{H^*}\mathcal{YD}$, and it is naturally isomorphic to $A^{*\mathrm{co}(\pi^*,\iota^*)}$.

$$\begin{array}{ccccccc} \mathscr{T}_H^{\mathsf{fd}} & \xrightarrow{\approx} & \mathscr{T}_{H^*}^{\mathsf{fd}} & \qquad & (A;\iota,\pi) & \longmapsto & (A^*;\pi^*,\iota^*) \\ \approx\downarrow & & \downarrow\approx & & \downarrow & & \downarrow \\ {}^H_H\mathcal{YDH}^{\mathsf{fd}} & \xrightarrow{\approx} & {}^{H^*}_{H^*}\mathcal{YDH}^{\mathsf{fd}}, & & \mathcal{B} = A^{\mathrm{co}(\iota,\pi)} & \longmapsto & \mathcal{B}^* \cong A^{*\mathrm{co}(\pi^*,\iota^*)}. \end{array}$$

For $\mathcal{B} \in {}^H_H\mathcal{YDH}^{\mathsf{fd}}$ with $(A;\iota,\pi) := \mathcal{G}(\mathcal{B})$, we get the following commutative diagram.

$$\begin{array}{ccccc} (\mathcal{B} \# H) \times (\mathcal{B}^* \# H^*) & & & \qquad (b \# h,\ f \# g) & \\ \cong\downarrow & \searrow & & \downarrow \quad \searrow & \\ A \times A^* & \xrightarrow[\langle\ ,\ \rangle]{} & \Bbbk, & (b\iota(h),\ f * \pi^*(g)) \longmapsto & f(b)g(h). \end{array}$$

Here, $\langle\ ,\ \rangle$ is the natural evaluation (the canonical pairing) of A and $f * \pi^*(g)$ is the convolution product of f and $\pi^*(g)$ ($= g \circ \pi$).

3. Hopf superalgebras

In this section, we suppose that the characteristic of $\Bbbk$ is not 2. Let $\mathbb{Z}_2 = \mathbb{Z}/2\mathbb{Z} = \{\bar{0}, \bar{1}\}$ denote the additive group of order two. We sometimes identify $\mathbb{Z}_2$ with the multiplicative group $\{\boldsymbol{e}, \boldsymbol{\sigma}\}$ of order two, where $\boldsymbol{e}$ is the identity element and $\boldsymbol{\sigma}^2 = \boldsymbol{e}$.

3.1. Superspaces. Let $\Bbbk\mathbb{Z}_2$ be the group algebra of $\mathbb{Z}_2$ over $\Bbbk$. An object of the category ${}^{\Bbbk\mathbb{Z}_2}\mathcal{M}$ of left $\Bbbk\mathbb{Z}_2$-comodules is just a $\mathbb{Z}_2$-graded vector space $V = V_{\bar{0}} \oplus V_{\bar{1}}$. The $\bar{0}$-component $V_{\bar{0}}$ (resp., $\bar{1}$-component $V_{\bar{1}}$) of V is called an *even part* (resp., *odd part*) of V. For a homogeneous element $0 \neq v \in V_{\bar{0}} \cup V_{\bar{1}}$, we denote its degree by $|v|$ ($\in \{0,1\}$). For simplicity, when we write $|v|$, v is always supposed to be homogeneous.

Since $\Bbbk\mathbb{Z}_2$ is a finite-dimensional Hopf algebra, ${}^{\Bbbk\mathbb{Z}_2}\mathcal{M}$ naturally becomes a tensor category. Note that our base field $\Bbbk$ is supposed to be a $\mathbb{Z}_2$-graded vector space with $\Bbbk_{\bar{0}} = \Bbbk$ and $\Bbbk_{\bar{1}} = 0$. Moreover, the following *supersymmetry* makes ${}^{\Bbbk\mathbb{Z}_2}\mathcal{M}$ into a symmetric tensor category.

$$\mathsf{s\text{-}sym}_{V,W} : V \otimes W \longrightarrow W \otimes V; \quad v \otimes w \longmapsto (-1)^{|v||w|} w \otimes v \quad (V, W \in {}^{\Bbbk\mathbb{Z}_2}\mathcal{M}). \tag{3.1}$$

We denote this nontrivial symmetric tensor category ${}^{\Bbbk\mathbb{Z}_2}\mathcal{M}$ by $\mathcal{SV}$. An object of $\mathcal{SV}$ is called a *superspace*. A superspace $V = V_{\bar{0}} \oplus V_{\bar{1}}$ is said to be *purely even* if the odd part $V_{\bar{1}}$ of V is zero. Obviously, $\Bbbk^{1|0} := \Bbbk \oplus 0$ and $\Bbbk^{0|1} := 0 \oplus \Bbbk$ are simple objects in $\mathcal{SV}$. Conversely, one sees that all simple objects are exhausted by $\Bbbk^{1|0}$ and $\Bbbk^{0|1}$.

In the following, we let $\boldsymbol{e}^*, \boldsymbol{\sigma}^* \in (\Bbbk\mathbb{Z}_2)^*$ denote the dual bases of $\boldsymbol{e}, \boldsymbol{\sigma} \in \Bbbk\mathbb{Z}_2$, respectively. The following is obvious.

LEMMA 3.1. *The linear map* $\Bbbk\mathbb{Z}_2 \to (\Bbbk\mathbb{Z}_2)^*; \boldsymbol{\sigma}^i \mapsto \boldsymbol{e}^* + (-1)^i \boldsymbol{\sigma}^*$ *is a Hopf algebra isomorphism. In particular,* $\Bbbk\mathbb{Z}_2$ *is self-dual.*

Let $V = V_{\bar{0}} \oplus V_{\bar{1}}$ be an object in $\mathcal{SV}$. By Lemma 3.1, V naturally becomes a left $\Bbbk\mathbb{Z}_2$-module with structure

$$\Bbbk\mathbb{Z}_2 \otimes V \longrightarrow V; \quad \boldsymbol{\sigma}^i \otimes v \longmapsto \boldsymbol{\sigma}^i.v := v_{\bar{0}} + (-1)^i v_{\bar{1}},$$

where $v = v_{\bar{0}} + v_{\bar{1}}$ with $v_{\bar{0}} \in V_{\bar{0}}$ and $v_{\bar{1}} \in V_{\bar{1}}$. If $v \in V$ is homogeneous, then we can simply write as $\boldsymbol{\sigma}^i.v = (-1)^{i|v|} v$ for $i \in \{0,1\}$.

The following result tells us that a Hopf algebraic approach can be employed in the study of the theory of super-mathematics (see Masuoka [**13**, Section 2] for example).

LEMMA 3.2. *The category* $\mathcal{SV}$ *is a monoidal full subcategory of* ${}^{\Bbbk\mathbb{Z}_2}_{\Bbbk\mathbb{Z}_2}\mathcal{YD}$*, and the braiding of* ${}^{\Bbbk\mathbb{Z}_2}_{\Bbbk\mathbb{Z}_2}\mathcal{YD}$ *restricts to the supersymmetry.*

PROOF. Let $V = V_{\bar{0}} \oplus V_{\bar{1}} \in \mathcal{SV}$. First, note that the left $\Bbbk\mathbb{Z}_2$-comodule structure $\delta : V \to \Bbbk\mathbb{Z}_2 \otimes V$ is explicitly given by $\delta(v) = \boldsymbol{\sigma}^i \otimes v$ for $v \in V_{\bar{i}}$. For $i \in \{0,1\}$ and $v = v_{\bar{0}} + v_{\bar{1}}$ with $v_{\bar{0}} \in V_{\bar{0}}$ and $v_{\bar{1}} \in V_{\bar{1}}$, we have

$$\delta(\boldsymbol{\sigma}^i.v) = \delta(v_{\bar{0}}) + (-1)^i \delta(v_{\bar{1}}) = \boldsymbol{e} \otimes v_{\bar{0}} + (-1)^i \boldsymbol{\sigma} \otimes v_{\bar{1}}.$$

This shows that V is an object of ${}^{\Bbbk\mathbb{Z}_2}_{\Bbbk\mathbb{Z}_2}\mathcal{YD}$. Take $V, W \in \mathcal{SV}$ and $v \in V, w \in W$ to be homogeneous. Since $\delta(v) = \boldsymbol{\sigma}^{|v|} \otimes v$ and $\boldsymbol{\sigma}^{|v|}.w = (-1)^{|v||w|} w$, the braiding given in (2.5) is calculated as

$$c_{V,W}(v \otimes w) = \boldsymbol{\sigma}^{|v|}.w \otimes v = (-1)^{|v||w|} w \otimes v = \mathsf{s\text{-}sym}_{V,W}(v \otimes w).$$

The proof is done. $\square$

A *superalgebra* (resp., *supercoalgebra*) is an algebra (resp., coalgebra) in $\mathcal{SV}$ ($\subset {}^{\Bbbk\mathbb{Z}_2}_{\Bbbk\mathbb{Z}_2}\mathcal{YD}$). Let $\mathcal{A} = (\mathcal{A}; m_{\mathcal{A}}, u_{\mathcal{A}})$ be a superalgebra. We say that $\mathcal{A}$ is *super-commutative* if it satisfies $ab = (-1)^{|a||b|} ba$ for all $a, b \in \mathcal{A}$. A superspace $V \in \mathcal{SV}$ is called a *left $\mathcal{A}$-supermodule* if it is equipped with a morphism $\rho : \mathcal{A} \otimes V \to V$ in $\mathcal{SV}$ such that $\rho \circ (m_A \otimes \mathsf{id}_V) = \rho \circ (\mathsf{id}_A \otimes \rho)$ and $\rho \circ (u_A \otimes \mathsf{id}_V) = \mathsf{id}_V$. The category of all left $\mathcal{A}$-supermodules is denoted by ${}_{\mathcal{A}}\mathcal{SM}$. We say that a finite-dimensional superalgebra $\mathcal{A}$ is *semisimple* if the category ${}_{\mathcal{A}}\mathcal{SM}$ is semisimple, that is, every object of ${}_{\mathcal{A}}\mathcal{SM}$ is direct sum of simple objects.

Let $\mathcal{C} = (\mathcal{C}; \Delta_{\mathcal{C}}, \varepsilon_{\mathcal{C}})$ be a supercoalgebra. We say that $\mathcal{C}$ is *super-cocommutative* if it satisfies $\Delta_{\mathcal{C}}(c) = (-1)^{|c_{(1)}||c_{(2)}|} c_{(2)} \otimes c_{(1)}$ for all $c \in \mathcal{C}$. The notion of right $\mathcal{C}$-supercomodule is defined in a similar manner. The category of all right $\mathcal{C}$-supercomodules is denoted by $\mathcal{SM}^{\mathcal{C}}$. We say that $\mathcal{C}$ is *pointed* if any simple right $\mathcal{C}$-supercomodule is one-dimensional.

3.2. Hopf superalgebras. A *Hopf superalgebra* is a Hopf algebra in $\mathcal{SV}$ ($\subset {}^{\Bbbk\mathbb{Z}_2}_{\Bbbk\mathbb{Z}_2}\mathcal{YD}$). One sees that the antipode $S_{\mathcal{H}}$ of a Hopf superalgebra $\mathcal{H}$ is a *super-anti-algebra map*, that is, it satisfies

$$S_{\mathcal{H}}(hh') = (-1)^{|h||h'|} S_{\mathcal{H}}(h') S_{\mathcal{H}}(h) \tag{3.2}$$

for $h, h' \in \mathcal{H}$. Also, one can show that $S_{\mathcal{H}}$ is a *super-anti-coalgebra map* (we omit the definition here since we do not use it in this paper).

Let $\mathcal{H} = (\mathcal{H}; m_{\mathcal{H}}, u_{\mathcal{H}}, \Delta_{\mathcal{H}}, \varepsilon_{\mathcal{H}}, S_{\mathcal{H}})$ be a Hopf superalgebra. Note that the comultiplication $\Delta_{\mathcal{H}}$ of $\mathcal{H}$ satisfies $\Delta_{\mathcal{H}}(\mathcal{H}_{\bar{\epsilon}}) \subset \sum_{\bar{\eta}+\bar{\eta}'=\bar{\epsilon}} \mathcal{H}_{\bar{\eta}} \otimes \mathcal{H}_{\bar{\eta}'}$ for $\epsilon \in \{0,1\}$. As in the nonsuper setting, the set

$$\mathbf{G}(\mathcal{H}) := \{g \in \mathcal{H}_{\bar{0}} \mid \varepsilon_{\mathcal{H}}(g) = 1,\ \Delta_{\mathcal{H}}(g) = g \otimes g\}$$

naturally becomes a group with $g^{-1} = S_{\mathcal{H}}(g)$ for $g \in \mathbf{G}(\mathcal{H})$. An element of $\mathbf{G}(\mathcal{H})$ is called a *group-like* element of $\mathcal{H}$ (cf. Remark 3.7).

For a fixed $g \in \mathbf{G}(\mathcal{H})$, an element $z \in \mathcal{H}$ is said to be *g-skew primitive* if it satisfies

$$\Delta_{\mathcal{H}}(z) = g \otimes z + z \otimes 1_{\mathcal{H}}.$$

In this case, we see that $\varepsilon_{\mathcal{H}}(z) = 0$ and $S_{\mathcal{H}}(z) = -g^{-1}z$. If a g-skew primitive element z belongs to $\mathcal{H}_{\bar{1}}$, then we shall call z an *odd g-skew primitive.* An (resp., odd) $1_{\mathcal{H}}$-skew primitive element is simply called an (resp., *odd*) *primitive* element. The set of all primitive elements in $\mathcal{H}$ is denoted by $\mathbf{P}(\mathcal{H})$. One easily sees that $\mathbf{P}(\mathcal{H})$ becomes a *Lie superalgebra* (i.e., a Lie algebra in $\mathcal{SV}$) with *superbracket* $[x, y] := xy - (-1)^{|x||y|}yx$ ($x, y \in \mathbf{P}(\mathcal{H})$).

EXAMPLE 3.3. Let V be a finite-dimensional vector space with basis $\{z_1, \ldots, z_\theta\}$. The exterior algebra

$$\mathcal{H} := \bigwedge(V) = \bigwedge(z_1, \ldots, z_\theta)$$

over V naturally becomes a super-commutative superalgebra, called the *exterior superalgebra.* Moreover, $\mathcal{H}$ becomes a super-commutative and super-cocommutative Hopf superalgebra by letting each $z_1, \ldots, z_\theta$ be odd primitive. One sees that $\mathbf{G}(\mathcal{H}) = \{1\}$, $\mathbf{P}(\mathcal{H}) = V$ and $\mathcal{H}$ is pointed. □

We assume our base field $\Bbbk$ is an algebraically closed field of characteristic zero throughout this Section 3.2.

Let $\mathcal{H}$ be a super-cocommutative Hopf superalgebra. Then the group $\Gamma := \mathbf{G}(\mathcal{H})$ acts on $\mathfrak{g} := \mathbf{P}(\mathcal{H})$ by the adjoint action as usual and the universal enveloping superalgebra $\mathcal{U}(\mathfrak{g})$ of the Lie superalgebra $\mathfrak{g}$ forms a left $\Bbbk\Gamma$-supermodule. The following result was first shown by Kostant [**9**, Theorem 3.3] (see also Andruskiewitsch, Etingof and Gelaki [**3**, Corollary 2.3.5] and Masuoka [**13**, Theorems 3.6 and 4.5]).

THEOREM 3.4. *$\mathcal{H}$ is isomorphic to the smash product $\Bbbk\Gamma \ltimes \mathcal{U}(\mathfrak{g})$ of $\Bbbk\Gamma$ and $\mathcal{U}(\mathfrak{g})$. In particular, if $\mathcal{H}$ is finite-dimensional, then $\mathcal{H} \cong \Bbbk\Gamma \ltimes \bigwedge(\mathfrak{g})$.*

Therefore, as in the nonsuper situation, in the classification theory of finite-dimensional Hopf superalgebras, the case of super-commutative or the case of super-cocommutative can be excluded as they are trivial.

EXAMPLES 3.5. Fix a primitive fourth root of unity $\zeta_4 \in \Bbbk$. According to Aissaoui and Makhlouf [**1**], the following is a complete list of pairwise nonisomorphic Hopf superalgebras of dimension 4 whose odd parts are nonzero.

- $\mathcal{H}_4^{(1)} := \bigwedge(z_1, z_2)$.

- $\mathcal{H}_4^{(2)} := \Bbbk\langle g, z \mid g^2 = 1, z^2 = 0, gz = zg\rangle$, where g is group-like and z is odd primitive. Note that $\mathcal{H}_4^{(2)} \cong \Bbbk\mathbb{Z}_2 \otimes \bigwedge(z)$.
- $\mathcal{H}_4^{(3)} := \Bbbk\langle g, z \mid g^2 = 1, z^2 = 0, gz = zg\rangle$, where g is group-like and z is odd g-skew primitive.
- $\mathcal{H}_4^{(4)} := \Bbbk\langle g, z \mid g^2 = 1, z^2 = 0, gz = -zg\rangle$, where g is group-like and z is odd primitive.
- $\mathcal{A}_4(\pm\zeta_4) := \Bbbk\langle x, z \mid x^2 + z^2 = 1, xz = zx = 0\rangle$, where x is even, z is odd and

$$\Delta(x) = x \otimes x \pm \zeta_4 z \otimes z, \quad \varepsilon(x) = 1, \quad S(x) = x,$$
$$\Delta(z) = x \otimes z + z \otimes x, \quad \varepsilon(z) = 0, \quad S(z) = \mp\zeta_4 z.$$

For the antipodes of $\mathcal{A}_4(\zeta_4)$ and $\mathcal{A}_4(-\zeta_4)$, one should note the formula (3.2). Also, one should note $\mathcal{A}_4(\zeta_4) \not\cong \mathcal{A}_4(-\zeta_4)$ since the eigenvalues of their antipodes are different. As we have mentioned, the exterior superalgebra $\mathcal{H}_4^{(1)} = \bigwedge(z_1, z_2)$ is pointed. We will see that $\mathcal{H}_4^{(2)}$, $\mathcal{H}_4^{(3)}$ and $\mathcal{H}_4^{(4)}$ are pointed, and $\mathcal{A}_4(\zeta_4)$ and $\mathcal{A}_4(-\zeta_4)$ are semisimple, see Theorem 5.5 and Example 4.5, respectively. □

3.3. A construction of Hopf superalgebras. We recall from [**3**] a construction of Hopf superalgebras. Let $H = (H; m_H, u_H, \Delta_H, \varepsilon_H, S_H)$ be a Hopf algebra, and let $c \in \mathbf{G}(H)$ be a group-like element such that $c^2 = 1_H$. From the pair (H, c), we can construct a Hopf superalgebra $\mathcal{H}$ as follows: As an algebra, $\mathcal{H} = H$. We make it into a $\mathbb{Z}_2$-graded vector space by letting

$$\mathcal{H}_{\bar{\epsilon}} := \{h \in \mathcal{H} \mid chc = (-1)^{\epsilon} h\} \qquad (\epsilon \in \{0, 1\}).$$

For each $h \in \mathcal{H}$, we define $\Delta_{\mathcal{H},\bar{0}}(h) \in \mathcal{H} \otimes \mathcal{H}_{\bar{0}}$ and $\Delta_{\mathcal{H},\bar{1}}(h) \in \mathcal{H} \otimes \mathcal{H}_{\bar{1}}$ satisfying $\Delta_H(h) = \Delta_{\mathcal{H},\bar{0}}(h) + \Delta_{\mathcal{H},\bar{1}}(h)$.

THEOREM 3.6 ([**3**, Theorem 3.1.1]). *The superalgebra $\mathcal{H}$ becomes a Hopf superalgebra together with the comultiplication, the counit and the antipode given by*

$$\Delta_{\mathcal{H}}(h) := \Delta_{\mathcal{H},\bar{0}}(h) - (-1)^{|h|}(c \otimes 1_H)\Delta_{\mathcal{H},\bar{1}}(h), \tag{3.3}$$

$\varepsilon_{\mathcal{H}} := \varepsilon_H$ *and* $S_{\mathcal{H}}(h) := c^{|h|} S_H(h)$ *for* $h \in \mathcal{H}$, *respectively. This construction establishes a bijective correspondence between:*

- *Pairs (H, c) consisting of a Hopf algebra H and an element $c \in \mathbf{G}(H)$ satisfying $c^2 = 1_H$.*
- *Pairs $(\mathcal{H}, g)$ consisting of a Hopf superalgebra $\mathcal{H}$ and en element $g \in \mathbf{G}(\mathcal{H})$ satisfying $g^2 = 1_{\mathcal{H}}$ and $gzg = (-1)^{|z|} z$ for all $z \in \mathcal{H}$.*

REMARK 3.7. For this theorem to hold, a group-like element of $\mathcal{H}$ needs to be homogeneous and this is one of reasons why we adopt our definition of $\mathbf{G}(\mathcal{H})$. We also note that there happen to exist a nonhomogeneous element $g \in \mathcal{H}$ such that $\varepsilon(g) = 1$ and $\Delta(g) = g \otimes g$. For example, we shall consider the Hopf superalgebra $\mathcal{H} = \mathcal{A}_4(-\zeta_4)$ given in Example 3.5. By definition, we have $\mathbf{G}(\mathcal{H}) = \{1_{\mathcal{H}}, x^2 - z^2\}$. Let ζ_8 be an element in $\Bbbk$ satisfying $\zeta_8^2 = \zeta_4$. Then $g := x + \zeta_8^3 z$ satisfies $\varepsilon(g) = 1$ and $\Delta(g) = g \otimes g$. However, g is nonhomogeneous, and hence $g \notin \mathcal{H}_{\bar{0}}$. □

We will see that a Hopf superalgebra obtained from the group algebra $\Bbbk\mathfrak{S}_3$ by Theorem 3.6 is, up to isomorphisms, a unique semisimple Hopf superalgebra of dimension 6 with nonzero odd part (see Section 5.4). One of four-dimensional Hopf superalgebra mentioned in Example 3.5 is also obtained by the construction of Theorem 3.6.

EXAMPLE 3.8. Let H_4 denote *Sweedler's four-dimensional Hopf algebra*

$$H_4 = \Bbbk\langle c, x \mid c^2 = 1, x^2 = 0, cx = -xc\rangle, \tag{3.4}$$

where c is group-like and x is c-skew primitive. One sees that the corresponding Hopf superalgebra of (H_4, c) is $(\mathcal{H}_4^{(4)}, g)$, where $\mathcal{H}_4^{(4)}$ is given in Example 3.5. □

3.4. Duals of Hopf superalgebras. For superspaces V, V' $(\in \mathcal{SV})$, a bilinear map $\langle\ ,\ \rangle : V' \times V \to \Bbbk$ is said to be a *pairing on V' and V* if it satisfies $\langle V'_{\bar\epsilon}, V_{\bar\eta}\rangle = 0$ if $\epsilon \neq \eta$ $(\epsilon, \eta \in \{0,1\})$. In other words, the induced map $\langle\ ,\ \rangle : V' \otimes V \to \Bbbk^{1|0}$ is a morphism in $\mathcal{SV}$. Let $\langle\ ,\ \rangle' : W' \times W \to \Bbbk$ be a paring on W' and W. Then we define a pairing

$$(V' \otimes W') \times (V \otimes W) \longrightarrow \Bbbk; \quad (f \otimes g, v \otimes w) \longmapsto \langle f, v\rangle\langle g, w\rangle'.$$

on $V' \otimes W'$ and $V \otimes W$. If $V' = W', V = W$ and $\langle\ ,\ \rangle' = \langle\ ,\ \rangle$, then we simply denote the pairing on $V' \otimes V'$ and $V \otimes V$ by the same symbol

$$\langle f \otimes g, v \otimes w\rangle = \langle f, v\rangle\langle g, w\rangle \quad (f, g \in V', v, w \in V)$$

as usual.

DEFINITION 3.9. For Hopf superalgebras $\mathcal{K} = (\mathcal{K}; m_{\mathcal{K}}, u_{\mathcal{K}}, \Delta_{\mathcal{K}}, \varepsilon_{\mathcal{K}}, S_{\mathcal{K}})$ and $\mathcal{H} = (\mathcal{H}; m_{\mathcal{H}}, u_{\mathcal{H}}, \Delta_{\mathcal{H}}, \varepsilon_{\mathcal{H}}, S_{\mathcal{H}})$, a pairing $\langle\ ,\ \rangle : \mathcal{K} \times \mathcal{H} \to \Bbbk$ on $\mathcal{K}$ and $\mathcal{H}$ is called a *Hopf pairing* if it satisfies the following conditions.

$$\langle k, hh'\rangle = \langle \Delta_{\mathcal{K}}(k), h \otimes h'\rangle\ (= \langle k_{(1)}, h\rangle\langle k_{(2)}, h\rangle), \quad \langle k, 1_{\mathcal{H}}\rangle = \varepsilon_{\mathcal{K}}(k),$$
$$\langle kk', h\rangle = \langle k \otimes k', \Delta_{\mathcal{H}}(h)\rangle\ (= \langle k, h_{(1)}\rangle\langle k', h_{(2)}\rangle), \quad \langle 1_{\mathcal{K}}, h\rangle = \varepsilon_{\mathcal{H}}(h)$$

for $k, k' \in \mathcal{K}$ and $h, h' \in \mathcal{H}$. □

In this case, we have $\langle k, S_{\mathcal{H}}(h)\rangle = \langle S_{\mathcal{K}}(k), h\rangle$ for $k \in \mathcal{K}, h \in \mathcal{H}$, as in the nonsuper setting.

Let $V \in \mathcal{SV}$, and let V^* denote the linear dual space of V (over $\Bbbk$) . By letting $(V^*)_{\bar\epsilon} := (V_{\bar\epsilon})^*$ $(\epsilon \in \{0,1\})$, we make V^* into an object of $\mathcal{SV}$. Then the evaluation map

$$V^* \times V \longrightarrow \Bbbk; \quad (f, v) \longmapsto f(v)$$

is a pairing on V^* and V.

For a finite-dimensional Hopf superalgebra $\mathcal{H}$, one can make $\mathcal{H}^*$ into a Hopf superalgebra, called the *dual Hopf superalgebra* of $\mathcal{H}$, so that the evaluation map $\langle\ ,\ \rangle : \mathcal{H}^* \times \mathcal{H} \to \Bbbk$ is a (nondegenerate) Hopf pairing. Since $\mathcal{H} \in {}^{\Bbbk\mathbb{Z}_2}_{\Bbbk\mathbb{Z}_2}\mathcal{YDH}^{\mathsf{fd}}$ and $(\Bbbk\mathbb{Z}_2)^* \cong \Bbbk\mathbb{Z}_2$, the linear dual $\mathcal{H}^*$ of $\mathcal{H}$ may be regarded as an object in ${}^{\Bbbk\mathbb{Z}_2}_{\Bbbk\mathbb{Z}_2}\mathcal{YDH}^{\mathsf{fd}}$, see Section 2.4. One easily sees that $\mathcal{H}^*$ actually is an object in $\mathcal{SV}$ and its Hopf superalgebra structure coincides with the one defined above.

REMARK 3.10. Some literature uses another definition of the dual Hopf superalgebras which we shall explain below. For pairings $\langle\ ,\ \rangle : V' \times V \to \Bbbk$ and $\langle\ ,\ \rangle' : W' \times W \to \Bbbk$, we note that there is another way to define a pairing on $V' \otimes W'$ and $V \otimes W$ as follows.

$$(V' \otimes W') \times (V \otimes W) \longrightarrow \Bbbk; \quad (f \otimes g, v \otimes w) \longmapsto (-1)^{|g||v|}\langle f, v\rangle\langle g, w\rangle'.$$

Let $\mathcal{H}$ be a finite-dimensional Hopf superalgebra. If we use this pairing, then we can also make the linear dual $\mathcal{H}^*$ of $\mathcal{H}$ into a Hopf superalgebra, which we denote by $\mathcal{H}^{\bar{*}}$ (just here), satisfying

$$\langle f, hh'\rangle = (-1)^{|f_{(2)}||h|}\langle f_{(1)}, h\rangle\langle f_{(2)}, h\rangle, \quad \langle f, 1_{\mathcal{H}}\rangle = f(1_{\mathcal{H}}),$$
$$\langle fg, h\rangle = (-1)^{|h_{(1)}||g|}\langle f, h_{(1)}\rangle\langle g, h_{(2)}\rangle, \quad \langle 1_{\mathcal{H}^{\bar{*}}}, h\rangle = \varepsilon_{\mathcal{H}}(h)$$

for $f, g \in \mathcal{H}^*$ and $h, h' \in \mathcal{H}$. Suppose that our base field $\Bbbk$ contains a primitive fourth root of unity ζ_4. Then one sees that the map $\mathcal{H}^* \to \mathcal{H}^{\bar{*}}; f \mapsto \zeta_4^{|f|} f$ is a Hopf superalgebra isomorphism. See [**16**, Sections 3.1 and 3.2] for the detail (see also [**14**, Section 1]). □

EXAMPLE 3.11. Let V be a finite-dimensional vector space with basis $\{z_1, \dots, z_\theta\}$. Then the evaluation map $V^* \times V \to \Bbbk$ extends to a nondegenerate Hopf pairing $\langle\ ,\ \rangle : \bigwedge(V^*) \times \bigwedge(V) \to \Bbbk$ defined by

$$\langle f_1 \wedge \cdots \wedge f_n, v_1 \wedge \cdots \wedge v_m\rangle = \delta_{n,m} \det\big(f_j(v_i)\big)_{i,j} \qquad (n, m \in \mathbb{N}),$$

where $\delta_{n,m}$ is the Kronecker symbol. In particular, $\bigwedge(z_1, \dots, z_\theta)$ is a self-dual Hopf superalgebra. □

EXAMPLE 3.12. Recall the Hopf superalgebras $\mathcal{H}_4^{(2)}, \mathcal{H}_4^{(3)}, \mathcal{H}_4^{(4)}$ given in Example 3.5. One sees that the pairings

$$\langle\ ,\ \rangle : \mathcal{H}_4^{(2)} \times \mathcal{H}_4^{(2)} \to \Bbbk; \quad \langle g, g\rangle = -1,\ \langle z, z\rangle = 1,\ \langle g, z\rangle = \langle z, g\rangle = 0$$

and

$$\langle\ ,\ \rangle : \mathcal{H}_4^{(3)} \times \mathcal{H}_4^{(4)} \to \Bbbk; \quad \langle g, g\rangle = -1,\ \langle z, z\rangle = 1,\ \langle g, z\rangle = \langle z, g\rangle = 0$$

are nondegenerate Hopf pairings. In particular, $\mathcal{H}_4^{(2)}$ is self-dual and the dual of $\mathcal{H}_4^{(3)}$ is isomorphic to $\mathcal{H}_4^{(4)}$. □

4. Bosonizations and super-forms

In this section, we also suppose that the characteristic of $\Bbbk$ is not 2.

4.1. Bosonization of Hopf superalgebras. In the following, we fix a Hopf superalgebra $\mathcal{H} = (\mathcal{H}; m_{\mathcal{H}}, u_{\mathcal{H}}, \Delta_{\mathcal{H}}, \varepsilon_{\mathcal{H}}, S_{\mathcal{H}})$. Since $\mathcal{H}$ ($\in \mathcal{SV}$) can be regarded as an object in the category of left $\Bbbk\mathbb{Z}_2$-Yetter–Drinfeld modules ${}^{\Bbbk\mathbb{Z}_2}_{\Bbbk\mathbb{Z}_2}\mathcal{YD}$ (see Lemma 3.2), we may consider the bosonization

$$\widehat{\mathcal{H}} := \mathcal{H} \# \Bbbk\mathbb{Z}_2$$

of $\mathcal{H}$ by $\Bbbk\mathbb{Z}_2$. The Hopf algebra structure of $\widehat{\mathcal{H}}$ is explicitly given as follows.

$$\begin{aligned}
(h \# \boldsymbol{\sigma}^i)(h' \# \boldsymbol{\sigma}^j) &= (-1)^{i|h'|} hh' \# \boldsymbol{\sigma}^{i+j}, \\
1_{\widehat{\mathcal{H}}} &= 1_{\mathcal{H}} \# \boldsymbol{e}, \\
\Delta_{\widehat{\mathcal{H}}}(h \# \boldsymbol{\sigma}^i) &= h_{(1)} \# \boldsymbol{\sigma}^{i+|h_{(2)}|} \otimes h_{(2)} \# \boldsymbol{\sigma}^i, \\
\varepsilon_{\widehat{\mathcal{H}}}(h \# \boldsymbol{\sigma}^i) &= \varepsilon_{\mathcal{H}}(h), \\
S_{\widehat{\mathcal{H}}}(h \# \boldsymbol{\sigma}^i) &= (-1)^{i+|h|} S_{\mathcal{H}}(h) \# \boldsymbol{\sigma}^{i+|h|}
\end{aligned}$$

for $h, h' \in \mathcal{H}$ and $i, j \in \{0, 1\}$. One easily sees that the map

$$\mathbf{G}(\mathcal{H}) \times \mathbb{Z}_2 \longrightarrow \mathbf{G}(\widehat{\mathcal{H}}); \quad (g, \boldsymbol{\sigma}^i) \longmapsto g \# \boldsymbol{\sigma}^i$$

is a group isomorphism.

EXAMPLE 4.1. The bosonization of the exterior superalgebra $\bigwedge(z_1, \dots, z_\theta)$ is isomorphic to the following Hopf algebra.

$$\Bbbk\langle c, x_i \mid c^2 = 1, x_i x_j = -x_j x_i, c x_i = -x_i c \ (i, j \in \{1, \dots, n\})\rangle,$$

where c is group-like and x_i is c-skew primitive ($i \in \{1, \dots, n\}$). In particular, the bosonization of $\bigwedge(z)$ is isomorphic to Sweedler's four-dimensional Hopf algebra H_4 (cf. Theorem 5.1). □

Any Hopf algebra H can be naturally regarded as a Hopf superalgebra by letting $H_{\bar{0}} := H$ and $H_{\bar{1}} := 0$. Thus, in the classification theory of Hopf superalgebras, we are interested in *non*-purely even Hopf superalgebras, that is, a Hopf superalgebra whose odd part is nonzero. The following is easy to see, however, it is remarkable in the classification theory of Hopf superalgebras (see Theorem 5.2).

LEMMA 4.2. *If $\mathcal{H}_{\bar{1}} \neq 0$, then $\widehat{\mathcal{H}}$ is neither commutative nor cocommutative.*

PROOF. By the assumption, we can take $0 \neq x \in \mathcal{H}_{\bar{1}}$. The we have

$$(1_{\mathcal{H}} \# \boldsymbol{\sigma})(x \# \boldsymbol{e}) = -x \# \boldsymbol{\sigma} \neq x \# \boldsymbol{\sigma} = (x \# \boldsymbol{e})(1_{\mathcal{H}} \# \boldsymbol{\sigma}),$$

and hence $\widehat{\mathcal{H}}$ is not commutative. We write $\Delta_{\mathcal{H}}(x) = \sum_{i=1}^m a_i \otimes b_i + \sum_{j=1}^n c_j \otimes d_j$, where $a_i, d_j \in \mathcal{H}_{\bar{0}}$, $b_i, c_j \in \mathcal{H}_{\bar{1}}$ and $\{c_j\}_{j=1}^n$ is linearly independent. Then

$$\Delta_{\widehat{\mathcal{H}}}(x \# \boldsymbol{e}) = \sum_{i=1}^m a_i \# \boldsymbol{e} \otimes b_i \# \boldsymbol{e} + \sum_{j=1}^n c_j \# \boldsymbol{\sigma} \otimes d_j \# \boldsymbol{e}.$$

If $d_j = 0$ for all j, then we have $x = \sum_{i=1}^m a_i \varepsilon_{\mathcal{H}}(b_i) = 0$, a contradiction. Thus $d_{j'} \neq 0$ for some j'. This implies that $x \# \boldsymbol{e}$ is not cocommutative. □

For a left $\mathcal{H}$-supermodule V, the action

$$\widehat{\mathcal{H}} \otimes V \longrightarrow V; \quad (h \# \boldsymbol{\sigma}^i) \otimes v \longmapsto (h \# \boldsymbol{\sigma}^i).v := (-1)^{i|v|} h.v$$

makes V into a left $\widehat{\mathcal{H}}$-module. This correspondence gives a category equivalence between the category ${}_{\mathcal{H}}\mathcal{SM}$ of left $\mathcal{H}$-supermodules and the category ${}_{\widehat{\mathcal{H}}}\mathcal{M}$ of left $\widehat{\mathcal{H}}$-modules. Dually, for a right $\mathcal{H}$-supercomodule V, the coaction

$$V \longrightarrow V \otimes \widehat{\mathcal{H}}; \quad v \longmapsto v_{(0)} \otimes (v_{(1)} \# \boldsymbol{\sigma}^{|v_{(0)}|})$$

makes V into a right $\widehat{\mathcal{H}}$-comodule, where the original right $\mathcal{H}$-comodule structure on V is written as $V \to V \otimes \mathcal{H}; v \mapsto v_{(0)} \otimes v_{(1)}$. This gives a category equivalence between the category $\mathcal{SM}^{\mathcal{H}}$ of right $\mathcal{H}$-supercomodules and the category $\mathcal{M}^{\widehat{\mathcal{H}}}$ of right $\widehat{\mathcal{H}}$-comodules.

Therefore, we have the following observation (cf. Andruskiewitsch, Angiono, and Yamane [**2**, Section 1.8]; see also Masuoka [**15**, Lemma 4]).

LEMMA 4.3. *$\mathcal{H}$ is semisimple (resp., pointed) if and only if $\widehat{\mathcal{H}}$ is semisimple (resp., pointed).*

DEFINITION 4.4. Let A be a Hopf algebra. We say that $\mathcal{H}$ is a *super-form* of A if A is isomorphic to $\widehat{\mathcal{H}}$ as a Hopf algebra. If $\mathcal{H}$ is purely even, then $\mathcal{H}$ is called a *trivial super-form* of A. □

By definition, all super-forms of a semisimple (resp., pointed) Hopf algebra is also semisimple (resp., pointed).

EXAMPLE 4.5. Suppose that $\Bbbk$ is an algebraically closed field of characteristic zero, and let $\zeta_4 \in \Bbbk$ be a fixed primitive fourth root of unity. According to Masuoka [**12**] (see also Kac and Paljutkin [**8**]), there is a unique (up to isomorphism) semisimple Hopf algebra of dimension 8 that is neither commutative nor cocommutative. We denote it by H_8. As noted in [**12**, Remark 2.14], the dual of H_8^* (which is actually isomorphic to H_8) has the following presentation:

$$H_8^* = \Bbbk\langle c, s, h \mid c^2 - s^2 = 1, sc = cs = 0, h^2 = 1, ch = hc, sh = -hs\rangle,$$

where

$$\begin{aligned} \Delta(c) &= c \otimes c - s \otimes s, \quad \varepsilon(c) = 1, \quad S(c) = c, \\ \Delta(s) &= c \otimes s + s \otimes c, \quad \varepsilon(s) = 0, \quad S(s) = s, \\ \Delta(h) &= h \otimes h + hs^2 \otimes h(1 - c - s), \quad \varepsilon(h) = 1, \quad S(h) = h(s^2 + s + 1). \end{aligned}$$

See also Section 5.3.1. Note that $\mathbf{G}(H_8^*) = \{1, c+\zeta_4 s, c-\zeta_4 s, c^2+s^2\}$ ($\cong \mathbb{Z}_2 \times \mathbb{Z}_2$). One sees that the algebra map

$$H_8^* \longrightarrow \widehat{\mathcal{A}_4(\zeta_4)}; \quad c \mapsto x^2 \# \boldsymbol{\sigma}, \quad s \mapsto -\zeta_4(x^2 - z^2) \# \boldsymbol{\sigma}, \quad h \mapsto x \# \boldsymbol{e} - \zeta_4 z \# \boldsymbol{\sigma}$$

is a Hopf algebra isomorphism, where $\mathcal{A}_4(\zeta_4)$ is given in Examples 3.5. Thus, we conclude that $\mathcal{A}_4(\zeta_4)$ and $\mathcal{A}_4(-\zeta_4)$ are nontrivial super-forms of H_8^* and are semisimple by Lemma 4.3. □

Suppose that $\mathcal{H}$ is finite-dimensional. Then for the dual Hopf superalgebra $\mathcal{H}^*$ of $\mathcal{H}$ (see Section 3.4), we also consider its bosonization $\widehat{\mathcal{H}^*} := \widehat{(\mathcal{H}^*)}$ of $\mathcal{H}^*$ by $\Bbbk\mathbb{Z}_2$. From the discussion in Section 2.4, we obtain the following.

LEMMA 4.6. *The bilinear map*

$$\widehat{\mathcal{H}^*} \times \widehat{\mathcal{H}} \longrightarrow \Bbbk; \quad (f \# \boldsymbol{\sigma}^i, h \# \boldsymbol{\sigma}^j) \longmapsto (-1)^{ij} f(h)$$

is a nondegenerate Hopf pairing and $\widehat{\mathcal{H}^*}$ *is isomorphic to the dual* $(\widehat{\mathcal{H}})^*$ *of* $\widehat{\mathcal{H}}$ *as a Hopf algebra.*

We summarize the above discussion and obtain the following result which is useful in the classification theory of finite-dimensional Hopf superalgebras.

PROPOSITION 4.7. *Let A be a Hopf algebra. If A has a nontrivial super-form $\mathcal{H}$, then we have the following.*

(1) *The dimension of A is an even number.*
(2) *A is neither commutative nor cocommutative.*
(3) *$\mathcal{H}$ is semisimple (resp., pointed) if and only if A is semisimple (resp., pointed).*
(4) *The group $\mathbf{G}(A)$ is decomposed into a direct product with $\mathbb{Z}_2$, that is, there exists a group Γ such that $\mathbf{G}(A) \cong \Gamma \times \mathbb{Z}_2$ as groups.*
(5) *If A is finite-dimensional, then the group $\mathbf{G}(A^*)$ is decomposed into a direct product with $\mathbb{Z}_2$.*

4.2. Admissible data for Hopf algebras. We fix a Hopf algebra $A = (A; m_A, u_A, \Delta_A, \varepsilon_A, S_A)$. Let $\mathsf{Aut}_{\mathsf{Hopf}}(A)$ denote the group of Hopf algebra automorphisms on A, and let A° denote the *finite dual* Hopf algebra of A. Recall that A° is given as $\bigcup_I (A/I)^*$, where I runs through the cofinite ideals of A. We regard $\alpha \in \mathbf{G}(A^\circ)$ as an algebra map $\alpha : A \to \Bbbk$ as usual.

It is easy to see that the map

$$\begin{array}{ccc} \{\Bbbk\mathbb{Z}_2 \hookrightarrow A^\circ\colon \text{Hopf inclusion}\} & \to & \{\alpha \in \mathbf{G}(A^\circ) \mid \mathsf{ord}(\alpha) = 2\}; \\ \iota & \mapsto & \iota(\boldsymbol{\sigma}) \end{array}$$

is bijective, where "Hopf inclusion" means an injective Hopf algebra map. Since the assignment $I \mapsto (A/I)^*$ gives a one-to-one correspondence between the the set of all cofinite Hopf ideals of A and set of all finite-dimensional Hopf subalgebras of A°, we see that the map

$$\begin{array}{ccc} \{A \twoheadrightarrow (\Bbbk\mathbb{Z}_2)^*\colon \text{Hopf surjection}\} & \to & \{\Bbbk\mathbb{Z}_2 \hookrightarrow A^\circ\colon \text{Hopf inclusion}\}; \\ \pi & \mapsto & \Bbbk\mathbb{Z}_2 \cong (\Bbbk\mathbb{Z}_2)^{**} \cong (A/\mathsf{Ker}(\pi))^* \hookrightarrow A^\circ \end{array}$$

is bijective, where "Hopf surjection" means a surjective Hopf algebra map.

Under the identification $\Bbbk\mathbb{Z}_2 \cong (\Bbbk\mathbb{Z}_2)^*$ given in Lemma 3.1, for a given order-two element $\alpha \in \mathbf{G}(A^\circ)$, the corresponding surjective Hopf algebra map is

$$\pi : A \longrightarrow \Bbbk\mathbb{Z}_2; \quad a \longmapsto \frac{\varepsilon_A(a)}{2}(\boldsymbol{e} + \boldsymbol{\sigma}) + \frac{\alpha(a)}{2}(\boldsymbol{e} - \boldsymbol{\sigma}). \tag{4.1}$$

On the other hand, for a given order-two element $g \in \mathbf{G}(A)$, the corresponding injective Hopf algebra map is

$$\iota : \Bbbk\mathbb{Z}_2 \longrightarrow A; \quad \boldsymbol{\sigma}^i \longmapsto g^i. \tag{4.2}$$

One easily sees that $\alpha(g) = -1$ if and only if $\pi \circ \iota = \mathsf{id}_{\Bbbk\mathbb{Z}_2}$ (i.e., π is a Hopf algebra split epimorphism with section ι). With this discussion in mind, we introduce the following notation.

Definition 4.8. A pair $(g, \alpha) \in \mathbf{G}(A) \times \mathbf{G}(A^\circ)$ is called an *admissible datum* for A if it satisfies $\mathsf{ord}(g) = 2$, $\mathsf{ord}(\alpha) = 2$ and $\alpha(g) = -1$. The set of all admissible data for A is denoted by $\mathsf{AD}(A)$. □

Let $\mathsf{Spl}(A)$ denote the set of all pairs (ι, π) of a Hopf algebra split epimorphism $\pi : A \to \Bbbk\mathbb{Z}_2$ with section $\iota : \Bbbk\mathbb{Z}_2 \to A$. In other words, $(A; \iota, \pi) \in \mathscr{T}_{\Bbbk\mathbb{Z}_2}$, see Section 2.2. The above argument shows that there is a one-to-one correspondence between $\mathsf{Spl}(A)$ and $\mathsf{AD}(A)$.

$$\begin{array}{ccc} \{\Bbbk\mathbb{Z}_2 \hookrightarrow A : \text{Hopf inclusion}\} \times \{A \twoheadrightarrow \Bbbk\mathbb{Z}_2 : \text{Hopf surjection}\} & \supset & \mathsf{Spl}(A) \\ \updownarrow & & \updownarrow \\ \{g \in \mathbf{G}(A) \mid \mathsf{ord}(g) = 2\} \times \{\alpha \in \mathbf{G}(A^\circ) \mid \mathsf{ord}(\alpha) = 2\} & \supset & \mathsf{AD}(A) \end{array}$$

For later use, for a given $(g, \alpha) \in \mathsf{AD}(A)$, we define $\pi_{(g,\alpha)} : A \to \Bbbk\mathbb{Z}_2$ by (4.1) and $\iota_{(g,\alpha)} : \Bbbk\mathbb{Z}_2 \to A$ by (4.2).

Remark 4.9. We note that if A is finite-dimensional, then the bijection $\mathsf{Spl}(A) \to \mathsf{Spl}(A^*); (\iota, \pi) \mapsto (\pi^*, \iota^*)$ induces a bijection $\mathsf{AD}(A) \to \mathsf{AD}(A^*); (g, \alpha) \mapsto (\alpha, g)$ under the canonical identification $A^{**} = A$. □

For $(g, \alpha), (h, \beta) \in \mathsf{AD}(A)$, if there exists $\varphi \in \mathsf{Aut}_{\mathrm{Hopf}}(A)$ such that $\varphi(g) = h$ and $\alpha = \beta \circ \varphi$, then we write $(g, \alpha) \sim (h, \beta)$. It is obvious that the relation $\sim$ forms an equivalence relation on $\mathsf{AD}(A)$.

Lemma 4.10. *For $(g, \alpha), (h, \beta) \in \mathsf{AD}(A)$, we have $(g, \alpha) \sim (h, \beta)$ if and only if $\big(\iota_{(g,\alpha)}, \pi_{(g,\alpha)}\big) \approx \big(\iota_{(h,\beta)}, \pi_{(h,\beta)}\big)$.*

PROOF. Suppose that $(g,\alpha) \sim (h,\beta)$. Then there exists $\varphi \in \mathsf{Aut}_{\mathrm{Hopf}}(A)$ such that $\varphi(g) = h$ and $\alpha = \beta \circ \varphi$. Since φ is an algebra (resp., coalgebra) map, one sees that $\pi_{(h,\beta)} \circ \varphi = \pi_{(g,\alpha)}$ (resp., $\varphi \circ \iota_{(g,\alpha)} = \iota_{(h,\beta)}$). Thus, this φ gives $(\iota_{(g,\alpha)}, \pi_{(g,\alpha)}) \approx (\iota_{(h,\beta)}, \pi_{(h,\beta)})$. The converse follows along the same argument. $\square$

For simplicity, we put

$$A^{\mathsf{coinv}(g,\alpha)} := A^{\mathsf{co}(\iota_{(g,\alpha)},\pi_{(g,\alpha)})}$$

for an admissible datum $(g,\alpha) \in \mathsf{AD}(A)$ for A. Then by Proposition 2.4, we have the following result.

PROPOSITION 4.11. *The assignment* $\mathsf{AD}(A) \to {}^{\Bbbk\mathbb{Z}_2}_{\Bbbk\mathbb{Z}_2}\mathcal{YDH}$; $(g,\alpha) \mapsto A^{\mathsf{coinv}(g,\alpha)}$ *gives a one-to-one correspondence between* $\mathsf{AD}(A)/\sim$ *and*

$$\{\mathcal{B} \in {}^{\Bbbk\mathbb{Z}_2}_{\Bbbk\mathbb{Z}_2}\mathcal{YDH} \mid \mathcal{B} \# \Bbbk\mathbb{Z}_2 \textit{ is isomorphic to } A \textit{ as a Hopf algebra}\}/\cong .$$

4.3. Super-data for Hopf algebras. In the following, we also fix a Hopf algebra $A = (A; m_A, u_A, \Delta_A, \varepsilon_A, S_A)$ and take an admissible datum $(g,\alpha) \in \mathsf{AD}(A)$ for A. Then by Proposition 4.11, the coinvariant subalgebra

$$\mathcal{B} := A^{\mathsf{coinv}(g,\alpha)}$$

of A is a Hopf algebra in ${}^{\Bbbk\mathbb{Z}_2}_{\Bbbk\mathbb{Z}_2}\mathcal{YD}$. In this section, we give a criterion for $\mathcal{B}$ to be an object of $\mathcal{SV}$. For $a \in A$, we use the following usual notations.

$$\alpha \rightharpoonup a := a_{(1)}\alpha(a_{(2)}), \quad a \leftharpoonup \alpha := \alpha(a_{(1)})a_{(2)}.$$

Note that $\alpha \rightharpoonup (-), (-) \leftharpoonup \alpha$ are algebra maps and $\alpha \rightharpoonup a \leftharpoonup \alpha := (\alpha \rightharpoonup a) \leftharpoonup \alpha = \alpha \rightharpoonup (a \leftharpoonup \alpha)$ for all $a \in A$.

PROPOSITION 4.12. *We have* $\mathcal{B} = \{b \in A \mid b = \alpha \rightharpoonup b\}$.

PROOF. For $a \in A$, a direct computation shows that

$$a_{(1)} \otimes \pi_{(g,\alpha)}(a_{(2)}) = \frac{1}{2}(a + (\alpha \rightharpoonup a)) \otimes \boldsymbol{e} + \frac{1}{2}(a - (\alpha \rightharpoonup a)) \otimes \boldsymbol{\sigma}.$$

Thus, $a \in \mathcal{B}$ if and only if $a = \alpha \rightharpoonup a$. $\square$

The left $\Bbbk\mathbb{Z}_2$-action on $\mathcal{B}$ is explicitly given by

$$\Bbbk\mathbb{Z}_2 \otimes \mathcal{B} \longrightarrow \mathcal{B}; \quad \boldsymbol{\sigma}^i \otimes b \longmapsto \boldsymbol{\sigma}^i \triangleright b = g^i b g^i,$$

see Proposition 2.1. The left $\Bbbk\mathbb{Z}_2$-coaction on $\mathcal{B}$ can be rephrased as follows.

LEMMA 4.13. *For* $a \in A$ *and* $i \in \{0,1\}$, *we have* $(\pi_{(g,\alpha)} \otimes \mathsf{id}_A)\Delta_A(a) = \boldsymbol{\sigma}^i \otimes a$ *if and only if* $a \leftharpoonup \alpha = (-1)^i a$.

Thus, if we let

$$\mathcal{B}_{i,j} := \{b \in \mathcal{B} \mid b \leftharpoonup \alpha = (-1)^i b \text{ and } gbg = (-1)^j b\} \quad (i,j \in \{0,1\}),$$

then $\mathcal{B}$ decomposes as $\mathcal{B} = \bigoplus_{i,j\in\{0,1\}} \mathcal{B}_{i,j}$. The following is a criterion for $\mathcal{B}$ to be a Hopf superalgebra such that $\widehat{\mathcal{B}} \cong A$.

PROPOSITION 4.14. *For the coinvariant subalgebra* $\mathcal{B} = A^{\mathsf{coinv}(g,\alpha)}$ *of* A, *the following assertions are equivalent.*

(1) $\mathcal{B}$ *is a super-form of* A.
(2) $b \leftharpoonup \alpha = gbg$ *for all* $b \in \mathcal{B}$.
(3) $\alpha \rightharpoonup a \leftharpoonup \alpha = gag$ *for all* $a \in A$.

PROOF. The decomposition of $\mathcal{B}$ shows that $\mathcal{B} \in \mathcal{SV}$ if and only if $\mathcal{B}_{0,1} = \mathcal{B}_{1,0} = 0$, and hence the conditions (i) and (ii) are equivalent. By Proposition 4.12, we see that

$$\alpha \rightharpoonup (bg^i) \leftharpoonup \alpha = (b \leftharpoonup \alpha)g^i \quad \text{and} \quad g(bg^i)g = (gbg)g^i$$

for all $b \in \mathcal{B}, i \in \{0,1\}$. Since $\mathcal{B} \# \Bbbk\mathbb{Z}_2 \to A; b \# \boldsymbol{\sigma}^i \mapsto bg^i$ is an isomorphism (see (2.6)), the conditions (ii) and (iii) are equivalent. The proof is done. □

If $\mathcal{B} = A^{\mathsf{coinv}(g,\alpha)}$ is a super-form of A, then $\mathcal{B} = \mathcal{B}_{\bar{0}} \oplus \mathcal{B}_{\bar{1}}$ with

$$\mathcal{B}_{\bar{\epsilon}} = \mathcal{B}_{\epsilon,\epsilon} = \{b \in \mathcal{B} \mid gbg = (-1)^{\epsilon} b\} \tag{4.3}$$

for $\epsilon \in \{0,1\}$. This implies the following criterion.

PROPOSITION 4.15. *If $\mathcal{B} = A^{\mathsf{coinv}(g,\alpha)}$ satisfies one of the conditions given in Proposition 4.14, then the following assertions are equivalent.*

1. *The super-form $\mathcal{B}$ of A is nontrivial.*
2. *There exists $b \in \mathcal{B}$ such that $b \neq 0$ and $b \leftharpoonup \alpha = -b$.*
3. *$g \notin Z(A)$, where $Z(A)$ is the center of A.*

DEFINITION 4.16. An admissible datum $(g, \alpha) \in \mathsf{AD}(A)$ for A is called a *super-datum* for A if it satisfies $g \notin Z(A)$ and $\alpha \rightharpoonup a \leftharpoonup \alpha = gag$ for all $a \in A$. The set of all super-datum for A is denoted by $\mathsf{SD}(A)$. □

Then by definition, the assignment $(g, \alpha) \mapsto A^{\mathsf{coinv}(g,\alpha)}$ induces a bijection

$$\mathsf{SD}(A)/\!\sim \longrightarrow \{\mathcal{H}\text{: Hopf superalgebra} \mid \mathcal{H}_{\bar{1}} \neq 0 \text{ and } \widehat{\mathcal{H}} \cong A\}/\!\cong .$$

Let $(g, \alpha) \in \mathsf{SD}(A)$. For $i \in \{0,1\}$ and $a \in A$, we put

$$\Delta_A^i(a) := \frac{1}{2} a_{(1)}(1_A + (-1)^i g) \otimes a_{(2)}.$$

The comultiplication and the antipode of the Hopf superalgebra structure can write down as follows (cf. (3.3)).

PROPOSITION 4.17. *Let $\epsilon \in \{0,1\}$, and let $\mathcal{B} := A^{\mathsf{coinv}(g,\alpha)}$. For $b \in \mathcal{B}_{\bar{\epsilon}}$, we get $\Delta_{\mathcal{B}}(b) = \Delta_A^0(b) - (-1)^{\epsilon}(g \otimes 1_A)\Delta_A^1(b)$ and $S_{\mathcal{B}}(b) = g^{\epsilon} S_A(b)$.*

PROOF. Since the antipode of $\Bbbk\mathbb{Z}_2$ is identical, we get

$$\Delta_{\mathcal{B}}(b) = b_{(1)}(\iota_{(g,\alpha)} \circ \pi_{(g,\alpha)})(b_{(2)}) \otimes b_{(3)} = \Delta_A^0(b) + \frac{1}{2} b_{(1)}(1_A - g) \otimes (b_{(2)} \leftharpoonup \alpha),$$

where we write $\Delta_A(b) = b_{(1)} \otimes b_{(2)}$. By Proposition 4.12, we have $\Delta_A(b) = \Delta_A(\alpha \rightharpoonup b) = b_{(1)} \otimes (\alpha \rightharpoonup b_{(2)})$. Combining this equation with Proposition 4.14(ii), we get

$$b_{(1)} \otimes (b_{(2)} \leftharpoonup \alpha) = b_{(1)} \otimes g b_{(2)} g. \tag{4.4}$$

On the other hand, by (4.3), we get $\Delta_A(b) = (-1)^{\epsilon} \Delta_A(gbg)$, and hence we have

$$g b_{(1)} g \otimes b_{(2)} = (-1)^{\epsilon} b_{(1)} \otimes g b_{(2)} g. \tag{4.5}$$

Thus, we obtain

$$\begin{aligned} \frac{1}{2} b_{(1)}(1_A - g) \otimes (b_{(2)} \leftharpoonup \alpha) &\overset{(4.4)}{=} \frac{1}{2} b_{(1)}(1_A - g) \otimes g b_{(2)} g \\ &\overset{(4.5)}{=} -(-1)^{\epsilon} \frac{1}{2} g b_{(1)}(1_A - g) \otimes b_{(2)}. \end{aligned}$$

This proves the first claim. The second claim follows directly. □

For reader's convince, we give a summary of our results as follows.

THEOREM 4.18. *Let* $(g, \alpha) \in \mathsf{SD}(A)$, *and let* $\mathcal{H} := A^{\mathsf{coinv}(g,\alpha)}$ *be the coinvariant subalgebra of* A *with respect to* (g, α). *Then*

$$\mathcal{H} = \{b \in A \mid \alpha \rightharpoonup b = b\}, \quad \mathcal{H}_{\bar{\epsilon}} = \{b \in \mathcal{H} \mid gbg = (-1)^{\epsilon} b\} \quad (\epsilon \in \{0, 1\})$$

and the Hopf superalgebra structure of $\mathcal{H}$ *is given as follows.*

- *(comultiplication)* $\Delta_{\mathcal{H}}(b) = \Delta_A^0(b) - (-1)^{|b|}(g \otimes 1_A)\Delta_A^1(b)$ *for* $b \in \mathcal{H}$.
- *(counit)* $\varepsilon_{\mathcal{H}} = \varepsilon_A$.
- *(antipode)* $S_{\mathcal{H}}(b) = g^{|b|} S_A(b)$ *for* $b \in \mathcal{H}$.

5. Classification of some classes of Hopf superalgebras

In the rest of this paper, our base field $\Bbbk$ is supposed to be an algebraically closed field of characteristic zero. In this section, we classify Hopf superalgebras of dimension up to 5 (Sections 5.1, 5.2 and 5.3). Also, we determine semisimple Hopf superalgebras of dimension 6 (Section 5.4) and give some examples of nonsemisimple nonpointed Hopf superalgebras of dimension 8 (Section 5.5).

5.1. Hopf superalgebras of prime dimension. We determine Hopf superalgebras of prime dimension. Two-dimensional Hopf superalgebras are given as follows.

THEOREM 5.1. *Up to isomorphism, the exterior superalgebra* $\bigwedge(z)$ *is the only two-dimensional Hopf superalgebra whose odd part is nonzero.*

PROOF. It is known that (up to isomorphism) Sweedler's four-dimensional Hopf algebra H_4 (see (3.4) for the definition) is the only Hopf algebra of dimension $4(= 2 \times 2)$ which is neither commutative nor cocommutative. Since $x^2 = 0$, we see that $\alpha \in \mathbf{G}(H_4^*)$ defined by $\alpha(c) = -1, \alpha(x) = 0$ is the only element in $\mathbf{G}(H_4^*)$ of order two, and hence we have $\mathsf{AD}(H_4) = \{(c, \alpha)\}$. Since $c \notin Z(H_4)$ and $\alpha \rightharpoonup x \leftharpoonup \alpha = -x = cxc$, we get $\mathsf{SD}(H_4) = \{(c, \alpha)\}$. Therefore, the coinvariant subalgebra $\mathcal{H} := H_4^{\mathsf{coinv}(c,\alpha)}$ of H_4 is generated by x and becomes a nontrivial super-form of H_4. The Hopf superalgebra structure of $\mathcal{H}$ is calculated as

$$\Delta_{\mathcal{H}}(x) = 1 \otimes x + x \otimes 1, \quad \varepsilon_{\mathcal{H}}(x) = 0, \quad S_{\mathcal{H}}(x) = -c^2 x = -x$$

by Theorem 4.18. Thus, the assignment $z \mapsto x$ gives a Hopf superalgebra isomorphism $\bigwedge(z) \cong \mathcal{H}$. □

We next show that any super-form of a Hopf superalgebra of an odd prime dimension is trivial.

THEOREM 5.2. *All Hopf superalgebras of odd prime dimensions are purely even.*

PROOF. Let p be an odd prime number, and let $\mathcal{H}$ be a p-dimensional Hopf superalgebra. Since the bosonization $\widehat{\mathcal{H}}$ of $\mathcal{H}$ is a Hopf algebra of dimension $2p$, by Masuoka [**11**] and Ng [**17**], there exists a finite group Γ such that $\widehat{\mathcal{H}}$ is isomorphic to the group algebra $\Bbbk\Gamma$ over Γ or its dual $(\Bbbk\Gamma)^*$. Hence by Lemma 4.2, the odd part $\mathcal{H}_{\bar{1}}$ of $\mathcal{H}$ must be zero. □

5.2. Pointed Hopf superalgebras of dimension 4. By [**1**] and Theorem 5.2, we now have a complete classification of Hopf superalgebras of dimension up to 5. As a demonstration of our method, we reproduce the classification of four-dimensional Hopf superalgebras by [**1**] and explore their properties in more detail. According to Masuoka [**11**] and Ştefan [**21**], Hopf algebras of dimension 8 are either semisimple or pointed. Here we address finding super-forms in the pointed case.

By Proposition 4.7 and the classification of nonsemisimple pointed Hopf algebras of dimension 8 given in [**21**], we find that all such Hopf algebras, with the exception listed below, do not admit nontrivial super-forms.

- $A_{C_2} = \Bbbk\langle c, x, y \mid c^2 = 1, cx = -xc, cy = -yc, xy = -yx, x^2 = y^2 = 0\rangle$, where c is group-like and x, y are c-skew primitive.
- $A_{C_2\times C_2} = \Bbbk\langle c, d, x \mid c^2 = d^2 = 1, cd = dc, cx = -xc, dx = -xd, x^2 = 0\rangle$, where d, c are group-like and x is c-skew primitive.

Super-forms of A_{C_2} are given as follows.

PROPOSITION 5.3. *Up to isomorphism, the exterior superalgebra* $\mathcal{H}_4^{(1)}=\bigwedge(z_1, z_2)$ *is the only Hopf superalgebra whose bosonization is isomorphic to* A_{C_2}.

PROOF. It is easy to see that $\alpha \in \mathbf{G}(A_{C_2}^*)$ defined by $\alpha(c) = -1, \alpha(x) = \alpha(y) = 0$ is the only algebra map $A_{C_2} \to \Bbbk$ of order two. Since $\alpha \rightharpoonup x \leftharpoonup \alpha = x$ and $\alpha \rightharpoonup y \leftharpoonup \alpha = y$, we obtain $\mathsf{AD}(A_{C_2}) = \mathsf{SD}(A_{C_2}) = \{(c, \alpha)\}$. By Proposition 4.12, one sees that the coinvariant subalgebra $A_{C_2}^{\mathsf{coinv}(c,\alpha)}$ of A_{C_2} is generated by x, y. By Theorem 4.18, these x, y are odd primitive. Thus, the assignment $z_1 \mapsto x, z_2 \mapsto y$ gives a Hopf superalgebra isomorphism $\bigwedge(z_1, z_2) \cong A_{C_2}^{\mathsf{coinv}(c,\alpha)}$. □

We will find super-forms of $A_{C_2\times C_2}$. It is easy to see

$$\mathbf{G}(A_{C_2\times C_2}) = \{1, c, d, cd\} \quad \text{and} \quad \mathbf{G}(A_{C_2\times C_2}^*) = \{\varepsilon, \alpha_1, \alpha_2, \alpha_3 := \alpha_1\alpha_2\},$$

where α_1 and α_2 are algebra maps $A_{C_2\times C_2} \to \Bbbk$ determined by $\alpha_1(c) = -1$, $\alpha_1(d) = 1$, $\alpha_2(c) = 1$, $\alpha_2(d) = -1$ and $\alpha_1(x) = \alpha_2(x) = 0$. Hence,

$$\mathsf{AD}(A_{C_2\times C_2}) = \{(c, \alpha_1), (cd, \alpha_1), (d, \alpha_2), (cd, \alpha_2), (c, \alpha_3), (d, \alpha_3)\}.$$

Since cd is central, we obtain

$$\mathsf{SD}(A_{C_2\times C_2}) = \{(c, \alpha_1), (c, \alpha_3), (d, \alpha_3)\}.$$

The group $\mathsf{Aut}_{\mathrm{Hopf}}(A_{C_2\times C_2})$ of Hopf algebra automorphisms on $A_{C_2\times C_2}$ is isomorphic to $\Bbbk^\times$. More precisely, we have

$$\mathsf{Aut}_{\mathrm{Hopf}}(A_{C_2\times C_2}) = \{\varphi_u \mid u \in \Bbbk^\times\},$$

where φ_u is the algebra automorphism on $A_{C_2\times C_2}$ determined by $\varphi_u|_{\mathbf{G}(A_{C_2\times C_2})} = \mathrm{id}$ and $\varphi_u(x) = ux$. Now we are ready to prove:

PROPOSITION 5.4. *Up to isomorphism, Hopf superalgebras* $\mathcal{H}_4^{(2)}$, $\mathcal{H}_4^{(3)}$ *and* $\mathcal{H}_4^{(4)}$ *of Example* 3.5 *are the only ones whose bosonization is isomorphic to* $A_{C_2\times C_2}$.

PROOF. By the above discussion, three elements of $\mathsf{SD}(A_{C_2\times C_2})$ are pairwise nonequivalent. We compute the coinvariant subalgebra for each element. First, we consider the coinvariant subalgebra $A_{C_2\times C_2}^{\mathsf{coinv}(c,\alpha_1)}$ of $A_{C_2\times C_2}$ with respect to (c, α_1). By Proposition 4.12, we see that $A_{C_2\times C_2}^{\mathsf{coinv}(c,\alpha_1)}$ is generated by d, x. By Theorem 4.18, these d is group-like and x is odd primitive. Thus, the assignment $g \mapsto d, z \mapsto x$ gives a Hopf superalgebra isomorphism $\mathcal{H}_4^{(2)} \cong A_{C_2\times C_2}^{\mathsf{coinv}(c,\alpha_1)}$.

Next, we consider the coinvariant subalgebra $A_{C_2\times C_2}^{\mathsf{coinv}(c,\alpha_3)}$ of $A_{C_2\times C_2}$ with respect to (c, α_3). Also, by Proposition 4.12 and Theorem 4.18, we see that $A_{C_2\times C_2}^{\mathsf{coinv}(c,\alpha_3)}$ is generated by cd, x, where cd is group-like and x is odd primitive. Thus, the assignment $g \mapsto cd, z \mapsto x$ gives a Hopf superalgebra isomorphism $\mathcal{H}_4^{(3)} \cong A_{C_2\times C_2}^{\mathsf{coinv}(c,\alpha_3)}$.

Finally, we consider the coinvariant subalgebra $A_{C_2\times C_2}^{\mathsf{coinv}(d,\alpha_3)}$ of $A_{C_2\times C_2}$ with respect to (d, α_3). We see that $A_{C_2\times C_2}^{\mathsf{coinv}(d,\alpha_3)}$ is generated by cd, x, where cd is group-like and x is odd cd-skew primitive. Thus, the assignment $g \mapsto cd, z \mapsto x$ gives a Hopf superalgebra isomorphism $\mathcal{H}_4^{(4)} \cong A_{C_2\times C_2}^{\mathsf{coinv}(d,\alpha_3)}$. □

Thus, we obtain the following result.

THEOREM 5.5. *Let $\mathcal{H}$ be a nonsemisimple pointed Hopf superalgebra of dimension 4 with $\mathcal{H}_{\bar{1}} \neq 0$. Then $\mathcal{H}$ is isomorphic to one of the Hopf superalgebras $\mathcal{H}_4^{(1)}, \mathcal{H}_4^{(2)}, \mathcal{H}_4^{(3)}, \mathcal{H}_4^{(4)}$, which are pairwise nonisomorphic. Moreover, $\mathcal{H}_4^{(1)}$ and $\mathcal{H}_4^{(2)}$ are self-dual and the dual of $\mathcal{H}_4^{(3)}$ is isomorphic to $\mathcal{H}_4^{(4)}$.*

5.3. Semisimple Hopf superalgebras of dimension 4. Next, we address finding super-forms of eight-dimensional semisimple Hopf algebras. In the following, we fix a primitive fourth root of unity $\zeta_4 \in \Bbbk$.

5.3.1. *Super-forms of H_8.* The classification of semisimple Hopf algebras of dimension 8 has been done by Masuoka [**12**]. By his result, the following is the only *nontrivial* (i.e., neither commutative nor cocommutative) semisimple Hopf algebra of dimension 8:

$$H_8 := \Bbbk \left\langle X, Y, Z \;\middle|\; \begin{array}{l} X^2 = Y^2 = 1,\ Z^2 = \frac{1}{2}(1 + X + Y - XY), \\ XY = YX,\ ZX = YZ,\ XZ = ZY \end{array} \right\rangle,$$

where X and Y are group-like and

$$\Delta(Z) = \frac{1}{2}(Z \otimes Z + Z \otimes XZ + YZ \otimes Z - YZ \otimes XZ),$$
$$\varepsilon(Z) = 1, \quad S(Z) = Z.$$

Both $\mathbf{G}(H_8)$ and $\mathbf{G}(H_8^*)$ are isomorphic to $\mathbb{Z}_2 \times \mathbb{Z}_2$ and are given by

$$\mathbf{G}(H_8) = \{1, X, Y, XY\}, \qquad \mathbf{G}(H_8^*) = \{\varepsilon, \alpha_+, \alpha_-, \alpha_+\alpha_-\},$$

where α_{s} for $\mathsf{s} \in \{+, -\}$ is given by $\alpha_{\mathsf{s}}(X) = \alpha_{\mathsf{s}}(Y) = -1$ and $\alpha_{\mathsf{s}}(Z) = \mathsf{s}\zeta_4$. Hence the set of all admissible data for H_8 is given as

$$\mathsf{AD}(H_8) = \{(X, \alpha_+), (X, \alpha_-), (Y, \alpha_+), (Y, \alpha_-)\}.$$

We also have $\mathsf{Aut}_{\mathrm{Hopf}}(H_8) = \{\mathsf{id}, \psi, \phi, \psi\phi\}$, where

$$\phi(X) = X, \quad \phi(Y) = Y, \quad \phi(Z) = XYZ,$$
$$\psi(X) = Y, \quad \psi(Y) = X, \quad \psi(Z) = \frac{1}{2}(1 + X + Y - XY)Z.$$

See Sage and Vega [**19**, Section 4.2] for example (see also Shi [**20**]). By a direct computation, we now can determine a set of complete representatives of equivalence classes of admissible data for H_8 as follows:

LEMMA 5.6. *We have* $(X,\alpha_+) \sim (Y,\alpha_-)$, $(X,\alpha_-) \sim (Y,\alpha_+)$ *and* $(X,\alpha_+) \not\sim (X,\alpha_-)$.

To determine super-data for H_8, we first calculate left/right actions of α_{s} for each $\mathsf{s} \in \{+,-\}$ as follows.

$$\alpha_{\mathsf{s}} \rightharpoonup X = -X, \quad \alpha_{\mathsf{s}} \rightharpoonup Y = -Y, \quad \alpha_{\mathsf{s}} \rightharpoonup Z = \mathsf{s}\zeta_4 YZ, \tag{5.1}$$

$$X \leftharpoonup \alpha_{\mathsf{s}} = -X, \quad Y \leftharpoonup \alpha_{\mathsf{s}} = -Y, \quad Z \leftharpoonup \alpha_{\mathsf{s}} = \mathsf{s}\zeta_4 YZ. \tag{5.2}$$

LEMMA 5.7. *The set* $\{(X,\alpha_+),(X,\alpha_-)\}$ *is a set of complete representatives of equivalence classes of* $\mathsf{SD}(H_8)$.

Therefore, up to isomorphism, there exists exactly two nontrivial super-forms of H_8. By Example 4.5, we see that such Hopf superalgebras are exhausted by $\mathcal{A}_4(\pm\zeta_4)$ given in Example 3.5. Thus, we get the following result.

THEOREM 5.8. *Let* $\mathcal{H}$ *be a semisimple Hopf superalgebra of dimension* 4 *with* $\mathcal{H}_{\bar{1}} \neq 0$. *Then* $\mathcal{H}$ *is isomorphic to one of the Hopf superalgebras* $\mathcal{A}_4(\zeta_4), \mathcal{A}_4(-\zeta_4)$. *Moreover, these Hopf superalgebras are nonisomorphic.*

5.3.2. *Coinvariant subalgebras.* In the following, we give an explicit isomorphism between $\mathcal{A}_4(-\zeta_4)$ and the coinvariant subalgebras $\mathcal{B} := H_8^{\mathsf{coinv}(X,\alpha_+)}$ of H_8. To accomplish this, we first determine generators and relations of the algebra $\mathcal{B}$ concretely. Set

$$\begin{aligned} g &:= XY, \\ v &:= \frac{1-\zeta_4}{4}(Z + \zeta_4 XZ + \zeta_4 YZ + XYZ), \\ w &:= \frac{\sqrt{2}}{4}(Z - \zeta_4 XZ + \zeta_4 YZ - XYZ). \end{aligned}$$

By (5.1) and Proposition 4.12, we see that $\mathcal{B}$ has the set $\{1,g,v,w\}$ as a basis. Moreover, v and w are homogeneous with $|v| = 0$ and $|w| = 1$, respectively. By direct computation, we have

$$\begin{gathered} g^2 = 1, \quad gv = v, \quad gw = -w, \\ wv = 0 = vw, \quad v^2 = \frac{1}{2}(1+g), \quad w^2 = \frac{1}{2}(1-g). \end{gathered}$$

In particular, we have $v^2 - w^2 = g$. Since $\mathcal{B}$ has dimension 4, it is now easy to see that $\mathcal{B}$ is generated by v and w with the relations $v^2 + w^2 = 1$ and $vw = wv = 0$.

By Theorem 4.18, one sees that the comultiplication $\Delta_{\mathcal{B}}$ of $\mathcal{B}$ is given as

$$\Delta_{\mathcal{B}}(v) = v \otimes v - \zeta_4 w \otimes w, \quad \Delta_{\mathcal{B}}(w) = v \otimes w + w \otimes v.$$

These results imply the following.

PROPOSITION 5.9. *There is an isomorphism of Hopf superalgebras determined by*

$$\mathcal{A}_4(-\zeta_4) \longrightarrow \mathcal{B} = H_8^{\mathsf{coinv}(X,\alpha_+)}; \quad x \longmapsto v, \quad z \longmapsto w.$$

An isomorphism $\mathcal{A}_4(\zeta_4) \cong H_8^{\mathsf{coinv}(X,\alpha_-)}$ is obtained by the same argument but with ζ_4 replaced with $-\zeta_4$.

5.3.3. *Duals.* In this section, we determine the dual of the Hopf superalgebra

$$\mathcal{H} := \mathcal{A}_4(-\zeta_4) = \Bbbk\langle x, z \mid x^2 + z^2 = 1, xz = zx = 0\rangle.$$

Recall that $|x| = 0$, $|z| = 1$, $\Delta_{\mathcal{H}}(x) = x \otimes x - \zeta_4 z \otimes z$ and $\Delta_{\mathcal{H}}(z) = x \otimes z + z \otimes x$.

Let $1^*, x^*, (x^2)^*, z^*$ denote the dual bases of $1, x, x^2, z$, respectively. Recall that the evaluation map $\langle\ ,\ \rangle : \mathcal{H}^* \times \mathcal{H} \to \Bbbk$ gives a Hopf superalgebra structure of $\mathcal{H}^*$, see Section 3.4. The multiplication table of $\mathcal{H}^*$ is given as follows.

	1^*	x^*	$(x^2)^*$	z^*
1^*	$1^* + (x^2)^*$	0	$-(x^2)^*$	0
x^*	0	x^*	0	z^*
$(x^2)^*$	$-(x^2)^*$	0	$2(x^2)^*$	0
z^*	0	z^*	0	$-\zeta_4 x^*$

Also, the comultiplication is

$$\begin{aligned}
\Delta_{\mathcal{H}^*}(1^*) &= 1^* \otimes 1^* + z^* \otimes z^*, \\
\Delta_{\mathcal{H}^*}(x^*) &= 1^* \otimes x^* + x^* \otimes 1^* + x^* \otimes (x^2)^* + (x^2)^* \otimes x^*, \\
\Delta_{\mathcal{H}^*}((x^2)^*) &= 1^* \otimes (x^2)^* + (x^2)^* \otimes 1^* + (x^2)^* \otimes (x^2)^* + z^* \otimes z^*, \\
\Delta_{\mathcal{H}^*}(z^*) &= 1^* \otimes z^* + z^* \otimes 1^*
\end{aligned}$$

and the counit is

$$\varepsilon_{\mathcal{H}^*}(1^*) = 1, \quad \varepsilon_{\mathcal{H}^*}(x^*) = 0, \quad \varepsilon_{\mathcal{H}^*}((x^2)^*) = 0, \quad \varepsilon_{\mathcal{H}^*}(z^*) = 0.$$

Then a direct computation shows that the algebra map

$$\mathcal{H} \longrightarrow \mathcal{H}^*; \quad x \longmapsto 1^*, \quad z \longmapsto \zeta_4 z^*$$

is an isomorphism of Hopf superalgebras. In other words, the pairing $\langle\ ,\ \rangle : \mathcal{H} \times \mathcal{H} \to \Bbbk$ defined by

$$\langle x, x\rangle = \langle x, z\rangle = \langle z, x\rangle = 0, \quad \langle z, z\rangle = \zeta_4$$

is a nondegenerate Hopf pairing. Thus, $\mathcal{H}$ is self-dual. In this way, we get the following result.

THEOREM 5.10. *Hopf superalgebras $\mathcal{A}_4(\zeta_4)$ and $\mathcal{A}_4(-\zeta_4)$ are self-dual.*

5.4. Semisimple Hopf superalgebras of dimension 6. In this section, we first construct a semisimple Hopf superalgebra of dimension 6 (whose odd part is not zero) using Theorem 3.6. Then next we show that it is (up to isomorphism) the only semisimple Hopf superalgebra of dimension 6.

5.4.1. *Existence.* By Masuoka [**12**, Theorem 1.10], any semisimple Hopf algebra of dimension 6 is isomorphic to either $\Bbbk\mathbb{Z}_6$, $(\Bbbk\mathfrak{S}_3)^*$ or $\Bbbk\mathfrak{S}_3$, where

$$\mathfrak{S}_3 = \langle s_1, s_2 \mid s_1^2 = s_2^2 = e,\ s_1 s_2 s_1 = s_2 s_1 s_2\rangle$$

is the symmetric group of degree three ($e \in \mathfrak{S}_3$ is the identity element). Note that $\Bbbk\mathfrak{S}_3$ is the only noncommutative Hopf algebra among them and has a noncentral group-like element

$$c := s_1 \quad \in \mathbf{G}(\Bbbk\mathfrak{S}_3)$$

of order two.

Therefore, by Theorem 3.6, we make the Hopf algebra $\mathcal{H} := \Bbbk\mathfrak{S}_3$ into a Hopf superalgebra such that $\mathcal{H}_{\bar{1}} \neq 0$. Moreover, $\mathcal{H}$ is semisimple as a Hopf superalgebra, since $\Bbbk\mathfrak{S}_3$ is semisimple as a Hopf algebra. In the following, we shall write down the Hopf superalgebra structure of $\mathcal{H}$. For simplicity, we set

$$x := \frac{1}{2}(s_1s_2 + s_2s_1), \quad y := \frac{1}{2}(s_2 + s_1s_2s_1),$$
$$z := \frac{1}{2}(s_1s_2 - s_2s_1), \quad w := \frac{1}{2}(s_2 - s_1s_2s_1).$$

Since $cxc = x, cyc = y, czc = -z$ and $cwc = -w$, we have

$$\mathcal{H}_{\bar{0}} = \Bbbk e \oplus \Bbbk c \oplus \Bbbk x \oplus \Bbbk y \quad \text{and} \quad \mathcal{H}_{\bar{1}} = \Bbbk z \oplus \Bbbk w.$$

The multiplication table of $\mathcal{H}$ is given as follows.

	e	c	x	y	z	w
e	e	c	x	y	z	w
c	c	e	y	x	w	z
x	x	y	$\frac{1}{2}(e+x)$	$\frac{1}{2}(c+y)$	$-\frac{1}{2}z$	$-\frac{1}{2}w$
y	y	x	$\frac{1}{2}(c+y)$	$\frac{1}{2}(e+x)$	$-\frac{1}{2}w$	$-\frac{1}{2}z$
z	z	$-w$	$-\frac{1}{2}z$	$\frac{1}{2}w$	$\frac{1}{2}(x-e)$	$\frac{1}{2}(c-y)$
w	w	$-z$	$-\frac{1}{2}w$	$\frac{1}{2}z$	$\frac{1}{2}(y-c)$	$\frac{1}{2}(e-x)$

Since $\Delta_{\Bbbk\mathfrak{S}_3}(x) = x \otimes x + z \otimes z$, we get $\Delta_{\mathcal{H},\bar{0}}(x) = x \otimes x$ and $\Delta_{\mathcal{H},\bar{1}}(x) = z \otimes z$ by definition. Thus, the comultiplication $\Delta_{\mathcal{H}}$ of x in $\mathcal{H}$ is given as

$$\Delta_{\mathcal{H}}(x) = \Delta_{\mathcal{H},\bar{0}}(x) - (-1)^{|x|}(c \otimes 1)\Delta_{\mathcal{H},\bar{1}}(x) = x \otimes x - w \otimes z.$$

The counit and the antipode are given as $\varepsilon_{\mathcal{H}}(x) = \varepsilon_{\Bbbk\mathfrak{S}_3}(x) = 1$ and $S_{\mathcal{H}}(x) = (-1)^{|x|}S_{\Bbbk\mathfrak{S}_3}(x) = x$, respectively. In this way, we get the following result.

PROPOSITION 5.11. *Let $\mathcal{A}_6$ be the 6-dimensional superalgebra generated by homogeneous elements x, y, z and w subject to*

$$|x| = |y| = 0, \quad |z| = |w| = 1,$$
$$x^2 = \frac{1}{2}(1 + x) = y^2, \quad z^2 = \frac{1}{2}(x - 1) = -w^2,$$
$$xy = yx, \quad xz = -\frac{1}{2}z = zx, \quad xw = -\frac{1}{2}w = wx,$$
$$yz = -\frac{1}{2}w = -zy, \quad yw = -\frac{1}{2}z = -wy, \quad zw = xy - y = -wz.$$

Then the following comultiplication Δ, counit ε and antipode S make $\mathcal{A}_6$ into a Hopf superalgebra.

$$\Delta(x) = x \otimes x - w \otimes z, \quad \Delta(y) = y \otimes y - z \otimes w,$$
$$\Delta(z) = z \otimes x + y \otimes z, \quad \Delta(w) = w \otimes y + x \otimes w,$$
$$\varepsilon(x) = \varepsilon(y) = 1, \quad \varepsilon(z) = \varepsilon(w) = 0,$$
$$S(x) = x, \quad S(y) = y, \quad S(z) = -w, \quad S(w) = z.$$

Moreover, $\mathcal{A}_6$ is semisimple as a Hopf superalgebra.

The set of all group-like elements is given as $\mathbf{G}(\mathcal{A}_6) = \{1, xy + zw\}$ $(\cong \mathbb{Z}_2)$.

5.4.2. *Uniqueness.* In this section, we show the following uniqueness result.

THEOREM 5.12. *Let $\mathcal{H}$ be a semisimple Hopf superalgebra of dimension 6 with $\mathcal{H}_{\bar{1}} \neq 0$. Then $\mathcal{H}$ is isomorphic to the Hopf superalgebra $\mathcal{A}_6$ given in Proposition 5.11. Moreover, $\mathcal{A}_6$ is self-dual.*

By Fukuda [**6**], any semisimple Hopf algebra of dimension 12 is isomorphic to either $\Bbbk\Gamma$, $(\Bbbk\Gamma)^*$, A_+ or A_-, where Γ is a group of order 12. It is also shown that $\mathbf{G}(A_+) \cong \mathbb{Z}_2 \times \mathbb{Z}_2$, $\mathbf{G}(A_-) \cong \mathbb{Z}_4$ and $A_\pm$ are self-dual, see [**6**, Remark 4.2]. Among these Hopf algebras, only A_+ has the possibility of having super-forms, see Proposition 4.7.

Let us recall the definition of the Hopf algebra $A := A_+$. In the following, we set $c := s_1 \in \mathfrak{S}_3$ and denote the dual base of $\sigma \in \mathfrak{S}_3$ by σ^*. As an algebra, A is generated by $(\Bbbk\mathfrak{S}_3)^*$ and ξ such that

$$\xi^2 = 1, \ \ \xi f = f^c \xi \ \text{ for all } f \in (\Bbbk\mathfrak{S}_3)^*, \ \text{ and } (\Bbbk\mathfrak{S}_3)^* \text{ is a subalgebra of } A,$$

where $f^c \in \Bbbk^{\mathfrak{S}_3}$ defined by $f^c(\sigma) := f(c\sigma c)$ for $\sigma \in \mathfrak{S}_3$. The Hopf algebra structure of A is given as

$$\Delta_A(\sigma^*) = \sum_{\tau\tau'=\sigma} \tau^* \otimes \tau'^*, \quad \varepsilon_A(\sigma^*) = \delta_{\sigma,e}, \quad S_A(\sigma^*) = (\sigma^{-1})^*$$

and $\Delta_A(\xi) = \xi \otimes \xi, \varepsilon_A(\xi) = 1, S_A(\xi) = \xi$ (i.e., ξ is group-like), where $\delta_{\sigma,e}$ is the Kronecker symbol and $e \in \mathfrak{S}_3$ is the identity element (as before).

Let $\mathsf{sgn} \in (\Bbbk\mathfrak{S}_3)^*$ be the signature map of $\mathfrak{S}_3$. Then we have

$$\mathbf{G}(A) = \{1, \xi, \mathsf{sgn}, \xi\,\mathsf{sgn}\}, \qquad \mathbf{G}(A^*) = \{\varepsilon, \alpha_1, \alpha_2, \alpha_3 := \alpha_1\alpha_2\},$$

where α_1, α_2 is given by $\alpha_1(\xi) = 1$, $\alpha_1(f) = f(c)$, $\alpha_2(\xi) = -1$ and $\alpha_2(f) = f(e)$ for $f \in (\Bbbk\mathfrak{S}_3)^*$.

The set of all admissible data for A is given as follows.

$$\mathsf{AD}(A) = \{(\xi, \alpha_2), (\xi, \alpha_3), (\mathsf{sgn}, \alpha_1), (\mathsf{sgn}, \alpha_3), (\xi\,\mathsf{sgn}, \alpha_1), (\xi\,\mathsf{sgn}, \alpha_2)\}.$$

LEMMA 5.13. *We have $(\xi, \alpha_2) \sim (\xi\,\mathsf{sgn}, \alpha_2)$ and $(\xi, \alpha_3) \sim (\xi\,\mathsf{sgn}, \alpha_1)$.*

PROOF. An algebra map $\varphi : A \to A$ determined by

$$\varphi|_{(\Bbbk\mathfrak{S}_3)^*} = \mathsf{id} \quad \text{and} \quad \varphi(\xi) = \xi\,\mathsf{sgn}$$

is a (well-defined) Hopf algebra automorphism on A. Using this φ, the claim easily follows. □

By this lemma, we conclude that the quotient set $\mathsf{SD}(A)/\sim$ is a singleton set.

LEMMA 5.14. *We have $\mathsf{SD}(A)/\sim\, = \{[(\xi, \alpha_3)]\}$.*

PROOF. Since the signature map sgn is central, we have $(\mathsf{sgn}, \alpha_1), (\mathsf{sgn}, \alpha_3) \notin \mathsf{SD}(A)$ by definition. Let us consider the case of (ξ, α_2). For all $\sigma \in \mathfrak{S}_3$, we get

$$\alpha_2 \rightharpoonup \xi = -\xi, \quad \alpha_2 \rightharpoonup \sigma^* = \sum_{\tau\tau'=\sigma} \tau^*\, \delta_{\tau',e} = \sigma^*, \quad \sigma^* \leftharpoonup \alpha_2 = \sigma^*.$$

This shows that the coinvariant subalgebra $A^{\mathsf{coinv}(\xi,\alpha_2)}$ of A coincides with $(\Bbbk\mathfrak{S}_3)^*$. However, for $s_2 \in \mathfrak{S}_3$, we have

$$\xi s_2^* \xi = (s_2^*)^c\, \xi^2 = (s_2^*)^c = (cs_2c)^* = (s_1s_2s_1)^* \neq s_2^* = \alpha_2 \rightharpoonup s_2^* \leftharpoonup \alpha_2,$$

and hence we conclude that $(\xi, \alpha_2) \notin \mathsf{SD}(A)$.

Next, we consider the case of (ξ, α_3). For all $\sigma \in \mathfrak{S}_3$, we get

$$(5.3)\quad \alpha_3 \rightharpoonup \xi = -\xi, \quad \alpha_3 \rightharpoonup \sigma^* = \sum_{\tau\tau'=\sigma} \tau^*(\tau'^*(c)) = (\sigma c)^*, \quad \sigma^* \leftharpoonup \alpha_3 = (c\sigma)^*.$$

Thus, for all $\sigma \in \mathfrak{S}_3$, we have

$$\xi\sigma^*\xi = (\sigma^*)^c\,\xi^2 = (c\sigma c)^* = \alpha_3 \rightharpoonup \sigma^* \leftharpoonup \alpha_3.$$

The proof is done. □

By Proposition 5.11 and Lemma 5.14, the proof of Theorem 5.12 is done.

5.4.3. *Coinvariant subalgebras.* Let $\mathcal{B}$ be the coinvariant subalgebra $A^{\mathsf{coinv}(\xi,\alpha_3)}$ of A. Since $\mathcal{B}$ is a semisimple Hopf superalgebra of dimension 6, it automatically follows that $\mathcal{B}$ is isomorphic to $\mathcal{A}_6$. In this section, we construct a Hopf superalgebra isomorphism from $\mathcal{A}_6$ to $\mathcal{B}$ explicitly.

Set

$$\begin{aligned}
x_1 &:= e^* + s_1^*,\\
x_2 &:= (e^* - s_1^*)\xi,\\
x_3 &:= s_2^* + (s_1s_2)^* + (s_2s_1)^* + (s_1s_2s_1)^*,\\
x_4 &:= (s_2^* - (s_1s_2)^* - (s_2s_1)^* + (s_1s_2s_1)^*)\xi,\\
w_1 &:= s_2^* - (s_1s_2)^* + (s_2s_1)^* - (s_1s_2s_1)^*,\\
w_2 &:= (s_2^* + (s_1s_2)^* - (s_2s_1)^* - (s_1s_2s_1)^*)\xi.
\end{aligned}$$

Then by (5.3), one sees that $\{x_1, x_2, x_3, x_4\}$ forms a basis of $\mathcal{B}_{\bar{0}}$ and $\{w_1, w_2\}$ forms a basis of $\mathcal{B}_{\bar{1}}$.

Let $M(1|1)$ be the set of all 2×2 square matrices with entries in $\Bbbk$. With the usual matrix multiplication and the following $\mathbb{Z}_2$-grading, this $M(1|1)$ becomes a superalgebra.

$$M(1|1)_{\bar{0}} := \{\left(\begin{array}{c|c} a & 0 \\ \hline 0 & d \end{array}\right) \mid a, d \in \Bbbk\}, \quad M(1|1)_{\bar{1}} := \{\left(\begin{array}{c|c} 0 & b \\ \hline c & 0 \end{array}\right) \mid b, c \in \Bbbk\}.$$

Then one sees that the linear map $\mathcal{B} \to \Bbbk^2 \oplus M(1|1)$ defined by

$$\begin{aligned}
&x_1 \longmapsto (1, 1, O), \quad x_2 \longmapsto (1, -1, O),\\
&x_3 \longmapsto (0, 0, \left(\begin{array}{c|c} 1 & 0 \\ \hline 0 & 1 \end{array}\right)), \quad x_4 \longmapsto (0, 0, \left(\begin{array}{c|c} 1 & 0 \\ \hline 0 & -1 \end{array}\right)),\\
&w_1 \longmapsto (0, 0, \left(\begin{array}{c|c} 0 & 1 \\ \hline 1 & 0 \end{array}\right)), \quad w_2 \longmapsto (0, 0, \left(\begin{array}{c|c} 0 & -1 \\ \hline 1 & 0 \end{array}\right)),
\end{aligned}$$

is a superalgebra isomorphism. Moreover, one can show that the following holds.

PROPOSITION 5.15. *The assignment*

$$x \longmapsto x_1 - \frac{1}{2}x_3, \quad y \longmapsto x_2 - \frac{1}{2}x_4, \quad z \longmapsto w_1, \quad w \longmapsto w_2$$

gives a Hopf superalgebra isomorphism $\mathcal{A}_6 \cong \mathcal{B}$.

5.5. Nonsemisimple nonpointed Hopf superalgebras of dimension 8. In this section, we fix a primitive fourth root of unity $\zeta_4 \in \Bbbk$. By Călinescu, Dăscălescu, Masuoka, and Menini [**5**], and García and Vay [**7**], we know that there exist precisely two isomorphism classes of nonsemisimple nonpointed self-dual Hopf algebras $H_{16}(\pm\zeta_4)$ of dimension 16. In the following, we determine (up to isomorphism) all super-forms of $H_{16}(\pm\zeta_4)$.

We concentrate on $H_{16} := H_{16}(\zeta_4)$ for simplicity. First of all, we recall the definition of H_{16}. As an algebra, H_{16} is generated by the subalgebra H_8 (for the definition and notations, see Section 5.3) and T subject to

$$T^2 = 0, \quad TX = -XT, \quad TY = -YT, \quad TZ = \zeta_4 XZT.$$

As a Hopf algebra $H_8 \subset H_{16}$ and T is a X-skew primitive element of H_{16}.

Both $\mathbf{G}(H_{16})$ and $\mathbf{G}(H_{16}^*)$ are isomorphic to $\mathbb{Z}_2 \times \mathbb{Z}_2$ and are given by

$$\mathbf{G}(H_{16}) = \{1, X, Y, XY\} \quad \text{and} \quad \mathbf{G}(H_{16}^*) = \{\varepsilon, \alpha_+, \alpha_-, \alpha_+\alpha_-\},$$

where α_{s} for $\mathsf{s} \in \{+,-\}$ is given by $\alpha_{\mathsf{s}}(X) = \alpha_{\mathsf{s}}(Y) = -1$, $\alpha_{\mathsf{s}}(Z) = \mathsf{s}\zeta_4$ and $\alpha_{\mathsf{s}}(T) = 0$. Hence, we get

$$\mathsf{AD}(H_{16}) = \mathsf{SD}(H_{16}) = \{(X,\alpha_+),(X,\alpha_-),(Y,\alpha_+),(Y,\alpha_-)\}.$$

In the contrast to the case of H_8 (Lemma 5.6), the number of isomorphism classes of super-data for H_{16} is 4.

LEMMA 5.16. *The set* $\{(X,\alpha_+),(X,\alpha_-),(Y,\alpha_+),(Y,\alpha_-)\}$ *is a complete set of representatives of* $\mathsf{SD}(H_{16})/\sim$.

PROOF. Let φ be a Hopf algebra automorphism on H_{16}. Since H_8 is the coradical of H_{16}, we see that

$$\varphi|_{H_8} \in \mathsf{Aut}_{\mathrm{Hopf}}(H_8) = \{\mathsf{id}, \psi, \phi, \psi\phi\},$$

see Section 5.3.1 for the notation. However, one easily sees that both ψ and $\psi\phi$ cannot extend to a Hopf algebra automorphism on H_{16}. The proof is done. □

Set $\mathcal{H}^{(1)} := H_{16}^{\mathsf{coinv}(X,\alpha_+)}$, $\mathcal{H}^{(2)} := H_{16}^{\mathsf{coinv}(X,\alpha_-)}$, $\mathcal{H}^{(3)} := H_{16}^{\mathsf{coinv}(Y,\alpha_+)}$, $\mathcal{H}^{(4)} := H_{16}^{\mathsf{coinv}(Y,\alpha_-)}$ for simplicity. Since $\alpha_\pm \rightharpoonup T = T$ and $XTX = YTY = -T$, we see that T is an odd element of $\mathcal{H}^{(i)}$ ($i \in \{1,2,3,4\}$), see Theorem 4.18. Set $g := XY$, $v := \frac{1-\zeta_4}{4}(Z + \zeta_4 XZ + \zeta_4 YZ + XYZ)$ and $w := \frac{\sqrt{2}}{4}(Z - \zeta_4 XZ + \zeta_4 YZ - XYZ)$ as before (see Section 5.3.2).

Then a direct computation shows that equations

$$Tg = gT, \quad Tv = vT, \quad Tw = -wT$$

hold in each $\mathcal{H}^{(i)}$ ($i \in \{1,2,3,4\}$). By Theorem 4.18, the comultiplications of $\mathcal{H}^{(i)}$ are given as $\Delta_{\mathcal{H}^{(1)}}(T) = \Delta_{\mathcal{H}^{(2)}}(T) = 1 \otimes T + T \otimes 1$ and $\Delta_{\mathcal{H}^{(3)}}(T) = \Delta_{\mathcal{H}^{(4)}}(T) = XY \otimes T + T \otimes 1 = g \otimes T + T \otimes 1$. The above argument shows the following.

PROPOSITION 5.17. *For* $\zeta \in \{\zeta_4, -\zeta_4\}$ *and* $\epsilon, \eta \in \{0,1\}$, *let* $\mathcal{K}_8(\zeta;\epsilon,\eta)$ *be the* 8*-dimensional superalgebra generated by* g, v, w *and* t *subject to*

$$|g| = |v| = 0, \quad |w| = |t| = 1,$$

$$g^2 = 1, \quad gv = v, \quad gw = -w, \quad vw = wv = 0, \quad v^2 = \frac{1}{2}(1+g), \quad w^2 = \frac{1}{2}(1-g),$$

$$t^2 = 0, \quad tg = gt, \quad tv = (-1)^\epsilon vt, \quad tw = -(-1)^\epsilon wt.$$

Then the following comultiplication Δ, counit ε and antipode S make $\mathcal{K}_8(\zeta;\epsilon,\eta)$ into a Hopf superalgebra.

$$\Delta(g) = g \otimes g, \quad \Delta(v) = v \otimes v - \zeta w \otimes w,$$
$$\Delta(w) = v \otimes w + w \otimes v, \quad \Delta(t) = g^\eta \otimes t + t \otimes 1,$$
$$\varepsilon(g) = \varepsilon(v) = 1, \quad \varepsilon(w) = \varepsilon(t) = 0,$$
$$S(g) = g, \quad S(v) = v, \quad S(w) = \zeta w, \quad S(t) = -g^\eta t.$$

Moreover, $\mathcal{K}_8(\zeta;\epsilon,\eta)$ is nonsemisimple and neither $\mathcal{K}_8(\zeta;\epsilon,\eta)$ nor $\mathcal{K}_8(\zeta;\epsilon,\eta)^*$ is pointed.

For $\zeta \in \{\zeta_4, -\zeta_4\}$ and $\epsilon, \eta \in \{0,1\}$, we may identify $\mathcal{A}_4(\zeta)$ as a Hopf subsuperalgebra of $\mathcal{K}_8(\zeta;\epsilon,\eta)$ and we see that the bosonization of $\mathcal{K}_8(\zeta;\epsilon,\eta)$ is isomorphic to $H_{16}(\zeta)$. As in Section 5.3.3, we can determine duals of $\mathcal{K}_8(\zeta;\epsilon,\eta)$. For example, the following gives a nondegenerate Hopf pairing $\langle\ ,\ \rangle : \mathcal{K}_8(\zeta;0,1) \times \mathcal{K}_8(\zeta;1,0) \to \Bbbk$.

$\langle\ ,\ \rangle$	g	v	w	t
g	1	-1	0	0
v	-1	0	0	0
w	0	0	ω	0
t	0	0	0	1

Here, ω denotes an element in $\Bbbk$ satisfying $\omega^2 = -\zeta$.

As a summary, we get the following result.

THEOREM 5.18. *Let $\mathcal{H}$ be a nonsemisimple Hopf superalgebra of dimension 8 such that $\mathcal{H}$ nor $\mathcal{H}^*$ is pointed and $\mathcal{H}_{\bar{1}} \neq 0$. Then $\mathcal{H}$ is isomorphic to one of the eight Hopf superalgebras $\mathcal{K}_8(\zeta;\epsilon,\eta)$ ($\zeta \in \{\zeta_4, -\zeta_4\}, \epsilon, \eta \in \{0,1\}$), which are pairwise nonisomorphic. Moreover, $\mathcal{K}_8(\zeta;0,0)$ and $\mathcal{K}_8(\zeta;1,1)$ are self-dual and the dual of $\mathcal{K}_8(\zeta;0,1)$ is isomorphic to $\mathcal{K}_8(\zeta;1,0)$.*

We say that a Hopf (resp., super)algebra $\mathcal{K}$ has the *Chevalley property* if the category of right $\mathcal{K}$-(resp., super)comodules has the Chevalley property, that is, the tensor product of any two simple objects is semisimple. It follows directly that the Hopf superalgebras $\mathcal{K}_8(\zeta;\epsilon,\eta)$ have the Chevalley property since the Hopf algebra $H_{16}(\zeta)$ has the Chevalley property ([**5**]).

Acknowledgments

We thank the organizers, Siu-Hung Ng and Susan Montgomery, of "the AMS Special Session on Quantum Groups, Hopf algebras, and Applications: In honor of Professor Earl J. Taft", held at University of Utah for giving us an opportunity to present the results. We also thank Kenichi Shimizu for his helpful comments.

The first author is supported by Japan Society of the Promotion of Science (JSPS) KAKENHI Grant Number JP22K13905. We are grateful for the referee's careful reading and valuable comments. Thanks to them, we were able to make some of the proofs more conceptual.

References

[1] S. Aissaoui and A. Makhlouf, *On classification of finite-dimensional superbialgebras and Hopf superalgebras*, SIGMA Symmetry Integrability Geom. Methods Appl. **10** (2014), Paper 001, 24, DOI 10.3842/SIGMA.2014.001. MR3210634

[2] N. Andruskiewitsch, I. Angiono, and H. Yamane, *On pointed Hopf superalgebras*, New developments in Lie theory and its applications, Contemp. Math., vol. 544, Amer. Math. Soc., Providence, RI, 2011, pp. 123–140, DOI 10.1090/conm/544/10752. MR2849717

[3] N. Andruskiewitsch, P. Etingof, and S. Gelaki, *Triangular Hopf algebras with the Chevalley property*, Michigan Math. J. **49** (2001), no. 2, 277–298, DOI 10.1307/mmj/1008719774. MR1852304

[4] M. Beattie and G. A. García, *Classifying Hopf algebras of a given dimension*, Hopf algebras and tensor categories, Contemp. Math., vol. 585, Amer. Math. Soc., Providence, RI, 2013, pp. 125–152, DOI 10.1090/conm/585/11615. MR3077235

[5] C. Călinescu, S. Dăscălescu, A. Masuoka, and C. Menini, *Quantum lines over non-cocommutative cosemisimple Hopf algebras*, J. Algebra **273** (2004), no. 2, 753–779, DOI 10.1016/j.jalgebra.2003.08.006. MR2037722

[6] N. Fukuda, *Semisimple Hopf algebras of dimension* 12, Tsukuba J. Math. **21** (1997), no. 1, 43–54, DOI 10.21099/tkbjm/1496163160. MR1467220

[7] G. A. García and C. Vay, *Hopf algebras of dimension 16*, Algebr. Represent. Theory **13** (2010), no. 4, 383–405, DOI 10.1007/s10468-009-9128-7. MR2660853

[8] G. I. Kac and V. G. Paljutkin, *Finite ring groups* (Russian), Trudy Moskov. Mat. Obšč. **15** (1966), 224–261. MR208401

[9] B. Kostant, *Graded manifolds, graded Lie theory, and prequantization*, Differential geometrical methods in mathematical physics (Proc. Sympos., Univ. Bonn, Bonn, 1975), Lecture Notes in Math., Vol. 570, Springer, Berlin-New York, 1977, pp. 177–306. MR580292

[10] S. Majid, *Cross products by braided groups and bosonization*, J. Algebra **163** (1994), no. 1, 165–190, DOI 10.1006/jabr.1994.1011. MR1257312

[11] A. Masuoka, *Semisimple Hopf algebras of dimension* $2p$, Comm. Algebra **23** (1995), no. 5, 1931–1940, DOI 10.1080/00927879508825319. MR1323710

[12] A. Masuoka, *Semisimple Hopf algebras of dimension* $6, 8$, Israel J. Math. **92** (1995), no. 1-3, 361–373, DOI 10.1007/BF02762089. MR1357764

[13] A. Masuoka, *The fundamental correspondences in super affine groups and super formal groups*, J. Pure Appl. Algebra **202** (2005), no. 1-3, 284–312, DOI 10.1016/j.jpaa.2005.02.010. MR2163412

[14] A. Masuoka, *Abelian and non-abelian second cohomologies of quantized enveloping algebras*, J. Algebra **320** (2008), no. 1, 1–47, DOI 10.1016/j.jalgebra.2008.03.034. MR2417975

[15] A. Masuoka, *Harish-Chandra pairs for algebraic affine supergroup schemes over an arbitrary field*, Transform. Groups **17** (2012), no. 4, 1085–1121, DOI 10.1007/s00031-012-9203-8. MR3000482

[16] A. Masuoka and T. Shibata, *On functor points of affine supergroups*, J. Algebra **503** (2018), 534–572, DOI 10.1016/j.jalgebra.2018.02.002. MR3780007

[17] S.-H. Ng, *Hopf algebras of dimension* $2p$, Proc. Amer. Math. Soc. **133** (2005), no. 8, 2237–2242, DOI 10.1090/S0002-9939-05-07804-4. MR2138865

[18] D. E. Radford, *The structure of Hopf algebras with a projection*, J. Algebra **92** (1985), no. 2, 322–347, DOI 10.1016/0021-8693(85)90124-3. MR778452

[19] D. S. Sage and M. D. Vega, *Twisted Frobenius-Schur indicators for Hopf algebras*, J. Algebra **354** (2012), 136–147, DOI 10.1016/j.jalgebra.2011.12.026. MR2879228

[20] Y. Shi, *Automorphism groups of the Suzuki Hopf algebras*, Preprint, `arXiv:2302.09773`, 2023.

[21] D. Ştefan, *Hopf algebras of low dimension*, J. Algebra **211** (1999), no. 1, 343–361, DOI 10.1006/jabr.1998.7602. MR1656583

[22] Y. Zhu, *Hopf algebras of prime dimension*, Internat. Math. Res. Notices **1** (1994), 53–59, DOI 10.1155/S1073792894000073. MR1255253

Department of Applied Mathematics, Okayama University of Science, 1-1 Ridai-cho Kita-ku Okayama-shi, Okayama 700-0005, Japan
Email address: `shibata@ous.ac.jp`

Department of Applied Mathematics, Okayama University of Science, 1-1 Ridai-cho Kita-ku Okayama-shi, Okayama 700-0005, Japan
Email address: `r23nda8mr@ous.jp`

Contemporary Mathematics
Volume **814**, 2025
https://doi.org/10.1090/conm/814/16307

Nakayama functor for coalgebras and a categorical perspective of the integral theory for Hopf algebras

Kenichi Shimizu

Abstract. We review basic properties of the Nakayama functor for coalgebras and introduce a number of applications to tensor categories. We also give equivalent conditions for a coquasi-bialgebra with preantipode to admit a nonzero integral on it.

1. Introduction

Given a finite-dimensional algebra A a field $\Bbbk$, we denote by ${}_A\mathfrak{M}_{\mathtt{fd}}$ the category of finite-dimensional left A-modules. The endofunctor on ${}_A\mathfrak{M}_{\mathtt{fd}}$ defined by $M \mapsto A^* \otimes_A M$ is called the *Nakayama functor* and has a prominent role in the representation theory of finite-dimensional algebras. Recently, this functor also attracts attention from a different perspective of the study of Hopf algebras, tensor categories and their applications. This rather expository paper aims to introduce numerous applications of the Nakayama functor to tensor categories mainly based on [**38, 39**].

The powerfulness of the use of the Nakayama functor for the study of tensor categories was demonstrated by Fuchs, Schaumann, and Schweigert in [**21**]. We recall that a finite abelian category is a linear category that is equivalent to ${}_A\mathfrak{M}_{\mathtt{fd}}$ for some finite-dimensional algebra A, and a finite tensor category is, roughly speaking, a finite abelian category equipped with a structure of a rigid monoidal category [**18**]. An important fact pointed out in [**21**] is the formula

$$A^* \otimes_A M = \int^{X \in {}_A\mathfrak{M}_{\mathtt{fd}}} \mathrm{Hom}_A(M, X)^* \otimes_{\Bbbk} X \tag{1.1}$$

of the Nakayama functor, where the integral means a coend [**29**]. By this formula, one can define the *right exact Nakayama functor* $\mathbb{N}^{\mathrm{r}}_{\mathcal{M}} : \mathcal{M} \to \mathcal{M}$ for a finite abelian category $\mathcal{M}$ as the same form of coend as (1.1). Applications of this functor given in [**21**] are based on the following two observations.

2020 *Mathematics Subject Classification.* Primary 18M05, 16T05.

Key words and phrases. Nakayama functor, locally finite abelian category, tensor category, coquasi-bialgebra.

The author is supported by JSPS KAKENHI Grant Number JP20K03520.

OBSERVATION 1.1. By the representation theory of finite-dimensional algebras, we know that $\mathbb{N}^{\mathrm{r}}_{\mathcal{M}}$ is an equivalence (respectively, the identity) if and only if $\mathcal{M} \approx {}_A\mathfrak{M}_{\mathtt{fd}}$ for some Frobenius (respectively, symmetric Frobenius) algebra A.

OBSERVATION 1.2. Given a functor F, we denote by F^{la} and F^{ra} a left and a right adjoint of F, respectively, if it exists. By the universal property of $\mathbb{N}^{\mathrm{r}}_{\mathcal{M}}$ as a coend, we have an isomorphism $F \circ \mathbb{N}^{\mathrm{r}}_{\mathcal{A}} \cong \mathbb{N}^{\mathrm{r}}_{\mathcal{B}} \circ F^{\mathrm{lla}}$ for a linear functor $F : \mathcal{A} \to \mathcal{B}$ between finite abelian categories such that F^{ra}, F^{la} and $F^{\mathrm{lla}} := (F^{\mathrm{la}})^{\mathrm{la}}$ exist.

Some nontrivial results on finite tensor categories now easily follow. To demonstrate, let $\mathcal{C}$ be a finite tensor category. By applying Observation 1.2 to the functor $F = (-) \otimes X$ or $F = X \otimes (-)$, we obtain natural isomorphisms

$$ {}^{\vee\vee}X \otimes \mathbb{N}^{\mathrm{r}}_{\mathcal{C}}(\mathbf{1}) \cong \mathbb{N}^{\mathrm{r}}_{\mathcal{C}}(X) \cong \mathbb{N}^{\mathrm{r}}_{\mathcal{C}}(\mathbf{1}) \otimes X^{\vee\vee} \quad (X \in \mathcal{C}), \tag{1.2} $$

where $\mathbf{1}$, $(-)^{\vee}$ and ${}^{\vee}(-)$ are the unit object, the left duality and the right duality, respectively. Since $\mathcal{C} \approx {}_A\mathfrak{M}_{\mathtt{fd}}$ for some Frobenius algebra A [**18**], Observation 1.1 implies that $\mathbb{N}^{\mathrm{r}}_{\mathcal{C}}$ is an equivalence. Hence $\alpha := \mathbb{N}^{\mathrm{r}}_{\mathcal{C}}(\mathbf{1})$ is invertible with respect to the tensor product. In particular, the Radford S^4-formula

$$ X^{\vee\vee\vee\vee} \cong \alpha^{\vee} \otimes X \otimes \alpha \quad (X \in \mathcal{C}) \tag{1.3} $$

follows from (1.2). Observation 1.1 also implies that $\mathcal{C} \approx {}_A\mathfrak{M}_{\mathtt{fd}}$ for some symmetric Frobenius algebra A if and only if there are an isomorphism $\alpha \cong \mathbf{1}$ and a natural isomorphism $X^{\vee\vee} \cong X$ $(X \in \mathcal{C})$.

After the work of Fuchs, Schaumann, and Schweigert [**21**], properties of the Nakayama functor are investigated and a lot of applications to finite tensor categories and their modules are found; see [**22, 23, 36, 37, 40–42, 46**]. Now, in view of these successful, it is natural to ask the following.

PROBLEM 1.3. Define the Nakayama functor by a coend of the form like (1.1) for a class of abelian categories and establish results describing homological properties of the category like Observation 1.2.

Given a coalgebra C, we denote by $\mathfrak{M}^C$ and $\mathfrak{M}^C_{\mathtt{fd}}$ the category of right C-comodules and its full subcategory consisting of finite-dimensional objects. A *locally finite abelian category* [**18**] is a linear category that is equivalent to $\mathfrak{M}^C_{\mathtt{fd}}$ for some coalgebra C. Our work [**39**] gives, in summary, a solution to the above problem for locally finite abelian categories and their ind-completions. More precisely, for a coalgebra C, we define the right exact Nakayama functor $\mathbb{N}^{\mathrm{r}}_C$ by

$$ \mathbb{N}^{\mathrm{r}}_C : \mathfrak{M}^C \to \mathfrak{M}^C, \quad \mathbb{N}^{\mathrm{r}}_C(M) = C \otimes_{C^*} M \quad (M \in \mathfrak{M}^C), $$

where C^* is the dual algebra of C, which naturally acts on a left (right) C-comodule from the right (left). We have the coend formula

$$ \mathbb{N}^{\mathrm{r}}_C(M) = \int^{X \in \mathfrak{M}^C_{\mathtt{fd}}} \mathrm{coHom}^C(X, M) \otimes_{\Bbbk} X \quad (M \in \mathfrak{M}^C), \tag{1.4} $$

where coHom^C is the coHom functor (see Subsection 2.3). Some properties of C are characterized through $\mathbb{N}^{\mathrm{r}}_C$ as follows.

THEOREM 1.4 (cf. [**38, 39**]). *For a coalgebra C, the following hold.*

(1) *C is right semiperfect if and only if $\mathbb{N}^{\mathrm{r}}_C$ induces an equivalence from the category of finite-dimensional projective right C-comodules to the category*

of injective right C-comodules whose socle is finite-dimensional. If these equivalent conditions are satisfied, then there is an isomorphism

$$\mathbb{N}^{\mathrm{r}}_{C}(P(S)) \cong E(S)$$

for each simple right C-comodule S, where $E(S)$ and $P(S)$ are an injective hull and a projective cover of S, respectively.

(2) *C is quasi-co-Frobenius (QcF) if and only if $\mathbb{N}^{\mathrm{r}}_{C}$ is an equivalence.*
(3) *C is co-Frobenius if and only if C is QcF and $\mathbb{N}^{\mathrm{r}}_{C}$ preserves dimensions of simple right C-comodules.*
(4) *C is symmetric co-Frobenius if and only if $\mathbb{N}^{\mathrm{r}}_{C}$ is the identity functor.*

Given a category $\mathcal{A}$, we denote its ind-completion [**28**] by $\mathrm{Ind}(\mathcal{A})$ (we may identify it with $\mathfrak{M}^{C}$ when $\mathcal{A} = \mathfrak{M}^{C}_{\mathtt{fd}}$ for some coalgebra C). One can define the right exact Nakayama functor $\mathbb{N}^{\mathrm{r}}_{\mathrm{Ind}(\mathcal{A})} : \mathrm{Ind}(\mathcal{A}) \to \mathrm{Ind}(\mathcal{A})$ for a locally finite abelian category $\mathcal{A}$ as the same form of coend as (1.4). Unlike the finite case, there is a technical difficulty that the functor $\mathbb{N}^{\mathrm{r}}_{C}$ does not preserve $\mathfrak{M}^{C}_{\mathtt{fd}}$ in general. The full subcategory $\mathfrak{M}^{C}_{\mathtt{fd}}$ is closed under $\mathbb{N}^{\mathrm{r}}_{C}$ if C is semiperfect. Thus, if $\mathcal{A} \approx \mathfrak{M}^{C}_{\mathtt{fd}}$ for some semiperfect C, then $\mathbb{N}^{\mathrm{r}}_{\mathrm{Ind}(\mathcal{A})}$ defines an endofunctor on $\mathcal{A}$.

The above results are applied to tensor categories as follows: In the same way as the finite case, we see that the Nakayama functor $\mathbb{N}^{\mathrm{r}}_{\mathrm{Ind}(\mathcal{C})}$ for a tensor category $\mathcal{C}$ has a similar expression as (1.2). An interesting result generalizing an existence criteria for non-zero integrals on Hopf algebras is that the functor $\mathbb{N}^{\mathrm{r}}_{\mathrm{Ind}(\mathcal{C})}$ is non-zero if and only if $\mathcal{C}$ is Frobenius. For a Frobenius tensor category $\mathcal{C}$, the functor $\mathbb{N}^{\mathrm{r}}_{\mathrm{Ind}(\mathcal{C})}$ is shown to be an equivalence. Radford's S^4-formula (1.3) for $\mathcal{C}$ is now easily established in the same way as the finite case. Like this, the Nakayama functor is useful for generalizing results on Hopf algebras that had been proved by using integrals. The Nakayama functor could be a categorical alternative for integrals on Hopf algebras.

1.1. Organization of this paper. We introduce further applications of the Nakayama functor as well as explaining the organization of this paper. In Section 2, we fix notations used throughout this paper and recall basic results on coalgebras and comodules from [**14, 17**].

In Section 3, for a coalgebra C, we introduce two endofunctors $\mathbb{N}^{\mathrm{l}}_{C}$ and $\mathbb{N}^{\mathrm{r}}_{C}$ on $\mathfrak{M}^{C}$, which we call the left exact and the right exact Nakayama functor, respectively (Definition 3.1). Apart from the results introduced in the above as Theorem 1.4, we show that $\mathbb{N}^{\mathrm{r}}_{C}$ is the zero functor if and only if the rational part of C^* is zero (Lemma 3.6). We also prove the following Calabi–Yau type property: For a semiperfect coalgebra C, there is a natural isomorphism

$$\mathrm{Hom}^{C}(P, M)^* \cong \mathrm{Hom}^{C}(M, \mathbb{N}^{\mathrm{r}}_{C}(P))$$

for $M, P \in \mathfrak{M}^{C}_{\mathtt{fd}}$ with P projective (Theorem 3.16).

In Section 4, we first give the coend formula of $\mathbb{N}^{\mathrm{r}}_{C}$ as in (1.4), which allows us to define the Nakayama functor $\mathbb{N}^{\mathrm{r}}_{\mathrm{Ind}(\mathcal{A})} : \mathrm{Ind}(\mathcal{A}) \to \mathrm{Ind}(\mathcal{A})$ for a locally finite abelian category $\mathcal{A}$ without referencing a coalgebra C such that $\mathcal{A} \approx \mathfrak{M}^{C}_{\mathtt{fd}}$ (Definition 4.8). We also observe that the left semiperfectness, the right semiperfectness, the semiperfectness and the QcF property of a coalgebra C are characterized in terms of the category $\mathfrak{M}^{C}_{\mathtt{fd}}$ (Lemmas 4.4–4.7). With the above preparation, we interpret results in Section 3 in the language of category theory. We also give a generalization of Observation 1.2 by the same way as in the finite case (Theorem 4.12).

In Section 5, we first give a characterization of Frobenius tensor categories (Theorem 5.5). As in the finite case, we obtain the Radford S^4-formula for Frobenius tensor categories (Theorem 5.11). We announce a generalization of the relationship between the Nakayama functor and modified traces pointed out in [**36**, **38**] in the finite setting (Theorem 5.19). Finally, we explain applications of the Nakayama functor for exact sequences of tensor categories. An important result is that the class of Frobenius tensor categories is closed under exact sequences (Theorem 5.21), as asked by Natale [**31**].

In Section 6, we give applications of our results to coquasi-bialgebras with preantipode. Bulacu and Caenepeel [**10**] gave several equivalent conditions for a coquasi-Hopf algebra to be co-Frobenius. We generalize this result to coquasi-bialgebras with preantipode (Theorem 6.2). Unlike the case of coquasi-Hopf algebras, there might be a QcF coquasi-bialgebra with preantipode that is not co-Frobenius, despite that we have no such examples (cf. Theorem 6.3). At the end of this paper, we discuss coquasi-bialgebras arising from the de-equivariantization process [**3**].

2. Preliminaries on coalgebras

2.1. Notation. Throughout this paper, we work over a field $\Bbbk$. The symbol **Vec** means the category of all vector spaces (over $\Bbbk$). Given a vector space X, we denote by $X^* := \mathrm{Hom}_{\Bbbk}(X, \Bbbk)$ the dual space. The symbol like x^* is sometimes used to express an element of X^*. For $x^* \in X^*$ and $x \in X$, we often write $\langle x^*, x\rangle$ to mean $x^*(x) \in \Bbbk$.

All (co)algebras are assumed to be (co)associative and (co)unital over $\Bbbk$. Given an algebra A, we denote by ${}_A\mathfrak{M}$ and $\mathfrak{M}_A$ the category of left and right A-modules, respectively. Similarly, given a coalgebra C, we denote by ${}^C\mathfrak{M}$ and $\mathfrak{M}^C$ the category of left and right C-comodules, respectively. The Hom functor for these four categories are written as ${}_A\mathrm{Hom}$, Hom_A, ${}^C\mathrm{Hom}$ and Hom^C, respectively. We use the subscript $\mathtt{fd}$ to mean the full subcategory of finite-dimensional objects. For example, $\mathfrak{M}^C_{\mathtt{fd}}$ is the category of finite-dimensional right C-comodules.

The comultiplication and the counit of a coalgebra C are denoted by Δ and ε, respectively. To express the comultiplication of $c \in C$, we adopt the Sweedler notation like $\Delta(c) = c_{(1)} \otimes c_{(2)}$. For a left C-comodule L and a right C-comodule R, the coactions of C are expressed by Sweedler-like notation as $\ell \mapsto \ell_{(-1)} \otimes \ell_{(0)}$ ($\ell \in L$) and $r \mapsto r_{(0)} \otimes r_{(1)}$ ($r \in R$), respectively.

If X is a finite-dimensional right C-comodule, then X^* is a left C-comodule by the coaction determined by $x^*_{(-1)}\langle x^*_{(0)}, x\rangle = \langle x^*, x_{(0)}\rangle x_{(1)}$ for all elements $x^* \in X^*$ and $x \in X$. Similarly, the dual space of a finite-dimensional left C-comodule has a natural structure of a right C-comodule. We note that these constructions give an anti-equivalence between $\mathfrak{M}^C_{\mathtt{fd}}$ and ${}^C\mathfrak{M}_{\mathtt{fd}}$.

2.2. Rational modules. If C is a coalgebra, then C^* is an algebra with respect to the convolution product defined by $\langle fg, c\rangle = \langle f, c_{(1)}\rangle\langle g, c_{(2)}\rangle$ for $f, g \in C^*$ and $c \in C$. The algebra C^* acts on a right C-comodule M from the left by

$$c^* \rightharpoonup m = m_{(0)}\langle f, m_{(1)}\rangle \quad (c^* \in C^*, m \in M)$$

and this construction gives rise to a fully faithful functor

$$\mathfrak{M}^C \to {}_{C^*}\mathfrak{M}, \quad M \mapsto (M, \rightharpoonup). \tag{2.1}$$

The image of this functor is the category of *rational* modules. Every left C^*-module M has the largest rational submodule, which is called *the rational part* of M and denoted by M^{rat}. We may, and do, regard the assignment $M \mapsto M^{\mathrm{rat}}$ as a functor $(-)^{\mathrm{rat}} : {}_{C^*}\mathfrak{M} \to \mathfrak{M}^C$, which is right adjoint to (2.1). The rational part functor for right C^*-modules is defined in a similar way.

In general, computing the rational part is a nontrivial problem. Here we note the following formula: If M is a left C-comodule, then M has a natural structure of a right C^*-module and hence its dual M^* is a left C^*-module. By [**17**, Corollary 2.2.16], the rational part of M^* is given by

$$M^{*\mathrm{rat}} = \{m^* \in M^* \mid \dim_{\Bbbk}(C^* m^*) < \infty\}. \tag{2.2}$$

2.3. Quasi-finite comodules and the coHom functor. Let C be a coalgebra. For each right C-comodule M, there is a linear functor

$$\mathbf{Vec} \to \mathfrak{M}^C, \quad W \mapsto W \otimes_{\Bbbk} M. \tag{2.3}$$

If the functor (2.3) has a left adjoint, then we denote it by $\mathrm{coHom}^C(M, -)$. Thus, by definition, we have a natural isomorphism

$$\mathrm{Hom}_{\Bbbk}(\mathrm{coHom}^C(M,N), W) \cong \mathrm{Hom}^C(N, W \otimes_{\Bbbk} M)$$

for $N \in \mathfrak{M}^C$ and $W \in \mathbf{Vec}$. In particular, by letting $W = \Bbbk$, we have

$$\mathrm{coHom}^C(M,N)^* \cong \mathrm{Hom}^C(N, M). \tag{2.4}$$

The functor $\mathrm{coHom}^C(M, -)$ is introduced by Takeuchi [**44**] in his study of Morita theory for coalgebras. According to Takeuchi [**44**], the functor (2.3) has a left adjoint if and only if M is *quasi-finite* in the sense that the vector space $\mathrm{Hom}(X, M)$ is finite-dimensional for all objects $X \in \mathfrak{M}^C_{\mathtt{fd}}$. Let $\mathfrak{M}^C_{\mathtt{qf}}$ denote the full subcategory of $\mathfrak{M}^C$ consisting of quasi-finite comodules. Then we have a functor

$$\mathrm{coHom}^C : (\mathfrak{M}^C_{\mathtt{qf}})^{\mathrm{op}} \times \mathfrak{M}^C \to \mathbf{Vec}, \quad (M,N) \mapsto \mathrm{coHom}^C(M,N),$$

which we call the *coHom functor*.

In [**44**], the coHom functor is given as a certain filtered colimit. For practical use, the following formula is useful: We have a natural isomorphism

$$\mathrm{coHom}^C(X,Y) \cong X^* \otimes_{C^*} Y \quad (X \in \mathfrak{M}^C_{\mathtt{fd}}, Y \in \mathfrak{M}^C) \tag{2.5}$$

by the tensor-Hom adjunction.

2.4. Injective and projective comodules. We recall that an object P of a category $\mathcal{A}$ is said to be *projective* if the functor $\mathrm{Hom}_{\mathcal{A}}(P, -) : \mathcal{A} \to \mathbf{Set}$ preserves epimorphisms, where **Set** is the category of all sets. An object of $\mathcal{A}$ is said to be *injective* if it is projective in $\mathcal{A}^{\mathrm{op}}$. To clarify, we introduce the following terminology:

DEFINITION 2.1. A right C-comodule X is said to be *projective* (*injective*) if it is a projective (injective) object of the category $\mathfrak{M}^C$.

Given a category $\mathcal{A}$, we denote by $\mathrm{Proj}(\mathcal{A})$ and $\mathrm{Inj}(\mathcal{A})$ the full subcategory of $\mathcal{A}$ consisting of projective and injective objects of $\mathcal{A}$, respectively. In this paper, we are interested in Proj or Inj of $\mathfrak{M}^C_{\mathtt{fd}}$ rather than $\mathfrak{M}^C$. For an algebra A, an object of $\mathrm{Proj}({}_A\mathfrak{M}_{\mathtt{fd}})$ may not be projective in ${}_A\mathfrak{M}$. However, the situation is easy for coalgebras.

LEMMA 2.2. *A finite-dimensional right C-comodule is projective (injective) in $\mathfrak{M}^C$ if and only if it is projective (injective) in $\mathfrak{M}^C_{\mathtt{fd}}$.*

This lemma can be obtained by assembling some fundamental results on projective and injective comodules found in [**17**] (see [**39**, Lemma 2.3] for details).

2.5. Injective comodules and idempotents. Let C be a coalgebra. An *idempotent* for a coalgebra C is an idempotent of the algebra C^*. Given an idempotent e for C and a right C-comodule M, we write $eM := \{e \rightharpoonup m \mid m \in M\}$. A similar notation will be used for left C-comodules. We note that eC and Ce are a left and a right C-comodule, respectively. It is easy to verify that the map

$$\mathrm{Hom}^C(M, W \otimes_{\Bbbk} Ce) \to \mathrm{Hom}_{\Bbbk}(eM, W), \quad f \mapsto (\mathrm{id}_W \otimes \varepsilon|_{Ce}) \circ (f|_{eM}) \tag{2.6}$$

is a natural isomorphism for $M \in \mathfrak{M}^C$ and $W \in \mathbf{Vec}$. Hence we have

$$\mathrm{coHom}^C(Ce, M) \cong eM \quad (M \in \mathfrak{M}^C) \tag{2.7}$$

as vector spaces. When M is moreover a B-C-bicomodule for some coalgebra B, (2.7) is in fact an isomorphism of left B-comodules.

According to [**14**, Proposition 1.13], an indecomposable injective left (right) C-comodule is isomorphic to eC (Ce) for some primitive idempotent e for C. Let $\mathfrak{I}^C_{\mathtt{sf}}$ be the full subcategory of $\mathfrak{M}^C$ consisting of injective objects whose socle is of finite length. Since an indecomposable injective comodule is the same thing as an injective hull of a simple comodule, $\mathfrak{I}^C_{\mathtt{sf}}$ coincides with the full subcategory of $\mathfrak{M}^C$ consisting of finite direct sums of injective hulls of simple comodules. We define the full subcategory ${}^C\mathfrak{I}_{\mathtt{sf}} \subset {}^C\mathfrak{M}$ in a similar way. Then we have the following lemma.

LEMMA 2.3. *Let C be a coalgebra. The contravariant functor*

$$D : \mathfrak{I}^C_{\mathtt{sf}} \to {}^C\mathfrak{I}_{\mathtt{sf}}, \quad E \mapsto \mathrm{coHom}^C(E, C) \tag{2.8}$$

is a well-defined anti-equivalence sending Ce to eC, where e is a primitive idempotent. Moreover, the anti-equivalence (2.8) *sends an injective hull of a simple right C-comodule S to an injective hull of $S^* \in {}^C\mathfrak{M}$.*

PROOF. By (2.7), we have an isomorphism $D(Ce) \cong eC$ for each idempotent e for C (as noted in [**14**, Corollary 1.6]). This shows that the functor D induces a bijective correspondence between the isomorphism classes of objects of $\mathfrak{I}^C_{\mathtt{sf}}$ and those of ${}^C\mathfrak{I}_{\mathtt{sf}}$. For idempotents e and f for C, we have isomorphisms

$$\mathrm{Hom}^C(Ce, Cf) \cong (eCf)^* \cong {}^C\mathrm{Hom}(fC, eC) \cong \mathrm{Hom}^C(D(Cf), D(Ce)) \tag{2.9}$$

by (2.6) and its variant for left comodules. One can check that (2.9) coincides with the linear map induced by the functor D. Thus D is fully faithful. We have proved that D is an anti-equivalence.

Now let S be a simple right C-comodule, and let e be a primitive idempotent for C such that Ce is an injective hull of S. Since eC ($\cong D(Ce)$) is an indecomposable injective left C-comodule, it is an injective hull of a simple left C-comodule. By the isomorphism (2.6), we have

$${}^C\mathrm{Hom}(S^*, D(Ce)) \cong {}^C\mathrm{Hom}(S^*, eC) \cong (S^*e)^* \cong (eS)^{**} \cong \mathrm{Hom}^C(S, Ce)^* \neq 0,$$

and therefore $D(Ce)$ is an injective hull of S^*. □

2.6. Special classes of coalgebras. Let C be a coalgebra. We recall from [**17**] the following classes of coalgebras:

(1) C is said to be *left semiperfect* if every finite-dimensional left C-comodule has a projective cover in ${}^{C}\mathfrak{M}$.
(2) C is said to be *left quasi-co-Frobenius* (or *left QcF* for short) if there are a cardinal κ and an injective homomorphism $C \to (C^*)^{\oplus\kappa}$ of left C^*-modules. If, moreover, the cardinal κ can be taken to be 1, then C is said to be *co-Frobenius*.
(3) For an adjective X $\in$ {semiperfect, quasi-co-Frobenius, co-Frobenius}, we say that C is right X if C^{cop} is left X. An X coalgebra means a coalgebra that is both left X and right X.

The book [**17**] is a comprehensive reference for results on these classes of coalgebras. One of important results is that a left (right) QcF coalgebra is left (right) semiperfect [**17**, Corollary 3.3.6].

2.7. Local units for $C^{*\mathrm{rat}}$. Given a coalgebra C, we denote by $C_\ell^{*\mathrm{rat}}$ and $C_r^{*\mathrm{rat}}$ the rational part of C^* as a left and a right C^*-module, respectively. By (2.2), both $C_\ell^{*\mathrm{rat}}$ and $C_r^{*\mathrm{rat}}$ are ideals of the algebra C^*.

We recall a technique of *local units* for semiperfect coalgebras. From now on, we assume that C is semiperfect. Since $C_\ell^{*\mathrm{rat}} = C_r^{*\mathrm{rat}}$ by the semiperfectness [**17**, Corollary 3.2.16], we write $C_\ell^{*\mathrm{rat}}$ and $C_r^{*\mathrm{rat}}$ simply as $C^{*\mathrm{rat}}$. By (2.2), the counit of C belongs to $C^{*\mathrm{rat}}$ only if C is finite-dimensional. Thus, in general, $C^{*\mathrm{rat}}$ is a "nonunital" ring. However, it is known that $C^{*\mathrm{rat}}$ has local units in the following sense: For any finite subset $X \subset C^{*\mathrm{rat}}$, there is an idempotent $e \in C^{*\mathrm{rat}}$ such that $ex = x = xe$ for all $x \in X$ [**17**, Corollary 3.2.17].

We note that local units can be extended for comodules as follows.

LEMMA 2.4. *Let C be a semiperfect coalgebra, and let M be a right C-comodule. For any finite subsets $X \subset C^{*\mathrm{rat}}$ and $Y \subset M$, there is an idempotent $e \in C^{*\mathrm{rat}}$ such that $ex = x = xe$ and $ey = y$ for all $x \in X$ and $y \in Y$.*

PROOF. Let N be a finite-dimensional subcomodule of M such that $Y \subset N$, and let D be the subspace spanned by $\langle n^*, n_{(0)}\rangle n_{(1)}$ ($n^* \in N^*$, $n \in N$). It is easy to see that D is a finite-dimensional subcoalgebra of C. The semiperfectness of C implies that $C^{*\mathrm{rat}}$ is dense in C^* with respect to the finite topology [**17**, Proposition 3.2.1]. Hence there is an element $\varepsilon' \in C^{*\mathrm{rat}}$ such that $\varepsilon'(d) = \varepsilon(d)$ for all $d \in D$. By the construction, we have

$$\langle n^*, \varepsilon' n\rangle = \varepsilon'(\langle n^*, n_{(0)}\rangle n_{(1)}) = \varepsilon(\langle n^*, n_{(0)}\rangle n_{(1)}) = \langle n^*, n\rangle$$

for all $n^* \in M^*$ and $n \in M$. Hence we have

$$\varepsilon' n = n \quad (n \in N). \tag{2.10}$$

Now let $e \in C^{*\mathrm{rat}}$ be a local unit for the finite set $X \cup \{\varepsilon'\}$. The element e fulfills the requirements. Indeed, it is trivial that the equations $ex = x = xe$ hold for all $x \in X$. By (2.10), we have $en = e(\varepsilon' n) = (e\varepsilon')n = \varepsilon' n = n$ for all $n \in N$. Hence, in particular, we have $ey = y$ for all $y \in Y$. The proof is done. □

By using this lemma, we prove the following technical result.

LEMMA 2.5. *Suppose that C is semiperfect. Then the map*

$$\kappa_M : C^{*\mathrm{rat}} \otimes_{C^*} M \to M, \quad \kappa_M(c^* \otimes_{C^*} m) = c^* m$$

is a natural isomorphism for $M \in \mathfrak{M}^C$.

PROOF. We first consider the case where M is finite-dimensional. By applying Lemma 2.4 to a basis of M, we obtain an idempotent $e \in C^{*\mathrm{rat}}$ such that $em = m$ for all $m \in M$. The space $X := C^{*\mathrm{rat}}e$ is finite-dimensional by (2.2). Again by Lemma 2.4, we get an idempotent $\tilde{e} \in C^{*\mathrm{rat}}$ such that $\tilde{e}x = x = x\tilde{e}$, $\tilde{e}e = e = e\tilde{e}$ and $\tilde{e}m = m$ for all $x \in X$ and $m \in M$. Now we define the linear map

$$\overline{\kappa}_M : M \to C^{*\mathrm{rat}} \otimes_{C^*} M, \quad \overline{\kappa}_M(m) = \tilde{e} \otimes_{C^*} m.$$

It is obvious that $\kappa_M \circ \overline{\kappa}_M$ is the identity. We also have

$$\begin{aligned} \overline{\kappa}_M \kappa_M (c^* \otimes_{C^*} m) &= \tilde{e} \otimes_{C^*} c^* m = \tilde{e} \otimes_{C^*} c^*(em) = \tilde{e} \otimes_{C^*} (c^* e)m \\ &= \tilde{e}(c^* e) \otimes_{C^*} m = (c^* e) \otimes_{C^*} m = c^* \otimes_{C^*} em = c^* \otimes_{C^*} m \end{aligned}$$

for $c^* \in C^{*\mathrm{rat}}$ and $m \in M$, where the fifth equality follows from $c^*e \in X$. Hence we have proved that κ_M is an isomorphism if M is finite-dimensional. The general case follows from the fundamental theorem for comodules. □

3. Nakayama functor for coalgebras

3.1. Nakayama functor for coalgebras. Let C be a coalgebra.

DEFINITION 3.1. We define the left exact Nakayama functor $\mathbb{N}^{\mathrm{l}}_C$ and the right exact Nakayama functor $\mathbb{N}^{\mathrm{r}}_C$ to be the endofunctors on $\mathfrak{M}^C$ given by

$$\mathbb{N}^{\mathrm{l}}_C(M) = \mathrm{Hom}^C(C, M)^{\mathrm{rat}} \quad \text{and} \quad \mathbb{N}^{\mathrm{r}}_C(M) = C \otimes_{C^*} M,$$

respectively, for $M \in \mathfrak{M}^C$. Here, the right C-comodule structure of $\mathbb{N}^{\mathrm{l}}_C(M)$ and $\mathbb{N}^{\mathrm{r}}_C(M)$ are given as follows:

(1) The vector space $\mathrm{Hom}^C(C, M)$ becomes a left C^*-module by the action given by $(c^* \cdot f)(c) = f(c \leftharpoonup c^*)$ for $c^* \in C^*$, $f \in \mathrm{Hom}^C(C, M)$ and $c \in C$. We take its rational part to obtain the right C-comodule $\mathbb{N}^{\mathrm{l}}_C(M)$.
(2) The right C-coaction on $\mathbb{N}^{\mathrm{r}}_C(M)$ is given by

$$c \otimes_{C^*} m \mapsto (c_{(1)} \otimes_{C^*} m) \otimes c_{(2)} \quad (c \in C, m \in M).$$

As their name suggest, $\mathbb{N}^{\mathrm{l}}_C$ and $\mathbb{N}^{\mathrm{r}}_C$ are left and right exact, respectively. This follows from Lemma 3.2 below and the fact that a functor is left (right) exact if it has a left (right) adjoint.

LEMMA 3.2. *$\mathbb{N}^{\mathrm{l}}_C$ is right adjoint to $\mathbb{N}^{\mathrm{r}}_C$.*

PROOF. For $M, M' \in \mathfrak{M}^C$, we have natural isomorphisms

$$\begin{aligned} \mathrm{Hom}^C(\mathbb{N}^{\mathrm{r}}_C(M), M') &\cong {}_{C^*}\mathrm{Hom}(C \otimes_{C^*} M, M') \\ &\cong {}_{C^*}\mathrm{Hom}(M, {}_{C^*}\mathrm{Hom}(C, M')) \\ &\cong \mathrm{Hom}^C(M, \mathrm{Hom}(C, M')^{\mathrm{rat}}) = \mathrm{Hom}^C(M, \mathbb{N}^{\mathrm{l}}_C(M')), \end{aligned}$$

where the second isomorphism is the tensor-Hom adjunction. □

The restriction of $\mathbb{N}^{\mathrm{l}}_C$ to $\mathfrak{M}^C_{\mathrm{qf}}$ has the following expression.

LEMMA 3.3. *There is a natural isomorphism*

$$\mathbb{N}^{\mathrm{l}}_C(M) \cong \mathrm{coHom}^C(M, C)^{*\mathrm{rat}} \quad (M \in \mathfrak{M}^C_{\mathrm{qf}}).$$

Chin and Simson [**12**, Equation 1.12] have introduced a functor ν_C for a coalgebra C and called it the Nakayama functor. The functor ν_C is defined on a particular class of quasi-finite comodules. By Lemma 3.3 and [**12**, Theorem 1.8 (a)], we see that ν_C is isomorphic to $\mathbb{N}_C^{\mathrm{l}}$ on the domain of ν_C.

PROOF. By (2.4) and the defining formula of $\mathbb{N}_C^{\mathrm{l}}(M)$, we establish an isomorphism of vector spaces as stated. One can check that it is indeed an isomorphism of right C-comodules. □

The restriction of $\mathbb{N}_C^{\mathrm{r}}$ to $\mathfrak{M}_{\mathrm{fd}}^C$ has the following expression.

LEMMA 3.4. *For $M \in \mathfrak{M}_{\mathrm{fd}}^C$, there are natural isomorphism*

$$\mathbb{N}_C^{\mathrm{r}}(M) \cong \mathrm{coHom}^{C^{\mathrm{cop}}}(M^*, C^{\mathrm{cop}}), \quad \mathbb{N}_{C^{\mathrm{cop}}}^{\mathrm{l}}(M^*) \cong \mathbb{N}_C^{\mathrm{r}}(M)^{*\mathrm{rat}}. \tag{3.1}$$

PROOF. The first isomorphism follows from (2.5). The second one is obtained from the first one by applying the functor $(-)^{*\mathrm{rat}}$ to the both sides. □

Two coalgebras C and D are said to be *Morita-Takeuchi equivalent* if $\mathfrak{M}^C$ and $\mathfrak{M}^D$ are equivalent as linear categories. The Nakayama functor is Morita-Takeuchi invariant in the following sense.

LEMMA 3.5. *Let C and D be coalgebras, and let $\Phi : \mathfrak{M}^C \to \mathfrak{M}^D$ be an equivalence of linear categories. Then there are isomorphisms*

$$\Phi \circ \mathbb{N}_C^{\mathrm{l}} \cong \mathbb{N}_D^{\mathrm{l}} \circ \Phi \quad \textit{and} \quad \Phi \circ \mathbb{N}_C^{\mathrm{r}} \cong \mathbb{N}_D^{\mathrm{r}} \circ \Phi.$$

We will give a proof of this lemma in Section 4.2 after recalling a categorical notion of coends. In fact, this lemma is trivial once the universal property of the Nakayama functor is formulated and proved in Subsection 4.2.

3.2. The Nakayama functor and the rational part of C^*. Let C be a coalgebra. As in Subsection 2.7, we denote by $C_\ell^{*\mathrm{rat}}$ the rational part of C^* as a left C^*-module. One of main purposes of this section is to characterize some homological properties of C in terms of the Nakayama functor. In preparation for this, we first observe:

LEMMA 3.6. *For a coalgebra C, the following are equivalent:*

(1) $C_\ell^{*\mathrm{rat}} = 0$.
(2) $\mathbb{N}_C^{\mathrm{l}}$ *is the zero functor.*
(3) $\mathbb{N}_C^{\mathrm{r}}$ *is the zero functor.*

PROOF. The tensor-Hom adjunction gives natural isomorphisms

$$\mathbb{N}_C^{\mathrm{r}}(M)^* \cong {}_{C^*}\mathrm{Hom}(M, \mathrm{Hom}_{\Bbbk}(C, \Bbbk)) \cong \mathrm{Hom}^C(M, C_\ell^{*\mathrm{rat}}) \tag{3.2}$$

for $M \in \mathfrak{M}^C$. Now (1) ⇔ (3) is obvious. The equivalence (2) ⇔ (3) follows from Lemma 3.2, which states that $\mathbb{N}_C^{\mathrm{r}}$ and $\mathbb{N}_C^{\mathrm{l}}$ are an adjoint pair. □

The natural isomorphism (3.2) is also used to prove:

LEMMA 3.7. *For a coalgebra C, the following are equivalent:*

(1) *The right C-comodule $C_\ell^{*\mathrm{rat}}$ is quasi-finite.*
(2) $\mathbb{N}_C^{\mathrm{l}}$ *preserves the class of quasi-finite comodules.*
(3) $\mathbb{N}_C^{\mathrm{r}}$ *preserves the class of finite-dimensional comodules.*

PROOF. The implication (1) ⇒ (3) follows immediately from (3.2). We suppose that (3) holds and let M be a quasi-finite right C-comodule. Then we have

$$\dim_{\Bbbk} \mathrm{Hom}^C(X, \mathbb{N}^{\mathrm{l}}_C(M)) = \dim_{\Bbbk} \mathrm{Hom}^C(\mathbb{N}^{\mathrm{r}}_C(X), M) < \infty$$

for all $X \in \mathfrak{M}^C_{\mathtt{fd}}$. This means that $\mathbb{N}^{\mathrm{l}}_C(M)$ is quasi-finite. Thus (2) holds. Finally, we assume that (2) holds. Then $C^{*\mathrm{rat}}_{\ell}$, which is isomorphic to $\mathbb{N}^{\mathrm{l}}_C(C)$, is quasi-finite since C is. Hence (1) holds. The proof is done. □

LEMMA 3.8. *For a coalgebra C, the following are equivalent:*

(1) *The right C-comodule $C^{*\mathrm{rat}}_{\ell}$ is injective.*
(2) *$\mathbb{N}^{\mathrm{l}}_C$ preserves the class of injective comodules.*
(3) *$\mathbb{N}^{\mathrm{r}}_C$ is exact.*

PROOF. The implication (1) ⇒ (3) follows from (3.2) and the fact that the duality functor preserves and reflects exact sequences. The implication (3) ⇒ (2) follows from Lemma 3.2. If (2) holds, then $\mathbb{N}^{\mathrm{l}}_C(C) \cong C^{*\mathrm{rat}}_{\ell}$ is injective since C is, and thus (1) holds. The proof is done. □

3.3. Semiperfect coalgebras. The Nakayama functor $A^* \otimes_A (-)$ for a finite-dimensional algebra A induces an equivalence from $\mathrm{Proj}({}_A\mathfrak{M}_{\mathtt{fd}})$ to $\mathrm{Inj}({}_A\mathfrak{M}_{\mathtt{fd}})$. We show that an analogous result holds for a coalgebra C if and only if C is right semiperfect. To be precise, we define the full subcategory $\mathfrak{I}^C_{\mathtt{sf}} \subset \mathfrak{M}^C$ as in Subsection 2.5 and write $\mathfrak{P}^C_{\mathtt{fd}} = \mathrm{Proj}(\mathfrak{M}^C_{\mathtt{fd}})$ for brevity. Then we have:

THEOREM 3.9. *For a coalgebra C, the following are equivalent:*

(1) *C is right semiperfect.*
(2) *The functor $\mathbb{N}^{\mathrm{l}}_C$ restricts to an equivalence from $\mathfrak{I}^C_{\mathtt{sf}}$ to $\mathfrak{P}^C_{\mathtt{fd}}$.*
(3) *The functor $\mathbb{N}^{\mathrm{r}}_C$ restricts to an equivalence from $\mathfrak{P}^C_{\mathtt{fd}}$ to $\mathfrak{I}^C_{\mathtt{sf}}$.*

Suppose that these equivalent conditions are satisfied. Then the equivalences between $\mathfrak{I}^C_{\mathtt{sf}}$ and $\mathfrak{P}^C_{\mathtt{fd}}$ induced by $\mathbb{N}^{\mathrm{l}}_C$ and $\mathbb{N}^{\mathrm{r}}_C$ are mutually quasi-inverse to each other. Moreover, for a simple right C-comodule S, there are isomorphisms

$$\mathbb{N}^{\mathrm{l}}_C(E(S)) \cong P(S) \quad \text{and} \quad \mathbb{N}^{\mathrm{r}}_C(P(S)) \cong E(S), \tag{3.3}$$

where $E(S)$ and $P(S)$ are an injective hull and a projective cover of S, respectively.

We note that $\mathbb{N}^{\mathrm{l}}_C$ does not even define a functor from $\mathfrak{I}^C_{\mathtt{sf}}$ to $\mathfrak{P}^C_{\mathtt{fd}}$ in general (see [**39**, Example 3.29]). The condition (2) of this theorem means that $\mathbb{N}^{\mathrm{l}}_C(E)$ belongs to $\mathfrak{P}^C_{\mathtt{fd}}$ for all objects $E \in \mathfrak{I}^C_{\mathtt{sf}}$ and, moreover, the induced functor $\mathbb{N}^{\mathrm{l}}_C : \mathfrak{I}^C_{\mathtt{sf}} \to \mathfrak{P}^C_{\mathtt{fd}}$ is an equivalence. The same remark applies to the condition (3).

PROOF. We first show that (1) implies both (2) and (3). For brevity, we write ${}^C\mathfrak{I}_{\mathtt{fd}} = \mathrm{Inj}({}^C\mathfrak{M}_{\mathtt{fd}})$. Suppose that (1) holds. Then, since every simple left C-comodule has a finite-dimensional injective hull [**17**, Theorem 3.2.3], we have ${}^C\mathfrak{I}_{\mathtt{sf}} = {}^C\mathfrak{I}_{\mathtt{fd}}$. The assignment $X \mapsto X^*$ induces an anti-equivalence between ${}^C\mathfrak{I}_{\mathtt{fd}}$ and $\mathfrak{P}^C_{\mathtt{fd}}$. By Lemmas 2.3 and 3.3, we see that the restriction of $\mathbb{N}^{\mathrm{l}}_C$ to $\mathfrak{I}^C_{\mathtt{sf}}$ factors into the composition

$$\mathbb{N}^{\mathrm{l}}_C : \mathfrak{I}^C_{\mathtt{sf}} \xrightarrow{\mathrm{coHom}(-,C)} {}^C\mathfrak{I}_{\mathtt{sf}} = {}^C\mathfrak{I}_{\mathtt{fd}} \xrightarrow{(-)^*} \mathfrak{P}^C_{\mathtt{fd}}$$

of anti-equivalences. Hence (2) holds. To show (3), we note that the above argument shows that every object of $\mathfrak{P}^C_{\mathtt{fd}}$ is a finite direct sum of objects of the form $(Ce)^*$

for some primitive idempotent e for C. Since we have

$$(3.4) \qquad \mathbb{N}^{\mathrm{r}}_C((Ce)^*) = C \otimes_{C^*} (Ce)^* = C \otimes_{C^*} C^*e \cong Ce$$

for a primitive idempotent e, the functor $\mathbb{N}^{\mathrm{r}}_C$ induces a functor from $\mathfrak{P}^C_{\mathtt{fd}}$ to $\mathfrak{I}^C_{\mathtt{sf}}$. It follows from Lemma 3.2 that the induced functor $\mathbb{N}^{\mathrm{r}}_C : \mathfrak{P}^C_{\mathtt{fd}} \to \mathfrak{I}^C_{\mathtt{sf}}$ is left adjoint to $\mathbb{N}^{\mathrm{l}}_C : \mathfrak{I}^C_{\mathtt{sf}} \to \mathfrak{P}^C_{\mathtt{fd}}$. As the latter one is an equivalence as we have proved, so is the former one. Namely, (3) holds.

Next, we prove that (2) implies (1). Suppose that (2) holds. Then there is an anti-equivalence $D : {}^C\mathfrak{I}_{\mathtt{sf}} \to \mathfrak{P}^C_{\mathtt{fd}}$ obtained by composing the quasi-inverse of (2.8) and the equivalence $\mathbb{N}^{\mathrm{l}}_C : \mathfrak{I}^C_{\mathtt{sf}} \to \mathfrak{P}^C_{\mathtt{fd}}$. By Lemma 3.3, we have $D(E) = E^{*\mathrm{rat}}$ for $E \in {}^C\mathfrak{I}_{\mathtt{sf}}$. We now consider the contravariant functor $\overline{D} : \mathfrak{P}^C_{\mathtt{fd}} \to {}^C\mathfrak{I}_{\mathtt{sf}}$ defined by $\overline{D}(P) = P^*$. It is easy to see that $D \circ \overline{D}$ is isomorphic to the identity functor on $\mathfrak{P}^C_{\mathtt{fd}}$. Since D is an anti-equivalence, so is $\overline{D}$. This implies that every object of ${}^C\mathfrak{I}_{\mathtt{sf}}$ is finite-dimensional. By a characterization of right semiperfect coalgebras [**17**, Theorem 3.2.3], we conclude that C is semiperfect.

One can prove that (3) implies (1) in a similar way: Suppose that (3) holds. Then, by applying Lemma 2.3 to C^{cop}, we obtain an anti-equivalence

$$(3.5) \qquad \mathrm{coHom}^{C^{\mathrm{cop}}}(-, C^{\mathrm{cop}}) : {}^C\mathfrak{I}_{\mathtt{sf}} \to \mathfrak{I}^C_{\mathtt{sf}}$$

of categories. Let $D : \mathfrak{P}^C_{\mathtt{fd}} \to {}^C\mathfrak{I}_{\mathtt{sf}}$ be the anti-equivalence obtained by composing the equivalence $\mathbb{N}^{\mathrm{r}}_C : \mathfrak{P}^C_{\mathtt{fd}} \to \mathfrak{I}^C_{\mathtt{sf}}$ and the quasi-inverse of (3.5). By Lemma 3.4, we have $D(E) \cong E^*$ for $E \in \mathfrak{P}^C_{\mathtt{fd}}$. This implies that every object of ${}^C\mathfrak{I}_{\mathtt{sf}}$ is finite-dimensional. Hence C is semiperfect.

We have proved that (1), (2), and (3) are equivalent. To complete the proof, we now assume that these equivalent conditions are satisfied. Then $\mathbb{N}^{\mathrm{l}}_C : \mathfrak{I}^C_{\mathtt{sf}} \to \mathfrak{P}^C_{\mathtt{fd}}$ and $\mathbb{N}^{\mathrm{r}}_C : \mathfrak{P}_{\mathtt{fd}} \to \mathfrak{I}^C_{\mathtt{sf}}$ are mutually quasi-inverse to each other as we have seen in the proof of (1) $\Rightarrow$ (3). By Lemmas 2.3 and 3.3, we have $\mathbb{N}^{\mathrm{l}}_C(E(S)) \cong E(S^*)^* \cong P(S)$ for simple $S \in \mathfrak{M}^C$. The latter isomorphism in (3.3) is obtained by applying the equivalence $\mathbb{N}^{\mathrm{r}}_C : \mathfrak{P}_{\mathtt{fd}} \to \mathfrak{I}^C_{\mathtt{sf}}$ to the former one. The proof is done. □

For our applications to locally finite abelian categories, it is important to know when the full subcategory $\mathfrak{M}^C_{\mathtt{fd}}$ is closed under $\mathbb{N}^{\mathrm{l}}_C$ and $\mathbb{N}^{\mathrm{r}}_C$. With the help of the above theorem, we now prove the following theorem.

THEOREM 3.10 ([**39**, Theorem 3.15]). *Let C be a semiperfect coalgebra. Then the adjunction $\mathbb{N}^{\mathrm{r}}_C \dashv \mathbb{N}^{\mathrm{l}}_C$ of Lemma 3.2 restricts to an adjunction on $\mathfrak{M}^C_{\mathtt{fd}}$. Moreover, the functors $\mathbb{N}^{\mathrm{l}}_C$ and $\mathbb{N}^{\mathrm{r}}_C$ induce equivalences*

$$(3.6) \qquad \mathbb{N}^{\mathrm{l}}_C : \mathfrak{I}^C_{\mathtt{fd}} \to \mathfrak{P}^C_{\mathtt{fd}} \quad \text{and} \quad \mathbb{N}^{\mathrm{r}}_C : \mathfrak{P}^C_{\mathtt{fd}} \to \mathfrak{I}^C_{\mathtt{fd}},$$

which are mutually quasi-inverse to each other.

PROOF. The semiperfect of C implies $\mathfrak{I}^C_{\mathtt{sf}} = \mathfrak{I}^C_{\mathtt{fd}}$. Thus, by Theorem 3.9, we have well-defined equivalences (3.6). It remains to show that $\mathfrak{M}^C_{\mathtt{fd}}$ is closed under $\mathbb{N}^{\mathrm{l}}_C$ and $\mathbb{N}^{\mathrm{r}}_C$. Let M be a finite-dimensional right C-comodule. By the semiperfectness of C, there are objects $E \in \mathfrak{I}^C_{\mathtt{fd}}$ and a monomorphism $M \hookrightarrow E$ of right C-comodules. Since $\mathbb{N}^{\mathrm{l}}_C$ is left exact, $\mathbb{N}^{\mathrm{l}}_C(M)$ is finite-dimensional as a subcomodule of $\mathbb{N}^{\mathrm{l}}_C(E) \in \mathfrak{P}^C_{\mathtt{fd}}$. A dual argument shows that $\mathbb{N}^{\mathrm{r}}_C$ preserves $\mathfrak{M}^C_{\mathtt{fd}}$. The proof is done. □

3.4. Quasi-co-Frobenius coalgebras. We investigate when $\mathbb{N}^{\mathrm{l}}_C$ and $\mathbb{N}^{\mathrm{r}}_C$ are equivalences. To begin with, we consider the case where C is co-Frobenius. The definition of a co-Frobenius coalgebra was recalled in Subsection 2.6. However, we use the following equivalent condition: According to Iovanov [**25**, Theorem 2.8], a coalgebra C is co-Frobenius if and only if it has a *Frobenius pairing*, that is, a nondegenerate C^*-balanced bilinear map $C \times C \to \Bbbk$. The following definition is due to [**11**, Section 6].

DEFINITION 3.11. Let C be a co-Frobenius coalgebra with Frobenius pairing β. Then there is a unique linear map $\nu_C : C \to C$ determined by

$$\beta(y, x) = \beta(\nu_C(x), y) \quad (x, y \in C),$$

which we call the *Nakayama automorphism* of C with respect to β.

It is known that ν_C is a coalgebra automorphism [**39**, Lemma 3.18]. Given a coalgebra automorphism f of C and a right C-comodule M, we denote by $M^{(f)}$ the right C-comodule obtained by twisting the coaction by f. The Nakayama functor for a co-Frobenius coalgebra is given as follows.

THEOREM 3.12 ([**39**, Lemma 3.20]). *For a co-Frobenius coalgebra C with Nakayama automorphism ν, there are natural isomorphisms*

$$\mathbb{N}^{\mathrm{l}}_C(M) \cong M^{(\nu^{-1})} \quad \textit{and} \quad \mathbb{N}^{\mathrm{r}}_C(M) \cong M^{(\nu)} \quad (M \in \mathfrak{M}^C).$$

PROOF. Let $\beta : C \times C \to \Bbbk$ be the Frobenius pairing of C. Then the map

$$\beta_\ell : C \to C^{*\mathrm{rat}}, \quad c \mapsto \beta(c, -)$$

is bijective [**11**, Section 1]. Thus we obtain an isomorphism

$$\mathbb{N}^{\mathrm{r}}_C(M) = C \otimes_{C^*} M \xrightarrow[\cong]{\beta_\ell \otimes_{C^*} \mathrm{id}_M} C^{*\mathrm{rat}} \otimes_{C^*} M \xrightarrow[\cong]{\text{Lemma 2.5}} M$$

of vector spaces. The formula for $\mathbb{N}^{\mathrm{r}}_C(M)$ is obtained by transporting the coaction of C through this isomorphism. The formula for $\mathbb{N}^{\mathrm{l}}_C(M)$ follows from that $\mathbb{N}^{\mathrm{l}}_C$ is right adjoint to $\mathbb{N}^{\mathrm{r}}_C$. $\square$

In particular, $\mathbb{N}^{\mathrm{l}}_C$ and $\mathbb{N}^{\mathrm{r}}_C$ are equivalences if C is co-Frobenius. The converse does not holds: Let C be a QcF coalgebra that is not co-Frobenius (see [**26**] for a concrete example). Then there is a co-Frobenius coalgebra D that is Morita-Takeuchi equivalent to C [**14**]. By Lemma 3.5 and Theorem 3.12, we see that $\mathbb{N}^{\mathrm{l}}_C$ and $\mathbb{N}^{\mathrm{r}}_C$ are equivalences, despite that C is not co-Frobenius.

The above argument shows that $\mathbb{N}^{\mathrm{l}}_C$ and $\mathbb{N}^{\mathrm{r}}_C$ are equivalences if C is QcF. With the help of Iovanov's result [**27**, Lemma 1.7] characterizing QcF coalgebras, one can prove that the converse holds. More specifically, Theorem 3.13.

THEOREM 3.13 (cf. [**39**, Theorems 3.22 and 3.23]). *For a coalgebra C, the following are equivalent:*

(1) *C is QcF.*
(2) *$\mathbb{N}^{\mathrm{l}}_C$ is an equivalence.*
(3) *$\mathbb{N}^{\mathrm{r}}_C$ is an equivalence.*
(4) *$\mathbb{N}^{\mathrm{r}}_C$ is exact and faithful.*

Suppose that these equivalent conditions are satisfied. Then we have

$$P(S) \cong E(\mathbb{N}^{\mathrm{l}}_C(S)) \quad \textit{and} \quad E(S) \cong P(\mathbb{N}^{\mathrm{r}}_C(S)) \tag{3.7}$$

for a simple right C-comodule S, where $E(S)$ and $P(S)$ are an injective hull and a projective cover of S, respectively.

PROOF. We have already proved (1) $\Rightarrow$ (2) and (1) $\Rightarrow$ (3). Since $\mathbb{N}^{\mathrm{r}}_C$ and $\mathbb{N}^{\mathrm{l}}_C$ are an adjoint pair, (2) and (3) are equivalent. It is trivial that (3) implies (4). Now we suppose that (4) holds and aim to show (1). Let $\{S_i\}_{i\in I}$ be a complete set of representatives of isomorphism classes of simple right C-comodules. The right C-comodule C is injective. By Lemma 3.8 and the assumption that $\mathbb{N}^{\mathrm{r}}_C$ is exact, $C^{*\mathrm{rat}}_{\ell}$ is also injective. Thus C and $C^{*\mathrm{rat}}_{\ell}$ are decomposed as

$$C \cong \bigoplus_{i\in I} E(S_i)^{\oplus a_i} \quad \text{and} \quad C^{*\mathrm{rat}}_{\ell} \cong \bigoplus_{i\in I} E(S_i)^{\oplus b_i}$$

for some cardinals a_i and b_i. Since $\mathrm{Hom}^C(S_i, C) \cong S_i^* \neq 0$, we have $a_i > 0$ for all i. By (3.2) and the assumption that $\mathbb{N}^{\mathrm{r}}_C$ is faithful, we have

$$\mathrm{Hom}^C(S_i, C^{*\mathrm{rat}}_{\ell}) \cong \mathbb{N}^{\mathrm{r}}_C(S_i)^* \neq 0$$

for all i. This implies that $b_i > 0$ for all i. Now let κ be an infinite cardinal bigger than all a_i's and all b_i's. Then we have $C^{\oplus\kappa} \cong (C^{*\mathrm{rat}}_{\ell})^{\oplus\kappa}$ by the above discussion. Thus, by [**27**, Lemma 1.7], we conclude that C is QcF, that is, (1) holds. We have proved that (1), (2), (3) and (4) are equivalent. The rest of the proof easily follows from Theorem 3.10. □

Let C be a QcF coalgebra, and let $\{S_i\}_{i\in I}$ be a complete set of representatives of isomorphism classes of simple right C-comodules. Then there is a permutation π of the index set I defined by $P(S_i) \cong E(S_{\pi(i)})$. Following the terminology used in the representation theory of finite-dimensional algebras, we may call π the *Nakayama permutation* of C. According to [**27**, Corollary 2.3], a QcF coalgebra is co-Frobenius if and only if its Nakayama permutation preserves the dimension of simple comodules. Theorem 3.13 says that the Nakayama permutation is induced by the Nakayama functor $\mathbb{N}^{\mathrm{l}}_C$. Thus we have Theorem 3.14.

THEOREM 3.14 ([**39**, Theorem 3.24]). *For a QcF coalgebra C, the following are equivalent:*

(1) *C is co-Frobenius.*
(2) $\dim_{\Bbbk} \mathbb{N}^{\mathrm{l}}_C(S) = \dim_{\Bbbk} S$ *for all simple right C-comodule S.*
(3) $\dim_{\Bbbk} \mathbb{N}^{\mathrm{r}}_C(S) = \dim_{\Bbbk} S$ *for all simple right C-comodule S.*

3.5. Symmetric coalgebras. A coalgebra C is said to be *symmetric* [**11**] if there is an injective homomorphism $C \to C^*$ of C^*-bimodules. This class of coalgebras is characterized by the Nakayama functor as follows.

THEOREM 3.15 ([**39**, Theorem 3.27]). *For a coalgebra C, the following are equivalent:*

(1) *C is a symmetric coalgebra;*
(2) *$\mathbb{N}^{\mathrm{l}}_C$ is isomorphic to the identity functor;*
(3) *$\mathbb{N}^{\mathrm{r}}_C$ is isomorphic to the identity functor.*

PROOF. A coalgebra automorphism f of C is said to be *coinner* if there is an invertible element $\alpha \in C^*$ such that $f(c) = \alpha \rightharpoonup c \leftharpoonup \alpha^{-1}$ for all $c \in C$. From a categorical point of view, a coalgebra automorphism f of C is coinner precisely if the autoequivalence $(-)^{(f)}$ on $\mathfrak{M}^C$ induced by f is isomorphic to the identity functor. It is known that a symmetric coalgebra is precisely a co-Frobenius coalgebra with coinner Nakayama automorphism [**11**, Proposition 6.2]. Thus (1) $\Rightarrow$ (2) and (1) $\Rightarrow$ (3) follow immediately from Theorem 3.12.

It is obvious from Lemma 3.2 that (2) and (3) are equivalent. To complete the proof, we prove (2) $\Rightarrow$ (1). We assume that (2) holds. Then, by Theorems 3.13 and 3.14, C is co-Frobenius. Moreover, by Theorem 3.12, the Nakayama automorphism of C is coinner. Thus C is symmetric. The proof is done. □

3.6. A Calabi–Yau type structure. In some literature, a Calabi-Yau structure on a linear category $\mathcal{A}$ refers to a natural isomorphism

$$\mathrm{Hom}_{\mathcal{A}}(X,Y)^* \cong \mathrm{Hom}_{\mathcal{A}}(Y,X) \quad (X, Y \in \mathcal{A})$$

or an equivalent structure. Here we point out that $\mathrm{Proj}(\mathfrak{M}^C_{\mathtt{fd}})$ for a semiperfect coalgebra C has a similar kind of structure.

THEOREM 3.16. *For a semiperfect coalgebra C, there is a natural isomorphism*

$$\mathrm{Hom}^C(P,M)^* \cong \mathrm{Hom}^C(M, \mathbb{N}^{\mathrm{r}}_C(P)) \quad (P \in \mathrm{Proj}(\mathfrak{M}^C_{\mathtt{fd}}), M \in \mathfrak{M}^C_{\mathtt{fd}}).$$

PROOF. We fix $P \in \mathrm{Proj}(\mathfrak{M}^C_{\mathtt{fd}})$ and set $P^\dagger = {}_{C^*}\mathrm{Hom}(P, C^*)$. Since P is projective as a left C^*-module [**17**, Corollary 2.4.19], the map

$$\theta_{P,M} : P^\dagger \otimes_{C^*} M \to {}_{C^*}\mathrm{Hom}(P,M), \quad \theta_{P,M}(f \otimes_{C^*} m)(p) = f(p)\cdot m$$

is a natural isomorphism for $M \in {}_{C^*}\mathfrak{M}$ with the inverse written by a pair of dual bases of P as a left C^*-module. There is a well-defined linear map

$$\phi_P : P^\dagger \otimes_{C^*} \mathbb{N}(P) \to \Bbbk, \quad \phi_P(f \otimes_{C^*} c \otimes_{C^*} p) = \langle f(p), c\rangle,$$

where $\mathbb{N} = \mathbb{N}^{\mathrm{r}}_C$. We use this map to define the bilinear pairing

$$\beta_{M,P} : \mathrm{Hom}^C(M, \mathbb{N}(P)) \times \mathrm{Hom}^C(P,M) \to \Bbbk, \quad \beta_M(f,g) = \phi_P \theta^{-1}_{P,\mathbb{N}(P)}(fg)$$

for $M \in \mathfrak{M}^C_{\mathtt{fd}}$. We note that every indecomposable projective right C-comodule is of the form $(Ce)^*$ for some primitive idempotent $e \in C$. If $P = (Ce)^*$ for some primitive idempotent e, then we have isomorphisms

$$\mathrm{Hom}^C(M, \mathbb{N}(P)) \cong \mathrm{Hom}^C(M, Ce) \cong (eM)^*,$$
$$\mathrm{Hom}^C(P,M) \cong {}_{C^*}\mathrm{Hom}(C^*e, M) \cong eM$$

by (2.6) and (3.4) and, through these isomorphisms, the pairing $\beta_{M,P}$ is identified with the canonical pairing between eM and $(eM)^*$. Thus $\beta_{M,P}$ is non-degenerate in this case. One can also verify the non-degeneracy of $\beta_{M,P}$ in the general case by decomposing P into the direct sum of indecomposable projective comodules. It is now routine to check that the linear map

$$\alpha_{M,P} : \mathrm{Hom}^C(M, \mathbb{N}(P)) \to \mathrm{Hom}^C(P,M)^*, \quad f \mapsto \beta_{M,P}(f,-)$$

induced by the pairing $\beta_{M,P}$ meets the required condition. □

By Theorems 3.15 and 3.16, we have the following corollary.

COROLLARY 3.17. *For a symmetric coalgebra C, there is a natural isomorphism*

$$\mathrm{Hom}^C(P,M)^* \cong \mathrm{Hom}^C(M,P) \quad (P \in \mathrm{Proj}(\mathfrak{M}^C_{\mathtt{fd}}), M \in \mathfrak{M}^C_{\mathtt{fd}}).$$

4. Categorical interpretation

4.1. The coHom functor for a linear category. In this section, we rephrase results of the previous section in the language of category theory for the purpose of broader applications. We first discuss what the coHom functor should be. Let $\mathcal{A}$ be a linear category, and let W be a vector space of dimension α (which is possibly infinite). The *copower* of $M \in \mathcal{A}$ by W, denoted by $W \otimes M$, is an object of $\mathcal{A}$ together with a natural isomorphism

$$\mathrm{Hom}_{\mathcal{A}}(W \otimes M, N) \cong \mathrm{Hom}_{\Bbbk}(W, \mathrm{Hom}_{\mathcal{A}}(M, N))$$

for $N \in \mathcal{A}$. By choosing a basis of W, we see that $W \otimes M$ exists if and only the direct sum $M^{\oplus \alpha} \in \mathcal{A}$ exists and, if they exist, they are isomorphic.

Now we fix an object $M \in \mathcal{A}$ and assume that the copower $W \otimes M$ exists for all objects $W \in \mathbf{Vec}$. Then the assignment $W \mapsto W \otimes M$ gives rise to a linear functor from $\mathbf{Vec}$ to $\mathcal{A}$. If a left adjoint of this functor exists, then we write it as $\mathrm{coHom}_{\mathcal{A}}(M, -)$. Thus we have a natural isomorphism

$$\mathrm{Hom}_{\mathcal{A}}(N, W \otimes M) \cong \mathrm{Hom}_{\Bbbk}(\mathrm{coHom}_{\mathcal{A}}(M, N), W)$$

for $N \in \mathcal{A}$. In view of the case of coalgebras, we say that an object $M \in \mathcal{A}$ is *quasi-finite* if the functor $\mathrm{coHom}_{\mathcal{A}}(M, -)$ exists. We denote by $\mathcal{A}_{\mathrm{qf}}$ the full subcategory of $\mathcal{A}$ consisting of quasi-finite objects. Then the *coHom functor*

$$\mathrm{coHom}_{\mathcal{A}} : (\mathcal{A}_{\mathrm{qf}})^{\mathrm{op}} \times \mathcal{A} \to \mathbf{Vec}$$

is defined. If $\mathcal{A} = \mathfrak{M}^C$ for some coalgebra C, then the functor $\mathrm{coHom}_{\mathcal{A}}$ is identical to the functor coHom^C introduced in Section 2.3.

Given a functor F, we denote by F^{la} and F^{ra} a left and a right adjoint of F, respectively, if they exist. For later use, we remark the following relation between the coHom functor and adjunctions.

LEMMA 4.1. *Let $\mathcal{A}$ and $\mathcal{B}$ be linear categories admitting arbitrary copowers, and let $F : \mathcal{A} \to \mathcal{B}$ be a linear functor. If F has a left adjoint and a right adjoint, then there is a natural isomorphism*

$$\mathrm{coHom}_{\mathcal{B}}(F(M), N) \cong \mathrm{coHom}_{\mathcal{A}}(M, F^{\mathrm{la}}(N)) \quad (M \in \mathcal{A}_{\mathrm{qf}}, N \in \mathcal{B}). \tag{4.1}$$

PROOF. We first note that F preserves copowers. Indeed, we have

$$\begin{aligned} \mathrm{Hom}_{\mathcal{B}}(F(W \otimes M), N) &\cong \mathrm{Hom}_{\mathcal{A}}(W \otimes M, F^{\mathrm{ra}}(N)) \\ &\cong \mathrm{Hom}_{\Bbbk}(W, \mathrm{Hom}_{\mathcal{A}}(M, F^{\mathrm{ra}}(N))) \\ &\cong \mathrm{Hom}_{\Bbbk}(W, \mathrm{Hom}_{\mathcal{B}}(F(M), N)) \\ &\cong \mathrm{Hom}_{\Bbbk}(W \otimes F(M), N) \end{aligned}$$

for all $W \in \mathbf{Vec}$, $M \in \mathcal{A}$ and $N \in \mathcal{B}$. The Yoneda lemma shows

$$F(W \otimes M) \cong W \otimes F(M) \quad (W \in \mathbf{Vec}, M \in \mathcal{A}). \tag{4.2}$$

For $W \in \mathbf{Vec}$, $M \in \mathcal{A}_{\mathrm{qf}}$ and $N \in \mathcal{B}$, we use the formula (4.2) to show:

$$\begin{aligned} \mathrm{Hom}_{\Bbbk}(\mathrm{coHom}_{\mathcal{B}}(F(M), N), W) &\cong \mathrm{Hom}_{\mathcal{B}}(N, W \otimes F(M)) \\ &\cong \mathrm{Hom}_{\mathcal{B}}(N, F(W \otimes M)) \\ &\cong \mathrm{Hom}_{\mathcal{A}}(F^{\mathrm{la}}(N), W \otimes M) \\ &\cong \mathrm{Hom}_{\Bbbk}(\mathrm{coHom}_{\mathcal{B}}(M, F^{\mathrm{la}}(N)), W). \end{aligned}$$

The proof is completed by the Yoneda lemma. □

4.2. The universal property of the Nakayama functor. The Nakayama functor has a certain universal property. To explain this, we first recall the notion of *coends* [**29**]. Let $\mathcal{A}$ and $\mathcal{V}$ be categories, and let $T : \mathcal{A}^{\mathrm{op}} \times \mathcal{A} \to \mathcal{V}$ be a functor. A *dinatural transformation* from T to an object $V \in \mathcal{V}$ is a family

$$\{i_X : T(X,X) \to V\}_{X \in \mathrm{Obj}(\mathcal{A})}$$

of morphisms in $\mathcal{V}$ such that the equation

$$i_X \circ T(f, \mathrm{id}_X) = i_Y \circ T(\mathrm{id}_Y, f)$$

holds for all morphisms $f : X \to Y$ in $\mathcal{V}$. A *coend* of T is an object of $\mathcal{V}$ equipped with a "universal" dinatural transformation from T. A coend of T is customary denoted by using an integral symbol as

$$\int^{X \in \mathcal{A}} T(X,X).$$

We note the following relation between coends and adjunctions.

LEMMA 4.2 ([**9**, Lemma 3.9]). *Let $T : \mathcal{B}^{\mathrm{op}} \times \mathcal{A} \to \mathcal{V}$ and $F : \mathcal{A} \to \mathcal{B}$ be functors, where $\mathcal{A}$, $\mathcal{B}$ and $\mathcal{V}$ are categories. If F has a right adjoint, then we have*

$$\int^{X \in \mathcal{A}} T(F(X), X) \cong \int^{Y \in \mathcal{B}} T(Y, F^{\mathrm{ra}}(Y)) \tag{4.3}$$

in the sense that the coend on the left-hand side exists if and only if the right one exists and, if they exist, they are canonically isomorphic.

A basic and important example of coends is found in Tannaka theory. Let C be a coalgebra. The Tannaka reconstruction theorem states that the coalgebra C is recovered from the category $\mathfrak{M}^C_{\mathtt{fd}}$ as a coend as

$$C = \int^{X \in \mathfrak{M}^C_{\mathtt{fd}}} X^* \otimes_{\Bbbk} X, \tag{4.4}$$

where the universal dinatural transformation for C is given by

$$i_X : X^* \otimes_{\Bbbk} X \to C, \quad i_X(x^* \otimes x) = \langle x^*, x_{(0)} \rangle x_{(1)} \tag{4.5}$$

for $x^* \in X^*$ and $x \in X \in \mathfrak{M}^C_{\mathtt{fd}}$.

We note that i_X is in fact a homomorphism of C-bicomodules. Let ${}^C\mathfrak{M}^C$ denote the category of C-bimodules. In the same way as we do in Tannaka theory, we can show that C is in fact a coend of the functor

$$(\mathfrak{M}^C_{\mathtt{fd}})^{\mathrm{op}} \times \mathfrak{M}^C_{\mathtt{fd}} \to {}^C\mathfrak{M}^C, \quad (X, Y) \mapsto X^* \otimes_{\Bbbk} Y \tag{4.6}$$

with the universal dinatural transformation (4.5).

Now we prove that $\mathbb{N}^{\mathrm{r}}_C(M)$ is a coend of a certain functor. The idea is simple: We know that C is a coend in ${}^C\mathfrak{M}^C$. If we apply a cocontinuous functor to the coend C, then the result is also a coend.

LEMMA 4.3 ([**39**, Lemma 2.11]). *Let C be a coalgebra. Then we have*

$$\mathbb{N}^{\mathrm{r}}_C(M) = \int^{X \in \mathfrak{M}^C_{\mathtt{fd}}} \mathrm{coHom}^C(X, M) \otimes_{\Bbbk} X \quad (M \in \mathfrak{M}^C).$$

Although we omit the details, we note that $\mathbb{N}^{\mathrm{l}}_C(M)$ is expressed as an end, that is, the dual notion of coends; see [**39**, Theorem 3.6].

PROOF. We fix $M \in \mathfrak{M}^C$ and consider the functor

$$ {}^C\mathfrak{M}^C \to \mathfrak{M}^C, \quad P \mapsto P \otimes_{C^*} M. \tag{4.7}$$

This functor has a right adjoint. Indeed, we have

$$\begin{aligned}\mathrm{Hom}^C(P \otimes_{C^*} M, N) &\cong {}_{C^*}\mathrm{Hom}_{C^*}(P, \mathrm{Hom}_{\Bbbk}(M, N)) \\ &\cong {}^C\mathrm{Hom}^C(P, \mathrm{Hom}_{\Bbbk}(M, N)^{\mathrm{rat}})\end{aligned}$$

for $P \in {}^C\mathfrak{M}^C$ and $N \in \mathfrak{M}^C$ by the tensor-Hom adjunction, where $(-)^{\mathrm{rat}}$ means the rational part as C-bicomodules. Since $C \in {}^C\mathfrak{M}^C$ is a coend of the functor (4.6), we see that $\mathbb{N}^{\mathrm{r}}_C(M)$ is a coend of the functor

$$(\mathfrak{M}^C_{\mathtt{fd}})^{\mathrm{op}} \times \mathfrak{M}^C_{\mathtt{fd}} \to \mathfrak{M}^C, \quad (X, Y) \mapsto (X^* \otimes_{\Bbbk} Y) \otimes_{C^*} M$$

by applying the functor (4.7) to C. The proof is completed by (2.5). □

We give a proof of Lemma 3.5, which we had skipped.

PROOF OF LEMMA 3.5. Let $\Phi : \mathfrak{M}^C \to \mathfrak{M}^D$ be an equivalence of linear categories, where C and D are coalgebras. Since a comodule is finite-dimensional if and only if it is of finite length, and since the length of an object is preserved by an equivalence, Φ induces an equivalence from $\mathfrak{M}^C_{\mathtt{fd}}$ to $\mathfrak{M}^D_{\mathtt{fd}}$. By the universal property given by Lemma 4.3, we have $\Phi \circ \mathbb{N}^{\mathrm{r}}_C \cong \mathbb{N}^{\mathrm{r}}_D \circ \Phi$. By taking right adjoints, we also have $\mathbb{N}^{\mathrm{l}}_C \circ \Phi^{-1} \cong \Phi^{-1} \circ \mathbb{N}^{\mathrm{l}}_D$. The proof is done. □

4.3. Locally finite abelian categories and their ind-completions. A *locally finite abelian category* [**18**, §1.8] is a linear abelian category $\mathcal{A}$ such that every object of $\mathcal{A}$ is of finite length and $\mathrm{Hom}_{\mathcal{A}}(X, Y)$ is finite-dimensional for all objects $X, Y \in \mathcal{A}$. By the result of Takeuchi [**44**] (see also [**18**, §1.8]), a linear category is locally finite abelian if and only if it is equivalent to $\mathfrak{M}^C_{\mathtt{fd}}$ for some coalgebra C.

Given a category $\mathcal{A}$, we denote by $\mathrm{Ind}(\mathcal{A})$ the ind-completion of $\mathcal{A}$ [**28**], that is, the category obtained from $\mathcal{A}$ by freely adjoining filtered colimits of objects of $\mathcal{A}$. There is a canonical functor $\iota : \mathcal{A} \to \mathrm{Ind}(\mathcal{A})$ by which we regard $\mathcal{A}$ as a full subcategory of $\mathrm{Ind}(\mathcal{A})$. If $\mathcal{A} = \mathfrak{M}^C_{\mathtt{fd}}$ for some coalgebra C, then $\mathrm{Ind}(\mathcal{A})$ and ι are identified with $\mathfrak{M}^C$ and the inclusion functor. Keeping this fact in mind, we give characterizations of some classes of coalgebras by properties of their category of finite-dimensional comodules. We first give a characterization of 'one-sided' semiperfect coalgebras.

LEMMA 4.4. *For a locally finite abelian category $\mathcal{A}$, the following are equivalent:*

(1) $\mathcal{A} \approx \mathfrak{M}^C_{\mathtt{fd}}$ *for some right semiperfect coalgebra C.*
(2) $\mathcal{A} \approx {}^C\mathfrak{M}_{\mathtt{fd}}$ *for some left semiperfect coalgebra C.*
(3) *Every object of $\mathcal{A}$ has a projective cover in $\mathcal{A}$.*
(4) *Every object of $\mathcal{A}$ has a projective cover in* $\mathrm{Ind}(\mathcal{A})$.
(5) $\mathcal{A}$ *has enough projective objects.*
(6) $\mathrm{Ind}(\mathcal{A})$ *has enough projective objects.*

PROOF. It is trivial that (1) and (2) are equivalent. By the locally finiteness of $\mathcal{A}$, there is a coalgebra C such that $\mathcal{A} \approx \mathfrak{M}^C_{\mathtt{fd}}$. We fix such a coalgebra C and identify $\mathcal{A}$ with $\mathfrak{M}^C_{\mathtt{fd}}$. Then the equivalence between (1), (4), (5), and (6) is obtained by rephrasing known characterizations of semiperfect coalgebras [**17**, Corollary 2.4.21 and Theorem 3.2.3]. The implication (3) ⇒ (4) is obvious.

To complete the proof, it remains to show (1) $\Rightarrow$ (3). We assume that C is right semiperfect. Let S be a simple right C-comodule. Then an injective hull of $E(S^*) \in {}^C\mathfrak{M}$ is finite-dimensional and $E(S^*)^* \in \mathfrak{M}^C$ is a projective cover of S (see the proof of [**17**, Theorem 3.2.3]). Thus (3) holds. The proof is done. $\square$

Let C be a coalgebra such that $\mathcal{A} \approx \mathfrak{M}^C_{\mathtt{fd}}$. Then the duality establishes equivalences ${}^C\mathfrak{M}_{\mathtt{fd}} \approx (\mathfrak{M}^C_{\mathtt{fd}})^{\mathrm{op}} \approx \mathcal{A}^{\mathrm{op}}$. By applying Lemma 4.4 to $\mathcal{A}^{\mathrm{op}}$, we have Lemma 4.5.

LEMMA 4.5. *For a locally finite abelian category $\mathcal{A}$, the following are equivalent:*

(1) $\mathcal{A} \approx \mathfrak{M}^C_{\mathtt{fd}}$ *for some left semiperfect coalgebra C.*
(2) $\mathcal{A} \approx {}^C\mathfrak{M}_{\mathtt{fd}}$ *for some right semiperfect coalgebra C.*
(3) *Every object of $\mathcal{A}$ has an injective hull in $\mathcal{A}$.*
(4) *$\mathcal{A}$ has enough injective objects.*

By Lemmas 4.4 and 4.5, semiperfect coalgebras are characterized as follows.

LEMMA 4.6. *For a locally finite abelian category $\mathcal{A}$, the following are equivalent:*

(1) $\mathcal{A} \approx \mathfrak{M}^C_{\mathtt{fd}}$ *for some semiperfect coalgebra C.*
(2) *Every object of $\mathcal{A}$ has a projective cover and an injective hull in $\mathcal{A}$.*
(3) *$\mathcal{A}$ has enough projective objects and enough injective objects.*

QcF coalgebras are characterized as follows.

LEMMA 4.7. *For a locally finite abelian category $\mathcal{A}$, the following are equivalent:*

(1) $\mathcal{A} \approx \mathfrak{M}^Q_{\mathtt{fd}}$ *for some QcF coalgebra Q.*
(2) $\mathcal{A} \approx \mathfrak{M}^Q_{\mathtt{fd}}$ *for some semiperfect coalgebra Q and* $\mathrm{Proj}(\mathcal{A}) = \mathrm{Ind}(\mathcal{A})$.

PROOF. This result is noted at the end of Section 3.1 of [**27**]. We explain in a bit more detail. First, let Q be a QcF coalgebra and set $\mathcal{A} = \mathfrak{M}^Q_{\mathtt{fd}}$. Then Q is semiperfect by [**17**, Corollary 3.3.6]. Since every injective right Q-comodule is projective [**17**, Theorem 3.3.4], we have $\mathrm{Inj}(\mathcal{A}) \subset \mathrm{Proj}(\mathcal{A})$. By applying the same argument to Q^{cop}, we also have an inclusion

$$\mathrm{Proj}(\mathcal{A}) = \mathrm{Inj}(\mathcal{A}^{\mathrm{op}}) \subset \mathrm{Proj}(\mathcal{A}^{\mathrm{op}}) = \mathrm{Inj}(\mathcal{A})$$

of classes of objects. Thus we have proved (1) $\Rightarrow$ (2).

To prove the converse, we let Q be a semiperfect coalgebra such that $\mathrm{Proj}(\mathcal{A}) = \mathrm{Ind}(\mathcal{A})$, where $\mathcal{A} := \mathfrak{M}^Q_{\mathtt{fd}}$. We decompose Q into a direct sum of indecomposable injective right Q-comodules, as $Q = \bigoplus_{i \in I} E_i$. The right semiperfectness of Q implies that each E_i is finite-dimensional [**17**, Theorem 3.2.3]. Thus Q is in fact a direct sum of projective objects. This shows that Q is projective, and therefore Q is left QcF [**17**, Theorem 3.3.4]. By applying the same argument to Q^{cop}, we see that Q is also right QcF. The proof is done. $\square$

4.4. Nakayama functor for locally finite abelian categories. Our categorical dictionary is now ready for translating the results on the Nakayama functor established in the previous section. In view of Lemma 4.3, we introduce the following definition.

DEFINITION 4.8. For a locally finite abelian category $\mathcal{A}$, we define the functor

$$\mathbb{N}^{\mathrm{r}}_{\mathrm{Ind}(\mathcal{A})} : \mathrm{Ind}(\mathcal{A}) \to \mathrm{Ind}(\mathcal{A}), \quad M \mapsto \int^{X \in \mathcal{A}} \mathrm{coHom}_{\mathrm{Ind}(\mathcal{A})}(X, M) \otimes X$$

and call $\mathbb{N}^{\mathrm{r}}_{\mathrm{Ind}(\mathcal{A})}$ the *right exact Nakayama functor* for $\mathcal{A}$. We also define the left exact Nakayama functor $\mathbb{N}^{\mathrm{l}}_{\mathrm{Ind}(\mathcal{A})}$ as a right adjoint of $\mathbb{N}^{\mathrm{r}}_{\mathrm{Ind}(\mathcal{A})}$.

Lemmas 3.2 and 4.3 imply that $\mathbb{N}^{\mathrm{r}}_{\mathrm{Ind}(\mathcal{A})}$ and $\mathbb{N}^{\mathrm{r}}_{\mathrm{Ind}(\mathcal{A})}$ are identified with $\mathbb{N}^{\mathrm{r}}_{C}$ and $\mathbb{N}^{\mathrm{l}}_{C}$, respectively, when $\mathcal{A} = \mathfrak{M}^{C}_{\mathtt{fd}}$ for a coalgebra C and $\mathrm{Ind}(\mathcal{A})$ is identified with $\mathfrak{M}^{C}$. Thus the functors $\mathbb{N}^{\mathrm{r}}_{\mathrm{Ind}(\mathcal{A})}$ and $\mathbb{N}^{\mathrm{r}}_{\mathrm{Ind}(\mathcal{A})}$ are defined for any locally finite abelian category $\mathcal{A}$. Theorem 3.10 is translated by Lemma 4.6 as follows:

THEOREM 4.9. *Let $\mathcal{A}$ be a locally finite abelian category with enough projective objects and enough injective objects. Then the functor $\mathbb{N}^{\mathrm{r}}_{\mathrm{Ind}(\mathcal{A})}$ restricts to an endofunctor on $\mathcal{A}$, which we denote by $\mathbb{N}^{\mathrm{r}}_{\mathcal{A}} : \mathcal{A} \to \mathcal{A}$. The functor $\mathbb{N}^{\mathrm{r}}_{\mathcal{A}}$ induces an equivalence $\mathbb{N}^{\mathrm{r}}_{\mathcal{A}} : \mathrm{Proj}(\mathcal{A}) \to \mathrm{Inj}(\mathcal{A})$.*

Based on Theorem 3.13, we prove Theorem 4.10.

THEOREM 4.10. *For a locally finite abelian category $\mathcal{A}$, the following are equivalent:*

(1) *$\mathcal{A} \approx \mathfrak{M}^{Q}_{\mathtt{fd}}$ for some QcF coalgebra Q.*
(2) *$\mathbb{N}^{\mathrm{r}}_{\mathrm{Ind}(\mathcal{A})}$ is an auto-equivalence of* $\mathrm{Ind}(A)$.
(3) *$\mathbb{N}^{\mathrm{r}}_{\mathrm{Ind}(\mathcal{A})}$ is exact and faithful.*
(4) *$\mathbb{N}^{\mathrm{r}}_{\mathrm{Ind}(\mathcal{A})}$ induces an auto-equivalence of $\mathcal{A}$.*

PROOF. Theorem 3.13 states that (1), (2,) and (3) are equivalent. It is trivial that (2) implies (4). To complete the proof, we shall show that (4) implies (3). We assume that $\mathbb{N}^{\mathrm{r}}_{\mathrm{Ind}(\mathcal{A})}$ induces an endofunctor on $\mathcal{A}$, say $\mathbb{N}^{\mathrm{r}}_{\mathcal{A}} : \mathcal{A} \to \mathcal{A}$. Then the endofunctor $\mathrm{Ind}(\mathbb{N}^{\mathrm{r}}_{\mathcal{A}})$ on $\mathrm{Ind}(\mathcal{A})$ induced by $\mathbb{N}^{\mathrm{r}}_{\mathcal{A}}$ is isomorphic to $\mathbb{N}^{\mathrm{r}}_{\mathrm{Ind}(\mathcal{A})}$. Thus, if $\mathbb{N}^{\mathrm{r}}_{\mathcal{A}}$ is an equivalence, so is the functor $\mathrm{Ind}(\mathbb{N}^{\mathrm{r}}_{\mathcal{A}})$. The proof is done. □

Theorem 3.15 implies Theorem 4.11.

THEOREM 4.11. *For a locally finite abelian category $\mathcal{A}$, the following are equivalent:*

(1) *$\mathcal{A} \approx \mathfrak{M}^{Q}_{\mathtt{fd}}$ for some symmetric coalgebra Q.*
(2) *$\mathbb{N}^{\mathrm{r}}_{\mathrm{Ind}(\mathcal{A})}$ is isomorphic to the identity functor on* $\mathrm{Ind}(\mathcal{A})$.
(3) *The restriction of $\mathbb{N}^{\mathrm{r}}_{\mathrm{Ind}(\mathcal{A})}$ to $\mathcal{A}$ is the identity functor in $\mathcal{A}$.*

4.5. Interaction with double adjoints. In the next section, we give some applications of the Nakayama functor to tensor categories. The following theorem is essential (cf. Observation 1.2).

THEOREM 4.12. *Let $\mathcal{A}$ and $\mathcal{B}$ be locally finite abelian categories, and let F : $\mathrm{Ind}(\mathcal{A}) \to \mathrm{Ind}(\mathcal{B})$ be a cocontinuous linear functor such that the double left adjoint $F^{\mathrm{lla}} = (F^{\mathrm{la}})^{\mathrm{la}}$ exists. Then there is an isomorphism*

$$F \circ \mathbb{N}^{\mathrm{r}}_{\mathrm{Ind}(\mathcal{A})} \cong \mathbb{N}^{\mathrm{r}}_{\mathrm{Ind}(\mathcal{B})} \circ F^{\mathrm{lla}}. \tag{4.8}$$

PROOF. Since $\mathfrak{M}^C$ for a coalgebra C is locally presentable, F has a right adjoint. For $M \in \mathrm{Ind}(\mathcal{A})$, we have natural isomorphisms

$$\begin{aligned} F(\mathbb{N}^{\mathrm{r}}_{\mathrm{Ind}(\mathcal{A})}(M)) &\cong \textstyle\int_{X \in \mathcal{A}} F(\mathrm{coHom}_{\mathrm{Ind}(\mathcal{A})}(X, M) \otimes X) \\ &\cong \textstyle\int_{X \in \mathcal{A}} \mathrm{coHom}_{\mathrm{Ind}(\mathcal{A})}(X, M) \otimes F(X) \\ &\cong \textstyle\int_{Y \in \mathcal{B}} \mathrm{coHom}_{\mathrm{Ind}(\mathcal{A})}(F^{\mathrm{la}}(Y), M) \otimes Y \\ &\cong \textstyle\int_{Y \in \mathcal{B}} \mathrm{coHom}_{\mathrm{Ind}(\mathcal{B})}(Y, F^{\mathrm{lla}}(M)) \otimes Y \\ &= \mathbb{N}^{\mathrm{r}}_{\mathrm{Ind}(\mathcal{B})}(F^{\mathrm{lla}}(M)), \end{aligned}$$

where the first isomorphism follows from the cocontinuity of F, the second from (4.2), the third from (4.3), and the fourth from (4.1). □

5. Applications to tensor categories

5.1. Frobenius tensor categories. In this section, we give some applications of our results to tensor categories. We mostly follow [**18**] for notation and terminology on monoidal categories. In view of Mac Lane's strictness theorem, we assume that all monoidal category are strict. Given a monoidal category $\mathcal{C}$, we usually denote by $\otimes$ and $\mathbf{1}$ the monoidal product and the unit object of $\mathcal{C}$, respectively. The category $\mathcal{C}^{\mathrm{rev}}$ is the monoidal category obtained from $\mathcal{C}$ by reversing the order of the monoidal product.

Following [**18**], a left dual object of $X \in \mathcal{C}$ is an object $X^{\vee} \in \mathcal{C}$ equipped with morphisms $\mathrm{ev}_X : X^{\vee} \otimes X \to \mathbf{1}$ and $\mathrm{coev}_X : \mathbf{1} \to X \otimes X^{\vee}$ satisfying certain equations. A right dual object of X is written as ${}^{\vee}X$. We recall that $\mathcal{C}$ is left (right) rigid if every object of $\mathcal{C}$ has a left (right) dual object. A rigid monoidal category is a left and right rigid monoidal category.

DEFINITION 5.1. A *tensor category* [**18**] is a locally finite abelian category $\mathcal{C}$ equipped with a structure of a rigid monoidal category such that the tensor product $\otimes : \mathcal{C} \times \mathcal{C} \to \mathcal{C}$ is $\Bbbk$-linear in each variable and the unit object $\mathbf{1} \in \mathcal{C}$ is a simple object whose endomorphism algebra is isomorphic to $\Bbbk$.

Our main interest in this section is the following class of tensor categories.

DEFINITION 5.2. A *Frobenius tensor category* [**2**] is a tensor category $\mathcal{C}$ such that every simple object of $\mathcal{C}$ has an injective hull in $\mathcal{C}$.

5.2. Equivalent conditions for Frobenius property. Let H be a Hopf algebra. Then the locally finite abelian category $\mathfrak{M}^H_{\mathtt{fd}}$ has a natural structure of a left rigid monoidal category, which may not be a tensor category. Indeed, $\mathfrak{M}^H_{\mathtt{fd}}$ is rigid if and only if the antipode of H is bijective [**45**] and there are some examples of Hopf algebras with non-bijective antipodes [**43**]. Thus, from a viewpoint of general theory of Hopf algebras, the following situation could be important.

ASSUMPTION 5.3. $\mathcal{C}$ is a locally finite abelian category equipped with a structure of a left rigid monoidal category such that the tensor product of $\mathcal{C}$ is $\Bbbk$-linear and exact in each variable, and the unit object $\mathbf{1}$ of $\mathcal{C}$ is a simple object whose endomorphism algebra is isomorphic to $\Bbbk$.

If $\mathcal{C}$ satisfies Assumption 5.3, then $\mathrm{Ind}(\mathcal{C})$ has a natural monoidal structure extending that of $\mathcal{C}$. The exactness of the tensor product of $\mathcal{C}$ implies that the tensor product of $\mathrm{Ind}(\mathcal{C})$ preserves colimits. Since the tensor product of $\mathcal{C}$ is faithful by

[**18**, Exercise 4.3.11], so is the tensor product of $\mathrm{Ind}(\mathcal{C})$. Now we prove the following formula of the Nakayama functor.

LEMMA 5.4. *Let $\mathcal{C}$ satisfy Assumption 5.3. Then we have*

$$\mathbb{N}^{\mathrm{r}}_{\mathrm{Ind}(\mathcal{C})}(X \otimes Y) \cong \mathbb{N}^{\mathrm{r}}_{\mathrm{Ind}(\mathcal{C})}(X) \otimes \mathbb{D}(Y), \tag{5.1}$$

$$\mathbb{N}^{\mathrm{l}}_{\mathrm{Ind}(\mathcal{C})}(X \otimes Y) \cong \mathbb{D}(X) \otimes \mathbb{N}^{\mathrm{l}}_{\mathrm{Ind}(\mathcal{C})}(Y) \tag{5.2}$$

for $X, Y \in \mathrm{Ind}(\mathcal{C})$, where $\mathbb{D} : \mathrm{Ind}(\mathcal{C}) \to \mathrm{Ind}(\mathcal{C})$ is the functor induced by the double left dual functor on the category $\mathcal{C}$. In particular, we have

$$\mathbb{N}^{\mathrm{r}}_{\mathrm{Ind}(\mathcal{C})}(X) \cong \mathrm{g} \otimes \mathbb{D}(X), \quad \mathbb{N}^{\mathrm{l}}_{\mathrm{Ind}(\mathcal{C})}(X) \cong \mathbb{D}(X) \otimes \overline{\mathrm{g}} \quad (X \in \mathrm{Ind}(\mathcal{C})), \tag{5.3}$$

where $\mathrm{g} = \mathbb{N}^{\mathrm{r}}_{\mathcal{C}}(\mathbf{1})$ and $\overline{\mathrm{g}} = \mathbb{N}^{\mathrm{l}}_{\mathcal{C}}(\mathbf{1})$.

PROOF. We write $\mathbb{N} = \mathbb{N}^{\mathrm{r}}_{\mathrm{Ind}(\mathcal{C})}$ for brevity. We fix an object $Y \in \mathcal{C}$. By applying Theorem 4.12 to $F = (-) \otimes Y^{\vee\vee}$, we have a natural isomorphism

$$\mathbb{N}(X \otimes Y) = \mathbb{N}F^{\mathrm{lla}}(X) \cong F(\mathbb{N}(X)) = \mathbb{N}(X) \otimes Y^{\vee\vee} \quad (X \in \mathrm{Ind}(\mathcal{C})).$$

Thus (5.1) holds for all $X \in \mathrm{Ind}(\mathcal{C})$ and $Y \in \mathcal{C}$. By the cocontinuity of $\mathbb{N}$ and $\otimes$, the formula (5.1) actually holds for all $X, Y \in \mathrm{Ind}(\mathcal{C})$. The isomorphism (5.2) is established by the universal property of $\mathbb{N}^{\mathrm{l}}_{\mathrm{Ind}(\mathcal{C})}$ as an end (see [**39**]). □

Let H be a Hopf algebra. A left (right) integral on H is a morphism $H \to \Bbbk$ of left (right) H-comodules, where $\Bbbk$ is the trivial comodule. It is known that H is co-Frobenius if and only if H has a non-zero (left or right) integral on H, if and only if $H^{*\mathrm{rat}}_{\ell} \neq 0$ [**17**, Corollary 5.2.4 and Theorem 5.3.2]. Moreover, the antipode of a co-Frobenius Hopf algebra is bijective [**17**, Corollary 5.4.6]. By Lemma 3.6, the condition $H^{*\mathrm{rat}}_{\ell} = 0$ is also equivalent to $\mathbb{N}^{\mathrm{r}}_{H} \neq 0$. As a categorical generalization of these results, we prove the following theorem.

THEOREM 5.5 (cf. [**39**, Theorem 5.2] and [**38**, Lemma 4.4]). *Let $\mathcal{C}$ satisfy Assumption 5.3, and let Q be a coalgebra such that $\mathcal{C} \approx \mathfrak{M}^{Q}_{\mathtt{fd}}$ as linear categories. Then the following are equivalent:*

(1) *Q is QcF.*
(2) *Q is right QcF.*
(3) *Q is right semiperfect.*
(4) $Q^{*\mathrm{rat}}_{\ell} \neq 0$.
(5) *Q is left QcF.*
(6) *Q is left semiperfect.*
(7) $Q^{*\mathrm{rat}}_{r} \neq 0$.
(8) *There is a non-zero object of $\mathcal{C}$ which is projective or injective.*
(9) *$\mathrm{Hom}_{\mathrm{Ind}(\mathcal{C})}(E, X) \neq 0$ for some $X \in \mathcal{C}$ and injective $E \in \mathrm{Ind}(\mathcal{C})$.*
(10) *The functor $\mathbb{N} := \mathbb{N}^{\mathrm{r}}_{\mathrm{Ind}(\mathcal{C})}$ is an auto-equivalence of $\mathrm{Ind}(\mathcal{C})$.*
(11) *The functor $\mathbb{N}$ induces an auto-equivalence of $\mathcal{C}$.*
(12) *The functor $\mathbb{N}$ is not zero.*

Suppose that these equivalent conditions are satisfied. Then $\mathcal{C}$ is rigid. Thus $\mathcal{C}$ is a Frobenius tensor category.

PROOF. (4) and (12) are equivalent by Lemma 3.6. (1), (10), and (11) are equivalent by Theorem 4.10. The general theory of coalgebras [**17**, Chapter 3] implies (1) ⇒ (2), (2) ⇒ (3), (3) ⇒ (4), (1) ⇒ (5), (5) ⇒ (6), (6) ⇒ (7), and (1)

$\Rightarrow$ (8). To show that all the conditions (1)–(12) are equivalent, it suffices to show (12) $\Rightarrow$ (1), (8) $\Rightarrow$ (9), (9) $\Rightarrow$ (7), and (7) $\Rightarrow$ (1).

(12) $\Rightarrow$ (1). This part was proved in [**38**, Lemma 4.4] under the assumption that $\mathcal{C}$ is a tensor category. We give a proof based on Lemma 4.10. Let $\mathcal{C}$ satisfy Assumption 5.3 and assume $\mathbb{N} \neq 0$. Then the functor $\mathbb{N}^{\mathrm{l}} := \mathbb{N}^{\mathrm{l}}_{\mathrm{Ind}(\mathcal{C})}$ is non-zero as well. Hence, by (5.3), the object $\overline{\mathtt{g}} = \mathbb{N}^{\mathrm{l}}(\mathbf{1})$ is nonzero.

We first show that the endofunctor $\mathbb{D} : \mathrm{Ind}(\mathcal{C}) \to \mathrm{Ind}(\mathcal{C})$ of Lemma 5.4 is left exact. Let $0 \to X \xrightarrow{p} Y \xrightarrow{q} Z$ be an exact sequence in $\mathrm{Ind}(\mathcal{C})$. By the left exactness of $\mathbb{N}^{\mathrm{l}}$ and (5.3), we have an exact sequence

$$0 \to \mathbb{D}(X) \otimes \overline{\mathtt{g}} \xrightarrow{\mathbb{D}(p)\otimes \mathrm{id}} \mathbb{D}(Y) \otimes \overline{\mathtt{g}} \xrightarrow{\mathbb{D}(q)\otimes \mathrm{id}} \mathbb{D}(Z) \otimes \overline{\mathtt{g}}$$

in $\mathrm{Ind}(\mathcal{C})$. By the exactness of the functor $(-) \otimes \overline{\mathtt{g}}$, we have

$$\mathrm{Ker}(\mathbb{D}(p)) \otimes \overline{\mathtt{g}} = 0 \quad \text{and} \quad \frac{\mathrm{Ker}(\mathbb{D}(q))}{\mathrm{Im}(\mathbb{D}(p))} \otimes \overline{\mathtt{g}} = 0.$$

Since the functor $(-) \otimes \overline{\mathtt{g}}$ is also faithful, we have $\mathrm{Ker}(\mathbb{D}(p)) = 0$ and $\mathrm{Ker}(\mathbb{D}(q)) = \mathrm{Im}(\mathbb{D}(p))$. We have proved that $\mathbb{D}$ is left exact.

The functor $\mathbb{N}$ is not only right exact, but also left exact by (5.3) and the left exactness of $\mathbb{D}$. Since the endofunctor $X \mapsto X^{\vee\vee}$ on $\mathcal{C}$ is faithful, so is $\mathbb{D}$. Thus, by (5.3), $\mathbb{N}$ is faithful. Now, by Lemma 4.10, we conclude that Q is QcF.

(8) $\Rightarrow$ (9). It is known that an object of a tensor category is projective if and only if it is injective [**18**, Proposition 6.1.3]. Although the rigidity is used in the proof presented in [**18**], one can prove that the same holds for categories satisfying Assumption 5.3 in the basically same manner (see [**39**, §5.2]). Hence (8) implies that $\mathcal{C}$ has a nonzero injective object. Now it is trivial that (9) holds.

(9) $\Rightarrow$ (7). We assume that there are an object $X \in \mathcal{C}$, an injective object $E \in \mathrm{Ind}(\mathcal{C})$ and a non-zero morphism $E \to X$ in $\mathrm{Ind}(\mathcal{C})$. For simplicity, we identify $\mathcal{C}$ and $\mathrm{Ind}(\mathcal{C})$ with $\mathfrak{M}^Q_{\mathrm{fd}}$ and $\mathfrak{M}^Q$, respectively. By replacing E with its direct summand, we may assume that E is indecomposable. Then E is a direct summand of Q. Hence we have a non-zero morphism $Q \to X$ of right Q-comodules. Since X is finite-dimensional, we have isomorphisms

$${}^{Q}\mathrm{Hom}(X^*, Q_r^{*\mathrm{rat}}) \cong \mathrm{Hom}_{Q^*}(X^*, Q^*) \cong \mathrm{Hom}^Q(Q, X) \neq 0,$$

which implies (7).

(7) $\Rightarrow$ (1). We have already proved that (1) and (4) of this theorem are equivalent. We note that $\mathcal{C}^{\mathrm{op,rev}}$ also satisfies Assumption 5.3 and there is an equivalence $\mathcal{C}^{\mathrm{op,rev}} \approx {}^{Q}\mathfrak{M}_{\mathrm{fd}}$ of linear categories. The equivalence (1) $\Leftrightarrow$ (7) is proved by applying (1) $\Leftrightarrow$ (4) to $\mathcal{C}^{\mathrm{op,rev}}$.

Rigidity of $\mathcal{C}$. We assume that (1)–(12) are satisfied. By Theorem 4.10, the right exact Nakayama functor for $\mathcal{C}$ and its right adjoint induce autoequivalences on $\mathcal{C}$, which we denote by $\mathbb{N}^{\mathrm{r}}_{\mathcal{C}}$ and $\mathbb{N}^{\mathrm{l}}_{\mathcal{C}}$, respectively. The functors $\mathbb{N}^{\mathrm{r}}_{\mathcal{C}}$ and $\mathbb{N}^{\mathrm{l}}_{\mathcal{C}}$ are mutually quasi-inverse. Now we set $\mathtt{g} = \mathbb{N}^{\mathrm{r}}_{\mathcal{C}}(\mathbf{1})$ and $\overline{\mathtt{g}} = \mathbb{N}^{\mathrm{l}}_{\mathcal{C}}(\mathbf{1})$. By (5.3) and $\mathbb{N}^{\mathrm{l}}_{\mathcal{C}}\mathbb{N}^{\mathrm{r}}_{\mathcal{C}} \cong \mathrm{id}_{\mathcal{C}} \cong \mathbb{N}^{\mathrm{r}}_{\mathcal{C}}\mathbb{N}^{\mathrm{l}}_{\mathcal{C}}$, we have isomorphisms $\mathtt{g} \otimes \overline{\mathtt{g}} \cong \mathbf{1} \cong \overline{\mathtt{g}} \otimes \mathtt{g}$ (this part is slight technical, though; see [**39**]). From this, we conclude that the double dual functor is an auto-equivalence on $\mathcal{C}$. Hence $\mathcal{C}$ is rigid. The proof is done. □

5.3. A categorical analogue of the modular function. From now on, we introduce interesting results on Frobenius tensor categories from [**38, 39**]. Let $\mathcal{C}$ be a Frobenius tensor category. By Theorems 3.13 and 5.5, the Nakayama functor $\mathbb{N}^{\mathrm{r}}_{\mathrm{Ind}(\mathcal{C})}$ restricts to an autoequivalence $\mathbb{N}^{\mathrm{r}}_{\mathcal{C}}$ on $\mathcal{C}$. Now we define

$$\mathrm{g}_{\mathcal{C}} := \mathbb{N}^{\mathrm{r}}_{\mathcal{C}}(\mathbf{1}) \in \mathcal{C}. \tag{5.4}$$

The object $\mathrm{g}_{\mathcal{C}}$ corresponds to the modular function of a co-Frobenius Hopf algebra (see [**39**]). According to the nomenclature in the Hopf algebra theory, we introduce the following terminology.

DEFINITION 5.6. We call $\mathrm{g}_{\mathcal{C}}$ the *modular object* of $\mathcal{C}$. We say that $\mathcal{C}$ is *unimodular* if $\mathrm{g}_{\mathcal{C}} \cong \mathbf{1}$.

By the same way as Lemma 5.4, we have the following theorem.

THEOREM 5.7 (cf. [**21**, Theorems 4.4 and 4.5]). *Let $\mathcal{C}$ be a Frobenius tensor category. Then there are natural isomorphisms*

$$^{\vee\vee}X \otimes \mathbb{N}^{\mathrm{r}}_{\mathcal{C}}(Y) \cong \mathbb{N}^{\mathrm{r}}_{\mathcal{C}}(X \otimes Y) \cong \mathbb{N}^{\mathrm{r}}_{\mathcal{C}}(X) \otimes Y^{\vee\vee}$$

for $X, Y \in \mathcal{C}$. In particular,

$$^{\vee\vee}X \otimes \mathrm{g}_{\mathcal{C}} \cong \mathbb{N}^{\mathrm{r}}_{\mathcal{C}}(X) \cong \mathrm{g}_{\mathcal{C}} \otimes X^{\vee\vee} \quad (X \in \mathcal{C}). \tag{5.5}$$

We give some immediate corollaries of this theorem. Let $\mathcal{C}$ be a Frobenius tensor category. Since $\mathbb{N}^{\mathrm{r}}_{\mathcal{C}}$ is an auto-equivalence of $\mathcal{C}$, we have Corollary 5.8.

COROLLARY 5.8. *The object $\mathrm{g}_{\mathcal{C}}$ is invertible, that is,*

$$\mathrm{g}_{\mathcal{C}} \otimes \mathrm{g}_{\mathcal{C}}^{\vee} \cong \mathbf{1} \cong \mathrm{g}_{\mathcal{C}}^{\vee} \otimes \mathrm{g}_{\mathcal{C}}.$$

Corollary 5.9 below has been proved in [**2**, Remark 2.10] under the assumption that $\mathcal{C}$ admits a dimension function. Given a simple object S, we denote by $E(S)$ and $P(S)$ an injective hull and a projective cover of S, respectively. By Theorems 3.13 and 5.7, we have Corollary 5.9.

COROLLARY 5.9 ([**39**, Corollary 5.10]). *There are isomorphisms*

$$E(S) \cong P(\mathrm{g}_{\mathcal{C}} \otimes S^{\vee\vee}), \quad P(S) \cong E(\mathrm{g}_{\mathcal{C}}^{\vee} \otimes {}^{\vee\vee}S)$$

for all simple objects $S \in \mathcal{C}$. In particular, $\mathrm{g}_{\mathcal{C}}$ is the top of $E(\mathbf{1})$ as well as the dual of the socle of $P(\mathbf{1})$.

By Theorems 3.15 and 5.7, we have the following corollary.

COROLLARY 5.10 ([**39**, Corollary 5.11]). *For a Frobenius tensor category $\mathcal{C}$, the following are equivalent:*

(1) *$\mathcal{C}$ is unimodular and the double dual functor $(-)^{\vee\vee}$ on $\mathcal{C}$ is isomorphic to the identity functor.*
(2) *$\mathcal{C} \approx \mathfrak{M}^{Q}_{\mathtt{fd}}$ for some symmetric coalgebra Q.*

5.4. A categorical analogue of Radford's S^4-formula. Radford's S^4-formula expresses the fourth power of the antipode of a finite-dimensional Hopf algebra in terms of certain elements defined by using integrals. This formula has been generalized to co-Frobenius Hopf algebras in [**6**, Theorem 2.8] and to finite tensor categories in [**19**]. It was pointed out that Radford's S^4-formula for finite tensor categories is obtained from a general result on the Nakayama functor for finite abelian categories in [**21**]. As we have established the theory of Nakayama functors for locally finite abelian categories, we obtain the following theorem in the same way as [**21**].

THEOREM 5.11. *For a Frobenius tensor category $\mathcal{C}$, there is an isomorphism*

$$X^{\vee\vee\vee\vee} \cong \mathrm{g}_{\mathcal{C}}^{\vee} \otimes X \otimes \mathrm{g}_{\mathcal{C}} \quad (X \in \mathcal{C}) \tag{5.6}$$

of tensor functors.

PROOF. There is a natural isomorphism as in (5.6) induced by

$$\mathfrak{r}_X := \left(X \otimes \mathrm{g}_{\mathcal{C}} \xrightarrow{(5.5)} \mathbb{N}_{\mathcal{C}}(X^{\vee\vee}) \xrightarrow{(5.5)} \mathrm{g}_{\mathcal{C}} \otimes X^{\vee\vee\vee\vee} \right) \quad (X \in \mathcal{C}). \tag{5.7}$$

We note that the isomorphism (5.5) used to define (5.7) comes from the isomorphism (4.8) given in Theorem 4.12. It follows from the coherence property of (4.8) that (5.6) is indeed an isomorphism of tensor functors. □

We call (5.7) the *Radford isomorphism.* In some situations, we desire to know an expression of the isomorphism (5.6) in much explicit way. For notational convenience, we rather discuss how the Radford isomorphism $\mathfrak{r}_X$ is given. Below we introduce some formulas of $\mathfrak{r}_X$ from [**39**] without expounding the detail of the proof. The following theorem generalizes [**18**, Theorem 8.10.7].

THEOREM 5.12 ([**39**, Theorem 5.14]). *For a braided Frobenius tensor category $\mathcal{C}$ with braiding σ, the Radford isomorphism is given by*

$$\mathfrak{r}_X = (\mathrm{id}_{\mathrm{g}} \otimes (u_{X^\vee}^{\vee})^{-1} u_X) \circ (\sigma_{\mathrm{g},X})^{-1}$$

for all $X \in \mathcal{C}$, where $\mathrm{g} = \mathrm{g}_{\mathcal{C}}$ and

$$u_X = (\mathrm{ev}_X \otimes \mathrm{id}_{X^{\vee\vee}}) \circ (\sigma_{X,X^\vee} \otimes \mathrm{id}_{X^{\vee\vee}}) \circ (\mathrm{id}_X \otimes \mathrm{coev}_{X^\vee}) \quad (X \in \mathcal{C}) \tag{5.8}$$

is the Drinfeld isomorphism associated to the braiding σ.

Let $\mathcal{C}$ be a braided tensor category with braiding σ. An object $T \in \mathcal{C}$ is said to be *transparent* if $\sigma_{X,T} \circ \sigma_{T,X} = \mathrm{id}_T \otimes \mathrm{id}_X$ for all objects $X \in \mathcal{C}$. The Müger center of $\mathcal{C}$ is the full subcategory of $\mathcal{C}$ consisting of all transparent objects of $\mathcal{C}$. We say that the Müger center of $\mathcal{C}$ is trivial if every transparent object of $\mathcal{C}$ is isomorphic to the direct sum of finitely many copies of $\mathbf{1}$. It is known that the modular object of a braided finite tensor category is transparent [**18**, Corollary 8.10.8]. With the help of Theorem 5.12, we can prove the following result by the same way as [**18**, Corollary 8.10.8].

COROLLARY 5.13. *The modular object of a braided Frobenius tensor category is transparent. Thus, a braided Frobenius tensor category with the trivial Müger center is unimodular.*

Since a cosemisimple coalgebra is a symmetric coalgebra [**11**], a semisimple tensor category is a unimodular Frobenius tensor category whose Nakayama functor is isomorphic to the identity. We note that the Radford isomorphism $\mathfrak{r}_X$ is viewed as an isomorphism from X to $X^{\vee\vee\vee\vee}$ in a unimodular Frobenius tensor category. For a semisimple tensor category, we have Theorem 5.14.

THEOREM 5.14 ([**39**, Theorem 5.18]). *Let $\mathcal{C}$ be a semisimple tensor category such that the endomorphism algebra of every simple object of $\mathcal{C}$ is isomorphic to $\Bbbk$. Then, for every simple object $X \in \mathcal{C}$, we have*

$$\mathfrak{r}_X = \frac{\mathsf{tr}((\phi^{-1})^\vee)}{\mathsf{tr}(\phi)} \cdot \phi^{\vee\vee} \circ \phi, \tag{5.9}$$

where $\phi : X \to X^{\vee\vee}$ is an arbitrary isomorphism in $\mathcal{C}$ and $\mathsf{tr}(f) \in \Bbbk$ is the "trace" for a morphism $f : X \to X^{\vee\vee}$ in $\mathcal{C}$ is defined by the equation

$$\mathsf{tr}(f) \cdot \mathrm{id}_{\mathbb{1}} = \mathrm{ev}_{X^\vee} \circ (f \otimes \mathrm{id}_{X^\vee}) \circ \mathrm{coev}_X.$$

5.5. Spherical tensor categories. A *pivotal structure* of a tensor category $\mathcal{C}$ is an isomorphism $\mathrm{id}_{\mathcal{C}} \to (-)^{\vee\vee}$ of tensor functors. Inspired by [**15**, Definition 3.5.2], we introduce the following definition.

DEFINITION 5.15. Let $\mathcal{C}$ be a unimodular Frobenius tensor category, and regard the Radford isomorphism $\mathfrak{r}_X$ ($X \in \mathcal{C}$) as an isomorphism from X to $X^{\vee\vee\vee\vee}$. We say that a pivotal structure p of $\mathcal{C}$ is *spherical* if the equation $\mathfrak{r}_X = p_{X^{\vee\vee}} \circ p_X$ holds for all $X \in \mathcal{C}$. A *spherical tensor category* is a unimodular Frobenius tensor category equipped with a spherical pivotal structure.

A motivation for introducing spherical *finite* tensor categories in [**15**] is to look for a 'correct' generalization of the notion of spherical fusion category [**5**] to a non-semisimple setting. As the importance of spherical finite tensor categories has become clearer in recent studies (see, e.g., [**13, 20, 37, 42**]), its "infinite" version Definition 5.15 could also be significant.

We here introduce some criteria for a tensor category to be spherical. The first one, which concerns semisimple tensor categories, is a generalization of the finite case proved in [**15**, Proposition 3.5.4]. Thanks to Theorem 5.14, one can prove Theorem 5.16 below in the completely same way as the finite case.

THEOREM 5.16 ([**39**, Theorem 5.20]). *Let $\mathcal{C}$ be a semisimple tensor category equipped with a pivotal structure p. Then $\mathcal{C}$ is spherical if and only if it is trace spherical* [**15**, Definition 3.5.3], *that is, the equation*

$$\mathsf{tr}(p_{X^\vee} \circ f^\vee) = \mathsf{tr}(p_X \circ f)$$

holds for all endomorphisms $f : X \to X$ in $\mathcal{C}$.

The second criteria concerns when a unimodular braided Frobenius tensor category is spherical. By Theorem 5.12 giving a formula of the Radford isomorphism for a braided Frobenius tensor category, one can prove the following theorem.

THEOREM 5.17 ([**39**, Theorem 5.21]). *Let $\mathcal{C}$ be a unimodular braided Frobenius tensor category equipped with a pivotal structure p. Then $\mathcal{C}$ is spherical if and only if the natural isomorphism $\theta := u^{-1} \circ p$ is a ribbon structure* [**18**, Definition 8.10.1] *on $\mathcal{C}$, where u is the Drinfeld isomorphism* (5.8).

5.6. Modified trace theory. In recent attempts to construct topological invariants from non-semisimple categories, a modified trace plays a central role.

DEFINITION 5.18. For a Frobenius tensor category $\mathcal{C}$ equipped with a pivotal structure p, a *right modified trace* on $\mathrm{Proj}(\mathcal{C})$ is a family of linear maps

$$t = \{t_P : \mathrm{Hom}_{\mathcal{C}}(P, P) \to \Bbbk\}_{P \in \mathrm{Proj}(\mathcal{C})}$$

satisfying the following two conditions:

(1) For any morphisms $f : P \to Q$ and $g : Q \to P$ in $\mathrm{Proj}(\mathcal{C})$, we have
$$t_Q(fg) = t_P(gf).$$
(2) For any morphism $f : P \otimes X \to P \otimes X$ in $\mathrm{Proj}(\mathcal{C})$ with $P \in \mathrm{Proj}(\mathcal{C})$ and $X \in \mathcal{C}$, we have $t_{P\otimes X}(f) = t_P(\mathrm{cl}_X(f))$, where
$$\mathrm{cl}_X(f) = (\mathrm{id}_P \otimes \mathrm{ev}_{X^\vee}) \circ (\mathrm{id}_P \otimes p_X \otimes \mathrm{id}_{X^\vee}) \circ (f \otimes \mathrm{id}_{X^\vee}) \circ (\mathrm{id}_P \otimes \mathrm{coev}_X)$$
is the morphism obtained from f by "closing" the X-strand in the string diagram representing f.

We note that $\mathcal{C}^{\mathrm{rev}}$ is also a Frobenius pivotal tensor category. A *left modified trace* on $\mathrm{Proj}(\mathcal{C})$ is a right modified trace on $\mathrm{Proj}(\mathcal{C}^{\mathrm{rev}})$. We say that a left or a right modified trace t on $\mathrm{Proj}(\mathcal{C})$ is *nondegenerate* if the pairing
$$\mathrm{Hom}_{\mathcal{C}}(P, Q) \times \mathrm{Hom}_{\mathcal{C}}(Q, P) \to \Bbbk, \quad (f, g) \mapsto t_Q(fg)$$
is nondegenerate for all $P, Q \in \mathrm{Proj}(\mathcal{C})$.

We denote by $t_\ell(\mathcal{C})$ and $t_r(\mathcal{C})$ the space of left and right modified traces on $\mathrm{Proj}(\mathcal{C})$, respectively. In the finite case, it has been observed in [**36**, **37**] that modified traces are closely related to the Nakayama functor. By Theorem 3.16 and the same argument as [**37**], one can prove Theorem 5.19.

THEOREM 5.19. *For a Frobenius tensor category equipped with a pivotal structure, the following are equivalent:*

(1) *There is a nonzero left modified trace on* $\mathrm{Proj}(\mathcal{C})$.
(2) *There is a nonzero right modified trace on* $\mathrm{Proj}(\mathcal{C})$.
(3) $\mathcal{C}$ *is unimodular.*

Suppose that the above equivalent conditions are satisfied. Then both $t_\ell(\mathcal{C})$ *and* $t_r(\mathcal{C})$ *are one-dimensional, and every nonzero left or right modified trace on* $\mathrm{Proj}(\mathcal{C})$ *is nondegenerate. Furthermore,* $\mathcal{C}$ *is spherical if and only if* $t_\ell(\mathcal{C}) = t_r(\mathcal{C})$.

The detail will be given in a forthcoming paper. A part of this theorem has been known. Indeed, (3) $\Rightarrow$ (1), (3) $\Rightarrow$ (2), and the uniqueness of modified traces (up to scalar) have been proved in [**24**].

5.7. Exact sequences of tensor categories. The notion of an exact sequence of Hopf algebras has been effectively used for the study of Hopf algebras. As a categorical counterpart of this notion, Bruguières and Natale [**7**, **8**, **31**] introduced and studied exact sequences of tensor categories. We exhibit an application of our results to exact sequences of tensor categories. To begin with, we prove Theorem 5.20.

THEOREM 5.20 ([**38**, Theorem 3.6]). *Let* $F : \mathcal{C} \to \mathcal{D}$ *be a tensor functor between tensor categories* $\mathcal{C}$ *and* $\mathcal{D}$*, and let* G *be a right adjoint of* $\mathrm{Ind}(F)$. *Then we have:*

(a) *If* $\mathcal{C}$ *is Frobenius and* G *is exact, then* $\mathcal{D}$ *is Frobenius.*
(b) *If* $\mathcal{D}$ *is Frobenius and* $G(\mathcal{D}) \subset \mathcal{C}$, *then* $\mathcal{C}$ *is Frobenius.*

PROOF.

(a) Suppose that $\mathcal{C}$ is Frobenius and G is exact. Then $\mathcal{C}$ has a non-zero projective object, say P. Since a tensor functor is faithful, we have $F(P) \neq 0$. Furthermore, $F(P)$ is a projective object of $\mathcal{D}$, since
$$\mathrm{Hom}_{\mathcal{D}}(F(P), -) \cong \mathrm{Hom}_{\mathrm{Ind}(\mathcal{C})}(P, -) \circ G$$

is exact. Hence $\mathcal{D}$ is Frobenius.

(b) Suppose that $\mathcal{D}$ is Frobenius and $G(\mathcal{D}) \subset \mathcal{C}$. Then the unit object $\mathbf{1} \in \mathcal{D}$ has an injective hull in $\mathcal{D}$, say E. A tensor functor is exact by definition. The object $G(E) \in \mathcal{C}$ is a non-zero injective object, since

$$\mathrm{Hom}_{\mathcal{C}}(-, G(E)) \cong \mathrm{Hom}_{\mathcal{D}}(-, E) \circ F$$

is exact and $\mathrm{Hom}_{\mathcal{C}}(\mathbf{1}, G(E)) \cong \mathrm{Hom}_{\mathcal{D}}(\mathbf{1}, E) \neq 0$. Hence $\mathcal{C}$ is Frobenius. □

Let $\mathcal{C}' \xrightarrow{\iota} \mathcal{C} \xrightarrow{F} \mathcal{D}$ be a sequence of tensor functors between tensor categories $\mathcal{C}$, $\mathcal{C}'$ and $\mathcal{D}$. We say that this sequence is *exact* [**7**,**8**,**31**] if the following conditions (1), (2), (3) and (4) are satisfied:

(1) The functor ι is fully faithful.
(2) The functor F is *dominant* in the sense that every object $Y \in \mathcal{D}$ is a subobject of an object of the form $F(X)$ for some $X \in \mathcal{C}$.
(3) The functor F is *normal* in the sense that every object $X \in \mathcal{C}$ has a subobject such that $F(X_0)$ is the maximal trivial subobject of $F(X)$ (here an object of $\mathcal{D}$ is said to be *trivial* if it is isomorphic to the direct sum of finitely many copies of the unit object).
(4) The kernel of F, defined by $\mathrm{Ker}(F) := \{X \in \mathcal{C} \mid F(X) \text{ is trivial}\}$, coincides with the essential image of the functor ι.

Let $G : \mathrm{Ind}(\mathcal{D}) \to \mathrm{Ind}(\mathcal{C})$ be a right adjoint of $\mathrm{Ind}(F)$. The conditions (2) and (3) above are characterized in terms of G as follows: The functor F is dominant if and only if G is faithful [**7**, Lemma 3.1]. The functor F is normal if and only if $G(\mathbf{1})$ is a filtered colimit of objects belonging to $\mathrm{Ker}(F)$ (see [**7**, Proposition 3.5] and [**38**, Lemma 4.2]).

It is known that the class of co-Frobenius Hopf algebras is closed under exact sequences [**1**]. Natale [**31**] asked whether the class of Frobenius tensor categories is closed under exact sequences. We have given an affirmative answer to this question as follows.

THEOREM 5.21 ([**38**, Theorem 3.9]). *Let $\mathcal{C}' \xrightarrow{\iota} \mathcal{C} \xrightarrow{F} \mathcal{C}''$ be an exact sequence of tensor categories such that a right adjoint of* $\mathrm{Ind}(F)$ *is exact. Then $\mathcal{C}$ is Frobenius if and only if both $\mathcal{C}'$ and $\mathcal{C}''$ are.*

6. Applications to coquasi-bialgebras

6.1. Coquasi-bialgebras with preantipode. A *coquasi-bialgebra* [**30**,**34**] is a coalgebra H equipped with two coalgebra maps $m : H \otimes_{\Bbbk} H \to H$ and $u : \Bbbk \to H$ and a linear map $\omega : H \otimes_{\Bbbk} H \otimes_{\Bbbk} H \to \Bbbk$ such that ω is invertible with respect to the convolution product and the equations

$$\omega(\mathrm{id}_H \otimes \mathrm{id}_H \otimes m) * \omega(m \otimes \mathrm{id}_H \otimes \mathrm{id}_H) = (\varepsilon \otimes \omega) * \omega(\mathrm{id}_H \otimes m \otimes \mathrm{id}_H) * (\omega \otimes \varepsilon),$$
$$\omega(\mathrm{id}_H \otimes u \otimes \mathrm{id}_H) = \varepsilon \otimes \varepsilon, \quad m(u \otimes \mathrm{id}_H) = \mathrm{id}_H = m(\mathrm{id}_H \otimes u),$$
$$m(\mathrm{id}_H \otimes m) * \omega = \omega * m(m \otimes \mathrm{id}_H)$$

hold, where $*$ denote the convolution product. This much complicated notion is natural from the viewpoint of Tannaka–Krein reconstruction: If H is a coquasi-bialgebra, then ${}^H\mathfrak{M}_{\mathtt{fd}}$ is a monoidal category with respect to the tensor product over $\Bbbk$ and the associator given by ω. Moreover, a coalgebra C has a structure of a coquasi-bialgebra if and only if ${}^C\mathfrak{M}_{\mathtt{fd}}$ has a structure of a monoidal category such that the forgetful functor ${}^C\mathfrak{M}_{\mathtt{fd}} \to \mathbf{Vec}$ is quasi-monoidal [**30**,**34**].

A coquasi-bialgebra is the dual notion of a quasi-bialgebra introduced by Drinfeld in [**16**]. As an analogue of the notion of a Hopf algebra, a *quasi-Hopf algebra* was also introduced in [**16**]. A *coquasi-Hopf algebra*, which is the dual notion of a quasi-Hopf algebra, is a coquasi-bialgebra (H, m, u) equipped with a triple (s, α, β) consisting of an anti-coalgebra map $s : H \to H$ and two linear maps $\alpha, \beta : H \to \Bbbk$ satisfying the equations

$$h_{(1)}\beta(h_{(2)})s(h_{(3)}) = \beta(h)1_H, \quad s(h_{(1)})\alpha(h_{(2)})h_{(3)} = \alpha(h)1_H,$$
$$\omega^{-1}(s(h_{(1)}) \otimes \alpha(h_{(2)})h_{(3)}\beta(h_{(4)}) \otimes s(h_{(5)}))$$
$$= \varepsilon(h) = \omega(h_{(1)} \otimes \beta(h_{(2)})s(h_{(3)})\alpha(h_{(4)}) \otimes h_{(5)})$$

for all $h \in H$, where $xy = m(x \otimes y)$ $(x, y \in H)$, $1_H = u(1_{\Bbbk})$ and ω^{-1} is the inverse of ω with respect to the convolution product. The triple (s, α, β) is called a *coquasi-antipode* of H. It is known that ${}^H\mathfrak{M}_{\mathtt{fd}}$ is right rigid if H is a coquasi-Hopf algebra. However, the converse does not hold in general, as has been pointed out by Schauenburg in [**33**].

According to Schauenburg [**35**, Theorem 2.6], the monoidal category ${}^H\mathfrak{M}_{\mathtt{fd}}$ for a coquasi-bialgebra H is right rigid if and only if a certain structure theorem for Hopf bimodules hold. Ardizzoni and Pavarin [**4**] showed that the latter condition for H is equivalent to that H has a *preantipode*, that is, a linear map $S : H \to H$ satisfying the equations

$$S(h_{(1)})_{(1)}h_{(2)} \otimes S(h_{(1)})_{(2)} = 1_H \otimes S(h),$$
$$S(h_{(2)})_{(1)} \otimes h_{(1)}S(h_{(2)})_{(2)} = S(h) \otimes 1_H,$$
$$\omega(h_{(1)} \otimes S(h_{(2)}) \otimes h_{(3)}) = \varepsilon(h)$$

for all $h \in H$. In summary, the monoidal category ${}^H\mathfrak{M}_{\mathtt{fd}}$ is right rigid if and only if H has a preantipode. We also remark that a coquasi-Hopf algebra H has a preantipode given by $S(h) = \beta(h_{(1)})s(h_{(2)})\alpha(h_{(3)})$ for $h \in H$, where (s, α, β) is the coquasi-antipode of H [**4**, Theorem 3.10].

Saracco [**32**] investigated how a preantipode is reconstructed from the forgetful functor ${}^H\mathfrak{M}_{\mathtt{fd}} \to \mathbf{Vec}$ when ${}^H\mathfrak{M}_{\mathtt{fd}}$ is right rigid. Moreover, an explicit construction of a dual object was given: Following [**32**, Remark 3.20], a right dual of $X \in {}^H\mathfrak{M}_{\mathtt{fd}}$ is given by ${}^{\vee}X := (X^* \otimes_{\Bbbk} H)^{\mathrm{co}\,H}$ as a vector space, where $(-)^{\mathrm{co}\,H}$ means the coinvariant as a right H-comodule. In general, ${}^{\vee}X$ may not be isomorphic to X^*. This is a crucial difference to the case where H is a coquasi-Hopf algebra.

6.2. QcF property of a coquasi-bialgebra with preantipode. Let H be a coquasi-bialgebra with preantipode. As in the case of Hopf algebras, we introduce the following definition.

DEFINITION 6.1. A left (right) *integral* on H is a homomorphism $H \to \Bbbk$ of left (right) comodules, where $\Bbbk$ is the trivial H-comodule.

As we have noted in the previous subsection, ${}^H\mathfrak{M}_{\mathtt{fd}}$ is a right rigid monoidal category with respect to the tensor product over $\Bbbk$. Thus the category $\mathcal{C} := ({}^H\mathfrak{M}_{\mathtt{fd}})^{\mathrm{rev}}$ satisfies Assumption 5.3. By applying our results to $\mathcal{C}$, we obtain Theorem 6.2.

THEOREM 6.2 (see [**17**] for Hopf algebras and [**10**] for coquasi-Hopf algebras). *For a coquasi-bialgebra H with preantipode, the following are equivalent:*

(1) *H is QcF (as a coalgebra; the same applies below).*

(2) *H is left semiperfect.*
(3) *H has a finite-dimensional nonzero projective left comodule.*
(4) *H has a finite-dimensional nonzero injective left comodule.*
(5) *The rational part of H^* as a left H^*-module is nonzero.*
(6) *There is a nonzero left integral on H.*
(7) *Left-right switched versions of (2), (3), (4), (5), and (6).*

PROOF. By applying Theorems 5.5 to $\mathcal{C} := ({}^H\mathfrak{M}_{\mathtt{fd}})^{\mathrm{rev}}$ or $\mathcal{C}^{\mathrm{op,rev}}$, we see that (1), (2), (3), (4), (5), and their left-right switched versions are equivalent. Moreover, Theorem 5.5 shows that (6) and its left-right switched version imply (1). By the fact that a QcF coalgebra Q is a generator both in ${}^Q\mathfrak{M}$ and $\mathfrak{M}^Q$ [**17**, Corollary 3.3.11], we see that (1) implies both (6) and its left-right switched version. The proof is done. □

By Theorems 3.14 and 6.2, we have Theorem 6.3.

THEOREM 6.3. *Let H be a coquasi-bialgebra with preantipode. We assume that H satisfies the equivalent conditions of Theorem 6.2. Then the following are equivalent:*

(1) *H is co-Frobenius.*
(2) $\dim_{\Bbbk}({}^{\vee\vee}X) = \dim_{\Bbbk}(X)$ *for all simple* $X \in {}^H\mathfrak{M}_{\mathtt{fd}}$.

In the case where H is a coquasi-Hopf algebra, the dual object of $X \in {}^H\mathfrak{M}_{\mathtt{fd}}$ is built on the dual space X^*. Thus we obtain the following consequence: If a coquasi-Hopf algebra H satisfies the equivalent conditions of Theorem 6.2, then H is co-Frobenius (cf. [**10**, Theorem 4.5]). A natural question arises.

QUESTION 6.4. Is there a QcF coquasi-bialgebra with preantipode that is not co-Frobenius?

6.3. De-equivariantization. The equivariantization is a kind of category of 'fixed points' of a category on which a group acts, and the de-equivariantization can be thought of as the inverse of equivariantization [**18**]. At the end of this paper, we discuss coquasi-bialgebras constructed by de-equivariantization by affine group schemes in [**3**]. A starting observation by Bruguières and Natale [**7**, **8**] is that the de-equivariantization of a finite tensor category by a finite group fits into a certain exact sequence of tensor categories. Their observation has been extended in [**38**] to de-equivariantization by affine group schemes as follows.

Given a monoidal category $\mathcal{M}$, we denote by $\mathcal{Z}(\mathcal{M})$ its Drinfeld center. Let G be an affine group scheme, and let $\mathcal{C}$ be a tensor category equipped with a central embedding of $\mathrm{Rep}(G) := {}^{\mathcal{O}(G)}\mathfrak{M}_{\mathtt{fd}}$ into $\mathcal{C}$, that is, a fully faithful tensor functor $\iota : \mathrm{Rep}(G) \to \mathcal{C}$ such that there exists a braided tensor functor $\tilde{\iota} : \mathrm{Rep}(G) \to \mathcal{Z}(\mathcal{C})$ such that $\iota = U \circ \tilde{\iota}$, where $U : \mathcal{Z}(\mathcal{C}) \to \mathcal{C}$ is the forgetful functor. Then the coordinate algebra $\mathcal{O}(G)$ becomes a commutative algebra in $\mathcal{Z}(\mathrm{Ind}(\mathcal{C}))$. The category of left $\mathcal{O}(G)$-modules in ${}_{\mathcal{O}(G)}\mathrm{Ind}(\mathcal{C})$ is an abelian monoidal category with respect to the tensor product over $\mathcal{O}(G)$. We recall that an object X of a category $\mathcal{A}$ is said to be *finitely generated* if the functor $\mathrm{Hom}_{\mathcal{A}}(X, -)$ preserves filtered colimits of monomorphisms. Now let $\mathcal{D}$ denote the full subcategory of ${}_{\mathcal{O}(G)}\mathrm{Ind}(\mathcal{C})$ consisting of finitely generated objects. It can be shown that $\mathcal{D}$ is closed under $\otimes_{\mathcal{O}(G)}$. Hence, $\mathcal{D}$ is a monoidal category.

THEOREM 6.5 ([**38**]). *Let G, $\mathcal{C}$ and $\mathcal{D}$ be as above. Suppose that the monoidal category $\mathcal{D}$ is a tensor category. Then there is an exact sequence*

$$\mathrm{Rep}(G) \to \mathcal{C} \to \mathcal{D}$$

of tensor categories. Hence, by Theorem 5.21, the tensor category $\mathcal{C}$ is Frobenius if and only if the Hopf algebra $\mathcal{O}(G)$ is co-Frobenius and the tensor category $\mathcal{D}$ is Frobenius.

We consider the case where $\mathcal{C} = {}^{H}\mathfrak{M}_{\mathtt{fd}}$ for some Hopf algebra H with bijective antipode. Let (K, r) be a braided central Hopf subalgebra of H [**3**, Definition 3.1], that is, a pair of a Hopf subalgebra K of H and a bilinear map $r : H \times K \to \Bbbk$ satisfying certain conditions similar to the axioms for universal r-forms. Since K is commutative [**3**, Remark 3.2], we may assume that $K = \mathcal{O}(G)$ for some affine group scheme G. As shown in [**3**], the pair (K, r) gives rise to a central embedding $\mathrm{Rep}(G) \to \mathcal{C}$.

We now assume that there is a convolution-invertible $\mathcal{O}(G)$-linear map from H to $\mathcal{O}(G)$ preserving the unit and the counit. The main result of [**3**] states that the quotient coalgebra $Q = H/\mathcal{O}(G)^{+}H$, where $\mathcal{O}(G)^{+}$ is the kernel of the counit of $\mathcal{O}(G)$, has a structure of a coquasi-bialgebra such that ${}^{Q}\mathfrak{M}$ is equivalent to ${}_{\mathcal{O}(G)}\mathrm{Ind}(\mathcal{C})$ as a monoidal category. Theorem 6.5 gives the following consequences: Suppose that the monoidal category ${}^{Q}\mathfrak{M}_{\mathtt{fd}}$ is rigid. Then the coquasi-bialgebra Q is QcF if and only if both $\mathcal{O}(G)$ and H are co-Frobenius.

Acknowledgments

The author thanks Taiki Shibata for discussions. The author also thanks the referee for the careful reading of the manuscript.

References

[1] N. Andruskiewitsch and J. Cuadra, *On the structure of (co-Frobenius) Hopf algebras*, J. Noncommut. Geom. **7** (2013), no. 1, 83–104, DOI 10.4171/JNCG/109. MR3032811

[2] N. Andruskiewitsch, J. Cuadra, and P. Etingof, *On two finiteness conditions for Hopf algebras with nonzero integral*, Ann. Sc. Norm. Super. Pisa Cl. Sci. (5) **14** (2015), no. 2, 401–440. MR3410615

[3] I. Angiono, C. Galindo, and M. Pereira, *De-equivariantization of Hopf algebras*, Algebr. Represent. Theory **17** (2014), no. 1, 161–180, DOI 10.1007/s10468-012-9392-9. MR3160718

[4] A. Ardizzoni and A. Pavarin, *Preantipodes for dual quasi-bialgebras*, Israel J. Math. **192** (2012), no. 1, 281–295, DOI 10.1007/s11856-012-0024-1. MR3004083

[5] J. W. Barrett and B. W. Westbury, *Spherical categories*, Adv. Math. **143** (1999), no. 2, 357–375, DOI 10.1006/aima.1998.1800. MR1686423

[6] M. Beattie, D. Bulacu, and B. Torrecillas, *Radford's S^4 formula for co-Frobenius Hopf algebras*, J. Algebra **307** (2007), no. 1, 330–342, DOI 10.1016/j.jalgebra.2006.06.004. MR2278058

[7] A. Bruguières and S. Natale, *Exact sequences of tensor categories*, Int. Math. Res. Not. IMRN **24** (2011), 5644–5705, DOI 10.1093/imrn/rnq294. MR2863377

[8] A. Bruguières and S. Natale, *Central exact sequences of tensor categories, equivariantization and applications*, J. Math. Soc. Japan **66** (2014), no. 1, 257–287, DOI 10.2969/jmsj/06610257. MR3161401

[9] A. Bruguières and A. Virelizier, *Quantum double of Hopf monads and categorical centers*, Trans. Amer. Math. Soc. **364** (2012), no. 3, 1225–1279, DOI 10.1090/S0002-9947-2011-05342-0. MR2869176

[10] D. Bulacu and S. Caenepeel, *Integrals for (dual) quasi-Hopf algebras. Applications*, J. Algebra **266** (2003), no. 2, 552–583, DOI 10.1016/S0021-8693(03)00175-3. MR1995128

[11] F. Castaño Iglesias, S. Dăscălescu, and C. Năstăsescu, *Symmetric coalgebras*, J. Algebra **279** (2004), no. 1, 326–344, DOI 10.1016/j.jalgebra.2004.05.007. MR2078404

[12] W. Chin and D. Simson, *Coxeter transformation and inverses of Cartan matrices for coalgebras*, J. Algebra **324** (2010), no. 9, 2223–2248, DOI 10.1016/j.jalgebra.2010.06.029. MR2684139

[13] F. Costantino, N. Geer, B. Patureau-Mirand, and A. Virelizier, *Non compact (2+1)-TQFTs from non-semisimple spherical categories*, Preprint, `arXiv:2302.04509`, (2023).

[14] J. Cuadra and J. Gómez-Torrecillas, *Idempotents and Morita-Takeuchi theory*, Comm. Algebra **30** (2002), no. 5, 2405–2426, DOI 10.1081/AGB-120003476. MR1904645

[15] C. L. Douglas, C. Schommer-Pries, and N. Snyder, *Dualizable tensor categories*, Mem. Amer. Math. Soc. **268** (2020), no. 1308, vii+88, DOI 10.1090/memo/1308. MR4254952

[16] V. G. Drinfel′d, *Almost cocommutative Hopf algebras* (Russian), Algebra i Analiz **1** (1989), no. 2, 30–46; English transl., Leningrad Math. J. **1** (1990), no. 2, 321–342. MR1025154

[17] S. Dăscălescu, C. Năstăsescu, and Ş. Raianu, *Hopf algebras*, Monographs and Textbooks in Pure and Applied Mathematics, vol. 235, Marcel Dekker, Inc., New York, 2001. An introduction. MR1786197

[18] P. Etingof, S. Gelaki, D. Nikshych, and V. Ostrik, *Tensor categories*, Mathematical Surveys and Monographs, vol. 205, American Mathematical Society, Providence, RI, 2015, DOI 10.1090/surv/205. MR3242743

[19] P. Etingof, D. Nikshych, and V. Ostrik, *An analogue of Radford's S^4 formula for finite tensor categories*, Int. Math. Res. Not. **54** (2004), 2915–2933, DOI 10.1155/S1073792804141445. MR2097289

[20] J. Fuchs, C. Galindo, D. Jaklitsch, and C. Schweigert, *Spherical Morita contexts and relative Serre functors*, Preprint, `arXiv:2207.07031`, (2022).

[21] J. Fuchs, G. Schaumann, and C. Schweigert, *Eilenberg-Watts calculus for finite categories and a bimodule Radford S^4 theorem*, Trans. Amer. Math. Soc. **373** (2020), no. 1, 1–40, DOI 10.1090/tran/7838. MR4042867

[22] J. Fuchs, G. Schaumann, and C. Schweigert, *A modular functor from state sums for finite tensor categories and their bimodules*, Theory Appl. Categ. **38** (2022), Paper No. 15, 436–594. MR4403279

[23] J. Fuchs and C. Schweigert, *Internal natural transformations and Frobenius algebras in the Drinfeld center*, Transform. Groups **28** (2023), no. 2, 733–768, DOI 10.1007/s00031-021-09678-5. MR4586249

[24] N. Geer, J. Kujawa, and B. Patureau-Mirand, *M-traces in (non-unimodular) pivotal categories*, Algebr. Represent. Theory **25** (2022), no. 3, 759–776, DOI 10.1007/s10468-021-10044-y. MR4421818

[25] M. C. Iovanov, *Co-Frobenius coalgebras*, J. Algebra **303** (2006), no. 1, 146–153, DOI 10.1016/j.jalgebra.2006.04.028. MR2253657

[26] M. C. Iovanov, *Abstract algebraic integrals and Frobenius categories*, Internat. J. Math. **24** (2013), no. 10, 1350081, 36, DOI 10.1142/S0129167X1350081X. MR3125851

[27] M. C. Iovanov, *Generalized Frobenius algebras and Hopf algebras*, Canad. J. Math. **66** (2014), no. 1, 205–240, DOI 10.4153/CJM-2012-060-7. MR3150709

[28] M. Kashiwara and P. Schapira, *Categories and sheaves*, Grundlehren der mathematischen Wissenschaften [Fundamental Principles of Mathematical Sciences], vol. 332, Springer-Verlag, Berlin, 2006, DOI 10.1007/3-540-27950-4. MR2182076

[29] S. Mac Lane, *Categories for the working mathematician*, 2nd ed., Graduate Texts in Mathematics, vol. 5, Springer-Verlag, New York, 1998. MR1712872

[30] S. Majid, *Tannaka-Kreĭn theorem for quasi-Hopf algebras and other results*, Deformation theory and quantum groups with applications to mathematical physics (Amherst, MA, 1990), Contemp. Math., vol. 134, Amer. Math. Soc., Providence, RI, 1992, pp. 219–232, DOI 10.1090/conm/134/1187289. MR1187289

[31] S. Natale, *On the notion of exact sequence: From Hopf algebras to tensor categories*, Hopf algebras, tensor categories and related topics, Contemp. Math., vol. 771, Amer. Math. Soc., [Providence], RI, [2021] ©2021, pp. 225–254, DOI 10.1090/conm/771/15514. MR4281372

[32] P. Saracco, *Coquasi-bialgebras with preantipode and rigid monoidal categories*, Algebr. Represent. Theory **24** (2021), no. 1, 55–80, DOI 10.1007/s10468-019-09931-2. MR4207391

[33] P. Schauenburg, *Hopf algebra extensions and monoidal categories*, New directions in Hopf algebras, Math. Sci. Res. Inst. Publ., vol. 43, Cambridge Univ. Press, Cambridge, 2002, pp. 321–381. MR1913442

[34] P. Schauenburg, *Hopf bimodules, coquasibialgebras, and an exact sequence of Kac*, Adv. Math. **165** (2002), no. 2, 194–263, DOI 10.1006/aima.2001.2016. MR1887584

[35] P. Schauenburg, *Two characterizations of finite quasi-Hopf algebras*, J. Algebra **273** (2004), no. 2, 538–550, DOI 10.1016/S0021-8693(03)00335-1. MR2037710

[36] C. Schweigert and L. Woike, *The trace field theory of a finite tensor category*, Algebr. Represent. Theory **26** (2023), no. 5, 1931–1949, DOI 10.1007/s10468-022-10147-0. MR4664538

[37] T. Shibata and K. Shimizu, *Modified traces and the Nakayama functor*, Algebr. Represent. Theory **26** (2023), no. 2, 513–551, DOI 10.1007/s10468-021-10102-5. MR4568854

[38] S. Taiki and K. Shimizu, *Exact sequences of Frobenius tensor categories*, Preprint, `arXiv:2303.14687`, (2023).

[39] T. Shibata and K. Shimizu, *Nakayama functors for coalgebras and their applications to Frobenius tensor categories*, Adv. Math. **419** (2023), Paper No. 108960, 61, DOI 10.1016/j.aim.2023.108960. MR4560996

[40] K. Shimizu, *Relative Serre functor for comodule algebras*, J. Algebra **634** (2023), 237–305, DOI 10.1016/j.jalgebra.2023.07.015. MR4624433

[41] K. Shimizu, *Nakayama functor for monads on finite abelian categories*, Preprint, `arXiv:2208.08203`, (2022).

[42] K. Shimizu, *Ribbon structures of the Drinfeld center of a finite tensor category*, Kodai Math. J. **46** (2023), no. 1, 75–114, DOI 10.2996/kmj46106. MR4560990

[43] M. Takeuchi, *Free Hopf algebras generated by coalgebras*, J. Math. Soc. Japan **23** (1971), 561–582, DOI 10.2969/jmsj/02340561. MR292876

[44] Y. Morita, *On the Hurwitz-Lerch L-functions*, J. Fac. Sci. Univ. Tokyo Sect. IA Math. **24** (1977), no. 1, 29–43. MR441924

[45] K.-H. Ulbrich, *On Hopf algebras and rigid monoidal categories*, Israel J. Math. **72** (1990), no. 1-2, 252–256, DOI 10.1007/BF02764622. Hopf algebras. MR1098991

[46] H. Yadav, *On unimodular module categories*, Adv. Math. **432** (2023), Paper No. 109264, 41, DOI 10.1016/j.aim.2023.109264. MR4632574

DEPARTMENT OF MATHEMATICAL SCIENCES, SHIBAURA INSTITUTE OF TECHNOLOGY, 307 FUKASAKU, MINUMA-KU, SAITAMA-SHI, SAITAMA 337-8570, JAPAN.

Email address: `kshimizu@shibaura-it.ac.jp`

Contemporary Mathematics
Volume **814**, 2025
https://doi.org/10.1090/conm/814/16308

The gamma function and a certain sequence of differences

David E. Radford

This paper is dedicated to the memories of Larry Lambe and Earl Taft.

ABSTRACT. We study a sequence of differences arising from the problem of finding the smallest factorial $n!$ greater than or equal to a^n, where $a > 1$, using the gamma function. This is Lalescu's sequence.

Introduction

Let $a > 1$. Then $a^n \leq n!$ for some $n \geq 2$. Let n_a be the smallest such positive integer. Then n_a is the unique solution n to $n \geq 2$ and a belongs to the half open interval $(\sqrt[n-1]{(n-1)!}\,, \sqrt[n]{n!}\,]$. An extensive discussion of n_a is found in [**Ra**].

Let $n \geq 1$. In this paper we study the sequence of differences $L_1, L_2, L_3, \ldots$, where $L_n = \sqrt[n+1]{(n+1)!} - \sqrt[n]{n!}$, the sequence of lengths of the half open intervals, by means of the gamma function $\Gamma(x)$. This sequence was introduced in [**L**] and has been referred to since as Lalescu's sequence.

It is well known that $\lim_{n\longrightarrow\infty} L_n = \frac{1}{e}$; see [**BS, F, T**] for example. Convergent generalizations of Lalescu's sequence are described and their limits are computed in [**F, T**]. Computations on pages 341–342 of [**F**] show the generalizations described therein are eventually strictly decreasing. In this paper we show, by elementary means, that Lalescu's sequence is strictly decreasing. The limit and decreasing nature of Lalescu's sequence play a central role in [**Ra**]

Let $G(x) = \Gamma(x+1)^{1/x}$, where $x > 0$. Then $L_n = G'(x)$ for some $n < x < n+1$. Our study of the sequence is based on results we obtain about $G(x)$ and its first and second derivatives. Our main result is Theorem 5.1 which states that the sequence $L_1, L_2, L_3, \ldots$ is strictly decreasing.

The main problem addressed in this paper is determining when $G'(x)$ is strictly decreasing. This we do by investigating when $G''(x)$ is negative. Section 1 sets up the machinery for this investigation, particularly the other functions used in the process. Sections 2–4 are devoted to the technical details about these functions. The approach we use to study Lalescu's sequence is that used in [**F**] to study a generalization. We relate the gamma function limits in [**F**] to sequences which

2020 *Mathematics Subject Classification.* Primary 33B15; Secondary 05A10.

Key words and phrases. Gamma function, Lalescu's sequence, factorials, and powers.

The author wishes to thank the referee for useful comments and for pointing out important references.

define the Euler–Mascheroni constant. Our main results are in Section 5. Properties of the sequence $S_1, S_2, S_3, \ldots$, where $S_n = eL_n$, are described in Theorem 5.2.

Let $A(x) = eG'(x)$, where $x > 0$, and let $n \geq 1$. It follows that $S_n = A(x)$ for some $n < x < n+1$. Since $\lim_{x \longrightarrow \infty} A(x) = 1$, we find it natural to study $A(x)$ in terms of the function $a(x) = (A(x) - 1)x$ since $A(x) = 1 + \frac{a(x)}{x}$. Theorem 5.2 states in particular that $1 + \frac{a(n+1)}{n+1} < S_n < 1 + \frac{a(n)}{n}$ for $n \geq 18$, $\lim_{n \longrightarrow \infty} a'(n) = 0$, and $\lim_{n \longrightarrow \infty} a(n) = \frac{1}{2}$. Many of the functions $F(x)$ of interest in [**Ra**] satisfy $\lim_{x \longrightarrow \infty} F(x) = 1$ and are treated as $F(x) = 1 + \frac{f(x)}{x}$.

1. $\Gamma(x)$ and the sequence $L_1, L_2, L_3, \ldots$

The gamma function $\Gamma(x)$ is infinitely differentiable and positive valued on $(0, \infty)$. Let $x > 0$. Since $\Gamma(x+1) = x\Gamma(x)$ and $\Gamma(1) = 1$, it follows $\Gamma(n) = (n-1)!$ for integers $n \geq 1$. See [**A**, §§1–2] for the properties of the gamma function we use.

Write $G(x) = \Gamma(x+1)^{1/x} = e^{f(x)}$, where $f(x) = \frac{\ln \Gamma(x+1)}{x}$. Then $G(x)$ and $f(x)$ are infinitely differentiable and $G(x)$ is positive valued on $(0, \infty)$. Let $n \geq 1$. Since $G(n) = \sqrt[n]{n!}$ we have that

$$L_n = \sqrt[n+1]{(n+1)!} - \sqrt[n]{n!} = G(n+1) - G(n) = G'(x) \tag{1.1}$$

for some $n < x < n+1$ by the mean value theorem.

Since

$$G''(x) = (f''(x) + f'(x)^2)G(x). \tag{1.2}$$

it follows that

$$G''(x) < 0 \text{ if and only if } f''(x) + f'(x)^2 < 0. \tag{1.3}$$

The function $h(x) = xf'(x)$ plays a central role in determining when $G''(x)$ is negative. First we describe it in terms of an infinite series and then reformulate (1.3) in terms of it.

As $\Gamma(x+1) = x\Gamma(x)$, we have

$$xf(x) = \ln x + \ln \Gamma(x). \tag{1.4}$$

Now

$$\ln \Gamma(x) = -Cx - \ln x + \sum_{n=1}^{\infty} \left(\frac{x}{n} - \ln(1 + \frac{x}{n}) \right),$$

where C is the Euler–Mascheroni constant, by (2.9) of [**A**]. Therefore,

$$f(x) = -C + \sum_{n=1}^{\infty} \left(\frac{1}{n} - \frac{1}{x} \ln(1 + \frac{x}{n}) \right).$$

As

$$(\ln \Gamma)'(x) = -C - \frac{1}{x} + \sum_{n=1}^{\infty} \left(\frac{1}{n} - \frac{1}{x+n} \right)$$

by (2.10) of [**A**], and

$$h(x) = \frac{1}{x} + (\ln \Gamma)'(x) - f(x) \tag{1.5}$$

by (1.4), the equation

$$h(x) = \sum_{n=1}^{\infty} \left(\frac{1}{x} \ln(1 + \frac{x}{n}) - \frac{1}{x+n} \right) \tag{1.6}$$

follows.

We can look at the terms of the series description of $h(x)$ in light of the usual estimates $\frac{a}{1+a} < \ln(1+a) < a$ for $\ln(1+a)$, where $a > 0$. The typical summand in (1.6) is

$$\frac{1}{x}\left(\ln(1+a) - \frac{a}{1+a} \right),$$

where $a = \frac{x}{n}$. Therefore $h(x) > 0$.

Observe that $xf''(x) = -\frac{1}{x^2} + (\ln \Gamma)''(x) - 2f'(x)$ by (1.5). By (2.11) of [**A**] we have $(\ln \Gamma)''(x) = \sum_{n=0}^{\infty} \frac{1}{(x+n)^2}$ and, therefore,

$$x^2(f''(x) + f'(x)^2) = \sum_{n=1}^{\infty} \frac{x}{(x+n)^2} + (1 - h(x))^2 - 1. \tag{1.7}$$

We have shown that (1.3) is equivalent to

$$G''(x) < 0 \text{ if and only if } \sum_{n=1}^{\infty} \frac{x}{(x+n)^2} + (1 - h(x))^2 < 1. \tag{1.8}$$

We briefly digress to discuss $x^2 f''(x)$ to show that $-f(x)$ is a convex function and relate $f'(x)$ and $f''(x)$. The equation of (1.6) and the two subsequent ones imply

$$x^2 f''(x) = 2\left(\sum_{n=1}^{\infty} \left(\frac{3x+2n}{2(x+n)^2} - \frac{1}{x}\ln(1+\frac{x}{n}) \right) \right). \tag{1.9}$$

Fix $x > 0$ and set

$$D(t) = \frac{1}{x}\ln(1+\frac{x}{t}) - \frac{3x+2t}{2(x+t)^2}$$

for $t > 0$. Then $\lim_{t\longrightarrow\infty} D(t) = 0$ and $D'(t) < 0$. Therefore $D(t)$ is a positive valued strictly decreasing function. Note $x^2 f''(x) = -2(\sum_{n=1}^{\infty} D(n))$.

An antiderivative of $D(t)$ is $\frac{t}{x}\ln(1+\frac{x}{t}) - \frac{t}{2(x+t)}$ which means

$$\int_1^{\infty} D(t)dt = \frac{1}{2} - \left(\frac{1}{x}\ln(1+x) - \frac{1}{2(x+1)} \right).$$

Using the integral test one can easily see that

$$\lim_{x\longrightarrow\infty} \left(\sum_{n=1}^{\infty} \left(\frac{3x+2n}{2(x+n)^2} - \frac{1}{x}\ln(1+\frac{x}{n}) \right) \right) = -\frac{1}{2}.$$

We have shown $x^2 f''(x) < 0$ and

$$\lim_{x\longrightarrow\infty} x^2 f''(x) = -1. \tag{1.10}$$

Since $-f''(x) > 0$ for all $x > 0$, it follows by the corollary to Theorem 1.4 of [**A**] that $-f(x)$ is a convex function on $(0, \infty)$.

Note that $D(1) > 0$ means

$$\frac{a}{1+a} < \frac{3a^2+2a}{2(1+a)^2} < \ln(1+a)$$

for all $a > 0$, which gives a better underestimate for $\ln(1+a)$ than the usual $\frac{a}{a+1} < \ln(1+a)$. Using (1.6) and (1.9), it is easy to see that $f'(x) > -\frac{x}{2}f''(x)$ for all $x > 0$. One way to prove this inequality is to think of the terms of the series involved in terms of the usual and better under estimates for $\ln(1+a)$.

Returning to (1.8), let

$$g(x) = \sum_{n=1}^{\infty} \frac{x}{(x+n)^2}$$

and let $d(x) = 1 - h(x)$. To determine when the second inequality of (1.8) holds, that is $g(x) + d(x)^2 < 1$, we derive estimates $g(x)$ and $d(x)$.

As for $g(x)$, fix $x > 0$ and set

$$D(t) = \frac{x}{(x+t)^2}$$

where $t > 0$. Observe that $D(t)$ is a positive valued strictly decreasing function and $\lim_{t\longrightarrow\infty} D(t) = 0$. As

$$\int_1^{\infty} D(t)dt = \frac{x}{x+1} \quad \text{and} \quad \frac{x}{x+1} + D(1) = 1 - \frac{1}{(x+1)^2},$$

the integral test for series yields

$$\frac{x}{x+1} < g(x) < 1 - \frac{1}{(x+1)^2}. \tag{1.11}$$

As a result $\lim_{x\longrightarrow\infty} g(x) = 1$. When m is positive integer, $g(m)$ has a particularly nice form. Since

$$\zeta(2) = \sum_{n=1}^{\infty} \frac{1}{n^2} = \frac{\pi^2}{6},$$

where $\zeta(x)$ is the Riemann zeta function,

$$g(m) = m(\frac{\pi^2}{6} - \sum_{n=1}^{m} \frac{1}{n^2}). \tag{1.12}$$

We consider $d(x)$ and its equivalent $h(x)$ in the next section.

2. $h(x)$ and applications to $G(x)$

Throughout $x > 0$. Fix x and set $D(t) = \frac{1}{x}\ln(1+\frac{x}{t}) - \frac{1}{x+t}$ for $t > 0$. Then using the usual underestimate for $\ln(1+a)$ one sees that $\frac{x}{x+t} < \ln(1+\frac{x}{t})$. Thus $D(t) > 0$. It is easy to see that $\lim_{t\longrightarrow\infty} D(t) = 0$. Since $D'(t) < 0$, we conclude $D(t)$ is a strictly decreasing function. An antiderivative of $D(t)$ is $\left(\frac{t}{x}\right)\ln\left(1+\frac{x}{t}\right)$. Thus we compute $\int_1^{\infty} D(t)dt = 1 - \frac{\ln(1+x)}{x}$. Using the integral test for series we deduce

$$1 - \frac{\ln(1+x)}{x} < h(x) < \frac{x}{x+1}. \tag{2.1}$$

Therefore

$$\lim_{x\longrightarrow\infty} h(x) = 1 \text{ and } \lim_{x\longrightarrow\infty} d(x) = 0, \tag{2.2}$$

$$\frac{1}{x+1} < d(x) < \frac{\ln(1+x)}{x},$$

and thus

$$\frac{1}{(x+1)^2} < d(x)^2 < \left(\frac{\ln(1+x)}{x}\right)^2. \tag{2.3}$$

The first limit of (2.2) follows by (1.10) and the discussion of $g(x)$ in Section 1 as well.

Our integral test approximations give us $L(x) < g(x) + d(x)^2 < R(x)$, where

$$L(x) = 1 - \frac{x}{(x+1)^2}$$

and where

$$R(x) = 1 + \left(\frac{\ln(1+x)}{x}\right)^2 - \frac{1}{(x+1)^2}.$$

See (1.11) and (2.3). Since $\frac{1}{x+1} < \frac{\ln(1+x)}{x}$, these approximations are not sufficient to give the second inequality of (1.8). We will use a better one for $g(x)$ implicit in part (a) of Corollary 3.2 which will be good enough for us.

We return to (2.2) for a moment. In light of (1.10) we have

$$\lim_{x\longrightarrow\infty} x^2(f''(x) + f'(x)^2) = 0 = \lim_{x\longrightarrow\infty}(f''(x) + f'(x)^2). \tag{2.4}$$

We now compute $h(x)$ when $x = m$ is a positive integer. In this case $h(x)$ is expressed in terms of the Euler–Mascheroni constant and terms of the sequence which defines it. Our calculations will lead to other properties of $G(x)$ and its first and second derivatives.

Let $N \geq 1$ and set

$$h_N(x) = \sum_{n=1}^{N}\left(\frac{1}{x}\ln(1+\frac{x}{n}) - \frac{1}{x+n}\right) = \frac{1}{x}\left(\sum_{n=1}^{N}\left(\ln(1+\frac{x}{n}) - \frac{x}{x+n}\right)\right).$$

Then $h(x) = \lim_{N\longrightarrow\infty} h_N(x)$. Let $m \geq 1$. Then

$$h_N(m) = \frac{1}{m}\left(\sum_{n=1}^{N}(\ln(m+n) - \ln n) - m\left(\sum_{n=1}^{N}\frac{1}{m+n}\right)\right).$$

To proceed we use a very simple summation formula.

LEMMA 2.1. *Let $a_1, a_2, a_3, \ldots$ be any sequence of real numbers and suppose $N \geq m \geq 1$. Then*

$$\sum_{n=1}^{N}(a_{m+n} - a_n) = \sum_{\ell=1}^{m}(a_{N+\ell} - a_\ell).$$

PROOF. The lemma follows from $\sum_{n=1}^{N} a_{m+n} = \sum_{\ell=1}^{N-m} a_{m+\ell} + \sum_{\ell=1}^{m} a_{N+\ell}$ and from $\sum_{n=1}^{N} a_n = \sum_{\ell=1}^{m} a_\ell + \sum_{\ell=1}^{N-m} a_{m+\ell}$. □

Assume further that $m \leq N$. In light of the preceding lemma

$$h_N(m) = \frac{1}{m}\left(\sum_{\ell=1}^{m}\left(\ln(N+\ell) - \ln\ell - \sum_{n=1}^{N}\frac{1}{m+n}\right)\right). \tag{2.5}$$

We consider each summand separately. Let $C_n = 1+1/2+1/3+\cdots+1/n-\ln n$ for $n \geq 1$. Then $C_1, C_2, C_3, \ldots$ is a strictly decreasing sequence of positive numbers which converges to C. See page 16 of [**A**]. Since

$$\ln(N+\ell) - \ln\ell - \sum_{n=1}^{N} \frac{1}{m+n} = -C_{N+\ell} + C_\ell - \sum_{j=\ell+1}^{m} \frac{1}{N+j} + \sum_{j=\ell+1}^{m} \frac{1}{j}$$

and $h(m) = \lim_{N\longrightarrow\infty} h_N(m)$, by (2.5) we have

$$h(m) = \frac{1}{m}\left(-mC + \sum_{\ell=1}^{m} C_\ell + \sum_{\ell=1}^{m}\left(\sum_{j=\ell+1}^{m}\frac{1}{j}\right)\right).$$

Now

$$\sum_{\ell=1}^{m}\left(\sum_{j=\ell+1}^{m}\frac{1}{j}\right) = m - \sum_{j=1}^{m}\frac{1}{j} = m - C_m - \ln m.$$

The first equation easily follows by induction on m. Therefore,

$$h(m) = 1 - C + \frac{C_1 + \cdots + C_{m-1} - \ln m}{m} \tag{2.6}$$

for $m \geq 1$, where $C_1 + \cdots + C_{m-1} = 0$ when $m = 1$ by convention. We simplify the expressions for $d(m)$ and $h(m)$. For $m \geq 1$ let $D_m = 1 + \frac{1}{2} + \cdots + \frac{1}{m} - \ln(\sqrt[m]{m!})$.

PROPOSITION 2.2. *$h(m) = D_m - C$ and $d(m) = C + 1 - D_m$ for all $m \geq 1$.*

PROOF. The second equation follows from the first. To establish the first, we set

$$D'_m = 1 + \frac{C_1 + \cdots + C_{m-1} - \ln m}{m}$$

for all $m \geq 1$. Observe that $D'_1 = D_1$ and $(m+1)D'_{m+1} + mD'_m = (m+1)D_{m+1} + mD_m$ for all $m \geq 1$. The first equation now follows by (2.6). □

Note (2.2) and Proposition 2.2 imply $\lim_{m\longrightarrow\infty} D_m = C + 1$. We examine the sequence $D_1, D_2, D_3, \ldots$ in more detail. Observe that

$$D_m = C_m + \ln m - \ln(\sqrt[m]{m!}) = C_m + \ln(\frac{m}{\sqrt[m]{m!}}).$$

Thus the well-known equation $\lim_{m\longrightarrow\infty} \frac{m}{\sqrt[m]{m!}} = e$, or, equivalently,

$$\lim_{m\longrightarrow\infty} \frac{\sqrt[m]{m!}}{m} = \frac{1}{e}$$

follows.

Now

$$D_{m+1} = D_m + \frac{1}{m+1}\left(1 + \ln\left(\frac{\sqrt[m]{m!}}{m+1}\right)\right).$$

Recall from Section 1 that $\sqrt[m]{m!} = G(m) = e^{f(m)}$. Thus

$$\ln\left(\frac{\sqrt[m]{m!}}{m+1}\right) = \ln\left(\frac{G(m)}{m+1}\right) = f(m) - \ln(m+1).$$

Let

$$D(x) = f(x) - \ln(x+1) = \ln\left(\frac{G(x)}{x+1}\right).$$

The second inequality of (2.1) can be expressed $xD'(x) < 0$. This means $D(x)$, and hence $\frac{G(x)}{x+1}$, is a strictly decreasing function. By virtue of the previous limit calculation

$$\lim_{m\longrightarrow\infty} \ln\left(\frac{\sqrt[m]{m!}}{m+1}\right) = -1.$$

We have shown that $D_1, D_2, D_3, \ldots$ is a strictly increasing sequence.

Since $D(x)$ is a strictly decreasing function and since $\lim_{m\longrightarrow\infty} D(m) = -1$, it follows that $\lim_{x\longrightarrow\infty} D(x) = -1$. We have observed in the discussion following by (1.6) that $h(x) > 0$. Therefore,

$$(x+1)f'(x) = \left(\frac{x+1}{x}\right) h(x) > 0$$

and hence $\lim_{x\longrightarrow\infty}(x+1)f'(x) = 1$ by (2.2). Since $G'(x) = f'(x)G(x) = e^{E(x)}$, where $E(x) = D(x) + \ln((x+1)f'(x))$, and $\lim_{x\longrightarrow\infty} E(x) = -1 = \lim_{x\longrightarrow\infty} D(x)$, we conclude that

$$\lim_{x\longrightarrow\infty} G'(x) = \frac{1}{e} = \lim_{x\longrightarrow\infty} \frac{G(x)}{x}. \tag{2.7}$$

We note the second equation of (2.7) is a special case of (1) of [**F**] and the first follows by the equation in the line after (6) of the same. A proof of $\lim_{n\longrightarrow\infty} L_n = \frac{1}{e}$ in [**F**] was based on (1).

Recall that $G''(x) = (f''(x) + f'(x)^2)G(x)$ by (1.2). Thus

$$\lim_{x\longrightarrow\infty} xG''(x) = 0 \tag{2.8}$$

follows by (2.4) and (2.7).

We end this section with estimates for $G'(x)$ in terms of $G(x)$.

$$\frac{G(x)}{x+1}R(x) \leq G'(x) \leq \frac{G(x)}{x+1}, \tag{2.9}$$

where

$$R(x) = \left(\frac{1+x}{x^2}\right)(x - \ln(1+x)).$$

It is easy to see that $0 < R(x) < 1$, that $R(x)$ is a strictly increasing function, and that $\lim_{x\longrightarrow\infty} R(x) = 1$. To establish the second inequality of (2.9), we let $h > 0$. Since $\frac{G(x)}{x+1}$ is a strictly decreasing function, we have that

$$\frac{G(x+h)}{x+h+1} < \frac{G(x)}{x+1}$$

from which

$$\frac{G(x+h) - G(x)}{h} < \frac{G(x)}{x+1}$$

follows. The second inequality is established.

Now let $F(x) = \frac{1}{x}\ln(1+x) + \ln(1+x)$. Then the first inequality of (2.1) can be written $xF'(x) < h(x)$. Thus

$$\ln\left(\frac{G(x)}{(x+1)\ell(x)}\right) = f(x) - F(x),$$

hence $\frac{G(x)}{(x+1)\ell(x)}$, is a strictly increasing function, where $\ell(x) = (1+x)^{1/x}$. At this point we leave it to the reader to establish the first inequality of (2.9) and assertions about $R(x)$, starting by mimicking our proof of the second inequality. The reader is encouraged to revisit (2.7) in light of (2.9).

3. $a(x)$ and $G''(x)$

Let $x > 0$ and set $A(x) = eG'(x)$. Now $\lim_{x\longrightarrow\infty} A(x) = 1$ by (2.7). Let $a(x) = (A(x)-1)x$. Then $A(x) = 1 + \frac{a(x)}{x}$. We find $a(x)$ is useful to associate with $A(x)$. As $a'(x) = A'(x)x + A(x) - 1 = e(xG''(x) + G'(x)) - 1$ we have

$$\lim_{x\longrightarrow\infty} a'(x) = 0 \tag{3.1}$$

by (2.7) and (2.8). To proceed we need estimates for a certain expression involving the derivative of the polygamma function $\psi(x) = (\ln\Gamma)'(x)$.

LEMMA 3.1. *Let $x > 0$. Then:*

(a) $\displaystyle \frac{1}{2} + \frac{x^2}{6(x+1/14)^3} < x^2\psi'(x) - x < \frac{1}{2} + \frac{1}{6x}$.

(b) $\displaystyle \lim_{x\longrightarrow\infty}(x^2\psi'(x) - x) = \frac{1}{2}$.

PROOF. Part (b) follows from part (a) which follows from the two inequalities

$$\frac{1}{x} + \frac{1}{2x^2} + \frac{1}{6(x+1/14)^3} < \psi'(x) < \frac{1}{x} + \frac{1}{2x^2} + \frac{1}{6x^3}$$

described in Theorem 4 of [**G**]. □

Since

$$g(x) = x\left(\psi'(x) - \frac{1}{x^2}\right),$$

it follows that $xg(x) - x = x^2\psi'(x) - x - 1$. We will need an equivalent formulation of the preceding lemma:

COROLLARY 3.2. *Let $x > 0$. Then:*

(a) $\displaystyle -\frac{1}{2} + \frac{x^2}{6(x+1/14)^3} < xg(x) - x < -\frac{1}{2} + \frac{1}{6x}$.

(b) $\displaystyle \lim_{x\longrightarrow\infty}(xg(x) - x) = -\frac{1}{2}$. □

Now

$$\begin{aligned} x^2G''(x) &= (g(x) + d(x)^2 - 1)G(x) \\ &= (xg(x) - x + xd(x)^2)\frac{G(x)}{x} \end{aligned}$$

follows by (1.2) and (1.7). Observe that $\lim_{x\longrightarrow\infty} xd(x)^2 = 0$ by (2.3) and recall that $\lim_{x\longrightarrow\infty}\frac{G(x)}{x} = \frac{1}{e}$ by (2.7). Hence

$$\lim_{x\longrightarrow\infty} x^2G''(x) = -\frac{1}{2e}. \tag{3.2}$$

now follows by part (b) of Corollary 3.2. We note that (3.2) is implied by the equation following (6) of [**F**]. Since $\lim_{x\longrightarrow\infty} A(x) = 1$, in light of (3.2) we have

$$\lim_{x\longrightarrow\infty} a(x) = \frac{1}{2} \tag{3.3}$$

by L'Hôspital's rule. Observe (3.2) implies $G'(x)$ is eventually decreasing.

4. $G''(x) < 0$ for $x \geq 18$

First of all, note (1.8) can be rewritten $G''(x) < 0$ if and only $g(x) + d(x)^2 < 1$; thus $G''(x) < 0$ if and only if $xg(x) - x + xd(x)^2 < 0$. Let $x > 0$. Now

$$xg(x) - x + xd(x)^2 < -\frac{1}{2} + \frac{1}{6x} + \frac{(\ln(1+x))^2}{x}$$

by part (a) of Corollary 3.2 and (2.3). The right hand side of this inequality is negative from some point on. When $D(x) = \frac{3x-1}{6} - (\ln(1+x))^2$ becomes and remains positive this is the case. Note that

$$D'(x) = \frac{1}{2} - 2\left(\frac{\ln(1+x)}{1+x}\right).$$

Since

$$D''(x) = -2\left(\frac{1 - \ln(1+x)}{(1+x)^2}\right),$$

it follows $D'(x)$ is strictly decreasing on $(0, e-1]$ and is strictly increasing on $[e-1, \infty)$. Since $D'(e-1) < 0$ and $\lim_{x \longrightarrow \infty} D'(x) = 1/2$, it follows that $D'(x)$ has a unique zero $a > e - 1$, which is the unique solution to this inequality and $\frac{\ln(1+a)}{1+a} = \frac{1}{4}$, and $D(x)$ is strictly increasing on $[a, \infty)$. As $\frac{\ln 9}{9} < \frac{1}{4} < \frac{\ln 8}{8}$ it follows that $7 < a < 8$. As $D(17) < 0 < D(18)$, we have shown the following proposition.

PROPOSITION 4.1. *Let $G(x) = \Gamma(x+1)^{1/x}$. Then there exists $17 < c < 18$ such that $G''(x) < 0$ for $x > c$. Thus $G'(x)$ is strictly decreasing on $[c, \infty)$.* □

We note that $7.61316 < a < 7.61317$. Our c is the unique solution to $c \geq a$ and $D(c) = 0$, satisfies $17.11650 < c < 17.11651$ as $D(17.11650) < 0 < D(17.11651)$.

We have shown $g(x) + d(x)^2 < 1$ for $x \geq c$. The reader might find it interesting to reformulate $g(m) + d(m)^2 < 1$ for $m \geq 18$ with the substitutions suggested by (1.12), (2.6), Proposition 2.2, and $g(m) = m\psi'(m) - \frac{1}{m}$.

5. The main results

We have noted that $\lim_{n \longrightarrow \infty} L_n = \frac{1}{e}$.

THEOREM 5.1. *Let $L_n = \sqrt[n+1]{(n+1)!} - \sqrt[n]{n!}$ for $n \geq 1$. Then:*

$$0.42 > L_1 > L_2 > L_3 > \cdots > \frac{1}{e}.$$

PROOF. Let $n \geq 1$. Then $L_n = G'(x)$ for some $n < x < n+1$ by (1.1). Since $G'(x)$ is strictly decreasing on $[18, \infty)$ by part (a) of Proposition 4.1, we have shown that $L_{18} > L_{19} > L_{20} > \cdots$. Using MATLAB [**M**], for example, one can show that $0.42 > L_1 > L_2 > L_3 > \cdots > L_{18}$. Thus the sequence $L_1, L_2, L_3, \ldots$ is strictly decreasing and the proof is complete by the statement preceding the theorem. □

The sequence $S_1, S_2, S_3, \ldots$ is fundamental to [**Ra**], where $S_n = eL_n$ for all $n \geq 1$. To study this sequence we use the functions $A(x)$ and $a(x)$ of Section 3. Let $x > 0$. Recall that $A(x) = eG'(x)$ and $A(x) = 1 + \frac{a(x)}{x}$. Now $A(x)$ is strictly decreasing on $[18, \infty)$ by part (a) of Proposition 4.1. Let $n \geq 18$. As $S_n = A(x)$ for some $n < x < n+1$ by (1.1); thus $A(n) > S_n > A(n+1)$ follows. These inequalities,Theorem 5.1, its preceding statement, (3.1), and (3.3) imply the following theorem.

THEOREM 5.2. *Let* $S_n = e\sqrt[n+1]{(n+1)!} - e\sqrt[n]{n!}$ *for* $n \geq 1$. *Then:*

(a) $\lim_{n\longrightarrow\infty} S_n = 1$.

(b) $1.15 > S_1 > S_2 > S_3 > \cdots > 1$.

(c) $1 + \dfrac{a(n+1)}{n+1} < S_n < 1 + \dfrac{a(n)}{n}$ *for all* $n \geq 18$.

(d) $\lim_{n\longrightarrow\infty} a'(n) = 0$ *and* $\lim_{n\longrightarrow\infty} a(n) = \dfrac{1}{2}$. □

In [**Ra**] we consider over estimates for S_n in detail. Our work based on the improvement on Sterling's approximation of factorials due to Robbins [**Ro**]. A few final comments based on parts (c) and (d) of Theorem 5.2.

Part (c) can be expressed in a different manner as $S_n < 1 + \frac{a(n)}{n}$ for $n = 18$ and $(S_n - 1)n < a(n) < (S_{n-1} - 1)n$ for $n > 18$. The last three inequalities imply $\lim_{n\longrightarrow\infty}(S_n - 1)n = \frac{1}{2}$. By part (b) there exists a sequence $a_1, a_2, a_3, \ldots$ such that $S_1 < 1 + a_1$ and $(S_n - 1)n < a_n < (S_{n-1} - 1)n$ for $n > 1$. For any such sequence $1 + \frac{a_{n+1}}{n+1} < S_n < 1 + \frac{a_n}{n}$ for all $n \geq 1$ and $\lim_{n\longrightarrow\infty} a_n = \frac{1}{2}$.

References

[A] E. Artin, *The gamma function*, Athena Series: Selected Topics in Mathematics, Holt, Rinehart and Winston, New York-Toronto-London, 1964. Translated by Michael Butler. MR165148

[BS] W. G. Boskoff and B. D. Suceavă, *A class of applications of AM-GM Inequality*, Austral. Math. Soc. Gaz., vol. **33** (2006), pp. 51–56.

[F] O. Furdui, *From Lalescu's sequence to a gamma function limit*, Austral. Math. Soc. Gaz. **35** (2008), no. 5, 339–344. MR2473205

[G] L. Gordon, *A stochastic approach to the gamma function*, Amer. Math. Monthly **101** (1994), no. 9, 858–865, DOI 10.2307/2975134. MR1300491

[L] N. Teodorescu, *Traian Lalescu, student and contributor to Gazeta Matematică* (Romanian), Gaz. Mat. (Bucharest) **87** (1982), no. 9-10, 339–346. MR698479

[M] MATLAB, Mathworks, Inc., Natick, MA, 2020.

[Ra] D. E. Radford, *Factorials and powers, a minimality result*, `arXiv:2106.02002v1 [math.NT]`, 3 June 2021, pp. 1–22.

[Ro] H. Robbins, *A remark on Stirling's formula*, Amer. Math. Monthly **62** (1955), 26–29, DOI 10.2307/2308012. MR69328

[T] Gh. Toader, *Lalescu sequences*, Univ. Beograd. Publ. Elektrotehn. Fak. Ser. Mat. **9** (1998), 25–28. MR1671291

DEPARTMENT OF MATHEMATICS, STATISTICS AND COMPUTER SCIENCE (M/C 249), 851 SOUTH MORGAN STREET, UNIVERSITY OF ILLINOIS AT CHICAGO, CHICAGO, ILLINOIS 60607-7045

Email address: `radford@uic.edu`

Selected Published Titles in This Series

814 **Susan Montgomery and Siu-Hung Ng, Editors,** Quantum Groups, Hopf Algebras, and Applications, 2025

810 **Robert J. Berman and Yanir A. Rubinstein, Editors,** Convex and Complex: Perspectives on Positivity in Geometry, 2025

809 **Fabrice Baudoin and Luca Rizzi, Editors,** New Trends in Sub-Riemannian Geometry, 2025

808 **Loring W. Tu, Editor,** Group Actions and Equivariant Cohomology, 2024

807 **Galina Filipuk, Editor,** Recent Progress in Special Functions, 2024

806 **Maximilian Engel, Hildeberto Jardón-Kojakhmetov, and Cinzia Soresina, Editors,** Topics in Multiple Time Scale Dynamics, 2024

805 **Anthony Iarrobino, Pedro Macias Marques, Maria Evelina Rossi, and Jean Vallès, Editors,** Deformation of Artinian Algebras and Jordan Type, 2024

804 **Mahir Bilen Can and Jörg Feldvoss, Editors,** A Glimpse into Geometric Representation Theory, 2024

803 **Peter Gothen, Margarida Melo, and Montserrat Teixidor i Bigas, Editors,** Moduli Spaces and Vector Bundles—New Trends, 2024

802 **Ralph M. Kaufmann, Martin Markl, and Alexander A. Voronov, Editors,** Higher Structures in Topology, Geometry, and Physics, 2024

801 **K. A. Brown, T. J. Hodges, M. Vancliff, and J. J. Zhang, Editors,** Recent Advances in Noncommutative Algebra and Geometry, 2024

800 **Avinoam Mann, Louis H. Rowen, David J. Saltman, Aner Shalev, Lance W. Small, and Uzi Vishne, Editors,** Amitsur Centennial Symposium, 2024

799 **Alberto A. Condori, Elodie Pozzi, William T. Ross, and Alan A. Sola, Editors,** Recent Progress in Function Theory and Operator Theory, 2024

798 **Marat V. Markin, Igor V. Nikolaev, and Carsten Trunk, Editors,** Advances in Functional Analysis and Operator Theory, 2024

797 **Sangita Jha, Mrinal Kanti Roychowdhury, and Saurabh Verma, Editors,** Recent Developments in Fractal Geometry and Dynamical Systems, 2024

796 **John Cremona, John Jones, Jennifer Paulhus, Andrew V. Sutherland, and John Voight, Editors,** LuCaNT: LMFDB, Computation, and Number Theory, 2024

795 **Michael A. Jones, David McCune, and Jennifer M. Wilson, Editors,** Mathematical Analyses of Decisions, Voting and Games, 2024

794 **Jean-Pierre Magnot, Editor,** Recent Advances in Diffeologies and Their Applications, 2024

793 **Abba Gumel, Editor,** Mathematical and Computational Modeling of Phenomena Arising in Population Biology and Nonlinear Oscillations, 2024

792 **Shaoming Guo, Zane Kun Li, and Brian Street, Editors,** Recent Developments in Harmonic Analysis and its Applications, 2024

791 **Mee Seong Im, Bach Nguyen, and Arik Wilbert, Editors,** Algebraic and Topological Aspects of Representation Theory, 2024

790 **Peter Crooks and Alexandru I. Suciu, Editors,** Compactifications, Configurations, and Cohomology, 2023

789 **I. S. Krasil′shchik, A. B. Sossinsky, and A. M. Verbovetsky, Editors,** The Diverse World of PDEs, 2023

788 **I. S. Krasil′shchik, A. B. Sossinsky, and A. M. Verbovetsky, Editors,** The Diverse World of PDEs, 2023

787 **Hemen Dutta, Editor,** Mathematical Modelling, 2023

786 **Hemen Dutta, Editor,** Mathematical Modelling, 2023

For a complete list of titles in this series, visit the
AMS Bookstore at **www.ams.org/bookstore/conmseries/**.